T0318422

Urban Ecology

URBAN ECOLOGY

EMERGING PATTERNS AND SOCIAL-ECOLOGICAL SYSTEMS

Edited by

PRAMIT VERMA

Integrative Ecology Laboratory (IEL), Institute of Environment & Sustainable Development (IESD),
Banaras Hindu University (BHU), Varanasi, Uttar Pradesh, India

PARDEEP SINGH

Department of Environmental Studies, PGDAV College, University of Delhi, New Delhi, India

RISHIKESH SINGH

Integrative Ecology Laboratory (IEL), Institute of Environment & Sustainable Development (IESD),
Banaras Hindu University (BHU), Varanasi, Uttar Pradesh, India

A.S. RAGHUBANSHI

Integrative Ecology Laboratory (IEL), Institute of Environment & Sustainable Development (IESD),
Banaras Hindu University (BHU), Varanasi, Uttar Pradesh, India

ELSEVIER

Elsevier
Radarweg 29, PO Box 211, 1000 AE Amsterdam, Netherlands
The Boulevard, Langford Lane, Kidlington, Oxford OX5 1GB, United Kingdom
50 Hampshire Street, 5th Floor, Cambridge, MA 02139, United States

Notices
Knowledge and best practice in this field are constantly changing. As new research and experience broaden our understanding, changes in research methods, professional practices, or medical treatment may become necessary.

Practitioners and researchers must always rely on their own experience and knowledge in evaluating and using any information, methods, compounds, or experiments described herein. In using such information or methods they should be mindful of their own safety and the safety of others, including parties for whom they have a professional responsibility.

To the fullest extent of the law, neither the Publisher nor the authors, contributors, or editors, assume any liability for any injury and/or damage to persons or property as a matter of products liability, negligence or otherwise, or from any use or operation of any methods, products, instructions, or ideas contained in the material herein.

Library of Congress Cataloging-in-Publication Data
A catalog record for this book is available from the Library of Congress

British Library Cataloguing-in-Publication Data
A catalogue record for this book is available from the British Library

ISBN: 978-0-12-820730-7

For information on all Elsevier publications visit our website at
https://www.elsevier.com/books-and-journals

Publisher: Joe Hayton
Acquisitions Editor: Candice Janco
Editorial Project Manager: Chiara Giglio
Production Project Manager: Vijayaraj Purushothaman
Cover Designer: Greg Harris

Typeset by TNQ Technologies

Contents

Theme I

Emerging facets of urban ecology — urban and urbanization, theory and concepts

1. Urban ecology — current state of research and concepts

Pramit Verma, Rishikesh Singh, Pardeep Singh and A.S. Raghubanshi

2. Urban metabolism: old challenges, new frontiers, and the research agenda ahead

Daniela Perrotti

Theme II

Urban land use land cover

3. Urban growth pattern detection and analysis

Mateo Gašparović

4. Exposition of spatial urban growth pattern using PSO-SLEUTH and identifying its effects on surface temperature

H.A. Bharath, G. Nimish and M.C. Chandan

Theme V
Urban material balance

Theme VI
Cities: healthy, smart and sustainable

Theme VII
Sustainable urban design

Contributors

Luca Afonso Centre for Invasion Biology, Department of Botany and Zoology, Stellenbosch University, Matieland, South Africa

Muhammad Akmal Water Research Laboratory, Department of Fisheries and Aquaculture, University of Veterinary and Animal Sciences, Lahore, Punjab, Pakistan

Waqas Ali Applied and Environmental Microbiology Laboratory, Department of Wildlife and Ecology, University of Veterinary and Animal Sciences, Lahore, Punjab, Pakistan

Carmen Antuña-Rozado VTT Research Centre of Finland Ltd., Espoo, Finland

Vidhu Bansal Research Scholar, Department of Architecture and Regional Planning (ARP), Indian Institute of Technology (IIT) Kharagpur, West Bengal, India

Sunny Bansal Research Scholar, Ranbir and Chitra Gupta School of Infrastructure Design and Management (RCGSIDM), Indian Institute of Technology (IIT) Kharagpur, West Bengal, India

André C.S. Batalhão Environmental Sciences, Center for Environmental and Sustainability Research — CENSE/Nova Lisbon University, Caparica, Portugal

Rahul Bhadouria Department of Botany, University of Delhi, Delhi, India

H.A. Bharath RCG School of Infrastructure Design and Management, Indian Institute of Technology Kharagpur, West Bengal, India

Antonia D. Bousbaine Département de Géographie, Laboratoire LAPLEC, Université de Liège, Liège, Belgium

Christopher Bryant Géographie, Université de Montréal, Canada & Adjunct Professor, School of Environmental Design and Rural Development, University of Guelph, Montréal, Québec, Canada

Syed Mohsin Bukhari Applied and Environmental Microbiology Laboratory, Department of Wildlife and Ecology, University of Veterinary and Animal Sciences, Lahore, Punjab, Pakistan

M.C. Chandan RCG School of Infrastructure Design and Management, Indian Institute of Technology Kharagpur, West Bengal, India

Ranit Chatterjee Kyoto University, Kyoto, Japan

Álvaro Corredor-Ochoa Tampere University, Tampere, Finland

Ambika Dabral Resilience Innovation Knowledge Academy, New Delhi, Delhi, India

Lalatendu Keshari Das IIT Bombay, Mumbai, Maharashtra, India

Rajkumari Sanayaima Devi Deen Dayal Upadhyaya College (University of Delhi), New Delhi, India

Juan Du Department of Architecture & Urban Ecologies Design Lab, Faculty of Architecture, The University of Hong Kong, Hong Kong Special Administrative Region, China

Karen J. Esler Department of Conservation Ecology and Entomology and Centre for Invasion Biology, Stellenbosch University, Matieland, South Africa

José Fariña-Tojo Universidad Politécnica de Madrid, Madrid, Spain

Mirijam Gaertner Nürtingen-Geislingen University of Applied Sciences (HFWU), Schelmenwasen 4-8, Nürtingen, Germany and Centre for Invasion Biology, Department of Botany and Zoology, Stellenbosch University, Matieland, South Africa

Mateo Gašparović Chair of Photogrammetry and Remote Sensing, Faculty of Geodesy, University of Zagreb, Zagreb, Croatia

Sjirk Geerts Department of Conservation and Marine Sciences, Cape Peninsula University of Technology, Cape Town, South Africa

Dilawar Husain Department of Mechanical Engineering, School of Engineering and Technology, Sandip University, Nashik, India

Ali Hussain Applied and Environmental Microbiology Laboratory, Department of Wildlife and Ecology, University of Veterinary and Animal Sciences, Lahore, Punjab, Pakistan

Syed Makhdoom Hussain Aquaculture Research Laboratory, Department of Zoology, Government College University, Faisalabad, Punjab, Pakistan

Arshad Javid Applied and Environmental Microbiology Laboratory, Department of Wildlife and Ecology, University of Veterinary and Animal Sciences, Lahore, Punjab, Pakistan

Vaishali Kapoor Deen Dayal Upadhyaya College (University of Delhi), New Delhi, India

Sushil Kumar School of Environmental Sciences, Jawaharlal Nehru University, New Delhi, India

Pyarimohan Maharana School of Environmental Sciences, Jawaharlal Nehru University, New Delhi, India

R.K. Mall DST-Mahamana Centre of Excellence in Climate Change Research, Institute of Environment and Sustainable Development, Banaras Hindu University, Varanasi, Uttar Pradesh, India

Y. Milshina National Research University Higher School of Economics, Moscow, Russia

Golam Morshed Department of Infrastructure Engineering, University of Innsbruck, Innsbruck, Austria

G. Nimish RCG School of Infrastructure Design and Management, Indian Institute of Technology Kharagpur, West Bengal, India

Tahir Noor Applied and Environmental Microbiology Laboratory, Department of Wildlife and Ecology, University of Veterinary and Animal Sciences, Lahore, Punjab, Pakistan

Piotr Nowakowski Silesian University of Technology, Katowice, Poland

Wenjian Pan Department of Architecture & Urban Ecologies Design Lab, Faculty of Architecture, The University of Hong Kong, Hong Kong Special Administrative Region, China

D. Pavlova National Research University Higher School of Economics, Moscow, Russia

Daniela Perrotti University of Louvain, Louvain-la-Neuve, Belgium

Ravi Prakash Department of Mechanical Engineering, Motilal Nehru National Institute of Technology, Allahabad, Uttar Pradesh, India

A.S. Raghubanshi Integrative Ecology Laboratory (IEL), Institute of Environment & Sustainable Development (IESD), Banaras Hindu University (BHU), Varanasi, Uttar Pradesh, India

Juho Rajaniemi Tampere University, Tampere, Finland

Rumana Islam Sarker Department of Infrastructure Engineering, University of Innsbruck, Innsbruck, Austria

Joy Sen Professor and Head, Department of Architecture and Regional Planning; Joint Faculty, Ranbir and Chitra Gupta School of Infrastructure Design and Management, Indian Institute of Technology (IIT) Kharagpur, West Bengal, India

Fariya Sharmeen Institute for Management Research, Radboud University, Nijmegen, the Netherlands; Faculty of Civil Engineering and Geosciences, Delft University of Technology, Delft, the Netherlands

Sujit kumar Sikder Leibniz Institute of Ecological Urban and Regional Development (IOER), Dresden, Germany

Nidhi Singh DST-Mahamana Centre of Excellence in Climate Change Research, Institute of Environment and Sustainable Development, Banaras Hindu University, Varanasi, Uttar Pradesh, India

Ravindra Pratap Singh Research Scholar, Integrative Ecology Laboratory, Institute of Environment and Sustainable Development, Banarasi Hindu University, Varanasi, Uttar Pradesh, India

Rishikesh Singh Integrative Ecology Laboratory (IEL), Institute of Environment & Sustainable Development (IESD), Banaras Hindu University (BHU), Varanasi, Uttar Pradesh, India

Saumya Singh DST-Mahamana Centre of Excellence in Climate Change Research, Institute of Environment and Sustainable Development, Banaras Hindu University, Varanasi, Uttar Pradesh, India

Anita Singh Department of Botany, Banaras Hindu University, Varanasi, Uttar Pradesh, India

Pardeep Singh Department of Environmental Studies, PGDAV College, University of Delhi, New Delhi, India

Rajeev Pratap Singh Department of Environment and Sustainable Development, Institute of Environment and Sustainable Development, Banaras Hindu University, Varanasi, Uttar Pradesh, India

Vaibhav Srivastava Department of Environment and Sustainable Development, Institute of Environment and Sustainable Development, Banaras Hindu University, Varanasi, Uttar Pradesh, India

Pratap Srivastava Shyama Prasad Mukherjee Post-graduate College, University of Allahabad, Allahabad, Uttar Pradesh, India

Nuala Stewart Master of Sustainable Development, Macquarie University, Sydney, NSW, Australia

Denilson Teixeira Environmental Engineering, Federal University of Goiás, Goiânia, Brazil

Sachchidanand Tripathi Deen Dayal Upadhyaya College (University of Delhi), New Delhi, India

Shweta Upadhyay Integrative Ecology Laboratory (IEL), Institute of Environment & Sustainable Development (IESD), Banaras Hindu University (BHU), Varanasi, Uttar Pradesh, India

Barkha Vaish Department of Environment and Sustainable Development, Institute of Environment and Sustainable Development, Banaras Hindu University, Varanasi, Uttar Pradesh, India

Pramit Verma Integrative Ecology Laboratory (IEL), Institute of Environment & Sustainable Development (IESD), Banaras Hindu University (BHU), Varanasi, Uttar Pradesh, India

Mariusz Wala PST Transgór S.A., Rybnik, Poland

Foreword

Harini Nagendra, School of Development, Azim Premji University, Pixel B, PES Campus, Electronic City, Hosur Road, Bangalore 560100, India.

In 2007, for the first time ever, more than half of the world's population lived and worked in urban areas. Cities occupy a relatively small fraction of the world's land cover but have an ecological, economic, social and cultural impact that is completely disproportionate to their size. Urban areas suck in resources including energy, water, food and people from the hinterland and from distant parts of the world and export their waste, creative ideas and money to far-flung regions. Understanding the ecological impact of cities is crucial in the era of the Anthropocene, if we are to learn how to move towards a more sustainable world (Seto and Pandey, 2019).

Urbanization has been criticized for its unsustainability. Yet the fact that much of the urban area projected to exist by 2050 is yet to be built also provides us with an opportunity to think differently about cities and to reimagine a different type of urban: one that is more sustainable, equitable and innovative (Parnell et al., 2018). That window of opportunity, if indeed it does exist, is closing fast. There is a real and urgent need for interdisciplinary research that examines the ecology of cities from diverse angles, using different disciplinary lenses, methodological approaches and drawing on case studies from all parts of the world.

'Urban Ecology: Emerging Patterns and Social-Ecological Systems' presents an ambitious attempt to examine a number of diverse facets of urban ecology, drawing on reviews, metaanalyses and case studies located in diverse parts of the world. Cities are, at their core, social-ecological systems (Wolfram et al., 2016), and this book appropriately treats them as such, combining research that looks at invasive species, urban metabolism, land cover change, air pollution and urban disaster management, as well as several other issues relevant to understanding the sustainability of urban social-ecological systems.

In far, too many reviews and books on the urban, fast-growing regions of the global South often get left out or underdeveloped, despite the fact that Southern cities are growing at much faster rates compared to their Northern counterparts (Nagendra et al., 2018). This edited volume presents a welcome departure from that trend, combining a number of case studies and reviews originating from South Asia with research from other parts of the world.

In this era of climate change, cities will be some of the worst hit in terms of urban sustainability and human well-being (Estrada et al., 2017). Ecological integrity, environmental quality and socioeconomic equity will play a major role in ensuring the resilience of cities to climate change and other shocks. Urban sustainability and resilience thus present important goals for the

21st century. Given the magnitude, intensity and interconnectedness of the challenge ahead, there is a pressing need for interdisciplinary research on urban social-ecological systems that investigate the challenges of urban sustainability and resilience from diverse angles. This book presents a welcome step in this direction.

References

Estrada, F., Botzen, W.W., Tol, R.S., 2017. A global economic assessment of city policies to reduce climate change impacts. Nature Climate Change 7, 403–406.

Nagendra, H., Bai, X., Brondizio, E.S., Lwasa, S., 2018. The urban south and the predicament of global sustainability. Nature Sustainability 1, 341–349.

Parnell, S., Elmqvist, T., McPhearson, T., Nagendra, H., Sörlin, S., 2018. Introduction - Situating knowledge and action for an urban planet. In: Elmqvist, T., et al. (Eds.), The Urban Planet. Cambridge University Press, pp. 1–16.

Seto, K.C., Pandey, B., 2019. Urban Land Use: Central to Building a Sustainable Future. One Earth 1, 168–170.

Wolfram, M., Frantzeskaki, N., Maschmeyer, S., 2016. Cities, systems and sustainability: Status and perspectives of research on urban transformations. Current Opinion in Environmental Sustainability 22, 18–25.

List of reviewers

S. No.	Name	Affiliation
1	Abhinav Yadav	Integrative Ecology Laboratory, Institute of Environment & Sustainable Development, Banaras Hindu University, Varanasi, India
2	Álvaro Corredor Ochoa	Faculty of Business and Built Environment, School of Architecture, Tempere University, Tempere, Finland
3	Mateo Gasparovic	Faculty of Geodesy, University of Zagreb, Croatia
4	Nidhi Singh	DST Mahamana Centre of Excellence in Climate Change Research, Institute of Environment & Sustainable Development, Banaras Hindu University, Varanasi, India
5	Rahul Bhadouria	University of Delhi, New Delhi, India
6	Ranit Chatterjee	Graduate School of Informatics, Kyoto University, Kyoto, Japan
7	S N Tripathi	Deen Dayal Upadhyaya College, University of Delhi, New Delhi, India
8	Saumya Singh	DST Mahamana Centre of Excellence in Climate Change Research, Institute of Environment & Sustainable Development, Banaras Hindu University, Varanasi, India
9	Shikha Singh	Integrative Ecology Laboratory, Institute of Environment & Sustainable Development, Banaras Hindu University, Varanasi, India
10	Sujit Kumar Sikder	Leibniz Institute of Ecological Urban and Regional Development (IOER), Germany
11	Tanu Kumari	Integrative Ecology Laboratory, Institute of Environment & Sustainable Development, Banaras Hindu University, Varanasi, India
12	Vidhu Bansal	Department of Architecture, School of Architecture, Central University of Rajasthan, Rajasthan, India

Emerging facets of urban ecology — urban and urbanization, theory and concepts

CHAPTER

1

Urban ecology – current state of research and concepts

Pramit Verma[1], Rishikesh Singh[1], Pardeep Singh[2], A.S. Raghubanshi[1]

[1]Integrative Ecology Laboratory (IEL), Institute of Environment & Sustainable Development (IESD), Banaras Hindu University (BHU), Varanasi, Uttar Pradesh, India; [2]Department of Environmental Studies, PGDAV College, University of Delhi, New Delhi, India

OUTLINE

3

1. Introduction

1.1 What is urban ecology?

Andrewartha and Birch (1954) considered ecology to be a study of the abundance and distribution of organisms. Odum (1975) gave the concept of ecosystem ecology, which focussed on the ecosystem. However, a better definition of ecology is given by the Carry Institute of Ecosystem Studies, focussing on the holistic and encompassing perspective of ecology, as 'the scientific study of the processes influencing the distribution and abundance of organisms, the interactions among organisms, and the interactions between organisms and the transformation and flux of energy and matter'. The important aspect of this definition is its emphasis on the 'processes' and 'interactions'.

Cities have become engines of development as well as drivers of environmental change. Drawing on the aforementioned description of ecology, *urban ecology* can be defined as the study of the relationship between living organisms and their environment, their distribution and abundance, the interactions between the organisms, and transformation and flux of energy and matter, in an urban area. An urban ecosystem is the growth of human population and its supporting infrastructure in the form of cities, towns, agglomerations and megacities.

An urban area consists of a number of processes and physical components, such as biodiversity in the form of parks, animals and trees, humans and their socioeconomic groups, built structures in the form of roads and buildings, transport, essential services such as finance, health and waste disposal, energy flow from different types of sources such as solar, electricity, coal, LPG, wood, and so on and material flow in the form of food supplies, building material (bricks, mortar, steel, etc.), waste generation, urban agriculture and biogeochemical cycles in urban areas. This is not an exhaustive list, but it gives an idea about the urban processes and components which constitute an urban ecosystem.

There are, however, two major aspects of this field that make it more important for the present times. First, the ecology of urban areas is not restricted to the urban boundary where the apparent indicator is observed (Verma and Raghubanshi, 2018), the indicator being urbanization. The meaning of 'urban area' and the boundary concept has been explored in greater details in section (3.1). Second, since human beings are the dominant organism in an urban area, urban ecology inevitably becomes a study focused on processes and interaction mediated by human actions. The resources, in the form of matter and energy, are not necessarily used where they are found, and the effects of human actions are felt at multiple scales and across system boundaries. The materials and energy may consist of hydropower energy transmitted from hydropower dams located at a remote location, built and other materials being carried into the cities for construction purposes, waste generated from urban areas getting dumped in landfills or other locations and finding its way to the oceans and rivers, waste produced during the manufacturing of food supplies to be consumed in urban areas, emission generated due to fuel consumption or changes in the biogeochemical cycles due to urban growth.

This input and output of material balance is the ex situ resource mobilization for urban growth due to trade and globalization and is mediated by anthropogenic subsidization of material and energy balance (Bai, 2016). However, the importance attributed to this phenomenon of urban growth, or the creation of urban ecosystems, is due to the scale at which it has developed and continues to do so with impunity. The urban population has increased from a mere '750 million (1951) to 4.2 billion (2018)' (Chapter 18; United Nations, 2018a, b) constituting more than 55% of the total world population. About 9.8 billion people will be living in urban areas by 2050 which will increase to 11.2 billion by 2100 (United Nations, 2018a, b). The resource base for such a massive population is made available at the cost of natural resources, environmental destruction and ecosystem services. Furthermore, apart from the environmental factors, the resource distribution and consumption in an urban ecosystem is not equitable since there is an influence of social factors, such as income, governance and policy, civic amenities, and so on.

Taking the example of CO_2 emission from electricity consumption from urban households in India, cities such as Allahabad had per capita emission of 12 kg CO_2 per capita per month, whereas Chennai emitted 81 kg CO_2 per capita per month (Ahmad et al., 2014). The reason has been attributed to the lifestyle and income disparities. Rural areas predominantly utilize traditionally solid fuels, which might be responsible for higher carbon emission from cooking activities. This kind of disparities within urban ecosystems also exist in different processes and components which determine the scale and magnitude of the effect of urban phenomena on its environment.

1.2 Social—ecological systems and urban metabolism

Cities and urban areas are human ecosystems where social, economic, biological and ecological components work together forming a system of feedback loops and interactions. These interactions in urban ecosystems are guided through human values and perceptions (Pickett and Cadenasso, 2013). Together, this forms the social—ecological system (SES) and determines the ecology of urban areas. Studies in the ecology of whole cities started in the 1970s centring on energy and nutrient cycling.

Energy flow through an ecosystem is considered unidirectional. It flows from the sun to the primary producers, consumers and decomposers and then to the nutrient pools across the food web. In urban ecosystems, the energy is consumed not only along the food chain but also to perform social and other basic activities, such as cooking, heating, cooling and travelling. As explained in an earlier example of urban electricity consumption, all activities using fuels and electricity contribute towards energy flow in an urban ecosystem which is different from the calorific content of food content. In a natural ecosystem, cycling of material also takes place, identified as the carbon, sulphur, phosphorus, nitrogen, oxygen and water cycle. Due to urbanization, these nutrient cycles are disturbed and modified. For example, due to input of fertilizers and pesticides, the natural flux across soil systems is modified, which leads to higher productivity as an immediate effect but lower fertility of soils over several years. This is one of the impacts of urban development. Urban areas are also considered the hotspot of consumption and waste generation.

The metabolic approach towards understanding the water supply and air and water pollution in cities originated from the biological concept of metabolism (Restrepo and Morales-Pinzon, 2018). The urban area is considered as an organism with dynamic functions maintaining the life of the urban system. The material and energy flow across its boundaries is compared to the way an organism or a cell takes in nutrients, converts them into energy and excretes the waste out of its body. In urban systems, material balance consists of natural resource requirements for activities such as housing and construction, transport, and so on (Schandl and Schaffartzik, 2015). It is simply how raw material or finished products are transferred to urban systems, and waste and transformed products come out. However, this flow of material takes place at economic and environmental costs.

Wolman's work in urban metabolism and ecosystems led to their recognition as an important area of research. Wolman's hypothetical city gave an estimate of material budget for food, fuel and water use, and sewage, refuse and air pollutants that a million US citizens would store and transform according to 1965 standard rates (Wolman, 1965). Such studies for whole cities are rare, and an in-depth analysis of a few cities by the UNESCO's Man and the Biosphere Programme in the 1970s gave further insight into urban metabolism studies (Bai, 2016). Urban metabolism is concerned with the flow and transformation of materials and energy in an urban setup. These are classified as inputs and outputs (Decker et al., 2000). It was found that the material balance of Hong Kong with a population of 5.5 million residents was approximately equal to that of Wolman's hypothetical US city (Decker et al. 2000). Furthermore, 3.65 million residents of Sydney, in 1990, metabolized as much as Wolman's hypothetical US city with the exception of high CO_2 levels. The cause would probably be the higher number of automobiles (Ibid).

Urban sustainability has a greater chance of success 'when the scales of ecological processes are well-matched with the human institutions charged with managing human−environment interactions' (Leslie et al., 2015). In the past few decades, the field of urban ecology transformed from studying ecology in the city to ecology of the city (Childers et al., 2014). This has led to the coupling of urban metabolism principles with human choices and preferences, giving rise to SES. Cities transform raw materials, fuel and water into the built structure, human biomass and waste. Energy flow and material transformation conceived as urban metabolism do not give a complete picture of these urban centres. The human aspect, when added to urban metabolism, provides a more holistic approach towards the study of these cities. Recently, this fact is being accepted and taken into account of urban system studies. The growth and development of cities is a process of organization in which human choices and preferences play a pivotal role, working in an ecological matrix.

Hence, this book deals with the emerging aspects of urban ecological studies from the perspective of SES. Urban ecology comprises a number of dimensions which have been outlined in this book, like, urban metabolism [Chapter 2], land use land cover change [Chapters 3 and 4], disaster risk management [Chapter 5], urban ecosystem services [Chapter 6], urban green space [Chapter 7], urban agriculture [Chapter 8], carbon emissions [Chapter 9], transport in cities [Chapter 10], urban air quality [Chapter 11], water management [Chapter 12], urban biodiversity [Chapter 13], waste management [Chapters 14, 15 and 23], climate change and human health [Chapter 17], urban heat island effect [Chapter 17], sustainable and smart cities [Chapters 18 and 19], urban design [Chapters 20 and 21], policy and management [Chapter 22] and nutrient fluxes [Chapter 16], among many others. This book discusses the conceptual undertakings and advances in the field of urban ecology. The next section gives a brief description of the state of research into urban ecology followed by a discussion on the major themes covered in this book.

2. State of research in urban ecology

2.1 Global trends in the past two decades (1999–2019)

'Urban ecology' was used as a keyword to search the Web of Science core database from 1999 to 2019. It was found that literature on urban ecology has grown from a mere 8 articles in 1999 to 158 articles in 2018. The period after 2008 experienced an exponential rise in the number of works of literature being published related to urban ecology (Fig. 1.1). The year 2009 also saw the publication of 'Planetary Boundaries: Exploring the Safe Operating Space for Humanity' by Rockstrom et al. (2009). It gave the concept of planetary boundaries for nine earth systems essential for humans to sustain themselves. However, the unprecedented growth of urban ecosystem with little regard to the ecological resilience has resulted in crossing over of some planetary boundaries. The latest research says that due to the development of society, certain systems, such as climate change, biodiversity loss, land and nutrient cycles (nitrogen and phosphorus), have 'gone beyond their boundary into unprecedented territories' (Steffen et al., 2015). This could be a possible reason for a large number of studies in this field now.

2.2 Country-wise division of urban ecology research (from 2009 to 2019)

Urbanization is expected to be led by the countries of Africa and Asia. India and China are expected to see an increase of one-third urban population by the end of 2020 (Shen et al., 2011). India and China, having the world's largest rural population, 893 and 578 million, respectively, will account for 35% of the urban population growth between 2018 and 2050 along with Nigeria (United Nations, 2018a, b). Asia houses 54% of the current world's urban population, followed by Europe and Africa (13% each). The pace of urbanization is expected to be the highest in low- and lower-middle-income countries (Singh et al., 2019). However, this is not reflected in the literature from the past 10 years. We found that the United States, England and Australia had the maximum number of publications in this field, followed by

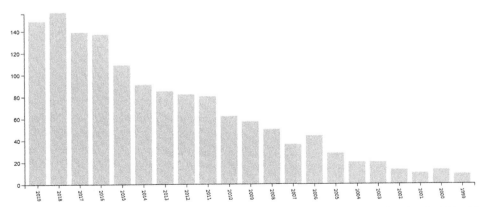

FIGURE 1.1   Publications related to 'urban ecology' for each year from 1999 to 2019 indexed in the Web of Science core collection (accessed on 05 December 2019).

China, Germany, Canada and Mexico (Fig. 1.2). African countries were represented by South Africa, and Asian countries were represented by China and Singapore in the top 25 results. This does not mean that various dimensions of urban ecology are not being researched in other countries; however, it does indicate that the transdisciplinary nature of urban ecology might be lacking in such studies.

The next section discusses the various themes covered in this book. It describes the conceptual undertakings and advances in the field of urban ecology covered in this book.

3. Urban ecology: concepts and definitions

The field of urban ecology can be approached in several ways, for example, from the perspective of material and energy balance, sustainable development in the form of economic, social and environmental sustainability and certain unique phenomena associated with urban growth, such as land-use land cover change and urban heat island effect, urban design and architecture and human-centric in the form of social equity and human health, leading to better resource management and sustainability, greenhouse gas (GHG) emission and climate change or ecosystem services and biodiversity. The transdisciplinary nature of this subject warrants understanding the nexus between human and ecological functions through the aforementioned mentioned lenses (Fig. 1.3). However, as pointed out by Simon et al. (2018), the transdisciplinary nature of coproduction is 'complex, time-consuming, and often unpredictable in terms of outcomes', and these issues gain greater importance when comparative studies are undertaken. More discussion on this aspect of urban ecology research has been done in the last chapter of this book. The following section describes the conceptual background, which would help the reader peruse through this book.

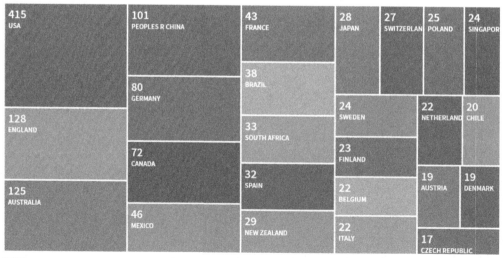

FIGURE 1.2 Tree diagram of the country and region-wise number of documents related to 'urban ecology' published between 2009 and 2019. *From Web of Science, accessed on 05 December 2019.*

3.1 The urban boundary

There is confusion of terminology used for describing the urban ecosystems, as the definition of a city boundary varies across countries making comparisons difficult. Due to advancements in geospatial and remote sensing technology, a growing scientific literature on the study of urban ecology is emerging, which warrants bringing forward the definition of the city at par with urban area boundary. In this section, we have first discussed the definitions of city and related terms, their inappropriateness in the implementation of urban ecological studies and data availability followed by suggestions.

Definitions for urban areas for city, town or any other administrative boundary are highly specific to the country. Political context determines these boundaries along with economic and social concerns (MacGregor-Fors, 2011). Urban land cover, material and energy balance, urban forestry and tree cover, urban planning, urban waste generation, pollution control and modelling, urban disaster mapping and many other fields require a geospatial boundary of constituent units to collect and analyze data. There have been attempts at defining terms used in the ecology of urban areas, but ultimately research from using such studies needs to be implemented on the ground, and thus it confronts the prevalent paucity of data and confusion in the functional, structural and administrative definitions of urban area boundaries. Acceleration and diversification of economic activities push the urban boundaries beyond their administrative or municipal limit. From an environmental point of view, the structure and function of the urban component in an urban ecosystem are more than that actually managed by the district or city administration.

FIGURE 1.3 Dimensions of urban ecology — a word of author keywords from each chapter in the book.

There are many definitions of the city according to different countries. How a city is defined generally depends on its population, presence of an administrative unit in the city and any other economic or social characteristic important for that country. Based on the number of people, cities have been defined by total population, population density or both. Some countries designated other terms for larger urban areas comprising more than one urban core such as urban agglomeration (India), urbanized area (United States) and conurbations (United Kingdom). A city has at least 50,000 population in Japan and the European Union, and 2500, 2000 and 200 population in the United States, Israel and Iceland, respectively. A list of some countries that have defined cities according to population is given in Fig. 1.4.

For a researcher, the first task before conducting any urban-based research is to identify and define the study area. Generally, district, city, urban agglomeration or block in case of rural areas is selected. The next step is to gather data and extract the boundary of the site. Information regarding the boundary of a city is generally not available in digital formats which makes the processing of data very difficult. If available, such digital information is out of date , for example, the urban boundaries of cities in India are expanding at a rate greater than that at which the administration works. This results in the presence of high-density urban patches classified as rural or outside the municipal boundary in local bodies' records and escaping full evaluation for urban landscapes. Satellite data and geographical information systems (GIS) play a vital role in landscape-based studies.

Confusion in data available for urban areas can be better understood by the following example, for example,. Data regarding population, literacy, number of households, area and employment sector are available at ward (sub-city) and village level for towns (urban) and blocks (rural), respectively. Calculation of secondary metrics and change in these quantities is possible for these categories at the aforementioned urban or rural units. Information regarding amenities and assets is available at subdistrict (tehsil) level. The boundary of a subdistrict is independent of the boundary of a city, town or village. Thus, metrics calculated from such data are applicable at different levels of urban areas, each having a different population, and thus pose a difficulty for researchers when calculating per capita metrics This example comes from India, where cities are constituted inside a district, however, it points out the confusing, often overlapping and sometimes absent data regarding urban areas. Other countries may have better systems of administrative demarcation, however, the point remains that in order to study urban systems, the data should reflect the ground reality.

Need for a uniform urban boundary becomes more apparent when we look at urban areas from a landscape perspective. The scale at which an urban area is perceived should match the scale at which it is expanding and information is available. The rate of urbanization should be taken into account to revise the definition of a city. Thus, the definition needs to be versatile and able to cope with rapid urban expansion as well as uniform to make comparisons across regions possible. A better way to define a city is by taking into account population, population density, employment and their concentration gradient identified through remote sensing and GIS. A GIS grid with these layers and a threshold value of concentration gradient for defining urban, semiurban and rural can be used similar to the methodology followed in a European Commission working paper (Dijkstra et al. 2018) but additionally having well-

City/Town by population

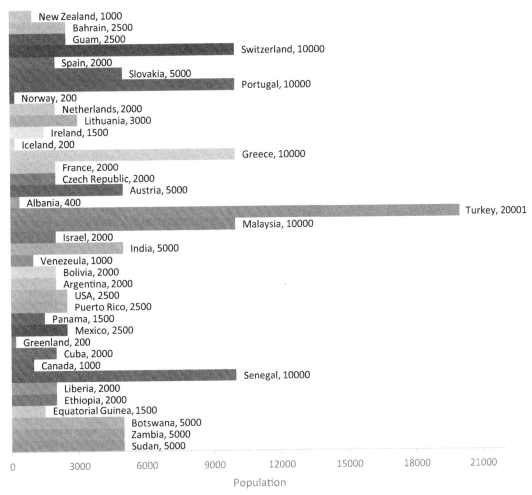

FIGURE 1.4   City/town by population in some countries (UNSD, 2015).

defined economic and employment thresholds similar to prerequisites already present in the definition of towns according to Census of India (2011). Threshold defining these values may differ from countries. This can result in a uniform definition of cities and needs to be investigated further.

3.2 Urban metabolism

Urban metabolism deals with urban sustainability indicators, GHG emission, policy analysis and their application to urban design (Kennedy et al., 2011). As mentioned earlier, the urban system depends on resources to sustain itself, in the form of a flow of materials and energy, and the various social and ecological interactions act like the 'metabolism' of living organisms. Urban metabolism is the study of the flow of matter and energy through a city providing a model for human and nature interactions.

3.3 Land use land cover change

Land cover change denotes a change in certain continuous characteristics of the land such as vegetation type, soil properties, and so on, whereas land-use change consists of an alteration in the way certain area of land is being used or managed by humans (Patel et al., 2019). This involves the transformation in the natural landscape due to urban growth. It is interesting to note that this change is responsible for a number of local and global effects, including biodiversity loss and its associated effects on human health, and the loss of habitat and ecosystem services (Patel et al., 2019). It is mainly driven by urban growth and is particularly important now for developing and underdeveloped countries. However, natural causes may result in land cover change, but land-use change requires human intervention (Joshi et al., 2016).

3.4 Modelling and remote sensing

To understand urban growth and quantify its impacts and future trajectories, certain mathematical tools are used, which is known as modelling. Modelling urban growth can provide better insights into managing urbanization and its related effects. Data collected from satellites and other sensors are used in modelling techniques. Development of modelling techniques which involve artificial neural network, fuzzy log and other nonparametric approaches have greatly increased the accuracy of mapping urban systems (Verma and Raghubanshi, 2019, 2020). Markov-chain and SLEUTH based on cellular automata are some of the models which help in the prediction of urban growth (Chandan and Bharath, 2018). Big data and crowdsourced data platforms are now increasing their impact on urban modelling (Johnson et al., 2017).

3.5 Disaster risk reduction

Rapid urban growth has resulted in unplanned settlements often with high population densities. It is found that the socially weaker sections of society inhabit these kinds of settlement (Chatterjee et al., 2015). Risk, due to natural and anthropogenic disasters, is increased in these places of unplanned built areas. Disaster-induced and rural-to-urban migration further puts a burden on urban resources (Chapter 5). Preventing this risk involves increasing the resilience of the people. This reliance includes a number of policy changes which invariably include urban design and social resilience in the form of education, income and demographics. These activities make up the disaster risk reduction strategies which have become more important due to the increase in the frequency of natural disasters due to climate change.

3.6 Economies of scale

This is an important concept in the field of urban ecology. It has been observed that cities follow scaling laws depending on their size. Bettencourt and West (2010) put forward three observations – (1) due to intense use of infrastructure, the space required per capita decreases;

(2) cocioeconomic activities increase leading to higher productivity; and (3) socioeconomic activities diversify leading to better opportunities. They showed that for doubling the population of a city, about 85% of infrastructure development is needed (Bettencourt and West, 2010). This indicates that cities essentially grow like an organism with some savings as the size increases. These savings are in the form of cost or material benefits which are made due to the increase in scale. However, this may not indicate that such growth is necessarily sustainable.

3.7 Urban ecological footprint

The urban ecological footprint is essentially the amount of earth needed to sustain and urban areas and recycle or absorb its waste and emissions. It denotes the number of resources needed to provide the necessary raw materials (natural resources, ecosystem services, etc.) and the earth's capacity to absorb or recycle the waste material generated including gaseous emission (like GHGs). The resource utilization by urban areas results in waste generation and emissions. The magnitude of this generation has crossed the critical threshold of planetary boundaries (Steffen et al., 2015).

3.8 Urban sustainability indicators

Indicators are an essential part of assessing the progress of any system. Urban sustainability indicators include a number of dimensions dealing with various aspects of urban systems, including policy and governance, demographics, economics, environment and energy. Indicators could be in the form of gross domestic product, Gini coefficient or ambient air quality. Indicators also provide an understanding of the phenomena being studied (Verma and Raghubanshi, 2018).

3.9 Smart city

According to the International Business Machines (IBM), a smart city is one that makes optimal use of the available information about various processes to better deliver and recognize its operations and make optimum use of resources available by balancing the social, commercial and environmental needs of the city (Nam and Pardo, 2011). This concept has grown to involve sustainability as a part of information and communication technology used to create a smart city. Efficiency and application of information and communication technology are the essential parts of a smart city.

3.10 Sustainable city

Urban ecosystems 'which are ethical, effective (healthy and equitable), zero-waste generating, self-regulating, resilient, self-renewing, flexible, psychologically fulfilling and cooperative' can be termed as sustainable (Newman and Jennings, 2012; Dizdaroglu, 2015).

3.11 Human health

Humans shape the ecology of cities as well as are influenced by the type of environment they create. Human health is an emerging aspect of urban ecology research, especially due to the effect of the urban ecosystem on human health, in the form of lack of green spaces, air quality, urban heat island effect, water and air pollution, psychological and mental health, and urban design (Giles-Corti et al., 2016).

3.12 Integrated approach

Ostrom (2009) suggested that the study of SES requires study of the 'complex, multivariable, nonlinear, cross-scale and changing systems'. Urban ecology when observed as an integrated and transdisciplinary subject would be able to offer better insights into urban sustainability, and thus, an integrated approach is required in this discipline.

3.13 Governance and planning

Implementation of sustainability practices to ensure a healthy urban ecosystem remains a challenge (Verma and Raghubanshi, 2018). Verma and Raghubanshi (2018) identified two major challenges in the application of sustainability monitoring programs in urban areas as the selection of relevant indicators followed by their application. However, sustaining such measures requires repeated assessments and policies tailored according to local conditions. Governance and planning play an important role in determining the nature of urban ecosystems, including education, urban design and planning, environmental laws and their implementation; hence, they remain one of the most essential features in this subject.

4. Conclusions

The preceding section described some of the concepts including the urban boundary, urban health, modelling and remote sensing, smart and sustainable city and indicators of sustainability, which have been used in this book. Since cities contain a number of components which are mainly created by humans, they function at different trajectory than natural ecosystems. Currently, cities are plagued by several problems such as biodiversity loss, air quality, green spaces, lack of open space, and so on. However, it is believed that cities are resilient ecosystems, and better performing cities continue to do so for several decades. Policy based on an understanding of the workings of different components in urban ecosystems would help in creating a sustainable future.

References

Ahmad, S., Mathai, M.V., Parayil, G., 2014. Household electricity access, availability and human well-being: evidence from India. Energy Policy 69, 308−315.
Andrewartha, H.G., Birch, L.C., 1954. The Distribution and Abundance of Animals. Univ. of Chicago Press.
Bai, X., 2016. Eight energy and material flow characteristics of urban ecosystems. Ambio 45 (7), 819−830.
Bettencourt, L., West, G., 2010. A unified theory of urban living. Nature 467 (7318), 912−913.
Chandan, M.C., Bharath, H.A., 2018. Modelling urban transition using cellular automata based Sleuth modelling. In: 2018 IEEE Symposium Series on Computational Intelligence (SSCI). IEEE, pp. 1656−1663.
Chatterjee, R., Shiwaku, K., Gupta, R.D., Nakano, G., Shaw, R., 2015. Bangkok to Sendai and beyond: implications for disaster risk reduction in Asia. International Journal of Disaster Risk Science 6 (2), 177−188.
Childers, D.L., Pickett, S.T., Grove, J.M., Ogden, L., Whitmer, A., 2014. Advancing urban sustainability theory and action: challenges and opportunities. Landscape and Urban Planning 125, 320−328.

Decker, E.H., Elliott, S., Smith, F.A., Blake, D.R., Rowland, F.S., 2000. Annual Review of Energy and the Environment 685—740.

Dijkstra, L., Florczyk, A., Freire, S., Kemper, T., Pesaresi, M., Schiavina, M., 2018. Applying the degree of urbanisation to the globe: A new harmonised definition reveals a different picture of global urbanisation. In: Proceedings of the 16th IAOS Conference: Better Statistics for Better Lives, pp. 19—20. Paris, France.

Dizdaroglu, D., 2015. Developing micro-level urban ecosystem indicators for sustainability assessment. Environmental Impact Assessment Review 54, 119—124.

Giles-Corti, B., Vernez-Moudon, A., Reis, R., Turrell, G., Dannenberg, A.L., Badland, H., Owen, N., 2016. City planning and population health: a global challenge. The lancet 388 (10062), 2912—2924.

Johnson, B.A., Iizuka, K., Bragais, M.A., Endo, I., Magcale-Macandog, D.B., 2017. Employing crowdsourced geographic data and multi-temporal/multi-sensor satellite imagery to monitor land cover change: a case study in an urbanizing region of the Philippines. Computers, Environment and Urban Systems 64, 184—193.

Joshi, N., Baumann, M., Ehammer, A., Fensholt, R., Grogan, K., Hostert, P., Reiche, J., 2016. A review of the application of optical and radar remote sensing data fusion to land use mapping and monitoring. Remote Sensing 8 (1), 70.

Kennedy, C., Pincetl, S., Bunje, P., 2011. The study of urban metabolism and its applications to urban planning and design. Environmental Pollution 159 (8—9), 1965—1973.

Leslie, H.M., Basurto, X., Nenadovic, M., Sievanen, L., Cavanaugh, K.C., Cota-Nieto, J.J., Nagavarapu, S., 2015. Operationalizing the social-ecological systems framework to assess sustainability. Proceedings of the National Academy of Sciences 112 (19), 5979—5984.

MacGregor-Fors, I., 2011. Misconceptions or misunderstandings? On the standardization of basic terms and definitions in urban ecology. Landscape and Urban Planning 100 (4), 347—349.

Nam, T., Pardo, T.A., June 2011. Conceptualizing smart city with dimensions of technology, people, and institutions. In: Proceedings of the 12th Annual International Digital Government Research Conference: Digital Government Innovation in Challenging Times. ACM, pp. 282—291.

Newman, P., Jennings, I., 2012. Cities as Sustainable Ecosystems: Principles and Practices. Island Press.

Odum, E.P., 1975. Ecology, the Link between the Natural and the Social Sciences. Oxford and IBH Publishing.

Ostrom, E., 2009. A general framework for analyzing sustainability of social-ecological systems. Science 325 (5939), 419—422.

Patel, S.K., Verma, P., Singh, G.S., 2019. Agricultural growth and land use land cover change in peri-urban India. Environmental Monitoring and Assessment 191 (9), 600.

Pickett, S.T.A., Cadenasso, M.L., 2013. In: Leemans, R. (Ed.), Ecological Systems. https://doi.org/10.1007/978-1-4614-5755-8.

Restrepo, J.D.C., Morales-Pinzon, T., 2018. Urban metabolism and sustainability: precedents, genesis and research perspectives. Resources, Conservation and Recycling 131, 216—224.

Rockström, J., Steffen, W., Noone, K., Persson, Å., Chapin, F.S., Lambin, E.F., Nykvist, B., 2009. A safe operating space for humanity. Nature 461 (7263), 472—475.

Schandl, H., Schaffartzik, A., 2015. Material Flow Analysis.

Shen, L.Y., Ochoa, J.J., Shah, M.N., Zhang, X., 2011. The application of urban sustainability indicators—A comparison between various practices. Habitat International 35 (1), 17—29.

Simon, D., Palmer, H., Riise, J., Smit, W., Valencia, S., 2018. The challenges of transdisciplinary knowledge production: from unilocal to comparative research. Environment and Urbanization 30 (2), 481—500.

Singh, N., Mhawish, A., Ghosh, S., Banerjee, T., Mall, R.K., 2019. Attributing mortality from temperature extremes: a time series analysis in Varanasi, India. The Science of the Total Environment 665, 453—464.

Steffen, W., Richardson, K., Rockström, J., Cornell, S.E., Fetzer, I., Bennett, E.M., Folke, C., 2015. Planetary boundaries: guiding human development on a changing planet. Science 347 (6223), 1259855.

United Nations, Department of Economic and Social Affairs, 2018a. World Population Projected to Reach 9.8 Billion in 2050, and 11.12 Billion in 2100. Retrieved on 31 December 2018. https://www.un.org/development/desa/en/news/population/world-population-prospects-2017.html.

United Nations, Department of Economic and Social Affairs, 2018b. 68% of the World Population Projected to Live in Urban Areas by 2050, Says UN. Retrieved on 31 December 2019. https://www.un.org/development/desa/en/news/population/2018-revision-of-world-urbanization-prospects.html.

UNSD — United Nations Statistics Division, 2015. Table 6 — Demographic Yearbook 2015. Available at: https://unstats.un.org/unsd/demographic-social/products/dyb/documents/dyb2015/Notes06.pdf. (Accessed 18 December 2019).

Verma, P., Raghubanshi, A.S., 2018. Urban sustainability indicators: challenges and opportunities. Ecological Indicators 93, 282—291.

Verma, P., Raghubanshi, A.S., 2019. Rural development and land use land cover change in a rapidly developing agrarian South Asian landscape. Remote Sensing Applications: Society and Environment 14, 138—147.

Verma, P., Raghubanshi, A., Srivastava, P.K., Raghubanshi, A.S., 2020. Appraisal of kappa-based metrics and disagreement indices of accuracy assessment for parametric and nonparametric techniques used in LULC classification and change detection. Modeling Earth Systems and Environment 1—15.

Wolman, A., 1965. The metabolism of cities. Scientific American 213 (3), 178—193.

Further reading

Ahmad, S., Baiocchi, G., Creutzig, F., 2015. CO_2 emissions from direct energy use of urban households in India. Environmental science & technology 49 (19), 11312—11320.

Urban metabolism: old challenges, new frontiers, and the research agenda ahead

Daniela Perrotti

University of Louvain, Louvain-la-Neuve, Belgium

1. Urban metabolism: from concept to methods

1.1 One concept, a variety of methods

Urban metabolism (UM) is an interdisciplinary research field, spanning across disciplines as different as industrial ecology, urban ecology, political ecology and political-industrial ecology (Wachsmuth, 2012; Castàn Broto et al., 2012; Newell et al., 2017). Each of these disciplines encompasses different schools, which focus on a wide range of methods and diversified scales of analysis. In quantitative terms, industrial ecology is the most influential research path in UM studies (Newell and Cousins, 2014). Industrial ecology approaches to resource accounting have extended beyond the original focus on the metabolism of industrial systems and industrial symbiosis to include the broader scale of cities (Bai, 2007; Kennedy et al., 2012). In industrial ecology, UM is defined as 'the sum total of the technical and socioeconomic processes that occur in cities, resulting in growth, production of energy, and elimination of waste' (Kennedy et al., 2007). In UM research, cities are studied as open systems whose metabolism is the result of the interactions with other (close or remote) anthropogenic systems and the natural environment. Beyond the concept, industrial ecology UM research provides analytical tools and methods to assess the resource intensity of urban systems and, when applied in policy and practice, to enable resource use optimization.

Material Flow Analysis (MFA) is the most used method for resource accounting in industrial ecology (Cui, 2018; Kennedy et al., 2011). Rather than a single model, it consists of a family of mass-balance models that can vary from national to local scales and includes aggregate materials and energy accounts as well as assessments of a single material or substance. The Eurostat's Economy-wide material flow accounting (EW-MFA) is the most widely spread method within the MFA family and represents the basis of standard statistical reporting in the EU (Eurostat, 2001). The EW-MFA was introduced in the late 1960s (Ayres and Kneese, 1969) and further developed in the 1990s (Baccini and Brunner, 1991; Bringezu, 1997). It was initially conceived as a standardized method for flow accounting at the scale of national economies. Hammer et al. (2003) adapted the EW-MFA at the city and regional level, opening the path to a still growing number of applications to urban systems (e.g., Hammer and Giljum, 2006; Barles, 2009; Voskamp et al., 2017; Bahers et al., 2019). In the EW-MFA, only material input and output flows entering or leaving the system are considered, excluding in-boundary processes and dynamics associated with resource use. Hence the model results in a 'black box' representation of the socioeconomic system itself. The method has in the past decades gained maturity thanks to considerable efforts by the scientific community to work consistently toward methodological harmonization and standardization across existing datasets (Fischer-Kowalski et al., 2011). In the past few years research efforts have concentrated on integrating the EW-MFA with other methods for the assessment of urban resource flows (Daigger et al., 2016) and with ecosystem services frameworks (Perrotti and Stremke, 2020). Substance Flow Analysis (SFA) belongs to the same mass-balance family of accounts as the Eurostat EW-MFA. It concentrates on the analysis of substance fluxes and key nutrients (primarily carbon, nitrogen and phosphorus), which are assessed either separately or coupled with other fluxes. SFA applications at the city level range from analysis of individual elements responsible for human-induced water, air, or soil contamination in urban ecosystems (or contamination risks) (Barles, 2010), to investigation into the mutual dependency of multiple elements in water-agro-food systems (Verger et al., 2018; Esculier et al., 2019), and to coupled material and energy fluxes in an 'urban nexus' perspective (Chen and Lu, 2015).

Beyond MFA and SFA, other popular methods for resource accounting within industrial ecology include Emergy-based analysis and Energy Flow Accounting (EFA). Emergy-based analysis is characterized by the use of the 'emergy' concept as a basis for resource accounting (Odum, 1996). Emergy is defined as the total amount of solar energy that is used directly and indirectly to deliver a product or a service. As exemplified in the study of Beijing from 1990 to 2004 (Zhang et al., 2009), in emergy-based analysis, the studied UM system includes the socioeconomic system and the natural subsystems (the natural capital included within the city's administrative boundaries). Solar equivalent joule (SEJ) is used as a common unit to account for all flows, including renewable sources (wind, rain, rivers, earth cycles), indigenous nonrenewable resources (e.g., coal, iron ore, sand, gravel) and all other resources imported from other systems (fuels, goods, services). The emergy issued from renewable and nonrenewable indigenous sources is considered regardless of the amount of final energy used in the socioeconomic subsystem. Despite the limits resulting from the use of a single unit for different energy flows and qualities (Hau and Bakshi, 2004), the emergy method provides a clear picture of the contribution of natural energy to the system's economy. Since the beginning of the 2000s, EFA has been established as an alternative method to the EW-MFA in industrial ecology research (Haberl, 2001a; Krausmann and Haberl, 2002). EFA is grounded on a socioeconomic perspective of the energy metabolism of human organizations and emphasizes the central role of energy flows in metabolic analysis, integrating technical energy (power and heating) with biomass flows (wood, food, feed and biomaterials). EFA adapts the set of indicators used in MFA to account for all streams of energy (including energy from renewable sources) and energy-rich materials that cross the system boundary based on their gross calorific value, regardless of the purpose for which they are used (Haberl, 2001a). All energy inputs that build up the biophysical structures of the societies are considered alongside the biomass combusted to generate heat and/or electricity. EFA also tracks the main energy conversion processes throughout the system (primary-final-useful energy) for both technical energy (used in artefacts) and nutritional energy (for humans and livestock). The EFA's focus on biomass is essentially due to the historical perspective adopted in these studies. This allows for comparison between different types of societal organizations, such as hunter-gatherer and agricultural societies, in which technical energy from fossil sources was not as dominant as in industrial societies. System boundaries are rarely restricted to a single administrative urban unit and EFA is mostly performed at the regional, national, or supranational scale (Krausmann, 2013). Moreover, through the introduction of additional indicators such as Human Appropriation of Net Primary Production (HANPP), EFA also offers valuable insights into the relations between land use and resource flows (Haberl et al., 2006), placing human social activities (e.g., economic production or the use of technologies) in a broader ecological context (Haberl, 2001b). EFA is regarded as a valuable tool to integrate socioeconomic and natural flows in a comprehensive UM framework (Golubiewsk, 2012) and can provide an interdisciplinary knowledge-base for industrial ecology, ecological economics and human ecology to jointly advance UM research (Barles, 2010).

Recent years have seen the rise of political-industrial ecology, an interdisciplinary field concerned with the cross-fertilization of epistemologies and the hybridization of qualitative and quantitative methodologies used in industrial and political ecology (Newell et al., 2017). In political ecology, socioeconomic flows of the UM are analyzed as the result of the interactions between power, institutional structures, politics, social and human capital (Swyngedouw and Heynen, 2003; Castàn Broto et al., 2012). Criticisms are addressed to 'orthodox' industrial ecological methods for their limited engagement with the social and political challenges arising from unequal access to resources across societies or stakeholder groups (Dalla Fontana and Boas, 2019) and uneven relations of power in the governance of natural resources (Gandy, 2004).

On the urban ecology side of the UM research spectrum, synergies and tradeoffs between environmental and social concerns were central to the studies carried out since the 1970s within the UNESCO—Man and the Biosphere Programme (UNESCO, UNEP, 1971). Follow-up research by the United Nations University Institute for the Advanced Study of Sustainability (UNU-IAS, former United Nations University Institute of Advanced Studies) has allowed fostering an 'urban ecosystems' approach to resource management and policy development in cities (Marcotullio and Boyle, 2003). Similar interdisciplinary questions were explored in experimental programmes such as the Long Term Ecological Research Program in the United States (LTER) that started in the 1980s and included 24 different projects (Hobbie et al., 2003). A long-lasting dialogue across disciplines was foreseen as a means to achieve shared theoretical bases to understand cities as dynamic, complex and adaptive systems linking ecology and society (Grimm et al., 2000; Pickett et al., 2011). More recently, interdisciplinary exploratory research initiatives were carried out within the US National Science Foundation's Urban Long-Term Research Areas (ULTRA-Ex) with pilot projects developed in over 20 US cities. The programme specifically tailored the opening of the 'black box' of city systems and the study of the dynamic interactions between people and natural ecosystems occurring in urban settings as well as their influence on cities' liveability and the health and functionality of nonurban systems. For example, in the *Ecosystem Study of Baltimore, Maryland, and the District of Columbia—Baltimore City ULTRA-Ex* project, biogeophysical models of water, carbon and nitrogen cycles from the watershed to the parcel scale were coupled with econometric and structural models simulating locational choices and patterns of land development; this aimed to inform sustainability policy scenarios for enhanced water quality and carbon sequestration (Grove et al., 2013). Other urban ecology experiments include the research, development and demonstration projects within the Public Interest Energy Research Program (PIER) established by the California Energy Commission in 1998. A research roadmap issued at the end of the programme proposed an expanded energy and resource use accounting framework including life cycle cost assessment, sociodemographic data and policy drivers underlying energy use. Outcomes of these projects provide evidence of the benefit of combining socioeconomic and environmental perspectives with UM research (Pincetl and Bunje, 2010; Pincetl et al., 2012).

When navigating through the aforementioned projects, one could argue that the strong interdisciplinary focus of contemporary ecological science links back to Eugene P. Odum's original idea of a 'new ecology' (Odum, 1977). E. P. Odum intended ecology as a conceptual framework and a project which should benefit society through linking physical and biological processes, combining holism and reductionism and bridging social and natural sciences. Ecology as a new 'integrative science' provides fertile soil to experiment with the integration of biophysical and socioeconomic parameters within the same system framework, bridging communication gaps and combining with economics methods (Barret and Odum, 1998; Barret, 2001). In line with this ambition, this chapter will focus on the interoperability of industrial and urban ecology methods and on the rise of new integrated approaches in ecological science as a way forward to face current challenges and identify opportunities for advancing UM research.

1.2 A variety of practical applications and end-users

Practical applications of UM research have developed in fields as varied as the definition of sustainability and environmental pressure indicators (e.g., Wood, 2012), the development of Decision Support Systems (e.g., Chrysoulakis et al., 2013), GHG emission accounting tools (e.g., Mohareb and Kennedy, 2012), and urban planning and design (e.g., Brugmans and Strien, 2014; Metabolic, 2015). There is nowadays a broad consensus that UM assessments can have significant impact on the work of professional designers and planners and help streamline urban planning strategies toward resource optimization (Galan and Perrotti, 2019). For urban planners and designers, an UM perspective is key to understanding the natural and anthropogenic processes that sustain socioeconomic activities and ecological cycles in humans' physical environments. These include both present and projected spatial and temporal dynamics through which cities gather, transform and use biotic and abiotic resources to ensure their functioning and, when relevant, growth. Applications of UM studies in design extend from mono- or multiprocess engineering systems and utility infrastructure (e.g., water, heat and electricity, waste cycling) to urban form at the neighborhood and city scale. If engineering applications represent a traditional focus of industrial ecology, spatial design strategies have gained increased attention across the industrial ecology research community over the last two decades (Ibañez and Katsikis, 2014; Galan and Perrotti, 2019). A better understanding of the interdependency between scopes and scales of these two design domains is a major challenge ahead and an essential condition for enhancing the applicability of UM models in policy and practice.

Following-up from the abovementioned practical applications, end-users of UM studies range from urban sustainability analysts (cf. previous section) to environmental and civil engineers, policymakers and urban planning and design professionals. The uptake of the UM concept and the applicability of UM studies may vary substantially according to the targeted end-users, their background, expertise, specialist and tacit knowledge, urban sustainability visions, as well as specific goals and priorities (Taleghani et al., 2020). Consequently, the design of UM assessment methods with the profile of the targeted end-users in mind, or in collaboration with them, is crucial for successful incorporation of the results of UM studies in policy and urban planning practice (Perrotti, 2019). Awareness of key characteristics and needs of end-user profiles and more structured collaborative approaches for codesigning resource accounting methods can increase the meaningfulness of UM research and its relevance for policy and practice. Moreover, the integration of end-users' input (on key characteristics of the studied urban system or specific policy orientations) since early stages of the research can favour successful UM knowledge exchange between science and practice. Knowledge coproduction in UM research is essential not only to favour the uptake of the UM concept by professionals but also for the development of assessment methods that can better serve practitioners' daily work and catalyze the move from research to action (Newell et al., 2019). With the considerable success known by the field in recent years, topics such as the actionability of the UM concept and UM theory and the delivery of more practice-relevant UM methods are constantly moving up in both industrial and urban ecology research agendas (Cui, 2018; Perrotti, 2019).

2. Challenges and new frontiers for urban metabolism research

2.1 Understanding distinct conceptual underpinnings and common methods across ecological sciences

There exist both conceptual differences and methodological commonalities between industrial ecology and urban ecology approaches in UM research. Understanding them can help identify common grounds and opportunities for cross-fertilization and more integrated work.

On the conceptual level, the main difference between industrial ecology and urban ecology approaches consists of their use of the 'organism' analogy (industrial ecology) and the 'ecosystem' concept (urban ecology) to refer to the city and to study its metabolism. A recurring tendency to use these two terms nearly interchangeably could be observed in industrial ecology studies until at least a few years ago (Golubiewsk, 2012; Kennedy, 2012). Interdisciplinary work across the two communities (Newell and Cousins, 2014; Bai, 2016) has favoured substantial progress and increased clarity and consensus on the fact that rather than organisms, cities are ecosystems and not analogous to them (Pickett et al., 1997; Grimm et al., 2008). Tansley (1935) originally coined the term 'ecosystem' to describe the constant interchange among organic and inorganic components in a biome, i.e., between the living organisms (individual plants and animals) and between these and all the inorganic elements that compose their environment. He did not attempt to translate these interchanges into energy and material flows, which was done later by Lindeman (1942), with the metabolism idea applied to ecosystems subsequently popularized by Odum (Fischer-Kowalski, 1998), nor did he refer to the human-dominated or urban ecosystems in particular. However, his contribution has the merit to have opened the discussion on which level is the most appropriate to study a city's degree of heterotrophy through consideration of the relationship with its environment and among its biotic and abiotic components (Golubiewsk, 2012). The urban ecosystem concept as defined in contemporary urban ecological science (Cadenasso and Pickett, 2013) is grounded in an understanding of cities as human-dominated systems (and, as such, different from 'natural' and said 'wild' ecosystems), characterized by interrelations and feedback loops among material cycles and energy flows (rather than linear input-output dynamics). Here regulating and governing mechanisms such as policy and planning play a crucial role in shaping social and ecological processes (Bai, 2016). Differentiating between the 'organism' and 'ecosystem' levels when referring to cities from an UM perspective is more than a question of semantics. Confining UM within the limits of the organismal analogy can limit the effective use of scientific principles and frameworks in analyzing how cities function and their relationship with the surrounding environment, as well as hamper the consolidation of a common knowledge base for more interdisciplinary work (Golubiewsk, 2012). In general, moving beyond organismic and biological metaphors to properly organize comprehensive knowledge frameworks and city-specific system analytics reflects a growing concern in systems thinking (Yang and Yamagata, 2019). A call for different urban disciplines to coalesce around a 'new science of cities' has been expressed in the past decade (Batty, 2013), stressing the need for streamlining theory and methods to better describe the intricate structure of the networks and flows that compose urban systems.

From a methodological standpoint, mass balance methods are used in both industrial and urban ecology to measure urban resource flows in order to understand drivers of human control upon them. Input-output studies of biogeochemistry fluxes (water, nutrients, carbon and other substances) and flows of useful and waste energy in ecosystems have a strong research tradition in urban ecology (McHale et al., 2015). Complementarily, industrial ecology EW-MFAs of cities concentrate on the account of flows of materials and technical energy that make up the economy of urban systems. MFA studies that consist of a mass-balance model for a single element (e.g., metals) normally concentrate on the pools and fluxes required for economic consumption, manufacturing and recycling process, as well as on waste generation.

In general, industrial ecology studies focus on applications in human-dominated processes and systems, although using the same mass-balance methods as in ecosystem ecology. The call for combining engineering with traditional biogeochemical models has been already expressed more than a decade ago, pointing, for example, to the need for the refinement of mechanical urban water modelling with data on soil and water processing through vegetation (Kaye et al., 2006). Building on common mass-balance methods, cross-fertilization between the two families of models can lead to the development of more comprehensive UM knowledge frameworks. Multielement mass-balance accounts can help identify the links between socioeconomic and biophysical UM dynamics, as well as help understand the coupling and decoupling between human-dominated flows and biogeochemical cycles in cities (Bai et al., 2016).

2.2 Expressing interdependency between biogeochemical cycles and socioeconomic flows of the urban metabolism

In industrial ecology, the need to expand traditional engineering accounting models and to couple them with urban ecology methods has been expressed in previous research (Pincetl et al., 2012; Perrotti and Iuorio, 2019). For example, the use of the ecosystem service concept and the integration of provisioning and regulating services in the Eurostat's EW-MFA can help express the contribution of urban green infrastructure towards the optimization of UM flows (Perrotti and Stremke, 2020). More generally, the incorporation of urban biogeochemical models in material and energy flow studies can help assess the biotic-abiotic interactions occurring heterogeneously within the urban natural capital (currently only scarcely accounted in industrial ecology research) as a part of the UM (Perrotti and Iuorio, 2019).

In urban ecology, an 'ecology *of* the city' approach (as complementary to an 'ecology *in* the city') has emerged in the 1990s as a project aiming at broadening the focus from just the vegetated areas to the entire urban social-ecological system and to reflect the multidisciplinary scope of urban ecological sciences (Pickett et al., 1997; Grimm et al., 2000). This approach strives for synthesizing human and social behaviours alongside purely ecological dynamics of specific species and organisms and for opening the 'black box' of urban systems to look into their physical, biological and social heterogeneities (Cadenasso et al., 2007, 2013).

Integrated socioeconomic and biogeochemical understanding of the UM in industrial and urban ecology becomes even more relevant when moving from analysis to policy implementation and action. Human-subsidized biogeochemical flows have been demonstrated to have a strong impact on biodiversity and the provision of ecosystem services in cities (DeStefano and DeGraaf, 2003). Hence their intensification as a consequence of strategies aiming at enhancing cities' self-sufficiency (e.g., intensification of urban agriculture initiatives) may result in ecological misfunctioning of both the urban system itself and the surrounding habitats (e.g., eutrophication of waterbodies) (Perrotti, 2012; Bai, 2016). This suggests that differentiation of resources based not only on quantity (mass) but also on the quality of the embodied substances (nutrients) is central to the design of effective resource optimization strategies. For example, in the case of water management policy, harmonization of quality and quantity accountings, linking analysis of the magnitude of wastewater flows with geochemical composition of nutrient flows and sediments (water quality parameters and pollutant concentration), can point to opportunities for water harvesting at the plot level and, potentially, sewage separation projects. Urban MFA studies in which water flows are accounted as a part of the total material balance traditionally do not include storm water in the wastewater flow and only concentrate on sewage from piped/drinking water (e.g., the MFA of Sydney, Newman, 1999). A recent MFA of Amsterdam is amongst the few exceptions (Voskamp et al., 2017). Storm water flows are accounted alongside piped drinking and nondrinking water, since both flows are treated at the main local, publicly held treatment plant (combined sewage), which implies higher operational, management and maintenance costs for the local authority. When coupled with geochemical analysis of storm water flows across seasons, this type of analyses can help identify the risk for chemical pollution of soils and water bodies from runoffs. Hence they can inform decisionmaking on whether sustainable drainage systems (SuDSs) may represent an ecologically and economically viable option in a given context. In general, it is paramount to interpret results of industrial ecology material balance studies (such as EW-MFA) based on consideration of reactants ratios in each flow analyzed (Kaye et al., 2006). Beyond the calculation of material flow mass, the relative proportion of nutrient elements in each mass unit per material flow is crucial to understand the behaviour of that flow when it becomes waste products and impacts downstream recipients both in-boundary and within other systems. For example, in the case of water management, consideration of quality aspects should involve studying the impacts of optimized and decentralized water retention/absorption solutions to reduce runoff and flood control (e.g., bioswales, rain gardens integrated in parking lots) on sediment transport, pollutant concentrations, and water quality at the watershed scale (Pataki et al., 2011). The same applies to human-controlled airflow regulation (e.g., windbreaks through screening vegetation, street trees, or, in some cases, urban form), which may represent an opportunity for passive cooling at both the building and open-space level. Consideration of air quality aspects is paramount to avoid exposure of urban populations to atmospheric pollutants and hence should be coupled with consideration of particulate matter (PM) sequestration capacity by street trees or other urban vegetation (Tiwary et al., 2009; Nowak et al., 2012). Quality aspects are indeed only rarely included in ecosystem service assessment and monitoring at the urban scale, which in general rest on GIS and remote sensing techniques and concentrate on quantification of biomass volumes (Wirion et al., 2017). Policy focusing only on reducing the magnitude of energy and material inflows to improve cities' self-sufficiency based on industrial ecology research may carry implications on ecosystem nutrients and energy provisioning structures, which can generate undesired side-effects at the urban and regional ecology scale (Bai, 2016). Cost-effectiveness at the urban system level (local authorities' budgets for urban infrastructure and utility management) needs to be coupled with ecological logics which in many cases extend beyond the administrative boundaries of the studied socioeconomic metabolism. Hence, coupling ecological and economics logics as argued in Odum's new integrative science requires cross-scale analysis to overcome methodological limitations of established models and rests on the willingness of public and private stakeholders to engage in alternative governance systems to change current resource management models.

Similar concerns can be raised on the other side of the UM research spectrum. Urban-ecology-extended mass balance studies have traditionally mainly focused on human-influenced ecological processes and the biogeochemical cycles of key nutrients such as carbon, nitrogen and phosphorous (Churkina, 2008). They need to be further expanded to incorporate also the study of other purely anthropogenic material and energy flows (including products and wastes) (Bai, 2016). A sign of relevant progress in this direction comes from the increasing number of studies incorporating both MFA and SFA methods as a means to establish a broader knowledge base on potentials for optimization across ecology and economy (Cui, 2018). They provide an accounting of anthropogenic flows and cycles linked to manufacturing and consumption, such as fossil fuels, water and construction materials and waste (e.g., excavated soil) while illustrating the biogeochemical breakdown of UM flows and level of contamination produced by the chemical substances incorporated in them. Further, working across urban scales (from the dwelling to the city and their surrounding habitats and landscapes) can help better understand the impact of human choice on ecosystems' functioning and vice versa. Results of multiscalar, integrative ecological research such as the above-mentioned LTER and ULTRA-Ex projects have shed new light on how the ecologies of an urban system and its biophysical structure react to human management, intervention, policy and design (Cadenasso and Pickett, 2013).

2.3 Linking urban metabolism research with design: systematizing the growing evidence-base on nature-based solutions

It has been widely acknowledged in the literature that the reference to the UM concept is of great interest for developing analytical tools that can broaden designers' understanding of the material and energy requirements of cities and support the optimization of resource demand through urban design strategies (Galan and Perrotti, 2019). In particular, the contribution of green and blue infrastructure design towards the cycling of material and biophysical flows and emission abatement has represented a strong focus in recent research (Hansen and Pauleit, 2014). Nature-based solutions (e.g., combined heat-and-power systems from biowaste, sustainable drainage systems for decentralized storm water management, PM capture by urban trees) provide an extensive reservoir of techniques and strategies to mitigate urban resource demand and to reduce the magnitude of associated carbon and pollutant emissions through the supply of provisioning and regulating ecosystem services (Hansen and Pauleit, 2014). Their fast growth in cities as well as the increasing uptake of the ecosystem service concept in urban policy (Hansen et al., 2015) points to the need to structurally address the role of green and blue infrastructure (and natural systems in general) in minimizing the resource-intensity of cities. Despite its relevance for UM research, there remains little evidence of how UM methods can help advance the scholarly discourse on nature-based solutions and their potential to improve resources management in cities. The EW-MFA model does not provide sufficient scope for systemic analysis of the potential of nature-based solutions to mitigate urban resource demand (Perrotti and Iuorio, 2019). Consequently, nature-based solutions are rarely addressed in UM-informed policy and circular economy action plans and often incorporated only in a restricted number of sectoral measures (e.g., energy recovery from organic waste). This limits the opportunities for systemic decisionmaking and the formulation of comprehensive resource-management strategies. As argued elsewhere, a new frontier in UM research consists of facilitating the application of industrial ecology mass-balance methods and urban ecology biogeochemical models in the planning and design of urban green infrastructure (Perrotti and Stremke, 2020). This involves studying how engineering solutions and spatial strategies for local sourcing and production (food, energy, water, bio-based materials) can be designed to make a clear and measurable contribution towards the optimization of urban resources flows and stocks and to be assessed as an integral part of the UM. A range of evidence-based approaches are well documented in the literature but still randomly integrated into practice, triggering the question as for why these approaches are not yet 'mainstream'. Beyond designers' growing enthusiasm for 'the fluidity of metabolic processes' (Ibañez and Katsikis, 2014), the agency of design could be deployed more strategically in order to benefit current scientific efforts towards the development of advanced and higher-impact UM accounts. A few existing nature-based initiatives already provide good examples of integrated energy and water management systems, in which recreational and ecological benefits by green infrastructure come into play (Lehmann, 2019). However, there remains little investigation on the implementation of monitoring systems that can inform a more systematic knowledge base on the metabolic gains that nature-based solutions can provide in different contexts. Additionally, designers' efforts towards metabolic optimization should not overview potential tradeoffs and environmental and social disservices that nature-based solutions can cause to urban populations. These include, for example, environmental disservices generated by resource demand and related carbon footprint of services provided by park and green space facilities. For example, a study of the Barcelona Montjuic Park showed that the CO_2-equivalent emissions produced by the life cycle of the energy consumed by the park's services would require a forest surface of more than 12 times the park's area to be absorbed (Oliver-Solà et al., 2008). Similar concerns can be raised for the social disservices and negative externalities on property value and community life in neighbourhoods affected by the development of green 'landmarks' projects (Lehmann, 2019). As argued for the New York City High Line, a greenway created on a former elevated railroad in Manhattan, often quoted as a paradigmatic ecological-urbanism example, nature-based solutions can be detrimental to social equity and trigger gentrification processes at the neighbourhood level, which can disrupt the delicate social-ecological balance in an urban system (Lang and Rothenberg, 2017). In sum, closing the loop from UM analysis to nature-based solution design and implementation requires more research into processes and strategies to systematize methods and connect findings across scientific domains, as well as research-practice joint efforts to monitor design outcomes. Such requirements reflect a set of fundamental steps towards the negotiation of a common UM research agenda, one that gives high priority to the translation of UM studies into actionable knowledge for policy and practice.

3. Conclusions: a tentative research agenda

This article has focused on three main challenges that UM research needs to face in order to serve as a conceptual and operational framework stimulating cross-fertilization among ecological sciences and to increase impact on policy and practice. The first challenge relates to the interdisciplinary nature of UM studies and the identification of silos across urban ecology and industrial ecology as the two main disciplines offering UM assessment methods. Although the two disciplines share the same mass-balance methods to quantify material and substance inflows and outflows and to measure the performance of urban systems, on a conceptual level they differ substantially in the way the urban system is defined. Moving beyond biological metaphors and engaging with the study of urban ecosystems and the biotic-abiotic interactions occurring heterogeneously within them is essential to properly organize comprehensive UM frameworks. The second challenge discussed in the article consists of broadening the UM interdisciplinary knowledge base to facilitate the understanding of the interdependency between biogeochemical cycles (e.g., ecosystem services) and socioeconomic flows of the UM. Linking engineering and biogeochemical modelling of the UM is crucial for bridging quality and quantity assessments of material resource flows and nutrient cycling in urban ecosystems. Industrial ecology resource-optimization strategies based on the regulation of human-controlled flows within the urban system boundary (e.g., local cycling of water, soil and excavated earth, airflow control through organic or inorganic structures for passive cooling) should be considered for their impacts on the health and functionality of other ecosystems at the level of watersheds, natural habitats and regional landscapes. This involves a more flexible approach to system boundary setting without neglecting methodological requirements for scientifically valid UM analysis and to avoid double counting. Reciprocally, urban ecology mass balance studies should be expanded beyond the accounting of human-influenced nutrient cycles (e.g., carbon, nitrogen, phosphorous) and incorporate the study of other purely anthropogenic material and energy flows (including products and wastes). The call for integrated UM accounting methods becomes even more compelling when research aims to inform more transversal urban policy and facilitate action across policy domains. Finally, the third challenge addressed in the article relates to the need for systematizing the increasing reservoir of design techniques and methods for the implementation of nature-based solutions in cities. This involves making sense of the growing evidence-base that landscape design research is gathering on the benefits that urban ecosystems can provide both in terms of resource-use optimization in urban ecosystems and the health and wellbeing of urban populations. Lessons learnt from the design and implementation of nature-based solutions through, for example, continued and longitudinal monitoring of impacts on material and biogeochemical cycles can shed new light on what kind of design solutions can lead to resource-use optimization and the quality and quantity of the metabolic gains provided. Systematization of efforts across analysts' and designers' communities can also benefit the search for innovative bottom-up modelling approaches that can bridge the divide between material flow and biogeochemical cycle accountings at multiple operational levels and scales of analysis.

Learning from the discussion of the three challenges identified, a tentative research agenda can be sketched to stimulate the adoption of UM as a mainstream, comprehensive framework for investigation and action on the ecology of the city. A first focal area would revolve around prioritizing cross-disciplinary efforts for the harmonization of languages and the establishment of a common vocabulary of concepts, which, beyond the use of metaphors, can express theoretical and methodological contributions of each scientific field to the UM discussion. A second point would aim at raising understanding of the UM 'physical nexus' linking resource flows and socioeconomic consumption patterns (e.g., water, energy, food, materials), on the one hand, and the ecological productivity capacity of natural habitats, on the other hand (based on water and air quality indicators and soil composition and properties). Finally, a third research frontier would consist of setting the ground for collaborative research efforts between science and design to work across analytical and operational frameworks and optimize (or potentially upscale) solutions. Establishing a more systematic dialogue among knowledge domains would also involve identifying levers to broaden current resource-management governance bases. The proposed interdisciplinary investigation into the UM 'physical nexus' could be expanded as a means to explore opportunities for a 'policy nexus', bridging UM stakeholders' competencies and policy domains. Last but not least, implementing such an ambitious UM research agenda would require addressing questions of stewardship systematically and comprehensively, for each stakeholder group in science, policy and practice to be given the opportunity and the responsibility to be part of the UM research to be.

References

Ayres, R., Kneese, A., 1969. Production, consumption and externalities. The American Economic Review 59 (3), 282–297.

Baccini, P., Brunner, P.H., 1991. Metabolism of the Anthroposphere. Springer-Verlag.

Bahers, J., Barles, S., Durand, M., 2019. Urban metabolism of intermediate cities: the material flow analysis, hinterlands and the logistics-hub function of rennes and le mans (France). Journal of Industrial Ecology 23, 686–698. https://doi.org/10.1111/jiec.12778.

Bai, X., 2007. Industrial ecology and the global impact of cities. Journal of Industrial Ecology 11 (2), 1–6. https://doi.org/10.1162/jie.2007.1296.

Bai, X., 2016. Eight energy and material flow characteristics of urban ecosystems. Ambio 45, 819–830. https://doi.org/10.1007/s13280-016-0785-6.

Bai, X., Surveyer, A., Elmqvist, T., et al., 2016. Defining and advancing a systems approach for sustainable cities. Current Opinion in Environmental Sustainability 23, 69–78. https://doi.org/10.1016/j.cosust.2016.11.010.

Barles, S., 2009. Urban metabolism of Paris and its region. Journal of Industrial Ecology 13 (6), 898–913. https://doi.org/10.1111/j.1530-9290.2009.00169.x.

Barles, S., 2010. Society, energy and materials: the contribution of urban metabolism studies to sustainable urban development issues. Journal of Environmental Planning and Management 53 (4), 439–455. https://doi.org/10.1080/09640561003703772.

Barrett, G.W., 2001. Closing the ecological cycle: the emergence of integrative science. Ecosystem Health 7 (2), 79–84.

Barrett, G.W., Odum, E.P., 1998. From the president: integrative science. BioScience 48, 970.

Batty, M., 2013. The New Science of Cities. MIT Press.

Bringezu, S., 1997. From quality to quantity: material flow analysis. In: Bringezu, S., Fischer-Kowalski, M., Kleijn, R., Palm, V. (Eds.), Regional and National Material Flow Accounting: From Paradigm to Practice of Sustainability. Proceedings of the ConAccount Workshop January 21–23. Wuppertal Institute, Wuppertal, pp. 43–57.

Brugmans, G., Strien, J., 2014. IABR-2014. Urban by Nature. Catalog of the 6th International Architecture Biennale Rotterdam (IABR). Rotterdam - IABR.

Cadenasso, M.L., Pickett, S.T.A., 2013. Three tides: the development and state of the art of urban ecological science. In: Pickett, S., Cadenasso, M., McGrath, B. (Eds.), Resilience in Ecology and Urban Design. Future City, vol. 3. Springer, Dordrecht. https://doi.org/10.1007/978-94-007-5341-9_2.

Cadenasso, M.L., Pickett, S.T.A., McGrath, B., Marshall, V., 2013. Ecological heterogeneity in urban ecosystems: reconceptualized land cover models as a bridge to urban design. In: Pickett, S., Cadenasso, M., McGrath, B. (Eds.), Resilience in Ecology and Urban Design. Future City, vol. 3. Springer, Dordrech. https://doi.org/10.1007/978-94-007-5341-9_6.

Cadenasso, M.L., Pickett, S.T., Schwarz, K., 2007. Spatial heterogeneity in urban ecosystems: reconceptualizing land cover and a framework for classification. Frontiers in Ecology and the Environment 5, 80–88. https://doi.org/10.1890/1540-9295(2007)5[80:SHIUER]2.0.CO;2.

Castán Broto, V., Allen, A., Rapoport, E., 2012. Interdisciplinary perspectives on urban metabolism. Journal of Industrial Ecology 16 (6), 851–861. https://doi.org/10.1111/j.1530-9290.2012.00556.x.

Chen, B., Lu, Y., 2015. Urban nexus: a new paradigm for urban studies. Ecological Modelling 318, 5–7. https://doi.org/10.1016/j.ecolmodel.2015.10.010.

Chrysoulakis, N., Lopez, M., San Josè, R., et al., 2013. Sustainable urban metabolism as a link between bio-physical sciences and urban planning: the BRIDGE project. Landscape and Urban Planning 112, 100–117. https://doi.org/10.1016/j.landurbplan.2012.12.005.

Churkina, G., 2008. Modeling the carbon cycle of urban systems. Ecological Modelling 216, 107–113. https://doi.org/10.1016/j.ecolmodel.2008.03.006.

Cui, X., 2018. How can cities support sustainability: a bibliometric analysis of urban metabolism. Ecological Indicators 93, 704–717. https://doi.org/10.1016/j.ecolind.2018.05.056.

Daigger, G., Newell, J.P., Love, N., 2016. Scaling up Agriculture in City-Regions to Mitigate FEW System impacts. In: White Paper NSF Workshop. University of Michigan, Ann Arbor, MI, USA. Available from: http://urbansustainability.snre.umich.edu/wp-content/uploads/2016/02/Umich_NSF-INFEWS-Workshop_Scaling-Up-Agriculture-in-City-Regions-to-Mitigate-Food-Energy-Water-Impacts.pdf. (Accessed 17 September 2019).

Dalla Fontana, M., Boas, I., 2019. The Politics of the Nexus in the City of Amsterdam, vol. 95, p. 102388. https://doi.org/10.1016/j.cities.2019.102388. Cities.

DeStefano, S., DeGraaf, R.M., 2003. Exploring the ecology of suburban wildlife. Frontiers in Ecology and the Environment 1, 95–101. https://doi.org/10.1890/1540-9295(2003)001[0095:ETEOSW]2.0.CO;2.

Esculier, F., Le Noë, J., Barles, S., et al., 2019. The biogeochemical imprint of human metabolism in Paris Megacity: a regionalized analysis of a water-agro-food system. Journal of Hydrology 573, 1028–1045. https://doi.org/10.1016/j.jhydrol.2018.02.043.

Eurostat, 2001. Economy-wide Material Flow Accounts and Derived Indicators. A Methodological Guide. Eurostat, European Commission, Office for Official Publications of the European Communities, Luxembourg. Available from: https://ec.europa.eu/eurostat/documents/3859598/5855193/KS-34-00-536-EN.PDF/411cd453-6d11-40a0-b65a-a33805327616. (Accessed 17 September 2019).

Fischer-Kowalski, M., 1998. Society's metabolism. Journal of Industrial Ecology 2 (1), 61–78. https://doi.org/10.1162/jiec.1998.2.1.61.

Fischer-Kowalski, M., Krausmann, F., Giljum, S., et al., 2011. Methodology and indicators of economy-wide material flow accounting. Journal of Industrial Ecology 15, 855–876. https://doi.org/10.1111/j.1530-9290.2011.00366.x.

Galan, J., Perrotti, D., 2019. Incorporating metabolic thinking into regional planning: the case of the Sierra Calderona Strategic Plan. Urban Planning 4 (1), 152–171. https://doi.org/10.17645/up.v4i1.1549.

Gandy, M., 2004. Rethinking urban metabolism: water, space and the modern City. City 8 (3), 363–379. https://doi.org/10.1080/1360481042000313509.

Golubiewski, N., 2012. Is there a metabolism of an urban ecosystem? An ecological critique. Ambio 41, 751. https://doi.org/10.1007/s13280-011-0232-7.

Grimm, N.B., Faeth, S.H., Golubiewski, N.E., et al., 2008. Global change and the ecology of cities. Science 319 (5864), 756–760. https://doi.org/10.1126/science.1150195.

Grimm, N., Morgan Grove, J., Pickett, S., Redman, C., 2000. Integrated Approaches to Long-Term Studies of Urban Ecological Systems: urban ecological systems present multiple challenges to ecologists— pervasive human impact and extreme heterogeneity of cities, and the need to integrate social and ecological approaches, concepts, and theory. BioScience 50 (7), 571–584. https://doi.org/10.1641/0006-3568(2000)050[0571:iatlto]2.0.CO;2.

Grove, J.M., Pickett, S.T.S., Whitmer, A., Cadenasso, M.L., 2013. Building an urban LTSER: the case of the Baltimore ecosystem study and the D.C./B.C. ULTRA-ex project. In: Singh, S., Haberl, H., Chertow, M., et al. (Eds.), Long Term Socio-Ecological Research. Human-Environment Interactions, vol. 2. Springer, Dordrecht. https://doi.org/10.1007/978-94-007-1177-8_16.

Haberl, H., 2001a. The energetic metabolism of societies. Part I: accounting concepts. Journal of Industrial Ecology 5 (1), 11–33. https://doi.org/10.1162/10881980152830141.

Haberl, H., 2001b. The energetic metabolism of societies. Part II: empirical examples. Journal of Industrial Ecology 5 (2), 71–88. https://doi.org/10.1162/10881980152830141.

Haberl, H., Weisz, H., Amann, C., et al., 2006. The energetic metabolism of the European union and the United States. Journal of Industrial Ecology 10 (4), 151–171. https://doi.org/10.1162/jiec.2006.10.4.151.

Hammer, M., Giljum, S., 2006. Material Flow Analysis of the City of Hamburg. NEDS Working Paper. No. 6, Hamburg. Available from: http://alt.seri.at/en/publications/other-working-papers/2009/09/19/materialflussanalysen-der-regionen-hamburg-wien-und-leipzig/. (Accessed 17 September 2019).

Hammer, M., Giljum, S., Bargigli, S., Hinterberger, F., 2003. Material flow analysis on the regional level: questions, problems, solutions. NEDS Working Papers 2 (04/2003). Wien: Sustainable Europe Research Institute (SERI). Available from: http://web205.vbox-01.inode.at/Data/personendaten/mh/NEDS%20WP%202_04_2003%20Druckvorst..pdf. (Accessed 17 September 2019).

Hansen, R., Pauleit, S., 2014. From multifunctionality to multiple ecosystem services? A conceptual framework for multifunctionality in green infrastructure planning for urban areas. Ambio 43 (4), 516–529. https://doi.org/10.1007/s13280-014-0510-2.

Hansen, R., Frantzeskaki, N., McPhearson, T., et al., 2015. The uptake of the ecosystem services concept in planning discourses of European and American cities. Ecosystem Services 12, 228–246. https://doi.org/10.1016/j.ecoser.2014.11.013.

Hau, J.L., Bakshi, B.R., 2004. Promise and problems of emergy analysis. Ecological Modelling 178 (1–2), 215–225. https://doi.org/10.1016/j.ecolmodel.2003.12.016.

Hobbie, J.E., Carpenter, S.R., Grimm, N.B., et al., 2003. The US long term ecological research program. BioScience 53 (1), 21–32. https://doi.org/10.1641/0006-3568(2003)053[0021:TULTER]2.0.CO;2.

Ibañez, D., Katsikis, N. (Eds.), 2014. Grounding Metabolism. New Geographies 6. Harvard University Press, Cambridge MA.

Kaye, J.P., Groffman, P.M., Grimm, N.B., et al., 2006. A distinct urban biogeochemistry? Trends in Ecology and Evolution 21 (4), 192–199. https://doi.org/10.1016/j.tree.2005.12.006.

Kennedy, C., 2012. Comment on article "Is there a metabolism of an urban ecosystem?" by golubiewski. Ambio 41, 765–766. https://doi.org/10.1111/j.1530-9290.2012.00564.x.

Kennedy, C., Baker, L., Dhakal, S., Ramaswami, A., 2012. Sustainable urban systems: an integrated approach. Journal of Industrial Ecology 16 (6), 775–779. https://doi.org/10.1111/j.1530-9290.2012.00564.x.

Kennedy, C., Cuddihy, J., Engel-Yan, J., 2007. The changing metabolism of cities. Journal of Industrial Ecology 11 (2), 43–59. https://doi.org/10.1162/jie.2007.1107.

Kennedy, C., Pincetl, S., Bunje, P., 2011. The study of urban metabolism and its applications to urban planning and design. Environmental Pollution 159 (8–9), 1965–1973. https://doi.org/10.1016/j.envpol.2010.10.022.

Krausmann, F., 2013. C. In: Singh, S., Haberl, H., Chertow, M., Mirtl, M., Schmid, M. (Eds.), Long Term Socio-Ecological Research. Human-Environment Interactions, vol. 2. Springer, Dordrecht. https://doi.org/10.1007/978-94-007-1177-8_11.

Krausmann, F., Haberl, H., 2002. The process of industrialization from the perspective of energetic metabolism. Socioeconomic energy flows in Austria 1830–1995. Ecological Economics 41, 177–201. https://doi.org/10.1016/S0921-8009(02)00032-0.

Lang, S., Rothenberg, J., 2017. Neoliberal urbanism, public space, and the greening of the growth machine: New York City's High Line park. Environment & Planning A: Economy and Space 49 (8), 1743–1761. https://doi.org/10.1177/0308518X16677969.

Lehmann, S., 2019. Reconnecting with nature: developing urban spaces in the age of climate change. Emerald Open Research 1, 2. https://doi.org/10.12688/emeraldopenres.12960.1.

Lindeman, n RL., 1942. The trophic-dynamic aspect of ecology. Ecology 23 (4), 399–417, 1942.

Marcotullio, P.J., Boyle, G., 2003. Defining an Ecosystem Approach to Urban Management and Policy Development. UNU/IAS Report. United Nations University Institute of Advanced Studies (UNU/IAS). March 2003. Available from: http://collections.unu.edu/eserv/UNU:3109/UNUIAS_UrbanReport.pdf. (Accessed 9 October 2019).

McHale, M.R., Pickett, S.T., Barbosa, O., et al., 2015. The new global urban realm: complex, connected, diffuse, and diverse social-ecological systems. Sustainability 7 (5), 211–240. https://doi.org/10.3390/su7055211.

Metabolic, Studioninedots and DELVA Landscape Architects, 2015. Circular Buiksloterham. Transitioning Amsterdam to a Circular City—Vision and Ambition. Final Report, English version. Available from: https://www.metabolic.nl/wp-content/uploads/2015/07/CircularBuiksloterham_ENG_FullReport.pdf. (Accessed 17 September 2019).

Mohareb, E., Kennedy, C., 2012. Greenhouse gas emission scenario modeling for cities using the PURGE model. Journal of Industrial Ecology 16 (6), 875–888. https://doi.org/10.1111/j.1530-9290.2012.00563.x.

Newell, J.P., Cousins, J.J., 2014. The boundaries of urban metabolism: towards a political–industrial ecology. Progress in Human Geography 39 (6), 702–728. https://doi.org/10.1177/0309132514558442.

Newell, J.P., Cousins, J.J., Baka, J.E., 2017. Political-industrial ecology: an introduction. Geoforum 85, 319–323. https://doi.org/10.1016/j.geoforum.2017.07.024.

Newell, J.P., Goldstein, B., Foster, A., 2019. A 40-year review of food–energy–water nexus literature and its application to the urban scale. Environmental Research Letters 14 (7). https://doi.org/10.1088/1748-9326/ab0767, 073003.

Newman, P.W.G., 1999. Sustainability and cities: extending the metabolism model. Landscape and Urban Planning 44, 219–226. https://doi.org/10.1016/S0169-2046(99)00009-2.

Nowak, D.J., Hirabayashi, S., Bodine, A., et al., 2012. Modeled PM2.5 removal by trees in ten U.S. cities and associated health effects. Environmental Pollution 178, 395–402. https://doi.org/10.1016/j.envpol.2013.03.050.

Odum, E.P., 1977. The emergence of Ecology as a new integrative discipline. Science 195 (4284), 1289–1293. https://doi.org/10.1126/science.195.4284.1289.

Odum, H.T., 1996. Environmental Accounting: Emergy and Environmental Decision Making. John Wiley and Sons, New York, 1996.

Oliver-Sola, J., Nuñez, M., Gabarrell, X., et al., 2008. Service sector metabolism: accounting for energy impacts of the montjuic urban park in Barcelona. Journal of Industrial Ecology 11 (2), 83–98. https://doi.org/10.1162/jie.2007.1193, 2007.

Pataki, D.E., Carreiro, M.M., Cherrier, J., et al., 2011. Coupling biogeochemical cycles in urban environments: ecosystem services, green solutions, and misconceptions. Frontiers in Ecology and the Environment 9 (1), 27–36. https://doi.org/10.1890/090220.

Perrotti, D., 2012. La ruralité urbaine: de plateforme d'expérimentation à lieu de la mise en scène d'un nouveau modèle de durabilité. Environnement Urbain/Urban Environment 6, 100–117. https://doi.org/10.7202/1013715ar.

Perrotti, D., Stremke, S., 2020. Can urban metabolism models advance green infrastructure planning? Insights from ecosystem services research. Environment and Planning B: Urban Analytics and City Science 47 (4), 380–397. https://doi.org/10.1177/2399808318797131.

Perrotti, D., 2019. Evaluating urban metabolism assessment methods and knowledge transfer between scientists and practitioners: a combined framework for supporting practice-relevant research. Environment and Planning B: Urban Analytics and City Science 46 (8), 1458–1479. https://doi.org/10.1177/2399808319832611.

Perrotti, D., Iuorio, O., 2019. Green infrastructure in the space of flows: an urban metabolism approach to bridge environmental performance and user's wellbeing. In: Lemes de Oliveira, F., Mell, I. (Eds.), Planning Cities with Nature: Theories, Strategies and Methods. Springer, pp. 265–277. https://doi.org/10.1007/978-3-030-01866-5_18.

Pickett, S., Burch Jr., W., Dalton, S., et al., 1997. Urban Ecosystems 1 (4), 185–199. https://doi.org/10.1023/A:1018531712889.

Pickett, S.T.A., Cadenasso, M.L., Grove, J.M., Boone, C.G., Irwin, E., Groffman, P.M., Kaushal, S.S., Marshall, V., McGrath, B.P., Nilon, C.H., Pouyat, R.V., Szlavecz, K., Troy, A., Warren, P., 2011. Urban ecological systems: foundations and a decade of progress. Journal of Environmental Management 92 (3), 331–362. https://doi.org/10.1016/j.jenvman.2010.08.022.

Pincetl, S., Bunje, P.M., 2010. Creating Sustainable Energy Systems in California Communities: A Research Roadmap. Consultant Report, California Energy Commission. Public Interest Energy Research Program. Available from: http://uc-ciee.org/downloads/UCLA%20Roadmap.pdf. (Accessed 17 September 2019).

Pincetl, S., Bunje, P., Holmes, T., 2012. An expanded urban metabolism method: toward a systems approach for assessing urban energy processes and causes. Landscape and Urban Planning 107 (3), 193–202. https://doi.org/10.1016/j.landurbplan.2012.06.006.

Swyngedouw, E., Heynen, N., 2003. Urban political ecology, justice and the politics of scale. Antipode 35, 898–918. https://doi.org/10.1111/j.1467-8330.2003.00364.x.

Taleghani, M., Montazami, A., Perrotti, D., 2020. Learning to Chill: the role of design schools and professional training to improve urban climate and urban metabolism. Energies 13 (9), 2243–2257. https://doi.org/10.3390/en13092243.

Tansley, A.G., 1935. The use and abuse of vegetational concepts and terms. Ecology 16 (3), 284–307.

Tiwary, A., Sinnett, D., Peachey, C., et al., 2009. An integrated tool to assess the role of new planting in PM10 capture and the human health benefits: a case study in London. Environmental Pollution 157, 2645–2653. https://doi.org/10.1016/j.envpol.2009.05.005.

Verger, Y., Petit, C., Barles, S., et al., 2018. A N, P, C, and water flows metabolism study in a peri-urban territory in France: the case-study of the Saclay plateau. Resources, Conservation and Recycling 137, 200–213. https://doi.org/10.1016/j.resconrec.2018.06.007.

Voskamp, I.M., Stremke, S., Spiller, M., et al., 2017. Enhanced performance of the Eurostat method for comprehensive assessment of urban metabolism: a material flow analysis of Amsterdam. Journal of Industrial Ecology 21 (4), 887–902. https://doi.org/10.1111/jiec.12461.

Wachsmuth, D., 2012. Three ecologies: urban metabolism and the society-nature opposition. The Sociological Quarterly 53, 506–523. https://doi.org/10.1111/j.1533-8525.2012.01247.x.

Wirion, C., Bauwens, W., Verbeiren, B., 2017. Location and time specific hydrological simulations with multi-resolution remote sensing data in urban areas. Remote Sensing 9 (7), 645.

Wood, J., 2012. Canadian Environmental Indicators: Air Quality. Studies in Environmental Policy, January 2012, Canadian Environmental Sustainability Indicators (CESI) Program. Environment Canada. Available from: http://publications.gc.ca/Collection/Statcan/16-251-X/16-251-XIE2006000.pdf. (Accessed 17 September 2019).

Yang, P.P.J., Yamagata, Y., 2019. Urban Systems Design: From "Science for Design" to "Design in Science". Environment and Planning B: Urban Analytics and City Science. Online First. https://doi.org/10.1177/2399808319877770.

Zhang, Y., Yanga, Z., Yub, X., 2009. Evaluation of urban metabolism based on emergy synthesis: a case study for Beijing (China). Ecological Modelling 220, 1690–1696. https://doi.org/10.1016/j.ecolmodel.2009.04.002.

Urban land use land cover

3

Urban growth pattern detection and analysis

Mateo Gašparović

Chair of Photogrammetry and Remote Sensing, Faculty of Geodesy, University of Zagreb, Zagreb, Croatia

OUTLINE

1. Introduction

In today's world of constant changes, the importance of accurate and timely information on the nature and extent of land resources and changes over time is constantly increasing. These changes are very important to scientists, planners, resource managers, and policy-makers. Urbanization is one of the most widespread anthropogenic causes of the loss of fertile land, habitat destruction, and the decrease of the natural vegetation cover (Gašparović et al., 2017; Dewan and Yamaguchi, 2009; Alphan, 2003; López et al., 2001). Much scientific research is concerned with the detection, analysis, and interpretation of changes in land-cover caused by the urbanization process and tries to create standardized parameters to assess the degree of urbanization (Verma and Raghubanshi, 2018; Fu and Weng, 2016). Geophysical data for obtaining and analyzing urban growth pattern can be obtained using remote sensing techniques. These data provide and describe land-cover changes in a given area over time. Although traditional techniques and measurements can follow these changes, satellite remote research provides a wider spectrum of information, with the advantage of saving time and money (Verma and Raghubanshi, 2019). This research aims to present an overview of the existing remote sensing methodology for collecting suitable data for detection, display, and analysis of urban growth pattern.

2. Satellite image selection

In order to monitor urban growth pattern detection and analysis, it is first necessary to define the following requirements:

(a) Define the study area.
(b) Define the study period (timeframe) within which the research was conducted. When it comes to monitoring urbanization in cities, that framework is usually larger than 10 years (Schneider, 2012; Taubenböck et al., 2009).
(c) Define the minimum size of the object on earth that we want to detect. Typically, the tendency always is the higher, the better. The higher spatial resolution also entails specific problems such as the need for a more powerful hardware system, longer data processing times, problem of shadows in cities, the cost of acquiring satellite images in high spatial resolution. Furthermore, high-resolution archival satellite images are often not available for a specific location. Historically, high-resolution satellite imagery has not been available for as long as mid-resolution satellite imagery such as Landsat (El Garouani et al., 2017; Taubenböck et al., 2009).
(d) Define the time resolution (frequency) of the data collected. That primarily depends on satellite revisit time. It should be emphasized that multi-annual analyses of urbanization monitoring uses from one image every year, up to one image every 2.5 or 10 years (Aburas et al., 2018; Hegazy and Kaloop, 2015).

Based on the requirements/criteria shown above, an adequate satellite platform for data collection for research was selected.

For this research, the study area of 242 km^2 urban part of the capital of the Republic of Croatia Zagreb was selected (Fig. 3.1). The goal is to monitor urbanization, i.e., the expansion of built-up land over the years. The research study period is 35 years (1984—2019). Given this is a significant timeframe, the time resolution of the data, i.e., satellite images used in the research was defined as 17—18 years. Because of the considerable timeframe of the research, the use of Landsat satellite imagery was considered. The Landsat-5 mission provided data from March 1984 to January 2013 in a 30 m spatial resolution and with a temporal resolution of approx. 16 days. Furthermore, Landsat-8 mission data has been selected for the research,

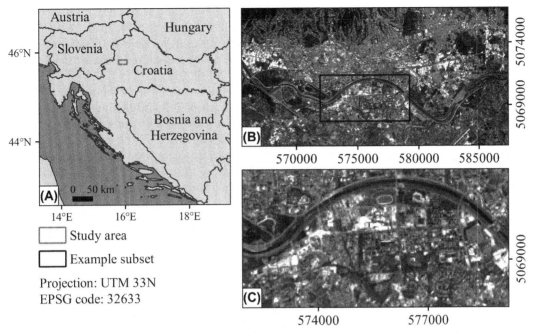

FIGURE 3.1 (A) study area location, (B) study area (22 km × 11 km) and example subset location, (C) example subset (7.3km × 3.6 km). Basemap Landsat-8 "true-color" composite (4-3-2).

which is available from February 2013 to the present. Landsat-8 has a 15 m panchromatic band that can be used for fusion (pansharpening) of the multispectral 30 m bands with the 15 m bands. In this research, 30 m multispectral Landsat-8 bands were used because Landsat-5 imagery has only 30 m multispectral bands (without 15 m panchromatic band).

3. Preprocessing of satellite imagery

After selecting the images, the preprocessing process follows. Preprocessing is an important process that enables better results of the image classification. Preprocessing can be divided into four steps: noise removal, radiometric correction, geometric correction, and fusion (e.g., pansharpening).

(a) Satellite imagery, especially in high resolution, has a common occurrence of noise. Noise on satellite imagery may occur at the time of observation, signal digitization, or in the process of data transfer. Also, noise can be caused by a malfunction of the satellite or satellite sensor. The most known example of such noise is the existence of a gap on the Landsat-7 satellite due to the failure of the Scan Line Corrector (SLC) device (Liu and Morgan, 2006; Zhang et al., 1999).

(b) Radiometric correction is an essential step in the preprocessing of satellite imagery since classification is performed based on radiometric values. The process of radiometric correction includes atmospheric correction (Liang et al., 2001; Chavez, 1996), sun elevation correction, and earth-sun distance correction (Richter, 1997), but also simpler processes such as radiometric normalization (Canty et al., 2004) and histogram equalization (Demirel et al., 2009). One of the essential processes for satellite images radiometric correction is the conversion of digital number to radiance (or reflectance) values. This process is necessary if measurements are taken from multiple sensing platforms such as the combination of Landsat-5, 7, 8 and Sentinel-2. Also, this conversion is necessary when satellite measurements are directly compared to physical measurements on earth (Lillesand et al., 2015).

(c) Satellite imagery can often contain geometrical distortions, so they cannot be used for direct terrain mapping, especially when it comes to spatio-temporal analysis. Such errors can be caused by various factors such as distortion, ground curvature, terrain topography influence, or terrain correction (Richter, 1998). The elimination of terrain correction requires a good knowledge of the terrain in the form of a digital terrain model. The orthorectification process eliminates the terrain impact on the satellite image and converts the raw image to an orthorectified one (Gašparović et al., 2018a). This issue has been addressed by many authors, notably the use of high-resolution satellite imagery such as WorldView (Gašparović et al., 2019a; Belfiore and Parente, 2015; Aguilar et al., 2013).

(d) Fusion or pansharpening is nowadays often used in satellite image processing to increase the spatial resolution of images. Fusion is a process in which the spatial resolution of images (multispectral bands) is increased by mathematical methods based on the higher spatial resolution images (panchromatic band). Numerous authors have engaged in the process of fusing satellite imagery (Wang et al., 2016; Alparone, 2007); some of them use the fusion process to increase the accuracy of land-cover classification (Gašparović and Jogun, 2018; Palsson et al., 2011). The fusion method can be applied to a combination of data from different satellites such as in the research of Gašparović et al. (2018b) where Sentinel-2 was sharpened with PlaneScope bands, or in the research (Gašparović et al., 2019b) where the Sentinel-2 was sharpened based on multispectral bands from WorldView-4 satellites.

4. Image classification methods

The main objective of image classification is to group all the pixels of a particular image into a specific land-cover class. There are numerous methods of classifying satellite images nowadays. Classification methods can be divided into the following (Abburu and Golla, 2015; Lillesand et al., 2015; Schowengerdt, 2012):

(a) Unsupervised classification that requires no samples and knowledge of the terrain but only the desired number of classes. The algorithm automatically classifies the entire satellite image by grouping pixels of similar radiometric values into groups, i.e., classes. Such classifications are carried out completely automatically and do not require any training samples, i.e., manual work. After classification based on unsupervised methods, the class must be assigned manually or automatically to the actual land-cover class on the earth's surface. Nowadays, there are many unsupervised classification methods and authors most commonly use ISODATA and k-means in their research (Gašparović et al., 2017; Li et al., 2016; Bandyopadhyay and Maulik, 2002; Melesse and Jordan, 2002).

(b) Supervised classification requires training samples collection, i.e., manual user operation. Such methods generally achieve more accurate results than unsupervised methods. The advantage of such methods is that, after the classification process has been completed, they correspond to the actual land-cover classes, i.e., the training samples upon which the classification was made. Nowadays, there are many supervised classification methods. It should be emphasized that traditional methods used in nowadays research include Maximum Likelihood (Bruzzone and Prieto, 2001), Minimum Distance (Yang et al., 2011), as well as, contemporary machine learning–based methods like Artificial Neural Network (Gašparović and Jogun, 2018), Random Forest (Thanh Noi and Kappas, 2018), Support Vector Machines (SVM, Qian et al., 2015).

(c) Hybrid methods are a combination of the above-mentioned methods. After the supervised classification is performed, it is possible to deepen a class, i.e., further classify it by the algorithms of the unsupervised classification. This method is often used when one is unable to define well the differences between types of subclasses, e.g., vegetation type, crop or soil type, type of material used as roof cover, etc. One type of hybrid methods can be rule-based classification methods. Such methods are based on conditions, e.g., vegetation represents the area where the NDVI (ormalized Difference Vegetation Index) value is greater than 0 or water is area where the MNDWI>0 (Modified Normalized Difference Water Index). An interesting hybrid method for fully automatic land-cover classification of the Landsat satellite imagery was presented in recent research (Gašparović et al., 2019c).

Classification methods can be divided according to the minimum classified feature, in the pixel-based and object-based classification (De Jong and Van der Meer, 2007; Congalton and Green, 2002).

(a) In pixel-based classifications, each pixel is assigned to a specific class according to its radiometric characteristics. With such a method, salt-and-paper noise is often presented, especially in classifications based on high spatial resolution satellite imagery (Salah, 2017; Lillesand et al., 2015).

(b) The first step in the object-based classification is the segmentation of satellite imagery (Csillik, 2017; Blaschke, 2010). Segmentation is the decomposition of a digital image into smaller homogenized parts (objects) of similar characteristics. Each object is made up of adjacent pixels of similar radiometric characteristics. Objects can vary in size and shape to better describe the real world, i.e., objects on earth. After classification, each object receives statistical indicators calculated based on the values of all the contained pixels in the object. The second step is to classify objects based on supervised, unsupervised, or hybrid methods.

5. Optimization of postclassification processing

After classification, especially those based on pixels, it is necessary to establish by objective visual analysis the presence level of a certain noise, e.g., paper-and-salt. Object-based classifications generally have no noise problem (Blaschke, 2010). The presence of noise corrupts classification itself and noise can and should be eliminated by filtering or other GIS-based approaches. The authors often use the so-called majority filter (Ghimire et al., 2010; Qian et al., 2005). It is a function where a moving window moves over the classification data. A majority class is calculated for each window position. If the central pixel of the window does not represent the majority class, it is identified and converted to the majority class. If the majority class cannot be detected, then the central pixel does not change. The window size for this type of filtering is variable, but the most commonly used sizes are 3×3 or 5×5 (Lillesand et al., 2015). Furthermore, some authors, in contrast to majority filters, recommend the use of a GIS-based approach. This approach is based on logical operations that can be used by many different types of filtering, e.g., expand and shrink filtering (Gašparović et al., 2018c; Ablameyko and Pridmore, 2012; Chambolle et al., 1998).

6. Accuracy assessment of the classification

After classifications and noise removal, the process of estimating accuracy follows. This process is essential because it gives objective indicators for the land-cover accuracy that is further used in the research (Congalton and Green, 2002). Inaccurate classifications can lead to misleading analyses, and thus to misleading conclusions. The basis for conducting the classification accuracy assessment is the confusion matrix, also known as the error matrix or contingency table. The confusion matrix is calculated based on the comparison of ground truth data (test samples) and classification results. Ground truth data used for accuracy assessment of the classification must be collected in the same timeframe as the classification. The ground truth data should be acquired on an independent, higher spatial resolution satellite imagery. Sometimes, ground truth data is collected or controlled on the field. Ground truth data can be collected as points or polygons, and their representation should correspond to the actual state of the satellite scene. Several measures that are directly used as indicators of overall classification accuracy assessment or individual class accuracy assessment can be calculated based on the confusion matrix (e.g., Overall Accuracy – OA, kappa coefficient – k; Figure of merit – F). The overall accuracy is defined by dividing the sum of the values on the main diagonal of the confusion matrix by the total number of samples (Mather and Tso, 2009). The concept of user's and producer's accuracy can be used to evaluate the accuracy of each class. The producer's accuracy can be calculated by dividing the cell (i, i) by the sum of column i in the confusion matrix, while the user's accuracy is dividing (i, i) by the sum of row i. Thus, producer accuracy shows the proportion of pixels in the test dataset that the classifier correctly recognizes. User accuracy measures the proportion of pixels identified by a classifier as belonging to a class that agrees with the test data (Mather and Tso, 2009). A commonly used unique coefficient for estimating accuracy, for which all confusion matrix data are used, is the kappa (k) coefficient (Foody, 1992). The Kappa coefficient is defined as:

$$k = \frac{N\sum_{i=1}^{r} x_{ii} - \sum_{i=1}^{r}(x_{i+} \cdot x_{+i})}{N^2 - \sum_{i=1}^{r}(x_{i+} \cdot x_{+i})}, \tag{3.1}$$

where

r − number of rows in the confusion matrix
x_{ii} − number of measurements in row i and column i (on the main diagonal)
x_{i+} − the total number of measurements in row i
x_{+i} − the total number of measurements in column i
N − the total number of measurements in the matrix.

Some authors (Gašparović et al., 2019c; Gomez et al., 2019; Charnock and Moss, 2017) proposed using the Figure of merit (F) instead of traditional methods for accuracy assessment, e.g., kappa. Pontius and Millones (2011) proposed F and presented certain limitations on the use of kappa as an accuracy assessment parameter. F was obtained by the following equation:

$$F = \frac{a}{o + a + c} 100\%, \tag{3.2}$$

where

a − number of agreements
o − number of omissions
c − the number of commissions.

7. Urbanization detection based on the image time-series

In order to detect changes in the land-cover, enough satellite images must be collected covering the entire study period. Through direct comparison (e.g., band subtraction) of images, spatial changes can be detected. A comparison of vegetation time-series or other indices is more commonly used. Such an application is very acceptable in forestry and agronomy for the rapid detection of changes in vegetation conditions (Zhu, 2017; Eckert et al., 2015; Forkel et al., 2013; Verbesselt et al., 2010). Further, spatial changes can be detected by comparing the time-series of land-cover classifications. This approach is most commonly used today for the detection and monitoring of urbanization growth pattern (Aburas et al., 2018; Garouani et al., 2017; Hegazy and Kaloop, 2015). In this way, it is straightforward detecting of which class has changed into which, e.g., deforestation for urban expansion, i.e., transition of forest class into build-up.

8. Results and analysis

For a better understanding of the presented methods, the results of urban growth pattern detection and analysis on the study area Zagreb, Croatia, are presented in this section. The observed timeframe was 35 years which were observed with two Landsat-5 satellite images from 1984 (first cloud-free Landsat-5 imagery of Zagreb), 2001, and one Landsat-8 satellite image from 2019 (Table 3.1). The 30 m optical Landsat-5 bands 1, 2, 3, 4, 5, 7 and Landsat-8 bands 2, 3, 4, 5, 6, 7 were used in the classification process. The supervised classification was based on the SVM algorithm in SAGA GIS (version 7.3.0). The SVM classification method was chosen according to many scientific studies, which showed that SVM achieved higher

TABLE 3.1 Satellite imagery used in this research.

Satellite	Sensing date	Satellite imagery ID
Landsat-5	27th June 1984	LT05_L1TP_189028_19840627_20170220_01_T1
Landsat-5	4th August 2001	LT05_L1TP_190028_20010804_20180430_01_T1
Landsat-8	31st August 2019	LC08_L1TP_189028_20190831_20190831_01_RT

accuracy for land-cover mapping than other classification methods (Thanh Noi and Kappas, 2018; Qian et al., 2015). Satellite imagery was classified into five land-cover classes: 1 — water; 2 — build-up; 3 — barren land; 4 — low vegetation; and 5 — high vegetation. The results of the land-cover classification for the entire study area and example subset are shown in Figs 3.2 and 3.3, respectively.

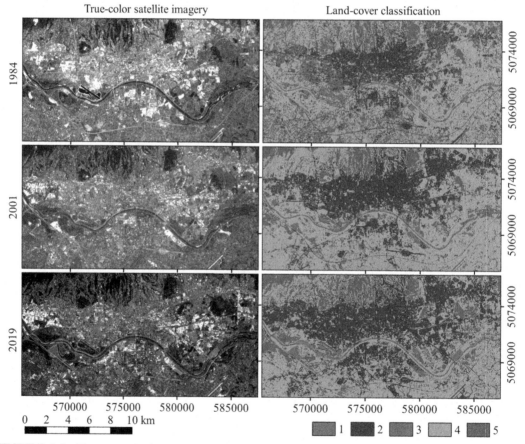

FIGURE 3.2 Time-series land-cover classification for growth pattern detection and analysis for the entire study area (1 — water; 2 — build-up; 3 — barren land; 4 — low vegetation; and 5 — high vegetation).

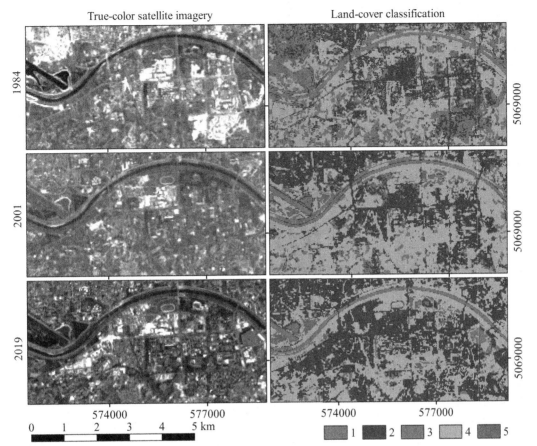

True-color satellite imagery

Land-cover classification

FIGURE 3.3   Time-series land-cover classification for growth pattern detection and analysis for the example subset (1 — water; 2 — build-up; 3 — barren land; 4 — low vegetation; and 5 — high vegetation).

For growth pattern detection and analysis, a time-series analysis of the representation of individual classes in the study area was made (Fig. 3.4). Fig. 3.4 clearly shows the increase in the built-up area at the expense of the low vegetation class and the barren land. The result indicates a process of urbanization over the past 35 years. In the study area of total 242 km², the built-up land grew almost linearly over the observed period and amounted to 50.9, 65.1, and 86.8 km² for the years 1984, 2001, and 2019, respectively. Urbanization, as well as growth pattern in urban areas, can be compared with the population increase. Fig. 3.4 shows a comparison of land-cover changes in all classes over the years as well as changes in the population. Official data from the censuses conducted every 10 years in the Republic of Croatia were used to track the population. These data were taken from the website of the Croatian Bureau of Statistics (CBS) (https://www.dzs.hr/default_e.htm). From Fig. 3.4, it is clear that the increase in population is accompanied by an increase in a built-up area in the study area.

For visualization of growth pattern and other environmental changes, a land-cover change detection analysis over a 35-year period was performed (Fig. 3.5). The analysis was made

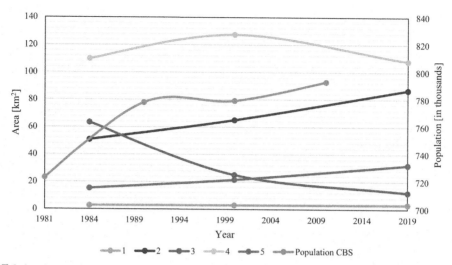

FIGURE 3.4 Time-series growth pattern analysis and comparison with population increase (1 — water; 2 — build-up; 3 — barren land; 4 — low vegetation; and 5 — high vegetation).

with a difference in land-cover classification for the years 1984 and 2019. The analysis shows all types of class changes for the stated period that were observed in the environment (e.g., water to barren land 1−3; low vegetation to built-up area 4−2; or high vegetation in barren land 5−3). Furthermore, an analysis of the representation of particular types of changes was expressed for the entire study area (Fig. 3.6).

The previous growth pattern analysis (Fig. 3.5) shows that changes are mostly in the city's outskirts in the expansion of the built-up area as opposed to the already built-up area in the old parts of the city, especially in the city center. Most of the changes (3−2 and 4−2, Fig. 3.6) indicate a high degree of urbanization. A change of 3−4 can be considered as a seasonal change of agricultural land (depends on the phenological state and the acquisition date of the satellite image) or on the overgrowth of agricultural land. Also, the overgrowth of agricultural land into the forest was clearly indicated by the fourth most frequent change 4−5 (Fig. 3.6).

9. Conclusions

This research presents the process of growth pattern detection and analysis in urban areas. The study area concerns the city of Zagreb, and the study period is 35 years. Satellite images of Landsat-5 and Landsat-8 were used for analysis. The change detection analysis was based on land-cover maps made using the SVM supervised classification method and 30 m multi-spectral bands. Based on the growth pattern analysis, a large degree of urbanization in the Zagreb area over the last 35 years is evident. For the most part, urbanization was based on a change in the class of barren land or low vegetation into the built-up area. Urbanization

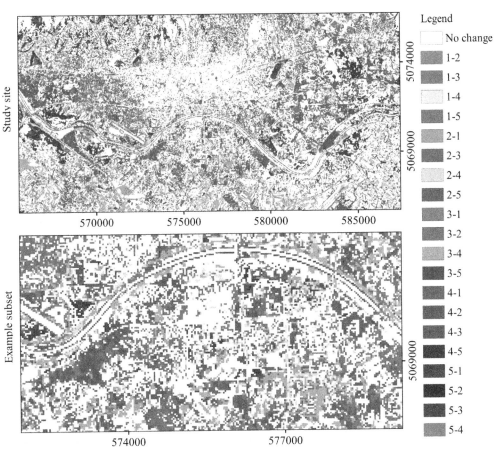

FIGURE 3.5 Time-series growth pattern change detection analysis in the period of 35 years (1984–2019). Raster classes represent the type of land-cover class change through 35 years (1 – water; 2 – build-up; 3 – barren land; 4 – low vegetation; and 5 – high vegetation).

was caused by population increases, which was clearly evidenced in the research results. Increasing population growth in big cities is a major problem today. The quality of life in cities is going down due to overcrowding. Increasing urbanization accelerates the increase of greenhouse gases. To prevent this, it is necessary to research and monitor the growing pattern in urban areas and to point out in time the problems of urban development. Urban planners and policymakers should put the quality of life and human health above profit and increasing urbanization. Nevertheless, it is important to emphasize that the methods and analyses presented in this research may be applicable in similar studies regardless of the study area or the type of satellite imagery. It should also be noted that open data and open-source software have been used and thus such procedures can be used easily and free of charge in the detection analysis of growing pattern for other urban areas all over the world.

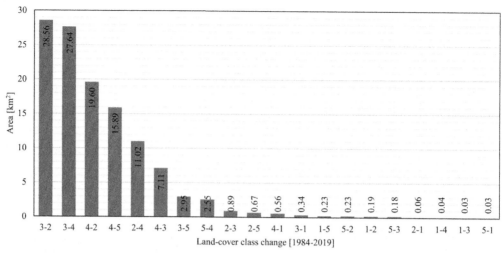

FIGURE 3.6   Sums of land-cover class change area expressed for the entire study area of 242 km² (1 − water; 2 − build-up; 3 − barren land; 4 − low vegetation; and 5 − high vegetation).

Acknowledgement

The author would like to thank the University of Zagreb that funded the RS4ENVIRO project entitled: "Advanced photogrammetry and remote sensing methods for environmental change monitoring" (Grant No. RS4ENVIRO) under which this research was conducted.

References

Abburu, S., Golla, S.B., 2015. Satellite image classification methods and techniques: a review. International Journal of Computer Applications 119 (8).

Ablameyko, S., Pridmore, T., 2012. Machine Interpretation of Line Drawing Images: Technical Drawings, Maps and Diagrams. Springer Science and Business Media.

Aburas, M.M., Ho, Y.M., Ramli, M.F., Ash'aari, Z.H., 2018. Monitoring and assessment of urban growth patterns using spatio-temporal built-up area analysis. Environmental Monitoring and Assessment 190 (3), 156.

Aguilar, M.A., del Mar Saldana, M., Aguilar, F.J., 2013. Assessing geometric accuracy of the orthorectification process from GeoEye-1 and WorldView-2 panchromatic images. International Journal of Applied Earth Observation and Geoinformation 21, 427−435.

Alparone, L., Wald, L., Chanussot, J., Thomas, C., Gamba, P., Bruce, L.M., 2007. Comparison of pansharpening algorithms: outcome of the 2006 GRS-S data-fusion contest. IEEE Transactions on Geoscience and Remote Sensing 45 (10), 3012−3021.

Alphan, H., 2003. Land-use change and urbanization of Adana, Turkey. Land Degradation and Development 14 (6), 575−586.

Bandyopadhyay, S., Maulik, U., 2002. An evolutionary technique based on K-means algorithm for optimal clustering in RN. Information Sciences 146 (1−4), 221−237.

Belfiore, O.R., Parente, C., 2015. Orthorectification and pan-sharpening of WorldView-2 satellite imagery to produce high resolution coloured ortho-photos. Modern Applied Science 9 (9), 122.

Blaschke, T., 2010. Object based image analysis for remote sensing. ISPRS Journal of Photogrammetry and Remote Sensing 65 (1), 2−16.

Bruzzone, L., Prieto, D.F., 2001. Unsupervised retraining of a maximum likelihood classifier for the analysis of multitemporal remote sensing images. IEEE Transactions on Geoscience and Remote Sensing 39 (2), 456−460.

Canty, M.J., Nielsen, A.A., Schmidt, M., 2004. Automatic radiometric normalization of multitemporal satellite imagery. Remote Sensing of Environment 91 (3–4), 441–451.

Chambolle, A., De Vore, R.A., Lee, N.Y., Lucier, B.J., 1998. Nonlinear wavelet image processing: variational problems, compression, and noise removal through wavelet shrinkage. IEEE Transactions on Image Processing 7 (3), 319–335.

Charnock, T., Moss, A., 2017. Deep recurrent neural networks for supernovae classification. The Astrophysical Journal Letters 837 (2), L28.

Chavez, P.S., 1996. Image-based atmospheric corrections-revisited and improved. Photogrammetric Engineering and Remote Sensing 62 (9), 1025–1035.

Congalton, R.G., Green, K., 2002. Assessing the Accuracy of Remotely Sensed Data: Principles and Practices. CRC press.

Csillik, O., 2017. Fast segmentation and classification of very high resolution remote sensing data using SLIC superpixels. Remote Sensing 9 (3), 243.

De Jong, S.M., Van der Meer, F.D. (Eds.), 2007. Remote Sensing Image Analysis: Including the Spatial Domain, vol. 5. Springer Science and Business Media.

Demirel, H., Ozcinar, C., Anbarjafari, G., 2009. Satellite image contrast enhancement using discrete wavelet transform and singular value decomposition. IEEE Geoscience and Remote Sensing Letters 7 (2), 333–337.

Dewan, A.M., Yamaguchi, Y., 2009. Land use and land cover change in Greater Dhaka, Bangladesh: using remote sensing to promote sustainable urbanization. Applied Geography 29 (3), 390–401.

Eckert, S., Hüsler, F., Liniger, H., Hodel, E., 2015. Trend analysis of MODIS NDVI time series for detecting land degradation and regeneration in Mongolia. Journal of Arid Environments 113, 16–28.

El Garouani, A., Mulla, D.J., El Garouani, S., Knight, J., 2017. Analysis of urban growth and sprawl from remote sensing data: case of Fez, Morocco. International Journal of Sustainable Built Environment 6 (1), 160–169.

Foody, G.M., 1992. On the compensation for chance agreement in image classification accuracy assessment. Photogrammetric Engineering and Remote Sensing 58 (10), 1459–1460.

Forkel, M., Carvalhais, N., Verbesselt, J., Mahecha, M., Neigh, C., Reichstein, M., 2013. Trend change detection in NDVI time series: effects of inter-annual variability and methodology. Remote Sensing 5 (5), 2113–2144.

Fu, P., Weng, Q., 2016. A time series analysis of urbanization induced land use and land cover change and its impact on land surface temperature with Landsat imagery. Remote Sensing of Environment 175, 205–214.

Gašparović, M., Jogun, T., 2018. The effect of fusing Sentinel-2 bands on land-cover classification. International Journal of Remote Sensing 39 (3), 822–841.

Gašparović, M., Zrinjski, M., Gudelj, M., 2017. Analysis of urbanization of Split. Geodetski List 71 (3), 189.

Gašparović, M., Dobrinić, D., Medak, D., 2018a. Spatial accuracy analysis of aerial and satellite imagery of Zagreb. Geodetski List 72 (1), 1–14.

Gašparović, M., Medak, D., Pilaš, I., Jurjević, L., Balenović, I., 2018b. Fusion of Sentinel-2 and PlanetScope imagery for vegetation detection and monitoring. In: Volumes ISPRS TC I Mid-term Symposium Innovative Sensing-From Sensors to Methods and Applications.

Gašparović, M., Simic Milas, A., Seletković, A., Balenović, I., 2018c. A novel automated method for the improvement of photogrammetric DTM accuracy in forests. Šumarski List 142 (11–12), 567–576.

Gašparović, M., Dobrinić, D., Medak, D., 2019a. Geometric accuracy improvement of WorldView-2 imagery using freely available DEM data. The Photogrammetric Record 34 (167) (Accepted for publication).

Gašparović, M., Rumora, L., Miler, M., Medak, D., 2019b. Effect of fusing Sentinel-2 and WorldView-4 imagery on the various vegetation indices. Journal of Applied Remote Sensing 13 (3), 036503.

Gašparović, M., Zrinjski, M., Gudelj, M., 2019c. Automatic cost-effective method for land cover classification (ALCC). Computers, Environment and Urban Systems 76, 1–10.

Ghimire, B., Rogan, J., Miller, J., 2010. Contextual land-cover classification: incorporating spatial dependence in land-cover classification models using random forests and the Getis statistic. Remote Sensing Letters 1 (1), 45–54.

Gomez, C., Dharumarajan, S., Féret, J.B., Lagacherie, P., Ruiz, L., Sekhar, M., 2019. Use of Sentinel-2 time-series images for classification and uncertainty analysis of inherent biophysical property: case of soil texture mapping. Remote Sensing 11 (5), 565.

Hegazy, I.R., Kaloop, M.R., 2015. Monitoring urban growth and land use change detection with GIS and remote sensing techniques in Daqahlia governorate Egypt. International Journal of Sustainable Built Environment 4 (1), 117–124.

II. Urban land use land cover

Li, Y., Tao, C., Tan, Y., Shang, K., Tian, J., 2016. Unsupervised multilayer feature learning for satellite image scene classification. IEEE Geoscience and Remote Sensing Letters 13 (2), 157–161.

Liang, S., Fang, H., Chen, M., 2001. Atmospheric correction of Landsat ETM+ land surface imagery. I. Methods. IEEE Transactions on Geoscience and Remote Sensing 39 (11), 2490–2498.

Lillesand, T., Kiefer, R.W., Chipman, J., 2015. Remote Sensing and Image Interpretation. John Wiley & Sons.

Liu, J.G., Morgan, G.L.K., 2006. FFT selective and adaptive filtering for removal of systematic noise in ETM+ image-odesy images. IEEE Transactions on Geoscience and Remote Sensing 44 (12), 3716–3724.

López, E., Bocco, G., Mendoza, M., Duhau, E., 2001. Predicting land-cover and land-use change in the urban fringe: a case in Morelia city, Mexico. Landscape and Urban Planning 55 (4), 271–285.

Mather, P., Tso, B., 2009. Classification methods for remotely sensed data. CRC Press.

Melesse, A.M., Jordan, J.D., 2002. A comparison of fuzzy vs. augmented-ISODATA classification algorithms for cloud-shadow discrimination from Landsat images. Photogrammetric Engineering and Remote Sensing 68 (9), 905–912.

Palsson, F., Sveinsson, J.R., Benediktsson, J.A., Aanaes, H., 2011. Classification of pansharpened urban satellite images. IEEE Journal of Selected Topics in Applied Earth Observations and Remote Sensing 5 (1), 281–297.

Pontius, R.G., Millones, M., 2011. Death to kappa: Birth of quantity disagreement and allocation disagreement for accuracy assessment. International Journal of Remote Sensing 32 (15), 4407–4429.

Qian, Y., Zhang, K., Qiu, F., March 2005. Spatial contextual noise removal for post classification smoothing of remotely sensed images. In: Proceedings of the 2005 ACM Symposium on Applied Computing. ACM, pp. 524–528.

Qian, Y., Zhou, W., Yan, J., Li, W., Han, L., 2015. Comparing machine learning classifiers for object-based land cover classification using very high resolution imagery. Remote Sensing 7 (1), 153–168.

Richter, R., 1997. Correction of atmospheric and topographic effects for high spatial resolution satellite imagery. International Journal of Remote Sensing 18 (5), 1099–1111.

Richter, R., 1998. Correction of satellite imagery over mountainous terrain. Applied optics 37 (18), 4004–4015.

Salah, M., 2017. A survey of modern classification techniques in remote sensing for improved image classification. Journal of Geomatics 11 (1), 21.

Schneider, A., 2012. Monitoring land cover change in urban and peri-urban areas using dense time stacks of Landsat satellite data and a data mining approach. Remote Sensing of Environment 124, 689–704.

Schowengerdt, R.A., 2012. Techniques for Image Processing and Classifications in Remote Sensing. Academic Press.

Taubenböck, H., Wegmann, M., Roth, A., Mehl, H., Dech, S., 2009. Urbanization in India–Spatiotemporal analysis using remote sensing data. Computers, Environment and Urban Systems 33 (3), 179–188.

Thanh Noi, P., Kappas, M., 2018. Comparison of random forest, k-nearest neighbor, and support vector machine classifiers for land cover classification using Sentinel-2 imagery. Sensors 18 (1), 18.

Verbesselt, J., Hyndman, R., Newnham, G., Culvenor, D., 2010. Detecting trend and seasonal changes in satellite image time series. Remote Sensing of Environment 114 (1), 106–115.

Verma, P., Raghubanshi, A.S., 2018. Urban sustainability indicators: challenges and opportunities. Ecological Indicators 93, 282–291.

Verma, P., Raghubanshi, A.S., 2019. Rural development and land use land cover change in a rapidly developing agrarian South Asian landscape. Remote Sensing Applications: Society and Environment 14, 138–147.

Wang, Q., Shi, W., Li, Z., Atkinson, P.M., 2016. Fusion of Sentinel-2 images. Remote Sensing of Environment 187, 241–252.

Yang, C., Everitt, J.H., Murden, D., 2011. Evaluating high resolution SPOT 5 satellite imagery for crop identification. Computers and Electronics in Agriculture 75 (2), 347–354.

Zhang, M., Carder, K., Muller-Karger, F.E., Lee, Z., Goldgof, D.B., 1999. Noise reduction and atmospheric correction for coastal applications of Landsat Thematic Mapper imagery. Remote Sensing of Environment 70 (2), 167–180.

Zhu, Z., 2017. Change detection using landsat time series: a review of frequencies, preprocessing, algorithms, and applications. ISPRS Journal of Photogrammetry and Remote Sensing 130, 370–384.

Exposition of spatial urban growth pattern using PSO-SLEUTH and identifying its effects on surface temperature

H.A. Bharath, G. Nimish, M.C. Chandan

RCG School of Infrastructure Design and Management, Indian Institute of Technology
Kharagpur, West Bengal, India

O U T L I N E

1. Background

Urbanization can be deduced as transformation of naturally occurring features to concrete impervious structures as a result of amplification in number of urban dwellers (Henderson and Wang, 2007; Ramachandra et al., 2012a). Increased urban population can be inferred to migrations from rural areas for better quality/standard of living and increased opportunities in terms of work profile. This results in hasty growth of cities and exerts substantial pressure on society to provide basic necessities such as adequate shelter, energy, water, health and education (Buhaug and Urdal, 2013). To provide infrastructure facilities for new urban dwellers, cities across the globe urbanize disproportionately and expand. Growth is inconsistent in developed nations and is even worse in developing nations (Farrell, 2017). As per a report published by UN DESA (2018), at present more than half of the global population resides in urban areas and it is predicted to increase to 68% by 2050. Out of this, Asia and Africa are predicted to house around 90% of the urban growth. India being one of the fastest developing countries (India Briefing, 2019) is facing some problems due to this.

Most of the urbanization in India started in early 1990s as a result of liberalization that allowed the expansion of private sector. Since then, India has witnessed numerous infrastructure developments to fulfil needs of residents and migrants. India being a growing economy is increasing its capacity in the form of industries, manufacturing and processing units, medical facilities, housing, infrastructure, etc., for better GDP (Gross Domestic Product). As a result of this, the rate of migration is swelling and core cities in India (Bangalore, Kolkata, Hyderabad, Chennai, Delhi, Mumbai, etc.) are gradually sprawling, giving rise to problems such as housing, lack of fresh treated water, inadequate infrastructure, and increased level of pollution (Ramachandra et al., 2012b). Two centres develop due to sprawling — occupational and satellite towns (Polidoro et al., 2012). The biggest problem is to provide basic amenities as it requires more resources along with energy intensive processes to develop infrastructure in these satellite towns. Urbanization not only creates problems related to infrastructure but also affects displacement of pollutants. Outskirts of cities acts as breathing spaces and help in displacing pollutants with heavy wind current and turbulence but satellite towns acts as an obstacle and do not allow both. Visualizing these changes along with interpreting results can play a vital role in minimizing the concerns related to rapid unplanned urbanization. Development in the field of remote sensing as well as algorithms to deduce land use has bought a revolution in all the fields including urban planning, climate change, etc. Land use can be defined as how physical features present on earth's surface are being utilized by humans for their benefits. Understanding urbanization using land-use analysis is important as it severely affects the environment (United Nations, 2018), biodiversity, ecology and existence of living being as it affects the four basic features important for humans — availability of water, air quality, land in terms of green and open spaces and ample energy production (Siedentop, 2005). Furthermore, changes in land use upsets the rate of evaporation, alters surface albedo, heat and moisture content in soil, wind turbulence and most importantly the surface temperature (Pal and Ziaul, 2017).

Land Surface Temperature (LST) is one of the most important parameters to quantify urban micro climate and can be defined as the radiative skin temperature of earth's surface (Ese Sentinel Online, 2018; Copernicus, 2018). It serves as the most important parameter in defining land processes and is the basic determinant of earthly thermal behaviour (Li et al., 2013). Albedo, emissivity, soil moisture, health and density of vegetation are some of the important parameters that define LST. Additionally, changes in land-use pattern alter the evapotranspiration rates and modify the latent and sensible heat flux (Mojolaoluwa et al., 2018). This results in increased thermal discomfort and development of stress due to heat. Increased construction activities escalate the atmospheric concentration of pollutants in terms of particulate matter ($PM_{10,\ 2.5,\ 1}$, Suspended PM) and greenhouse gases such as CO_2, CH_4, CO, SO_x, NO_x, O_3, etc. These pollutants trap the heat reflected from the earth's surface and reradiate it towards surface, thus, enhancing greenhouse gas effect and causing rise in surface and ambient temperature. Moreover, LST also affects natural atmospheric cycles (water cycle, energy cycle, bio-geo-chemical cycle, etc.), micro-meso-macro climate, crop and wind patterns, biodiversity, rainfall patterns, etc. (Bharath et al., 2013; Jin et al., 2015). It not only affects the environment but increased LST also results in some serious insinuations on health and welfare of dwellers such as difficulty in breathing, increased pulmonary diseases, cramps and exhaustion due to heat, lethal and nonlethal heat strokes and alteration in mortality rate (Yang, 2016; Lal, 2017; EPA, 2019). In addition, heat waves along with increased emissions (pollution and greenhouse gases) generate Urban Heat Island (UHI) effect that misbalances the natural energy exchange causing increased heat waves and the vicious cycle continues (Liang and Shi, 2009; Founda and Santamouris, 2017). These issues make it important to understand how urban growth in the future might affect the earth. Urban growth models can efficiently predict the land use for future using agents and constraints as inputs.

Urban growth models have been into existence since 1960s. Scientists report two major types of urban growth models: (a) Cellular Automata (CA) and (b) Agent Based Model (ABM) (Heppenstall et al., 2011). CA is a dynamic two-dimensional modelling approach where simple changes in the local neighbourhood produces complex global changes (Wolfram, 1984). CA is made up of five basic elements: cell, lattice or a group of cells, neighbourhood (defined by either von-Neumann or Moore), transition rules and temporal aspect defined by Markov chains (Torrens and O'Sullivan, 2001). Due to its simple implementation procedure and flexibility to handle geospatial datasets, CA is often regarded as a base module for various other land-use change models such as: CLEU-S, GEOMOD, CA-Markov Chain (Kamusoko et al., 2009; Pan et al., 2010; Hyandye and Martz, 2016). SLEUTH is one of the finest CA-based open-source land-use change simulation model, first conceptualized by Prof. Clarke in the year 1997. Acronym of the model name stands for its six raster layers used in simulation namely: Slope, Land use, Exclusion, Urban, Transportation, Hillshade (Sakieh et al., 2016; Chandan et al., 2019). To calibrate urban area and predict, the model requires four different time period historical, urban and road data, a minimum of two different time period land-use data, single time period slope, exclusion and Hillshade data even though Hillshade layer is used only for visualization and does not have any other significant role (Chaudhuri and Foley, 2019). Model calibration routine is governed by a set of five control parameters namely: dispersion, breed, spread, slope resistance and road gravity along with four growth transition rules: spontaneous, diffusive, organic and road influenced growth (Gazulis and Clarke, 2006; Nimish et al., 2018). SLEUTH involves a rigorous calibration procedure to estimate optimum value for each control parameter. Traditional brute force method (BFM) takes days together to arrive at three phases of calibration, i.e., coarse, fine, full and shortlisted values for prediction. Most researchers have made a successful attempt to reduce SLEUTH computation time and therefore to increase its efficiency. For instance, efforts like optimal SLEUTH metric (OSM), pSLEUTH, SLEUTH-3r, SLEUTH-Genetic Algorithm and distributed SLEUTH have been reported by global urban modellers (Guan & Clarke, 2010; Jantz et al., 2010; Chaudhuri and Foley, 2019). However, very few researchers have incorporated CA with emerging artificial swarm intelligence and optimization techniques such as ant colony optimization, artificial bee colony, particle swarm optimization (PSO), grey wolf algorithm, etc., and this research area still remains unexplored (Feng et al., 2011; Naghibi et al., 2016; Lu et al., 2018). PSO is a population-based stochastic optimization technique inspired by social behaviour of birds flocking and fish schooling (Eberhart & Kennedy, 1995). Feng et al. (2011) conducted a study to explore advantages of integrated CA and PSO approach. Objective of the study was to stochastically optimize transition rules and therefore reduce computation time, prediction uncertainties as well as improving location accuracy, a case of Fengxian District, Shanghai, China. Authors claim similarity between PSO and CA drives to search global optimum parameters of CA rules. Further, this concept can be applied to SLEUTH-PSO spread optimization to reduce computation time and predict accurate transitions. PSO optimizes allocation of pixels by considering 3×3 kernel and returns a set of eight values that represent Moore neighbourhood. In this communication we attempt to improvize routine BFM with the aid of PSO to reduce computation time, as well as to achieve better modelled results.

2. Study area

Indian metropolitan of Bangalore was chosen as study area as shown in Fig. 4.1 to demonstrate the urban growth pattern, its linkage with LST and simulating urban models for future

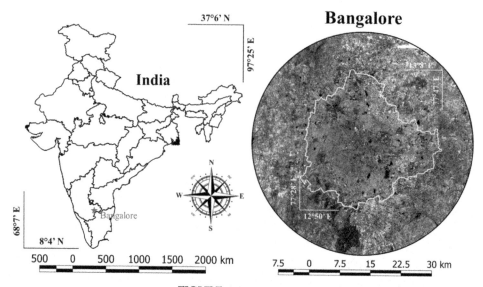

FIGURE 4.1   Study area.

predictions. Bangalore is located in Karnataka state of Southern India on the Deccan Plateau and has an elevation of 900 m above MSL (mean sea level). It comprises of 198 wards spread over 741 km^2 area and is the fifth largest metropolitan city of India. Since the beginning of the 20th century, there has been a gigantic upsurge (almost two folds) in the population of Bangalore. This can be inferred to the advent of Information Technology (IT) sector and setting up of International organizations such as Infosys, Wipro, Cognizant, IBM, Accenture, etc., along with other public sector organizations including Bharat Electrical Limited, Hindustan Aeronautics Limited, National Aeronautics Laboratories, Indian Space Research Organization, Defence Research and Development Organization, etc. Bangalore houses numerous educational institutes from primary school to doctoral level institutions that act as an important agent in its increased demography. The city is also known for its textile industry and is one of the major producers of cotton. Bangalore emerged as the most vivacious city and attracted foreign investment, thus becoming the fourth largest technology cluster across the globe. The city since beginning has a good transportation network, other infrastructure facilities and an excellent connectivity with other Indian states and across the borders. The geographic and demographic profile of Bangalore is as shown in Table 4.1.

In terms of climate, Bangalore is classified as tropical savanna climate under Koppen climate classification. The city usually experiences a moderate and pleasant climate with occasional heat waves during summers. The annual mean temperature varies from 14°C (average low) to 36°C (average high). Bangalore receives rainfall from Northeast as well as Southwest monsoons.

TABLE 4.1 Bangalore — geographic and demographic information.

Area	741 km^2		
Latitude	12° 59′ N		
Longitude	77° 35′ E		
Altitude	740−960 m		
Mean annual rainfall	859 mm		
Climate	• Tropical savanna climate (according to koppen climate classification) with diverse wet and dry seasons • City usually experiences moderate climate with few occasional heat waves during summer • Receives rainfall from both Southwest and Northeast monsoons		
Air temperature	Winters minimum — January: 15.1°C Summer maximum — April: 35°C		
Population	**Year**	**Population**	**Decadal growth rate in percentage**
	1981	2,922,000	
	1991	4,130,000	41.34
	2001	5,101,000	23.51
	2011	8,425,970	65.18
	2019	11,882,666 (predicted)	41.02

3. Method

A structured process with six phases was performed in the study as illustrated in Fig. 4.2 that includes (1) Data acquisition, (2) Extraction of various features, (3) Preprocessing, (4) Land-use analysis, (5) Estimation of land surface temperature, (6) Landscape modelling and prediction using particle swarm optimization − SLEUTH (PSO-SLEUTH) model.

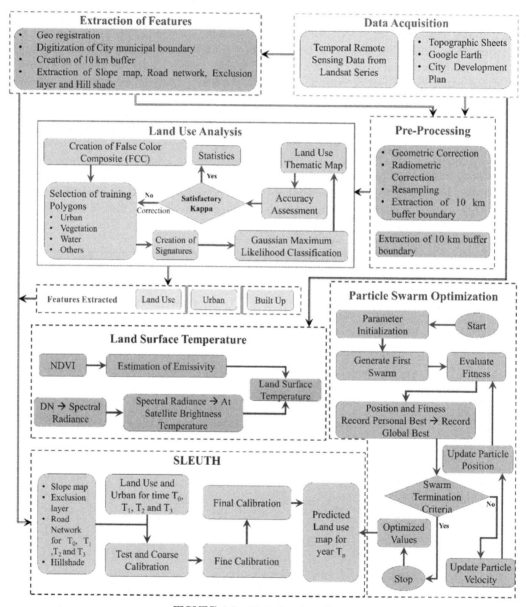

FIGURE 4.2   Method − flow chart.

3.1 Data acquisition

Data for analysis was acquired from various sources as mentioned in Table 4.2.

Satellite data used in the study was captured from Landsat series that has a good resolution (Table 4.3) and is available for no cost on public domains. It provides historical data since 1973, and is thus very useful to detect changes over the time. Post third Landsat mission, temperature sensitive band(s) in Thermal Infrared region of EM waves was added that contributed to a significant rise in applications in the form of surface temperature, GHG estimations, etc. DEM that serves as one of the inputs to SLEUTH model and that defines the local elevation of the region was captured from ASTER.

3.2 Extraction of features

City development plan was obtained from BBMP and was georegistered (assigned coordinates) using various Ground Control Points (GCPs). Places that do not change over time such as highway junctions, railway and highway crossings, historical buildings were chosen as GCPs. City municipal boundary was then digitized with the georegistered map, topographic sheets and other ancillary datasets. Centroid was then created for the city boundary and considering that as centre, a 10 km circular buffer was demarcated.

Other factors influencing the growth such as road network were digitized with the help of classified map and auxiliary data such as Open Street Maps, Bhuvan, Google Earth. Slope map and hillshade were extracted from ASTER DEM. City development plan and other datasets were used for deriving exclusion layer (areas resisting the growth) incorporating protected areas, water bodies, etc.

3.3 Preprocessing

Remotely sensed data can have various errors including geometric distortions, radiometric error, etc. These errors reduce the quality of information. The raw satellite data was corrected for any geometric or radiometric errors using topographic maps, Google Earth and ground truth data. The data was then resampled and all the data (satellite) was cropped relating to the study area.

TABLE 4.2 Data acquired and their sources.

Data	Source
Temporal remote sensing data	USGS earth explorer (https://earthexplorer.usgs.gov)
Data elevation model	
Topographic sheets (1:50,000 and 1:250,000)	Survey of India
City development plan	Bruhat Bengaluru Mahanagara Palike
Ancillary data	Open street maps, Google Earth and Bhuvan

TABLE 4.3   Landsat — resolution and sensor on board.

Landsat series	Sensor	Resolution			
		Spatial	Spectral	Temporal	Radiometric
Landsat 5	Thematic mapper	Optical - 30 m Thermal — 120m[a]	7 bands	16 days	8 bits
Landsat 8	Operational land imager/ thermal Infrared	Optical - 30m PAN — 15 m Thermal — 100m[a]	11 bands	16 days	16 bits[b]

[a]Resampled to 30 m.
[b]collected at 12 bits but later converted to 16 bits.

3.4 Land-use analysis: Gaussian maximum likelihood classifier

Land-use analysis was performed using a four-step process — (1) Generation of False Color Composite (FCC), (2) Training the classifier, (3) Classification, (4) Accuracy Assessment.

Generation of FCC: Classification of image is efficient if standard FCC is considered as multiple classes such as urban, vegetation, water, etc., can be easily separated. FCC can be created by assigning wrong combination of color to three bands (i.e., blue color to red band, green color to blue band and red color to green band, etc.). A standard FCC was created by assigning red color to NIR band, green color to red band and blue color to green band.

Training the classifier: Representative and heterogeneous polygons from each land-use class were digitized. This step was performed by overlaying FCC on Google Earth for cross-validation of the class to which the polygon is assigned. Training polygons were uniformly distributed across the study area and covered at least 15% of the total study area.

Classification: Polygons selected in previous step were assigned with statistical values (mean, covariance and variance) by converting them into raster form, thus, making them unique signatures representing heterogeneous feature on the map. These signatures were then provided as input to Gaussian Maximum Likelihood Classifier (GMLC) which is a supervised classification algorithm. GMLC estimates probability of pixel under consideration belonging to a specific class by creating probability density functions (PDF) (Duda et al. 2012; Ramachandra et al., 2014). Pixels are assigned as per these PDF and mathematically it can be represented as shown in Eq. (4.1). As per literature and other requirements of the study four land-use classes were chosen as shown in Table 4.4.

$$X \varepsilon C_j \text{ if } p(C_j / X) = \max[p(C_1 / X), p(C_2 / X), ..., p(C_m / X)] \tag{4.1}$$

Here $p(C_j/X)$ represents the conditional probability of pixel X being a member of class C_j.

Accuracy Assessment: A classification is inadequate without assessing its accuracy and can be defined as the precision by which a classifier processes image classification with respect to the reference (ground truth). Standard procedure involves construction of error/confusion matrix that includes all correctly and incorrectly classified pixels. Using this matrix descriptive statistical measurements such as overall accuracy, kappa coefficient, producer's accuracy and user accuracy are estimated. Overall accuracy is calculated by ratio of sum of correctly

TABLE 4.4 Land-use classes.

S. No.	Land use Class	Features included in the class
1.	Urban	Impervious/Paved surfaces including residential and industrial buildings, mixed pixels containing more than 50% built-up
2.	Vegetation	All the vegetated areas including forest, parks, nurseries, agricultural fields, etc.
3.	Water	Lakes, ponds, tanks, reservoirs
4.	Others	Barren agricultural fields, open grounds, rocks, mining areas, unconstructed roads, etc.

classified pixels to total number of pixels. Kappa coefficient provides robust information as it considers the misclassified pixels as well as a chance agreement between the pixel in the reference map and the classified map. Classified map was then compared with validation map to generate confusion/error matrix using which overall accuracy and coefficient of kappa were quantified.

3.5 Land surface temperature: single window algorithm

The range 10.4–12.5 μm in the electromagnetic spectrum is thermal infrared that obtains reflectance and emissivity of various objects on earth's surface. The bands in this range were used to extract surface temperature. For Landsat 5, band 6, and for Landsat 8, band 10 were used. The process involved three steps – (1) Generation of thermal maps, (2) Quantification of emissivity, (3) Calculation of LST.

Generation of thermal maps: Landsat series post Landsat 3 has introduced thermal bands to capture the emissivity and radiation from materials on earth's surface. Landsat 8 has two thermal infrared sensors (band 10 and band 11) and Landsat 5 has one. These sensors store thermal data in terms of DN values. Since radiometric resolution of Landsat 5 is 8 bits thus, the DN value ranges from 0–255 and for Landsat 8 it is 16 bits; thus, it ranges from 0–65535. For Landsat 5, band 6 and for Landsat 8, band 10 was used. Band 10 is preferred over band 11 because the range of EM radiation for band 10 (10.60–11.19 μm) has less attenuation from atmosphere when compared to band 11 (11.50–12.51 μm).

Quantification of emissivity: Emissivity was quantified using NDVI threshold method as it is considered to be one of the most efficient methods and takes into effect the diversity of land. A threshold value of NDVI was chosen (using FCC, NDVI map and Google Earth) and based on that various classes were defined as shown in Table 4.5 (NOTE: Threshold values are location dependent).

LSE for water, soil and vegetation were directly chosen as single values (Landsat 8 LST Analysis, 2016), while LST for mixture of soil and vegetation was obtained using Eq. (4.2).

$$\varepsilon_{SV} = \varepsilon_V P_V + \varepsilon_S (1 - P_V) + C \tag{4.2}$$

TABLE 4.5   NDVI thresholds for land-use classes.

S. No.	NDVI thresholds	Land-use class
1.	$NDVI_{min} - NDVI_W$	Water
2.	$NDVI_W - NDVI_S$	Soil
3.	$NDVI_S - NDVI_V$	Soil + vegetation
4.	$NDVI_V - NDVI_{max}$	Vegetation

Here ε_{sv}: emissivity of soil and vegetation ε_s: emissivity of soil ε_v: emissivity of vegetation P_V: calculated as shown in Eq. (4.3) C: calculated as shown in Eq. (4.4)

$$P_{V1} = \left(\frac{NDVI - NDVI_S}{NDVI_V - NDVI_S}\right)^2 \qquad (4.3)$$

$$C = (1 - \varepsilon_S)\varepsilon_V F(1 - P_V) \qquad (4.4)$$

F is geometrical factor ranging from 0 to 1 depending on surface geometry (usually F = 0.55).
Calculation of LST: It included three steps

• *Conversion of DN into Radiance*

$$L_\lambda = (Gain \ X \ DN) + Offset \qquad (4.5)$$

Gain and offset was calculated using Metadata

• *Conversion of Radiance into At-Sat brightness temperature*

$$T_B = \frac{K_2}{\ln\left(\frac{K_1}{L_\lambda} + 1\right)} \qquad (4.6)$$

Here T_B: at-satellite brightness temperature K_1: calibration constant 1; K_2: calibration constant 2 and were obtained from metadata.

• *Conversion of At-Sat brightness temperature into LST*

$$LST = \frac{T_B}{1 + \left(\frac{\lambda T_B}{\rho} \ X \ \ln(\varepsilon)\right)} \qquad (4.7)$$

Here λ is the wavelength of the thermal band at which the relative response in maximum with respect to wavelength (μm). ε is the emissivity of the element $\rho = \frac{hc}{\sigma} = 1.438 \times 10^{-2} mK$ (where, h is planck's constant, c is speed of light, σ is Stefan Boltzmann constant).

3.6 Landscape modelling and prediction: PSO-SLEUTH

SLEUTH analysis consists of three phases: Input data and verification phase, calibration and prediction phase. Analysis of SLEUTH starts with input data preparation and data standardization. Verification phase is carried out to ensure compatibility of datasets prepared, before calibration and prediction. Scenario file is modified according to standard layers along with random coefficients. Calibration and prediction phase involves exploring coefficient range of parameters varying from 1 to 100, 1 indicating nil or least influence and 100 indicating highest influence on urban growth. Overall goal of calibration phase is to achieve optimized and best fit statistic of each coefficient for prediction. Calibration is carried out in three modes: coarse, fine and final. Monte Carlo iterations need to be set during each calibration stage where as literature suggests 5, 8 and 10 iterations for coarse, fine and final calibration, respectively (Silva & Clarke, 2002). Top Lee-Salee metric are selected to obtain five unique coefficient values responsible for urban growth. The scenario file will be modified accordingly to predict urban growth in prediction phase along with changes in source code based on PSO values.

PSO is popular among various nature-inspired techniques for its simplicity, minimum mathematical computation, achieving improved optimization in less computation time and ease of application to general engineering problems (Hu et al., 2018). Integrating concepts of PSO with SLEUTH includes modification in the eight Moore neighbourhood cell values that are to be automatically updated after each swarm iteration and storing best values. PSO provides optimum spread solution ranging between 1 and 99 for these cell values. These best values obtained are then adopted in spread file of SLEUTH 3.0 code with p01 patch. Input layers used for the model are depicted in Fig. 4.3. Routine full calibration was performed using brute force method (BFM), output controlstat.log file was carefully examined to obtain model fit statistics and optimum SLEUTH metric (OSM), calculated by multiplying metrics such as compare population, edges, cluster, slope, X-mean and Y-mean. Best OSM was chosen and corresponding coefficient values were replaced in the scenario file during prediction phase.

4. Results and discussion

4.1 Land-use analysis

Temporal land-use analysis for Bangalore (Greater Bangalore) with 10 km buffer was performed and the maps are as shown in Fig. 4.5. Fig. 4.4 represents variation in percentage cover of each land-use class during the study period. Impervious surfaces in study area have increased threefold since 1991. In this short duration, study area has experienced a haphazard and unplanned growth that led the city to expand in all the directions from the core at the cost of agricultural fields and barren areas that serve as breathing space. Urban

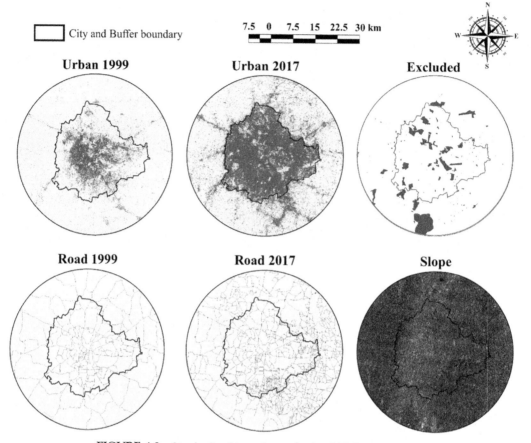

FIGURE 4.3   Standardized input layers used in PSO-SLEUTH model.

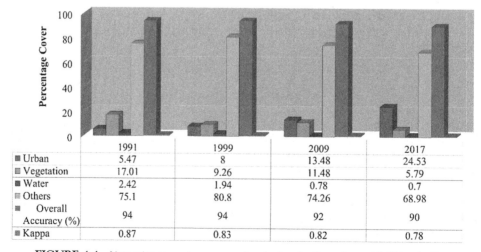

	1991	1999	2009	2017
■ Urban	5.47	8	13.48	24.53
▨ Vegetation	17.01	9.26	11.48	5.79
■ Water	2.42	1.94	0.78	0.7
▨ Others	75.1	80.8	74.26	68.98
■ Overall Accuracy (%)	94	94	92	90
▨ Kappa	0.87	0.83	0.82	0.78

FIGURE 4.4   Year-wise percentage cover of each class for Bangalore with 10 km buffer.

II.   Urban land use land cover

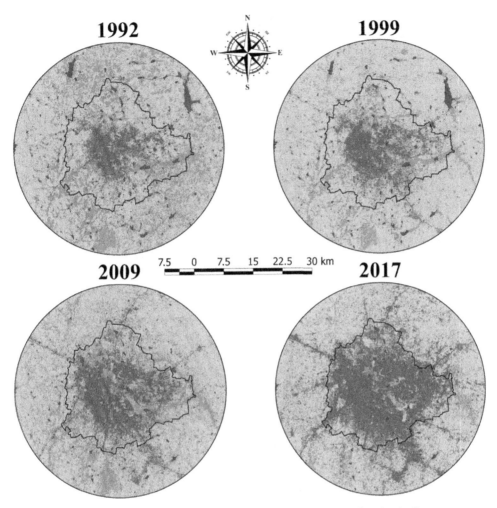

FIGURE 4.5 Temporal land-use analysis for Bangalore with 10 km buffer.

structures have been constructed on wetlands and floodplains as well, causing reduction in water bodies such as wetlands, lakes, reservoirs, etc., and increasing the problem of urban floods. All these agents have led to a severe reduction in vegetation (17.01% to 5.79%) and other categories including barren space, open areas (75.10% to 68.98%) within a span of three decades. The increase in vegetation from 9.26% in 1999 to 11.48 in 2009 can be attributed to extensive afforestation drives in and around Turahalli Reserve Forest, Ragihalli State Forest, Jarkabandi Forest, etc.

Bangalore in early 1990s was blessed with more than 180 lakes that ominously reduced to merely few in the present scenario, causing the class water bodies to reduce from 2.42% to 0.70%. Reduction in green space as well as water bodies is impacting the ground water regime and already Bangalore is grieving over shortage of water. One of the most important agents of this growth is the booming of IT sector and headquarters of various private as well as public sector organizations. To sustain the increased population influx, the city has been growing since 1990s. As per a study by Ramachandra and Bharath (2016), since 1973, the urban jungle in Bangalore has increased by 1005% along with decline in vegetation by 88% and water bodies by 79%.

4.2 Land surface temperature

Temporal Land Surface Temperature was estimated for the study area as shown in Fig. 4.6. Statistical analysis in terms of mean, range and standard deviation with respect to individual land-use class was carried out as represented in Table 4.6 and it was observed that mean surface temperature of urban category has increased from 33.07°C to 41.14°C. This upsurge in surface temperature can be inferred to increased impervious surfaces that alter the latent and sensible heat content. An increment of 7.5°C was witnessed for others category as these areas have lost the moisture content due to altering neighbourhood. Reduction in number of wells and depth and contamination of water bodies due to increased human intervention have caused mean surface temperature of water class to increase by 9.13°C. A rise of 7.68°C in mean surface temperature was found for vegetation category as a result of depleting green spaces in the city. Cumulatively analysis shows that the mean surface temperature of the study area has increased to 41°C in 2017 from 33°C in 1992. The result of our analysis is in line with other similar studies (Chakraborty et al., 2015; Deng et al., 2018).

Location-wise analysis was performed and the locations that have the highest temperatures (44−51°C) in 2017 were found to be airfields in HAL and Jakkur region, Kempegowda International airport, agricultural fields near Shankanipura, dried lakes of Hesaraghatta. Green spaces such as Cubbon park, golf course, dense vegetation at Indian Institute of Science (IISc), etc., show moderate temperature of 33−35°C. Minimal surface temperature range was obtained in water bodies such as Sankey tank, Bellandur lake, Ulsoor lake that varies from 30°C to 33°C.

4.3 Landscape modelling and prediction

SLEUTH codes were modified with respect to spread optimization obtained from PSO. Spread optimization gave eight best-fit values indicating eight directions of optimum growth as shown in Moore neighbourhood network. Best PSO results were achieved with an OSM value of 0.623.

Results obtained from PSO were compared with traditional brute force method (BFM). Corresponding statistics are pictorially shown in Fig. 4.7. Substantial enhancement in terms of metrics such as Product, Compare, Population, Clusters, Size, Slope, Percentage urban and Y-Mean were observed by adopting PSO technique, while urban edges, Lee-Salee and Rad showed minimum difference between these two methods. Lee-salee metric was higher in BFM method with a value of 0.4158 and lesser in PSO with a value of 0.3626. OSM was calculated, achieved OSM was 0.3229 during BFM and 0.6238 during PSO-SLEUTH. This clearly shows significant improvement in calibration of coefficients with PSO improvized, spread optimized pixel values. Another important development was observed in computation time. Post full calibration process, PSO took 1day, 1 h, 37 mins and 3 s to yield model fit

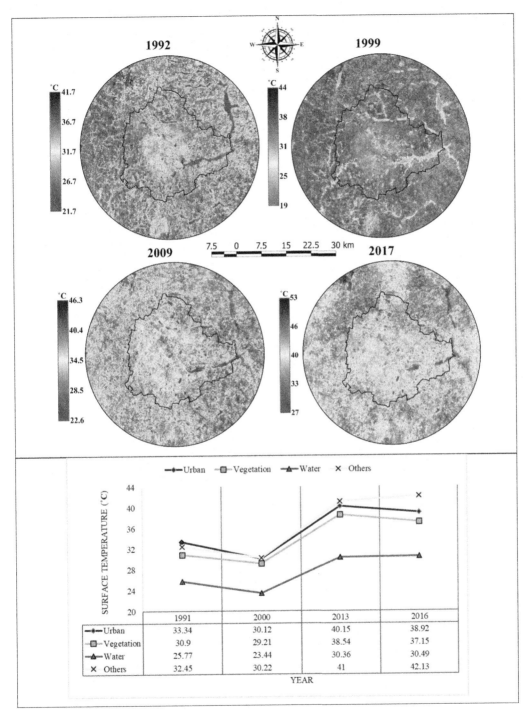

FIGURE 4.6 Temporal Land Surface Temperature for Bangalore with 10 km buffer.

II. Urban land use land cover

TABLE 4.6 Temperature statistics for Bangalore with 10 km buffer.

Year	Land-use class	Min. Temp. (°C)	Max. Temp. (°C)	Mean temp. (°C)	Standard deviation
1992	Urban	22.99	41.74	33.07	1.82
	Vegetation	22.94	41.74	31.09	2.27
	Water	21.73	39.78	25.40	2.70
	Others	22.17	41.35	34.57	2.02
1999	Urban	22.57	42.90	33.56	1.63
	Vegetation	21.68	40.57	29.95	2.02
	Water	22.61	40.17	27.37	3.42
	Others	18.58	44.06	34.41	2.03
2009	Urban	25.21	44.44	35.26	1.70
	Vegetation	23.38	46.34	33.16	3.02
	Water	22.61	44.82	28.11	3.70
	Others	23.38	45.20	36.56	2.43
2017	Urban	27.00	52.50	41.14	1.95
	Vegetation	30.77	49.72	38.77	2.22
	Water	30.08	47.15	34.53	2.88
	Others	30.71	52.49	42.12	2.51

statistics. These results not only indicate the model robustness by integrating PSO, but also it can deliver excellent fit statistics without compromising optimum coefficients.

Fig. 4.8 represents urban growth pattern predicted for the year 2025. After three phase calibration, optimum growth coefficients were derived. Slope resistance (14) can be observed in fewer parts such as: areas lying between SH9 and NH7 due north of Bangalore city, IISc and

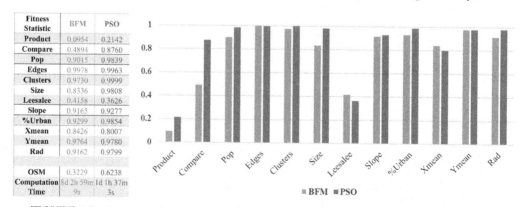

FIGURE 4.7   Summary of model fit statistics, comparison of metric values between BFM and PSO.

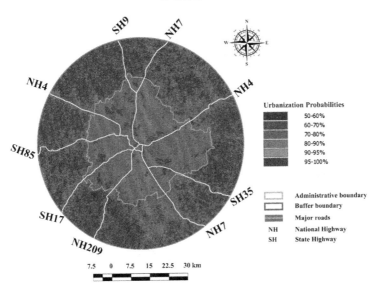

FIGURE 4.8 PSO-SLEUTH modeled output for the year 2025, Bangalore region.

Airforce training command near Yeshwanthpur due west of the city, Bannerghatta national park and surroundings due south of the city. Medium road gravity (43) and slightly higher values observed for spread (80) suggested the possibility of spreading urban growth along major roads found in all directions of the city.

Road gravity values reflect highways and transport corridors acting as key urbanization factors. To name a few, Bangalore region has major connecting roads NH4 towards NW and NE directions connecting to Tumkur and Kolar, respectively, interstate NH7 (Hosur road) connecting to neighbour state Tamilnadu in the SW direction from city interstate, NH209 in SW direction towards Kanakapura and SH17 in SW direction connecting to erstwhile capital of Karnataka, Mysore. It is evident that Bangalore has followed historical urban growth, especially infilling type within the administrative boundary. Future predictions also suggest that growth will be spreading in nature and along major transport corridors. Urban growth has occurred at the cost of intruding other categories. Statistically, urban cover in the year 2017 was 24.53% has dramatically increased to 57.72% in the year 2025, insisting on immediate policy interventions.

5. Conclusion

Temporal remote sensing analysis of Bangalore city with 10 km buffer reveals a rise in builtup area from 5.47% in 1992 to 24.53% in 2017. The density and health of green spaces in the study area has reduced by 193%. These changes lead to increased pollution level causing respiratory and pulmonary-related issues. Water bodies in the region have shown a drastic fall and if the same trend continues, it would not be a surprise to experience water crisis within a span of few years. As a result of increased migration, the city has experienced a disaggregated and ribbon type of growth. The study area has experienced an average rise of 8°C in surface temperature that has caused increased thermal discomfort and increased heat stress among the residents. The results obtained signify that there will be a rise in surface temperature with increased fraction of urban and others category as construction materials, dry soil, exposed rocks have a tendency of absorbing more heat, thus, showing a positive correlation. On the other hand, surface temperature will have negative correlation with vegetated surface and water bodies as these tend to alter the latent heat flux while maintaining the sensible heat flux. All these changes can be inferred to altering land-use pattern and replacement of manmade structures at the cost of natural landscape. The situation seems to be even worse as the prediction results show that the city might experience a twofold rise in concrete area within a span of next 8 years.

The study thus signifies how remote sensing data can be useful to understand the dynamics of city with respect to urban growth as well as climate in terms of surface temperature and it also demonstrates a new and optimized method for prediction of urban area. In this study PSO and SLEUTH were integrated to optimize spread using Moore neighbourhood, calibration and prediction of Bangalore region's urban expansion for the year 2025. PSO turned out to be faster in terms of computation time, efficient, reliable and has lesser parameters for human intervention. Regular BFM took around 8 days to complete entire calibration and prediction phase whereas PSO-SLEUTH took only 1 day. PSO-SLEUTH compared with BFM has produced excellent improvement in results. OSM from BFM was observed to be 0.3229 while integration of PSO has helped in achieving an OSM of 0.6238 which is significant. Enhancement of other model fit statistics also suggests that SLEUTH is one of the robust CA-based models. The chapter can be a significant input for creating policies to mitigate the effect of urbanization on the residents.

Acknowledgement

We are grateful to SERB, India, The Ministry of Science and Technology, Government of India, Ranbir and Chitra Gupta School of Infrastructure Design and Management, Sponsored research in Consultancy cell, Indian Institute of Technology Kharagpur for the financial and infrastructure support. We thank United States Geological Survey, Bhuvan, Google Earth for providing temporal remote sensing data.

References

Bharath, S., Rajan, K.S., Ramachandra, T.V., 2013. Land surface temperature responses to land use land cover dynamics. Geoinformatics Geostatistics: An Overview 1 (4), 1—10. https://doi.org/10.4172/2327-4581.1000112.

Briefing, I., 2019. Indian cities in top 10 list of world's fastest growing cities. Retrieved from: https://www.india-briefing.com/news/india-tops-list-fastest-growing-cities-world-18175.html/. (Accessed 24 September 2019).

Buhaug, H., Urdal, H., 2013. An urbanization bomb? Population growth and social disorder in cities. Global Environmental Change 23 (1), 1—10. https://doi.org/10.1016/j.gloenvcha.2012.10.016.

Chakraborty, S.D., Kant, Y., Mitra, D., 2015. Assessment of land surface temperature and heat fluxes over Delhi using remote sensing data. Journal of Environmental Management 148, 143—152. https://doi.org/10.1016/j.jenvman.2013.11.034.

Chandan, M.C., Nimish, G., Bharath, H.A., 2019. Analysing spatial patterns and trend of future urban expansion using SLEUTH. Spatial Information Research 7, 1—13. https://doi.org/10.1007/s41324-019-00262-4.

Chaudhuri, G., Foley, S., 2019. April. *DSLEUTH: a distributed version of SLEUTH urban growth model. Paper presented at the Spring Sim.* In: ANSS (Society for Modelling and Simulation International. Tucson, Arizona. Retrieved from: https://www.researchgate.net/publication/332910528_DSLEUTH_a_distributed_version_of_SLEUTH_urban_growth_model.

Clarke, K.C., Hoppen, S., Gaydos, L., 1997. A self-modifying cellular automaton model of historical urbanization in the san Francisco Bay area. Environment and Planning B: Planning and Design 24 (2), 247–261. https://doi.org/10.1068/b240247.

Copernicus, 2018. Land Surface Temperature. Retrieved from: https://land.copernicus.eu/global/products/lst. (Accessed 16 July 2019).

Deng, Y., Wang, S., Bai, X., Tian, Y., Wu, L., Xiao, J., et al., 2018. Relationship among land surface temperature and LUCC, NDVI in typical karst area. Scientific Reports 8 (1), 641. https://doi.org/10.1038/s41598-017-19088-x.

Duda, R.O., Hart, P.E., Stork, D.G., 2012. Pattern classification. John Wiley & Sons.

Eberhart, R., Kennedy, J., 1995. October. A new optimizer using particle swarm theory. In: MHS'95. Proceedings of the Sixth International Symposium on Micro Machine and Human Science. Ieee, pp. 39–43.

EPA, 2019. Heat Island Impacts. Retrieved from: https://www.epa.gov/heat-islands/heat-island-impacts. (Accessed 16 July 2019).

Ese Sentinel Online, 2018. Land Surface Temperature. Retrieved from: https://sentinel.esa.int/. (Accessed 16 July 2019).

Farrell, K., 2017. The rapid urban growth triad: a new conceptual framework for examining the urban transition in developing countries. Sustainability 9 (8), 1407. https://doi:10.3390/su9081407.

Feng, Y., Liu, Y., Tong, X., Liu, M., Deng, S., 2011. Modeling dynamic urban growth using cellular automata and particle swarm optimization rules. Landscape and Urban Planning 102 (3), 188–196. https://doi.org/10.1016/j.landurbplan.2011.04.004.

Founda, D., Santamouris, M., 2017. Synergies between urban heat island and heat waves in athens (Greece), during an extremely hot summer (2012). Scientific Reports 7 (1), 1–11. https://doi.org/10.1038/s41598-017-11407-6 (Article number: 10973).

Gazulis, N., Clarke, K.C., September 2006. Exploring the DNA of our regions: classification of outputs from the SLEUTH model. In: Paper Presented at the 7th International Conference on Cellular Automata for Research and Industry. https://doi.org/10.1007/11861201_54. Perpignan, France. Retrieved from.

Guan, Q., Clarke, K.C., 2010. A general-purpose parallel raster processing programming library test application using a geographic cellular automata model. International Journal of Geographical Information Science 24 (5), 695–722.

Henderson, J.V., Wang, H.G., 2007. Urbanization and city growth: the role of institutions. Regional Science and Urban Economics 37 (3), 283–313. https://doi.org/10.1016/j.regsciurbeco.2006.11.008.

Heppenstall, A.J., Crooks, A.T., See, L.M., Batty, M. (Eds.), 2011. Agent-based Models of Geographical Systems. Springer Science & Business Media, Maryland.

Hu, W., Wang, H., Yan, L., Du, B., 2018. A hybrid cellular swarm optimization method for traffic-light scheduling. Chinese Journal of Electronics 27 (3), 611–616. https://doi.org/10.1049/cje.2018.02.002.

Hyandye, C., Martz, L.W., 2016. A Markovian and cellular automata land-use change predictive model of the Usangu Catchment. International Journal of Remote Sensing 38 (1), 64–81. https://doi.org/10.1080/01431161.2016.1259675.

Jantz, C.A., Goetz, S.J., Donato, D., Claggett, P., 2010. Designing and implementing a regional urban modeling system using the SLEUTH cellular urban model. Computers, Environment and Urban Systems 34 (1), 1–16.

Jin, M., Li, J., Wang, C., Shang, R., 2015. A practical split-window algorithm for retrieving land surface temperature from Landsat-8 data and a case study of an urban area in China. Remote Sensing 7 (4), 4371–4390. https://doi.org/10.3390/rs70404371.

Kamusoko, C., Aniya, M., Adi, B., Manjoro, M., 2009. Rural sustainability under threat in Zimbabwe — simulation of future land use/cover changes in the Bindura district based on the Markov-cellular automata model. Applied Geography 29 (3), 435–447. https://doi.org/10.1016/j.apgeog.2008.10.002.

Lal, D.S., 2017. Climatology (Revised Edition: 2017). Sharda Pustak Bhawan, Allahabad.

Li, Z.-L., Tang, B.-H., Wu, H., Ren, H., Yan, G., Wan, Z., et al., 2013. Satellite-derived land surface temperature: current status and perspectives. Remote Sensing of Environment 131, 14–37. https://doi.org/10.1016/j.rse.2012.12.008.

Liang, S., Shi, P., May 2009. Analysis of the relationship between urban heat island and vegetation cover through Landsat ETM+. In: A Case Study of Shenyang. Paper Presented at 2009 Joint Urban Remote Sensing Event https://doi.org/10.1109/urs.2009.5137474. Shanghai, China.

Lu, C., Gao, L., Yi, J., 2018. Grey wolf optimizer with cellular topological structure. Expert Systems with Applications 107, 89–114. https://doi.org/10.1016/j.eswa.2018.04.012.

Mojolaoluwa, T.D., Emmanuel, O.E., Kazeem, A.I., 2018. Assessment of thermal response of variation in land surface around an urban area. Modelling Earth Systems and Environment 4 (2), 535−553. https://doi.org/10.1007/s40808-018-0463-8.

Naghibi, F., Delavar, M., Pijanowski, B., 2016. Urban growth modeling using cellular automata with multi-temporal remote sensing images calibrated by the artificial bee colony optimization algorithm. Sensors 16 (12), 2122. https://doi.org/10.3390/s16122122.

Nimish, G., Chandan, M.C., Bharath, H.A., 2018. November). Understanding current and future landuse dynamics with land surface temperature alterations: a case study of Chandigarh. In: Paper Presented at ISPRS Annals of Photogrammetry, Remote Sensing and Spatial Information Sciences, vols. IV-5. Dehradun, ISPRS, pp. 79−86. https://doi.org/10.5194/isprs-annals-iv-5-79-2018.

Pal, S., Ziaul, S., 2017. Detection of land use and land cover change and land surface temperature in English Bazar urban centre. The Egyptian Journal of Remote Sensing and Space Science 20 (1), 125−145. https://doi.org/10.1016/j.ejrs.2016.11.003.

Pan, Y., Roth, A., Yu, Z., Doluschitz, R., 2010. The impact of variation in scale on the behavior of a cellular automata used for land use change modeling. Computers, Environment and Urban Systems 34 (5), 400−408. https://doi.org/10.1016/j.compenvurbsys.2010.03.003.

Polidoro, M., De Lollo, J.A., Barros, M.V.F., 2012. Urban sprawl and the Challenges for urban planning. Journal of Environmental Protection 3 (9), 1010−1019. https://doi.org/10.4236/jep.2012.39117.

Ramachandra, T.V., Aithal, B.H., Sreekantha, S., 2012a. Spatial metrics based landscape structure and dynamics assessment for an emerging Indian megalopolis. International Journal of Advanced Research in Artificial Intelligence 1 (1), 48−57. https://doi.org/10.14569/ijarai.2012.010109.

Ramachandra, T.V., Aithal, B.H., Sanna, D.D., 2012b. Insights to urban dynamics through landscape spatial pattern analysis. International Journal of Applied Earth Observation and Geoinformation 18, 329−343. https://doi.org/10.1016/j.jag.2012.03.005.

Ramachandra, T.V., Aithal, B.H., Sowmyashree, M.V., 2014. Urban structure in Kolkata: metrics and modelling through geo-informatics. Applied Geomatics 6 (4), 229−244.

Ramachandra, T.V., Aithal, B.H., 2016. Bengaluru's reality: towards unlivable status with unplanned urban trajectory. Current Science 110 (12), pp. 2207−2208.

Sakieh, Y., Salmanmahiny, A., Mirkarimi, S.H., 2016. Rules versus layers: which side wins the battle of model calibration? Environmental Monitoring and Assessment 188 (11), 1−26. https://doi.org/10.1007/s10661-016-5643-2.

Siedentop, S., 2005. Understanding, measuring and controlling urban sprawl: starting points for an empirical measurement and evaluation concept of urban settlement development. disP - The Planning Review 41 (160), 23−35. https://doi.org/10.1080/02513625.2005.10556903.

Silva, E.A., Clarke, K.C., 2002. Calibration of the SLEUTH urban growth model for Lisbon and Porto. Portugal. Computers, Environment and Urban Systems 26 (6), 525−552.

Suresh, S., 2017. Will Bangalore Kill Itself by 2020? This Study Says So!. Retrieved from. https://timesofindia.indiatimes.com/. (Accessed 5 September 2019).

Torrens, P.M., O'Sullivan, D., 2001. Cellular automata and urban simulation: where do we Go from here? Environment and Planning B: Planning and Design 28 (2), 163−168. https://doi.org/10.1068/b2802ed.

UN, D.E.S.A., 2018. 2018 Revision of World Urbanization Prospects. Retrieved from: https://www.un.org/. (Accessed 16 June 2019).

United Nations, 2018. Climate Change. Retrieved from: http://www.un.org/en/sections/issues-depth/climate-change/index.html. (Accessed 16 June 2019).

Wolfram, S., 1984. Cellular automata as models of complexity. Nature 311 (5985), 419−424. https://doi.org/10.1038/311419a0.

Yang, L., Qian, F., Song, D., Zheng, K., 2016. Research on urban heat-island effect. In: Paper Presented at the 4th International Conference on Countermeasures to Urban Heat Island. Elsevier, Singapore. https://doi.org/10.1016/j.proeng.2016.10.002. Retrieved from:

Social ecological systems

CHAPTER

5

Stressors of disaster-induced displacement and migration in India

Ranit Chatterjee[1], Lalatendu Keshari Das[2], Ambika Dabral[3]

[1]Kyoto University, Kyoto, Japan; [2]IIT Bombay, Mumbai, Maharashtra, India; [3]Resilience
Innovation Knowledge Academy, New Delhi, Delhi, India

1. Introduction

The recent Global Report on Internal Displacement[1] shows a greater number of people being displaced internally by disasters in comparison to those due to conflict (IDMC Report,

---

[1]Displacement in this context means movement from one's home or place of habitual residence.

2018). The Philippines, China and India top the chart of having the highest number of Internal Displacements (IDPs) in 2018. Meanwhile, India has recorded 2.7 million people displaced internally due to naturally triggered disasters in the year 2018 (UNHCR, 2019). With studies predicting an increase in frequency and severity of natural hazards in Asia, one can easily expect these figures to rise in the coming years. A look back into the 2004 Indian Ocean Tsunami: concerns were raised on issues of sexual and gender-based violence; discrimination in access to assistance on ethnic, caste and religious grounds; recruitment of children into fighting forces; lack of safety in areas of displacement and return areas; and inequities in dealing with property and compensation (Cohen, 2009). The 2013 Cyclone Phailin relief and reconstruction efforts in Odisha, India, highlight the mobilization of women's unremunerated social reproductive labor, particularly through their role as primary caregivers (Tanyag, 2018).

But beyond issues of protection, Disaster Induced Displacements (DIDs) lead to other stressors on the urban ecology. The implications of such displacements often go beyond individual lives and lead to change in sociocultural trends, land use, services and ecosystem services.

DIDs stem from a complex mix of sociocultural, economic, political and environmental push and pull factors forcing communities to make a life-changing decision. Terminski (2012) classifies four categories of displacements, namely 1. development-induced displacement, 2. conflict-induced displacement, 3. disaster-induced displacement and 4. environmentally induced displacement. In 2007, at the international level, the Representative of the UN Secretary General on the Human Rights of IDPs added IDPs uprooted by disasters to the concerns of his mandate (Cohen, 2009). The UN General Assembly in its New York Declaration on refugee and migration suggests that many a time an interplay of more than one of the above four categories leads to displacement (UN, 2016). Belcher and Bates (1982) theorize DID as an opportunity to break the cycle of poverty. They further point out the increase in existing squatter settlements in the prime urban locations immediately after the disaster as a result of DID. Based on such instances, it is arguable if the disaster risk is reduced or reinforced due to such displacement. Further, Srivastav and Shaw (2014) point at shift of the intraurban balance, especially in infrastructure capacity as a result of DID.

India leads disaster-linked displacements in South Asia. Justin Ginnetti and Chris Lavell (2015) in their report share that in 2013 alone more than one million people were displaced in India due to cyclone and flooding. The displacements due to 2004 Indian Ocean Tsunami and Chernobyl disaster are mostly irreversible in nature (Terminski, 2012). Szynkowska (2015) taking the case of 2015 Chennai floods argues for a shift in trend from development to DID. The reason for such high numbers can be annotated to India's high hazard exposure in conjunction with high population density, poverty levels, rapid urbanization and environmental degradation (IDMC website, 2019). Having said that, India is one of the few countries in the world to officially recognize the risks and impacts of DID with policies and legal frameworks on the issue. For DIDS, the National Disaster Management Plan advices against secondary displacement as a part of relocation (NDMP, 2016), but doesn't spell out the possible approach to tackle such issues, leading this study to deepen the understanding on the impacts of DIDs, specially in the urban socioeconomic environment.

This chapter is divided into six parts where the first part introduces the topic followed by a theoretical construct of DID and migration. The third section focuses on disaster and related displacements in India. The fourth section delves into two case studies of Odisha and Bihar floods.[2] The following section discusses the various thematic heads that emerge from literature and case studies leading to the conclusion.

2. Disaster and climate change induced displacements: a theoretical construct

Three terms have been found to be used interchangeably while referring to mobility of community due to disaster, namely displacement, migration and permanent relocation. Displacement is due to a disaster event where risk is intensive and sudden in nature and mobility is for a short distance, whereas migration has certain degree of voluntarism in it and is due to extensive risk. Lastly, planned relocation is mostly done by the government or sometimes by the community and is accompanied by resettlement of the community (Wilkinson et al., 2016).

A close look at the world history reveals disasters to be a major factor of conflict, shift in political power and rise and fall of empires. But in the scholarship of displacement, the causality of DID gained importance post 1980s (Oliver-Smith, 2018). Gray et al. (2014) reports increase in scepticism in the recent times over the potential impact of environmental displacement due to its temporary nature and short distance traversed from impact zone. The inclusion of DID as a term in UN convention is to safeguard the rights of minority from political parties during a disaster. From field experience of various disasters, a general trend is to either shift to a safe shelter or a temporary relief camp or move with a relative or friend close to the area of impact mostly as a function of social capital. From example, owing to a severe drought in 2000 in the Bolangir district of Odisha, India, approximately 60,000 people got displaced to the neighbouring state of Andhra Pradesh in search of employment and food. Such displacement serves as a means of survival and hope for a better and stable future and is induced by poverty and lack of livelihood and not just the hazard itself (Naik et al., 2007).

From various scholarships, three distinct categories of environmental displacement, environmental emergency migration, environmentally forced migration and environmentally motivated migration, have been proposed by Renaud et al. (2008). Of these, the first type is synonymous to DID where people are displaced due to sudden disaster events. The second is due to long-term impact of climate change and the last one links to declining economic opportunities. Agustoni and Maretti (2019) add one more layer of environmentally conditioned migration of a totally voluntary nature to the above identified categories. In India, such migrations have been prevalent for over a period of time. The shifting of summer and winter capital during the British colonial period is one such case. Each of these categories have been studied in depth and various factors have been attributed based on field evidence (Table 5.1).

The majority of climate-induced migrants and displaced people move to urban centers (Mosel and Jackson, 2013; UNHCR, 2016). This may be attributed to the growth of cities in Asia which has been attracting a chunk of this population with lucrative option of

---

[2]Bihar's case study is written by the Ranit Chatterjee, for which fieldwork was done during 2008. While, for the case study on Odisha, the Lalatendu Keshari Das did fieldwork in Odisha during 2014––18. The latter was part of the author's doctoral project to understand the issue of (under-)development in Odisha.

TABLE 5.1   Contributing factors to types of displacements.

1. Mobility that results from irreversible or long-term changes in the surrounding ecosystem.	• Land degradation, inappropriate agricultural practices, desertification of soil, • Consequences of deforestation, • Salinity rise of water bodies and soil, • Temperature rise in certain territories (which prevents the maintenance of agriculture), • Rising sea levels and coastal erosion, irreversible consequences of major natural and technological disasters
2. Mobility that results from cyclical environmental factors that hinder normal human function in a particular territory.	• Periodic droughts, floods, etc. • Migration in the Asian monsoon season, • Periodic migrations caused by the threat of forest fire
3. Mobility caused by natural disasters or manmade catastrophes.	• Disasters of natural cause flood, wildfire or bushfire, volcanic eruption, earthquake, tsunami, cyclones, storms, landslides, heat and cold waves. • Technological disasters
4. Environmentally conditioned migration of a totally voluntary nature.	• Health tourism • Summer and winter vacation travels • Permanent shifting of workspace

good life and livelihood. In most of the developing Asian economics, cities are already overburdened in providing public services and utilities and are not equipped to accommodate such large influx of people (The Rockefeller Foundation, 2017). Out of the above four categories listed in Table 5.1, the influx due to a disaster is the most worrisome as to its small lead time for preparation. Such pressures often lead to conflict of services, rise in crime rates and failure of governance, which are mostly not flexible. The IDMC report of 2018 finds that displacement to urban areas contributes to the proliferation of informal settlements and puts pressure on land in periurban areas if there is limited affordable housing stock. This is of serious consequence to the open and green spaces, land-use planning and future response strategies of the cities in face of a disaster. Kirbyshire et al. (2017) stress on the need for a focus on urban resilience through effective policies taking into account these dynamic conditions. On the other hand, IDMC report (2018) shares that it is not always easy to estimate the number of new displacements in case of events such as seasonal floods that have multiple waves of displacement. For India it has been noticed that most affected population stay with their relatives or friends instead of relief camps. The most common source of DID data comes from National and State Disaster Management Authorities (NDMA and SDMAs).

3. Disasters and displacement in India

India owing to its multihazard profile has faced disasters over time and again. IDMC reports on an average 2,300,999 annually for all hazards in India (Fig. 5.1).

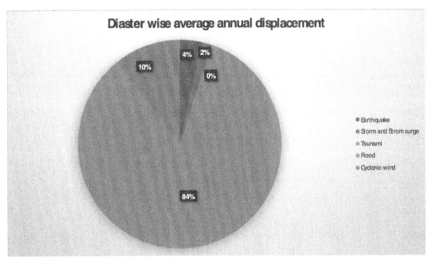

FIGURE 5.1 Disaster-wise average annual displacement. *Source: IDMC report, 2018. UnSettlement: Urban Displacement in the 21st Century, Thematic Series.*

This is a big number when compared to the list of total affected population in India in the last decade. Analyzing the disaster affected population data (Table 5.2), it can be said that majority of the affected and displaced population is due to disaster of natural causes in comparison to technological disasters.

Comparing Table 5.1 and Fig. 5.2, it is observed that the rate of disaster induced new displacements was higher in 2012, though the affected population was comparatively lower. There could possibly be more than one reason attributed to it. For example, 1. there were

TABLE 5.2 List of affected population in India in the last decade.

Year of occurrence	Affected population due to natural cause	Affected population due to technical cause
2009	11,096,639	2,401
2010	4,279,488	695
2011	12,829,319	486
2012	4,280,860	333
2013	16,708,827	332
2014	6,154,264	475
2015	346,558,129	3,613
2016	3,816,813	893
2017	22,395,195	466
2018	32,361,348	407
2019	12,400,928	35

Source: EM-DAT (2018).

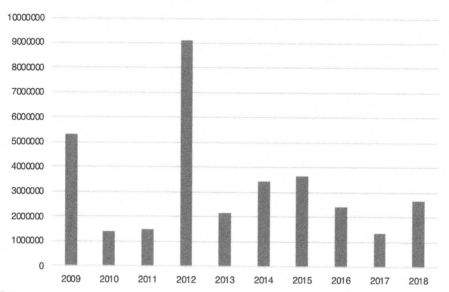

FIGURE 5.2   Disaster induced displacement in the last decade. *Source: IDMC report, 2018. UnSettlement: Urban Displacement in the 21st Century, Thematic Series.*

small-scale disasters displacing community; 2. The disasters struck new geographic locations; 3. Most affected population relied on government relief shelters and could be accounted for. This reinforces the earlier argument on lack of data on disaster induced displacement.

4. Case studies

This section delves in two case studies to draw parallel and derive a framework for stressor on urban and periurban ecology due to DID and migration. The two cases taken are from Bihar and Odisha. The Bihar case focuses on 2008 Kosi floods and is a first-hand account of the first author's fieldwork in Saharsa. The case study of Odisha looks at historic data on the displacement and migration and various socioeconomic stressors.

4.1 Case study of Kosi floods 2008

The Kosi River floods of August 2008 caused huge loss to lives, livelihoods, infrastructure and property in northeastern Bihar. Although floods in Bihar have been a recurring feature, the 2008 floods were not usual. The Kosi burst its embankments and changed course, inundating areas of Bihar that had not experienced such flooding for half a century. About 1700 villages in five districts (Araria, Madhepura, Purnia, Saharsa and Supaul) were affected, involving three million people. About one million were evacuated to various districts headquarters (UNDP, 2009). The first author has been part of the Tata Institute of Social Sciences team working in Saharsa in the first 2 months of the flood. A profile of the affected districts is as given below (Table 5.3).

TABLE 5.3 Distribution of working groups in Kosi flood affected districts.

District	% Working population	Agricultural labour	Agriculture cultivators	Household industry	Others
Araria	40.3	64.7	26.2	1.9	7.2
Madhepura	45.5	57.4	34.0	1.7	6.9
Purnia	38.9	66.3	24.0	1.6	8.1
Supaul	42.7	55.9	35.2	1.8	7.1
Saharsa	40.4	54.7	32.9	2.0	10.4

Source: Census (2001).

It can be seen that majority of the population either were agricultural labours or cultivators who are affected. Among all affected population, 57% fell under the officially defined below-poverty line. The displacement trend as documented by UNDP report shows a trend in which 33% of households shifted to government camps close by during the flood, while 4% took shelter in the camps run by NGOs/charitable organizations. About 25% of the households made their own arrangements to move out of flood-affected villages, while 25% of the people belonging to medium and large cultivators and salaried employees stayed back in their villages, perhaps because they possessed houses that were comparatively safe and secure. Mobility to outside of Bihar was higher among the male population who could find work as casual agricultural labourers, construction labourers or other informal employment.

An influx of about 0.5 million people happened in Saharsa due to 2008 Kosi floods in addition to its existing population of about 125,167 as per 2001 census. In Saharsa, relief camps were set up by the government, NGOs and political parties mostly in open spaces or occupied school premises. But due to their inability to house so many people, public places like railway stations, bus stops, etc., became a common space for the displaced to put up. This also led to various international and national humanitarian aid organizations moving into Saharsa. The immediate effect was an increase in the price of housing and daily essential commodities.

4.1.1 Water, sanitation and hygiene

During and post floods, the contamination of ground water is common and hence one has to be careful about the drinking water and solid waste management. In and around relief camps, as shown in Fig. 5.3, contamination of ground water due to open defaecation was a big risk. The arrangement of mobile toilets in such huge numbers was not possible in such a short span of time. Further, there were no dedicated bathing areas in many of the temporary shelters. Various waterborne diseases were prevalent. This has been reported by various other reports and studies (Disaster in Bihar: A Report from the TISS Assessment report, 2008; UNDP, 2009; Kumar and Ahlawat, 2008).

The non-availability of safe drinking water was the main reason for waterborne diseases leading to many deaths in the relief camps mainly among the children.

4.1.2 Healthcare and infrastructure

The health infrastructure gets unduly stressed due to the influx of such huge number of displaced populations. The availability of doctors and hospital beds are mostly limited in

FIGURE 5.3   A temporary relief shelter in Saharsa damaged by storm. *Source: first Author.*

number with little possibility of expansion especially for critical care and Out Patient Department (OPDs). The World Health Organisation (WHO) provided emergency medical supplies and equipment for almost 200,000 people in Bihar. Hundreds of cases of diarrhea and many deaths were reported from the relief camps as also from several parts of the flood affected area (TISS Assessment report, 2008). In Saharsa the district hospital was flooded with patients mainly suffering from waterborne diseases. The humanitarian aid agencies like Indian red cross, All Indian Institute of medical Sciences (AIIMS) and other medical colleges volunteered but mostly in the rural areas while the condition in Saharsa was grim.

4.1.3 Security, law and order issues

The five affected districts of Kosi floods had low literacy rates, especially among women. Roy (2010) establishes correlation between rise in crime rates and disasters in Indian context. Post 2008 Bihar floods due to location of flood inundated districts along the Nepal border witnessed higher cases of women trafficking. Due to family disorganization, extreme low literacy and male migrations, the women were left to the aspirations of better law and order situation in Bihar. TISS Assessment report (2008) highlight the need to focus on trafficking, law and order issues, Social exclusion relief camps based on caste hierarchy (Fig. 5.4).

The TISS assessment team reported that several relief camps and distribution centres were not treating flood victims with dignity. Many of the relief camps deployed security team to protect belongings and check on other crimes in the relief camps. Due to sudden influx of humanitarian workers from different parts of India and abroad, the antisocial elements in the society were in full swing. The place of residence used by the first author along with his batchmates from TISS disaster management had to report to the local police about possible trespassing incidences and issues of safety of students.

FIGURE 5.4   District hospital in Saharsa with patients queuing for checkup. *Source: Ranit Chatterjee.*

4.1.4 Livelihood

The lands outside the embankment along Kosi became waterlogged and the low-lying areas got inundated for 3 to 6 months, while in some areas water remained throughout the year (Kumar and Jha, 2018). This hampered the livelihood of the local community dependent on agricultural production and led to migration to distant states in the north and west of India. Although, the lost livelihoods of the displaced population had been well documented, but what was left unexplored was the livelihood stress that got generated in the new location due to restart of income generation activities. In 2008 Kosi floods, male population's migration to outside Bihar was reported but what has not been reported is the new livelihood options taken up by the displaced population who stayed back in Bihar. The migrating population outside Bihar sent remittance which formed a source of family sustenance. The study by Kumar and Jha (2018) document the type of jobs taken up in Delhi by the 2008 flood migrants. As salaried jobs are limited, most people who migrated to Delhi worked as helpers in construction, in shops and dairy, head loading, rickshaw and cart pulling, among others (Fig. 5.5).

4.1.5 Land-use planning

A time series data analysis of Google Earth images of the refugee camp area in Saharsa shows a gradual rise in population around that area (Fig. 5.6). The area is in the periphery of Saharsa's main nerve centre.

4.2 Natural disaster induced migration in/from Odisha

The colonial Bengal, which included the present-day states of Odisha, Bihar and Jharkhand, witnessed famines on a large scale both during the late 19th century as well as the 20th century. It not only affected lives of millions of human beings, but also their livestock

FIGURE 5.5   Chaos in relief distribution often leads to violence. *Source: Ranit Chatterjee-First author.*

and a general desire to live. As Sen (1982) succinctly argues, these famines resulted not merely because of the failure of the southwest monsoons, but because of the statist policies of the colonial regime that gave more importance to commercial exploitation of food grain stocks than channelizing them to the needy population. The agrarian crisis created by the colonial state brought suitable conditions due to which people from rural areas migrated to the city of Kolkata to populate the latter's slums, so graphically described in Dominique Lapierre's novel, *The City of Joy*. The depressing effects of this large-scale rural to urban

FIGURE 5.6   Growth of settlements in the refugee camp area of Saharsa. *Source: Google Earth.*

migration had grave implications on the ecology of Kolkata, a majorly swampy landmass on the fringes of the Indian Sundarbans. Not only immense areas under forest cover were decimated to build houses, townships and slums to accommodate the burgeoning population, but also swamps and other wetlands were drained to make space for construction activities.

As it was during the regime of the colonial state, the post colonial states in the east coast of India, namely, West Bengal, Odisha, Andhra Pradesh and Tamil Nadu in the south, have historically been prone to drought, floods and cyclonic rainfalls. Although, large-scale famines, caused after the aftermath of a major natural disaster, no longer decimate millions of people, nevertheless, distress migration to cities have become far more severe than before. According to Census of India 2011 (within South India) the rural to urban migration in the state of undivided Andhra Pradesh remains the highest, with a great deal of population migrating from coastal Andhra to cities like Hyderabad, Visakhapatnam and more recently Vijayawada. Climate change has only hastened this process of rural-urban migration by increasing the severity of droughts, floods and cyclones.

In the past two decades, the state of Odisha has faced the brunt of four major cyclones in the years 1999 (Odisha Super Cyclone), 2013 (Phailin), 2014 (Hudhud) and 2019 (Fani). The 1999 and 2019 cyclones are considered to be 'super cyclones'[3] which not only uprooted hundreds of thousands of trees, but also destroyed livelihoods in the state, with estimates ranging from Rs. 12,000 (India Today 2019) to Rs. 24,000 crores (The Hindu 2019) in loss of property. According to the Government of Odisha, in the Super Cyclone of 1999, more than 10,000 persons lost their lives, while nongovernment sources put this figure as three times more. Since then, the state machinery has been able to reduce the number of human and animal casualties to the extent that in Cyclone Fani only 40 persons reportedly died. Nevertheless, there is no substantial decrease in loss to property. For example, Dash (2013) reports that in Cyclone Phailin, Paddy in 5,43,587 ha, Pulses in 47,742 ha and horticultural crops in 34,079 ha got destroyed, leaving the small and marginal farmers with no other choice than to migrate to urban areas in search of basic livelihood opportunities. Cyclones, big and small, not only destroy standing crops but also dwellings and other means of production of the most marginalized sections of society.

If we divide the state of Odisha into four geographical regions, namely, coastal east, highland north, plateau west and hilly south, then migration due to natural disasters like floods and cyclones is higher in the coastal regions, whereas in the west, it is due to chronic draughts. The latter incident is not due to failure in rainfalls, rather is is due to a general lack of irrigation facilities that could mitigate cyclical reduction in median rainfall in the region. Although, countries like Japan and the south east coast of the USA witness climate induced disasters with a frequency comparable to the Odisha case, however, the level of migration is not so desperate in the former countries as it is in Odisha. Therefore, it is important that we understand the other reasons that accentuate the natural disasters' led distress migration from rural areas in Odisha. There are three interrelated reasons, (A) a lagging human development index (HDI), (B) infrastructure and agriculture, and (C) political awareness and assertion, due to which natural disasters have been tragic to the Odia people.

[3]Super cyclones are category 5 and above hurricanes with wind speeds ranging from 250 to 350 km per hour accompanied by heavy to very heavy rainfall.

4.2.1 A lagging HDI

Out of the total population of Odisha almost 39% of its inhabitants claim to belong to one or the other subcastes of scheduled castes (SC) (16.5%) and its corresponding scheduled tribes (STs) (23%) (Census of India, 2011). While the SCs are concentrated in the coastal districts of the state, a large number of STs live in the western Odisha districts of Mayurbhanj, Kendujhar, Sundargarh, Kandhamal, Gajapati, Koraput, Rayagada, Malkangiri and Nabarangpur. The population of STs in these districts are close to 50% of the total population. Incidentally, these districts are also the poorest ones in the state, particularly in terms of HDI (Census of India, 2011).[4]As most of the households coming under the categories of SC and ST fare rather low in HDI, so, more than 75% of them are under desperate conditions, also known as 'below poverty line' (Padhi undated).

A World Bank report states that between 1994 and 2012, the percentage of people in Odisha living under the officially sanctioned 'poverty line' has reduced from 59% to 33% with the corresponding figures for rural areas being 63% and 36%, and for urban areas being 35% and 17%, respectively. Incidentally, the Census of India 2011 reports that during the decade 2001−11 close to 143,672 farmers left active cultivation in the state (with its total percent increasing to 61.8% from 64.8% in the 2001 census) and joined along with others into an ever-growing population of agricultural labourers (which increased by 1,740,889 individuals to 6,739,993).

4.2.2 Infrastructure and agriculture

The second parameter by which backwardness of Odisha is measured is in terms of infrastructure and agriculture. As in the case of the Chandaka Industrial Estate which was established by the Nehruvian regime to create downstream manufacturing units both for employment generation and holistic industrialization of the state, the neoliberal state of Odisha has continued to open new industrial estates. At present, there are about 86 industrial estates spread over 4000 acres of land in Odisha. On paper, most of these estates are operated under the quasistate agency Odisha Industrial Infrastructure Development Corporation (IDCO). However, the arrangement since the postliberalisation period has been to give precedence to privately developed and held industrial estates.

Like their special economic zone (SEZ) counterparts, these privately held estates are created not only to compete with the already existing ones but they also receive land, water, electricity and taxation at concessional rates. Relaxed/rationalized labour laws in these estates are also on par with the SEZs. The state government claimed that nine steel and ancillary manufacturing industrial estates would come up in the state, and almost 4,90,000 personnel would get employment (direct and indirect) in them (Government of Odisha, 2016). Of this the project at Kalinganagar in Jajpur district of Odisha itself would provide employment to about 4,50,000 persons. Nevertheless, historically privately owned and operated industrial estates in Odisha have not been known to provide a great number of jobs to local people as is the case across different states (Mishra, 2010). For example, despite all the promises of

---

[4]HDI takes care of different capabilities that humans have and how through various welfare state measures these can be bettered. A few of the parameters at which HDI is calculated are premature mortality rate, life expectancy, fertility, safe motherhood and safe methods of family planning, level of female/male literacy, access to primary healthcare and primary education.http://censusindia.gov.in/Tables_Published/SCST/dh_sc_orissa.pdf.

development of these industrial estates (rechristened as 'industrial parks'), as the case of Kalinganagar shows in which a number of adivasi who were protesting against dispossession and a few of them were shot dead by the state armed police, shows neither the proposed infrastructure has come up in the state nor the projected employment generated.

The story is not much different in terms of agriculture. Unlike many other states in India, Odisha, except a few pockets in the coastal districts, did not witness the green revolution of the 1960s and 70s. The irrigation facilities that were supposed to bring relief to drought-prone areas of northern and western Odisha after the construction of multiple river valley projects had little impact on the ground situation (Meher, 1996). To cite an example, the state government argues that out of an estimated 49.90 lakh hectares of land that has the potential of getting irrigated, by 2017 it had already provided irrigation facilities to almost 38.15 lakh or 76.45% of the land.[5] However, an audit report by the Comptroller and Auditor General (CAG) of India reported that more than 65% of the cultivable land area does not have any irrigation facility (Indian Express, 2014).[6] In fact, growth of agriculture in Odisha as percent of gross state domestic product (GSDP) between 2005 and 2012 has never crossed 1%. In fact during 2005, 2010 and 2012 agriculture witnessed a negative growth rate (World Bank, 2016). This has resulted in the absolute decrease in number of cultivators in the state, as we saw in the Census of India 2011 data.

4.2.3 Political awareness and assertion

The third and final parameter to judge Odisha's backwardness is regarding political awareness of the people. The state of Odisha, at least since the late 1980s has seen a great number of protest movements targeting the state government and opposing dispossession of one form or the other (Kumar, 2014). At a much smaller scale there have been conflicts between different religious groups for control over natural resources and land (Kanungo, 2003, 2014). However, not since the late colonial period, when they rose against the princely states, known as Praja Mandal movement (Mishra, 2008), the people of Odisha, particularly its lower castes and classes, have risen for social, economic and political justice (Jena, 2016).[7] Meanwhile, despite the scheduled castes and scheduled tribes 28 constituting close to 40% of Odisha's population, their actual participation in the decisionmaking bodies, whether political or bureaucratic, has been negligible. In fact, while the neighbouring states of Jharkhand, Bihar and Chattisgarh have seen the emergence of tribal, Dalit and intermediate caste leaders of stature, there are no such examples to be found in Odisha. The two tribal chief ministers, Hemananda Biswal and Giridhar Gamang, who came to govern the state did so for only a brief period of time. Listing out the caste backgrounds of all the chief ministers that Odisha had, Mohanty (2014) notes that of the 17 chief ministers that served Odisha since its formation, seven were of Karan caste, five of Brahmin, two from Kshatriya caste, one was a Khandayat and two were tribals.

The continued domination of the two castes of Karan and Brahmin in the polity, economy and culture of the state has been intriguing. Historically, Karan, Brahmin and the Khandayat

[5]http://www.dowrorissa.gov.in/Irrigation/IrrigationScenario.pdf.

[6]http://indianexpress.com/article/india/india-others/two-out-of-3-blocks-in-orissa-lack-irrigation-facilities-cag/.

[7]http://www.dailypioneer.com/state-editions/bhubaneswar/social-reform-movements-needed-to-builddemocraic-odisha.html.

castes have been the landholders in the coastal districts of Odisha. While Brahmins in Odisha, like their counterparts in other regions of India, had maintained the cultural and ritual superiority, their closeness to the ruling elites engendered their material wellbeing too. In lieu of providing ritual and hierarchical justification for the emerging princes, many of them coming from tribal backgrounds, Brahmins received large tracts of revenue-free land grants. They also enjoyed the fruits of the production from lands granted to temples by the ruling establishment. The Khandayats also hold large tracts of revenue-free lands granted to them by the princely states to provide the latter with military assistance. But during most of the peace time, Khandayats worked as agriculturists, either on their own or with share cropping or a mix of both. In the coastal districts of the state, in terms of landed property, Khandayats still form the largest group of people. But this control over land has not translated into effective control over state power. Looking at this scenario, Mohanty (2014) has argued that the Khandayats within the ruling coalition in Odisha are seen more as appendages of the hitherto powerful castes of Karan and Brahmin without an autonomous claim to state power. The former, unlike the case of Kamma and Reddys in undivided Andhra Pradesh, who also happened to be landed castes, could not translate their control over land in spreading to other branches of the economy and polity. In regions like Chilika, while the Karan and Brahmin caste groups are considered shrewd in dealing with things by the local fishers, the Khandayats are seen more as people prone to violence to achieve their goals. The legend of being a martial caste hangs heavily on the shoulders of the Khandayat caste groups than their other two counterparts.

Of the three dominant castes in Odisha, the Karans were the only ones whose emergence as a caste dates back only to the coming of Muslim rulers to India. Known as Kayasthas in northern India and Bengal, the Karanam in the erstwhile Hyderabad state, Karns came to symbolize a term to identify a constellation of occupations related to book-keeping and education (Rao, 2004; Leonard, 1978). In its earlier period of consolidation, the Kayasthas relied heavily on serving the rulers in various administrative positions for their status and hierarchy. In the case of Odisha this caste was not only active in the field of education alongside the Brahmins in the precolonial period, but was also active during the anticolonial struggles in Odisha. Their holding of administrative positions during both precolonial and colonial periods endowed them with enough leverage to usurp lands legally and illegally, to attain the status of landlords in the countryside. The domination of the three castes of Karan, Brahmin and Khandayats in the fields of economy, polity and culture perpetuate to this day, without much overt challenge from the lower caste groups (Padhy and Muni, 1987; Mohanty, 1990, 2014; Adduci, 2012).

Owing to the noncompetitive politics regarding allocation of resources and political one-upmanship between different castes and classes in Odisha, particularly the ones lower in the socioeconomic order, Odisha has largely remained socially backward with an upper caste, patriarchal order of things.[8] This is as true of coastal region of the 30 states as it is of the western districts, who have been demanding a separate state of Koshala stating regional

---

[8]This is as true of the coastal tracts of the state as it is of the western districts, who have been demanding a separate state of Koshala stating regional imbalance in development. For a sectoral report on the nature of 'regional imbalances' see Sahoo et al. (2004).

imbalance in development. As a result, natural disasters are leading an increasing number of people to migrate to urban locations out of the state in search of livelihood opportunities.

5. Discussion

The recent UN Global Assessment Report 2019 recognizes displacement as one of the stressors for systematic failure, aggravation of vulnerabilities and addition of new risks and other socioeconomic challenges such as inequality, climate change, poverty, under/unemployment and fast-paced urbanization. Considering that in the next three decades majority of the world population will be in the urban and periurban areas, there is need to make cities resilient and sustainable to future shocks. Considering that ecological arguments are never socioeconomically neutral, it is important to understand urban ecosystems as an interaction of sociocultural and economic dynamics intrinsic to human nature. Belcher and Bates (1982) counter the existing narrative that after a disaster people often return to their predisaster homes and locations (Bolton, 1979; Bolin, 1982). Jones and Jones (1991) point out at the limited studies focusing on people who do not return and their tendency to recreate the old environment in the new place.

Disaster induced displacements often lead to stress which in turn is passed on to the new location and its immediate environment by means of human-nature interactions. The stresses most often written about are the environmental, health-related and financial of the displaced population, but hardly any study on the impact of the mass displacement on the new location. Such disaster induced mobility has potential to put pressure on shelter, security, service provision, local economy and social relations. The urban ecology to be disaster resilient needs to account for not only the shocks from disasters but also from these dynamic stresses mostly beyond its own control. DID is a kind of forced and involuntary mobility of the affected population whereby in the aftermath of disasters, the socioeconomic and ecological conditions of affected population push them for relocation and outward migration to nearby or far-flung areas providing better opportunities of survival to them. It is not an adaptation strategy an attempt of survival resulting from lack of other adaptation measures.

The probability of marginalized households moving out of a disaster area and returning is less in comparison to households which are economically sound and have access to resources (Jones and Jones, 1991). The Report on Indian Urban Infrastructure and Services (2011) highlights the stress on the urban services mainly in the smaller cities and towns due to limited capacity of the municipal authorities, inadequate investment and poor asset maintenance. The sudden influx in the population exerts further pressure on the urban services leading to systematic failure. States like Odisha and Bihar do not have strong healthcare infrastructure and the Planning Commission of India report of 1999 and NITI Ayog 2019 report highlight the constant decline in healthcare in both these States. Under such circumstances, the provision of medical support to the displaced population needs to be planned.

Solid waste management, drinking water, health and hygiene are chronic problems in many urban centres of India. Availability of budgets to cover the costs associated with developing proper waste collection, storage, treatment and disposal remains an issue to be addressed (Kumar et al., 2017). Daley et al. (2001) in their analysis of the post turkey earthquake of 1999 points at the shortage of drinking water, water for other daily usage like bathing, washing, among others. Open defaecations and spells of rains lead to contamination of pond water in and around temporary relief camps. Gupta (2007) in the backdrop of Indian Ocean Tsunami 2004 highlights the high presence of *E.Coli* in drinking water 6 months after the disaster. Needless to say, this becomes more severe in case of floods and cyclones due to higher chance of water contamination and spread of waterborne diseases.

Cultural interpretations and representations of the affected community play an important role in post disaster displacement. In words of Button (2010) cultural background and practices, which are quite specific for a community, especially in a country like India, play an important role. The mixing of two cultural traits due to sudden displacement and the level of acceptance vary from one location to the other. This often leads to conflicts of culture that snowballs into law and order issue. The inclusion of displaced community is of serious concern, as most of them have lost their physical assets and in many cases identity cards, property documents, etc. Srivastav and Shaw (2014) studied the impact of floods on rural, periurban and urban population in Ahmedabad due to 2011 floods. Migration from rural area is due to loss of assets and majority of the displaced are male. This creates a dearth of male population in the rural area while male population increases in the urban areas. Education and job prospects are the pull factors. Disaster induced displacement shifts the onus on the city to expand its informal base and accommodate the sudden increase in the number of unskilled labourers.

Land-use planning and safe housing become a regulatory measure for disaster risk reduction. Kirbyshire et al. (2017) suggests revisiting the land-use zoning and planning, monitoring and data management, plan approval processes and upgrading of informal settlements. One way of doing it is to better the infrastructure facility as is done by the European Bank for Reconstruction and Development project in Jordan and Turkey. The investments have gone into the following sector: water, wastewater, solid waste management and urban transport systems in the refugee camps in the cities (Multilateral Development Banks, 2014). Considering the case of cities like Mumbai, Delhi and others, the displaced population occupies areas which are hotspots of future disasters. When cities respond to displacement or migration by being less willing to accommodate marginalized people, development of overcrowded and underserviced informal settlements happens in pockets (IOM, 2014). Any future natural hazard event in the urban scenario will put more lives at risk due the unsafe conditions. Based on the case studies and above discussion, a list of various stressors for urban ecology due to disaster induced displacement and migration is developed (Fig. 5.7).

6. Conclusion

The impact of DID and migration on urban ecology has limited scholarships but is of importance considering sustainable development of cities in future. Further there are only a few urban resilience plans and policies that have been inclusive of the issues of mass displacement and migration to urban areas. The risk assessment at various levels needs to take such past trends and future possibilities for supporting the local government to design the plan. The displaced community brings with them various skill sets, culture and traditional practices which may be beneficial for the city growth. In future studies can be taken

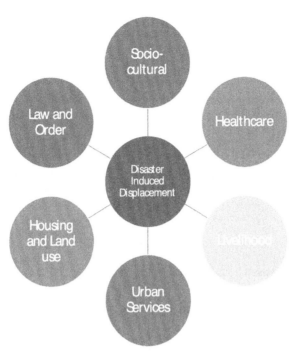

FIGURE 5.7 A framework for the stressors on urban ecology due to DID.

up to detail out each of the above dimensions and provide evidences from field to make informed policies and actions.

References

Adduci, M., 2012. Neoliberalism and class reproduction in India: the political economy of privatisation of the mineral sector in the Indian state of Orissa. Forum for Social Economics 41 (1), 68–96.

Agustoni, A., Maretti, M., 2019. Towards a global ecology of migration: an introduction to climatic-environmental migration. International Review of Sociology 29 (2), 125–141. https://doi.org/10.1080/03906701.2019.1641262.

Bank, World, 2016. Doing Business: Measuring Regulatory Quality and Efficiency. World Bank Group, Washington, DC.

Belcher, J.C., Bates, F.L., 1982. Aftermath of natural disasters: Coping through residential mobility. Disasters 7 (2). Retrieved from. https://onlinelibrary.wiley.com/doi/abs/10.1111/j.1467-7717.1983.tb00805.x.

Bolton, P., 1979. Long Term Recovery from Disaster: The Case of Managua. University of Colorado, Unpublished Dissertation. Boulder.

Bolin, R., 1982. Long-Term Family Recovery from Disaster. University of Colorado;-institute of Behavioral Science, Boulder. Monograph #36.

Button, G., 2010. Disaster Culture Knowledge and Uncertainty in the Wake of Human and Environmental Catastrophe. Left Coast Press, Walnut Creek, CA.

Cohen, R., 2009. An Institutional Gap for Disaster IDPs. FMR32. Accessed at: https://core.ac.uk/download/pdf/25998584.pdf on 18.10.2019.

Dash, S.P., 2013. How Odisha managed the Phailin disaster. Economic and Political Weekly 48 (44), 15–18.

Daley, W.R., Karpati, A., Sheik, M., 2001. Needs assessment of the displaced population following the August 1999 earthquake in Turkey. Disasters 25 (1), 67–75. Retrieved from. https://www.researchgate.net/publication/

12086921_Needs_Assessment_of_the_Displaced_Population_Following_the_August_1999_Earthquake_in_ Turkey.

EM-DAT, 2018. In: Guha-Sapir, D. (Ed.), The Emergency Events Database - Université catholique de Louvain (UCL) - CRED. Brussels, Belgium. www.emdat.be.

Gray, C., Frankenberg, E., Gillespie, T., Sumantri, C., Thomas, D., 2014. Studying displacement after a disaster using large scale survey methods: Sumatra after the 2004 Tsunami.

Ginnetti, J., Lavell, C., 2015. The Risk of Disaster-Induced Displacement in South Asia, Technical Paper. IDMC.

Government of Odisha, 2016. Compendium of Investment Projects: Industrial Parks/Infrastructure. Government of Odisha, Bhubaneswar.

Gupta, S.K., Suantio, A., Gray, A., Widyastuti, E., Jain, N., Rolos, R., Quick, R., 2007. Factors associated with E. Coli contamination of household drinking water among tsunami and earthquake survivors, Indonesia. The American Journal of Tropical Medicine and Hygiene 1158–1162.

IDMC report, 2018. UnSettlement: Urban Displacement in the 21st Century, Thematic Series.

Indian Express (2014). https://indianexpress.com/article/india/india-others/two-out-of-3blocks-in-orissa-lack-irrigation-facilities-cag/.

IOM, December 2014. Migration Report 2015: Urbanisation, Rural-Urban Migration and Urban Poverty. Background Paper. IOM, Geneva.

Jena (2016). http://www.dailypioneer.com/state-editions/bhubaneswar/social-reform-movements-needed-to-builddemocraic-odisha.html.

Jones, H.A., Jones, C.R., 1991. Mobility Due to Natural Disaster: Theoretical Considerations and Preliminary Analyses. Disasters 15 (2). Retrieved from. https://onlinelibrary.wiley.com/doi/abs/10.1111/j.1467-7717.1991.tb00441.x.

Kanungo, P., 2003. Hindutva's Entry into a 'Hindu Province': Early Years of RSS in Orissa. Economic and Political Weekly 38 (31), 3293–3303.

Kanungo, P., 2014. Shift from Syncretism to Communalism. Economic and Political Weekly 49 (14), 48–55.

Kirbyshire, A., Wilkinson, E., Le Masson, V., Batra, P., 2017. Mass displacement and the challenge for urban resilience. ODI. Retrieved from. https://www.odi.org/sites/odi.org.uk/files/resource-documents/11202.pdf.

Kumar, K., 2014. Confronting extractive capital: social and environmental movements in Odisha. Economic and Political Weekly 49 (14), 66–73.

Kumar, P., Ahalwat, M., 2008. Kosi floods 2008: devastation, displacement and migration experience. Geophysical Research Abstracts 19. EGU2017-1488, 2019.

Kumar, P., Jha, M., 2018. From Kosi to Delhi: Life and Labour of the Migrants. Accessed online at: file:///Users/Ron/Downloads/KositoDelhi-LifeandLabourofMigrantsPre-publication.pdf.

Kumar, S., Smith, S.R., Fowler, G., Velis, C., Jyoti Kumar, S., Arya, S., Rena, R.K., Cheeseman, C., 2017. Challenges and Opportunities Associated with Waste Management in India, vol. 4. Royal Society Open Science.

Leonard, K.I., 1978. Social History of an Indian Caste: The Kayasthas of Hyderabad. University of California Press, Berkeley.

Meher, R., 1996. Inter-district Disparities in Orissa in Development, Investment and Performance. Nabakrushna Chaudhury Centre of Development Studies, Bhubaneswar.

Ministry of Urban Development, Governemnt of India, 2011. Report on Indian Urban Infrastructure and Services. Retrieved from. https://icrier.org/pdf/FinalReport-hpec.pdf.

Mishra, K.C., 2008. Prajamandal movements in the feudatory states of Western Orissa. Proceedings of the Indian History Congress 69, 543–553.

Mishra, B., 2010. Agriculture, industry and mining in Orissa in the post-liberalisation era: an inter-district and inter-state panel analysis. Economic and Political Weekly 45 (20), 49–68.

Mohanty, M., 1990. Caste, class and dominance in a backward state: Orissa. In: Frankel, F.R., Rao, M.S.A. (Eds.), Dominance and State Power in Modern India: Decline of a Social Order. Oxford University Press, Delhi, pp. 321–366.

Mohanty, M., 2014. Persisting dominance: crisis in democracy in a resource-rich region. Economic and Political Weekly 49 (14), 39–47.

Mosel, I., Jackson, A., 2013. Sanctuary in the City? Urban Displacement and Vulnerability in Peshawar, Pakistan. Humanitarian Protection Group, Working Paper.

Multilateral Development Banks. (2014). The Forced Displacement Crisis. Retrieved from https://www.adb.org/sites/default/files/related/43279/Joint%20MDB%20paper%20displacement%20%5Bfinal%.pdf.

Naik, A., Stigter, E., Laczko, F., 2007. Migration, development and environment. In: International Organizatiuon for Migration Series, vol. 35.

National Disaster Management Authority, 2016. National Disaster Management Plan. New Delhi. Retrieved from. https://ndma.gov.in/images/policyplan/dmplan/National%20Disaster%20Management%20Plan%20May%202016.pdf.

Office of the Registrar General & Census Commissioner, India, 2001. Population Enumeration Data. Retrieved from Census India : Retrieved from Census India. https://censusindia.gov.in/Metadata/Metada.htm.

Office of the Registrar General & Census Commissioner, India, 2011. Population Enumeration Data. Retrieved from Census India. http://censusindia.gov.in/2011census/population_enumeration.html.

Oliver Smith, A., 2018. Disasters and Large-Scale Population Dislocations: International and National Responses, Oxford Research Encyclopedia of Natural Hazard Science. Oxford press.

Padhy, K.S., Muni, P.K., 1987. Corruption in Indian Politics: a Case Study of an Indian State. Discovery Publishing House, Delhi.

Rao, V.N., 2004. Print and prose: pundits, karanams, and the East India company in the making of modern telugu. In: Dalmia, V., Blackburn, S.H. (Eds.), India's Literary History: Essays on the Nineteenth Century. Permanent Black, Delhi.

Renaud, E., et al., 2008. Deciphering the importance of environmental factors in human migration. In: Environment, Forced Migration and Social Vulnerability Conference. UNU-EHS, Bonn, Germany.

Roy, S., 2010. The Impact of Natural Disasters on Crime. WORKING PAPER No. 57/2010. Department of Economics and Finance, College of Business and Economics University of Canterbury.

Sahoo, M., Abraham, V., Mishra, R.K., Pradhan, J.P., 2004. Interpreting the Demand for Koshala State in Orissa: Development Versus Underdevelopment KDF Working Paper No. 1, 2004. Available at SSRN. https://ssrn.com/abstract=1523640 or https://doi.org/10.2139/ssrn.1523640.

Sen, A., 1982. Poverty and Famines: an Essay on Entitlement and Deprivation. Oxford university press.

Srivastava, N., Shaw, R., 2014. Establishing parameters for identification of vulnerable occupations in a disaster scenario in Gujarat, India. Risk Hazards & Crisis in Public Policy 5 (2). Retrieved from. https://www.researchgate.net/publication/268882036_Establishing_Parameters_for_Identification_of_Vulnerable_Occupations_in_a_Disaster_Scenario_in_Gujarat_India.

Szynkowska, M., 2015. Flood in Tamil Nadu, India Disaster-Induced Displacement.

Tanyag, M, 2018. Resilience, Female Altruism, and Bodily Autonomy: Disaster-Induced Displacement in Post-Haiyan Philippines. Journal of Women in Culture and Society 43 (3).

Terminski, B., 2012. Development-induced Displacement and Human Security: a Very Short Introduction, Geneva. https://nbn-resolving.org/urn:nbn:de:0168-ssoar-359788.

The Rockefeller Foundation, 9 May 2017. Global Migration: Resilient Cities at the Forefront - Strategic Actions to Adapt and Transform Our Cities in an Age of Migration. Available at. https://reliefweb.int/report/world/global-migration-resilient-cities-forefront-strategic-actions-adapt-and-transform-our.

TISS Assessment report, 2008. Disaster in Bihar: a Report From the TISS Assessment Team. Accessed at: http://www.mcrg.ac.in/development/p_writing/manish1.pdf.

United Nations Development Programme, 2009. Kosi Floods 2008: How we coped! What we need? Perception Survey on Impact and Recovery Strategies. Retrieved from. http://www.undp.org/content/dam/india/docs/kosi_floods_2008.pdf.

UNHCR, 2016. Global Trends: Forced Displacement in 2015. UNHCR, Geneva.

Wilkinson, E., Kirbyshire, A., Mayhew, L., Batra, P., Andrea, M., 2016. Climate-induced migration and displacement: closing the policy gap. ODI. Retrieved from. https://www.odi.org/sites/odi.org.uk/files/resource-documents/10996.pdf.

III. Social ecological systems

Ecological economics of an urban settlement: an overview

Vaishali Kapoor[1], Sachchidanand Tripathi[1],
Rajkumari Sanayaima Devi[1], Pratap Srivastava[2],
Rahul Bhadouria[3]

[1]Deen Dayal Upadhyaya College (University of Delhi), New Delhi, India; [2]Shyama Prasad
Mukherjee Post-graduate College, University of Allahabad, Allahabad, Uttar Pradesh, India;
[3]Department of Botany, University of Delhi, Delhi, India

OUTLINE

1. Introduction

The world is witnessing a rapid pace of urbanization. These builtup centres of economic growth support a large population (Malthus, 1986). The colossal rise in urbanization is synonymous with continuous degradation of the environment, greater consumption of energy and hampering biodiversity, which has impact on biodiversity, human health and even its own survival. Because of the interplay between urban growth centres embedded within ecological systems, there is a need to understand whether the two ecosystems work in isolation or they are dependent (and if they are, to what extent) and also how ecological economics better addresses the nexus of these issues.

1.1 Ecological economics

The prices in the market are believed to reflect true value of resources and play allocative and distributive roles, thus bringing equilibrium in the society. But in 1960s, it was realized that markets failed on two accounts. First, there are ecosystem services (like control of climate change, protection from harmful UV rays, oxygen production and recreational experiences) and common property resources (like common grazing grounds, ponds, etc.) that are not valued in markets as they are not traded (Costanza and Wainger, 1991). Second, there is evidence of the degradation of environment and ecosystems that could not have been the case of efficient utilization of resources given that these resources are indispensable for human survival. So, if it is perceived to be important, why it was not preserved by providing it a higher value. Thus, valuation of ecosystem services has either been flawed or markets have not been working as they ought to (Gómez-Baggethun and Martín-López, 2015; Nobel Media AB, 2017).

Further, conventional economics failed to cater the valuation of ecosystem services on two accounts. The first failure ground is being 'Newtonian' because of too much generalizing human behaviour and finding patterns and regularities. Second, since it conceals complexities and makes oversimplistic assumptions about the real world, i.e., being 'Hermetic', it is only helpful in giving an approximate picture of reality (Gowdy, 2000). Furthermore, extending mainstream economic tools to environment will lead to faulty decisions. Rather, the basic premise in which mainstream economic tools operate has serious shortcomings.

Ecological economics differs from both conventional economics and conventional ecology, rather it 'encompasses and transcends these disciplinary boundaries' (Costanza et al., 1991). At one extreme is conventional economics whose sole aim is growth of national product and at the other end is conventional ecology with the goal of survival of species. Ecological economics aims at sustainability of economic system assuming that it is a subsystem of natural ecosystem. Ecological economics uses the tools of either of the two conventional subjects as appropriate. The first logical fallacy of conventional economics pertains to the definition of 'sustainability'. In conventional economics, sustainability is used in weaker sense of nondeclining or stable growth rates. It only focuses on rise in GNP. But this concept, referred to as 'Economic Growth', is distinguishable from economic development and sustainable development. Development is a broader term that includes growth and is a multidimensional measure of standard of living with other indicators being health, education, eradication of poverty, sanitation, etc. According to Bruntland Commission Report (1987),

'Sustainable development is development that meets the needs of the present without compromising the ability of future generations to meet their own needs'. These two terms are not to be confused with weak and strong sustainability where the former assumes substitutability of natural capital with man-made capital and latter does not.

The ecological economists believe in strong sustainability that ensures the preservation of natural capital since it is considered to be a steady source of inputs for production. Not only does it emphasize the indispensable nature of essential ecosystem services but also indicates the threats to the very survival of human. In the wake of rapid growth of urban builtup structures, a check on the rate of exploitation of resources is mandatory such that stock of natural capital is maintained at desirable level. To put it in practice, we need to keep record of economic value of ecosystem goods and services or natural capital stock. Also, including ecosystem into our economic accounting is a must. The use of extended input—output social accounting is advised (Gowdy and Erickson, 2005) so that environmental resources' uses and costs of economic impact on environment are taken into consideration while making key policies.

Since urbanization is highly dependent on energy consumption and has rising population density as a pressing issue, we need to understand the arguments relating to the energy crisis and carrying capacity of the earth. With the publication of 'The Limits to Growth' by Meadows et al. (1972), the world was warned about the energy crisis and explained how ever-rising growth rates cannot be maintained forever. The ecological economist Daly (1994) also emphasized that growth faces limits and the limiting factor in development is no longer man-made capital but remaining natural capital (Costanza et al., 1991). On the other hand, technology optimists believe that 'Earth's carrying capacity cannot be measured scientifically because it is the function or artifact of the state of knowledge and technology' (Sagoff, 1995). Thus, there are no constraints as such, and resource constraints, if any, can be curbed by technological inventions/innovations and new ideas.

Sagoff (1995) presented three reasons why mainstream economists do not get anxious about perils of collapsing ecosystem. He explained that they believe that more resources will be discovered, cheap sources of energy will substitute nonrenewable sources and energy requirement per unit of production will fall, due to the advent of technology. All these arguments were supported by substantial data by Sagoff (1995).

Technology pessimists argue that the energy crisis, which is still not a reality, does not imply that we can circumvent it infinitely. Daly and Townsend (1993) claim that labelling growth 'green' or 'sustainable' would not translate into growth for infinite period, rather greater the delay in inevitable transition to a changed ecosystem, more painful the process will be. The basis for energy crisis can be traced in second law of thermodynamics as explained by Georgescu-Roegen (1971). According to second law of thermodynamics, there is always loss of energy (in entropy terms) in process of exchange of energy. Since this is universal and inevitable, thus, it does not require much attention. What needs to be addressed is that stock of energy is fixed and low entropy resource base shrinks for an isolated system (Rees and Wackernagel, 1996). But technology optimists maintain a view that use of energy other than fossil fuels along with updated technology makes limits to growth of a delusion (Lovins and Lovins, 1991; Sagoff, 1995).

Now the dilemma is 'who is correct?'. Can urban centres carry on economic growth infinitely while maintaining the sustenance of mankind? Or, are there certain checks and steps to be taken? To understand this let us consider the view of technological optimists and pessimists on it. Costanza (1989) presented a payoff matrix in his paper relating to choice of technological optimist policy versus pessimist policy. In the game theory framework, he proved that optimal strategy is to adopt technological pessimist policies because if pessimist proposition is true then in that case even the worst outcome would be tolerable for the world. This is widely known as 'Precautionary Principle', which recommends that safe minimum standards should be adopted to tackle the unpleasant surprises by nature (Ehrlich, 1994). Ecological economists, maintaining pessimist view, argue that technology may hinder the natural processes in such a way that quality of environment may deteriorate to a level where human survival is threatened. Also, in contrast to technology optimists' view of substitutability of natural capital with man-made capital, technological pessimists argue that 'the basic relation between man-made and natural capital is one of complementarity, not substitutability' (Daly, 1994).

In recent years, the international platforms like the Economics of Ecosystem Services and Biodiversity (TEEB) and the Intergovernmental Platform on Biodiversity and Ecosystem Services were witnessed, and the new discipline 'ecological economics' emerged. The ecological economics is transdisciplinary area that aims at resolving cross-cutting issues in economics and ecology.

This chapter deals with the ecological economics of urban settlements, elaborating upon the urbanization, its impact on the urban ecosystem and its functions. The focus is also on urban ecosystem services and various tools applied to their valuation. A brief discussion is provided for ecosystem disservices (EDSs) of an urban area and its impact on human well-being.

2. Urbanization

2.1 Overview

By the year 2030, more than 60% of the estimated world population will live in urban settlements (Seto et al., 2012). Economic development has been claimed to be interconnected to urbanization since long. However, stimulating economic growth has posed serious environmental concerns across the world, and therefore, serious consequences are apparent towards sustainable development (Peters et al., 2007; Yansui et al., 2014). Urbanization is also associated with the change in land-use patterns, particularly in terms of quantity, type and spatial distribution. Further, mass conversion of agrarian lands to nonagrarian is severely jeopardizing the land-use/earth surface cover patterns and carbon source/sink processes. It has been observed in recent decades that rapid urbanization has also led to an increase in CO_2 emissions (Peters et al., 2007; Raupach et al., 2007). Urbanization has critically influenced the carbon balance in regional and terrestrial ecosystems, and therefore, in recent years, studies have been focused on this concern (Seto et al., 2012). Furthermore, studies have also been undertaken to generate a comprehensive understanding of carbon cycle mechanisms in urban ecosystems (Seto et al., 2012).

Interaction between urbanization and biodiversity is very complex and has various dimensions (McKinney, 2002). It is now well established that both the size and distribution pattern of urban areas significantly influence biodiversity (Tratalos et al., 2007; Müller et al., 2013). Further, development of urban infrastructure and expansion of cities to accommodate the burgeoning population has put tremendous pressure on local biodiversity (Pauchard et al., 2006; Grimm et al., 2008). Moreover, such rapid urban expansion has been directly linked to genetic isolation of native species owing to habitat fragmentation (Ricketts et al., 2001; Concepción et al., 2015). Also, the replacement of native with exotic species is one of the repercussions of urbanization (Hope et al., 2003). This has severely restricted dispersal of propagules and posed challenges to endemic species (Bierwagen, 2007; McKinney, 2006, 2008; Seto et al., 2012).

There are various direct and interdict connotations of urbanization. Directly, it is linked to habitat fragmentation/modification and loss, alteration in soil properties and physical transformation of other components of erstwhile nonurbanized landscape, while indirectly urbanization has been associated to enhancement in abiotic pressures like air and noise pollution, alteration in water and nutrient availability and spurt in competition between native and nonnative species, apart from altered rates of herbivory and predation (Pickett et al., 2009).

Rapidly expanding cities are associated with greater changes in land cover. It has been observed across the world that low-elevation coastal zones are prone to urbanization owing to their higher biodiversity (Elmqvist et al., 2013). Though urban expansion is comparatively very less on terrestrial ecosystems, it has severe consequences for various ecoregions, particularly coastal and islands. It has been estimated that by 2030, more than 25% of endangered and critically endangered species will be negatively affected, either directly or indirectly (Güneralp and Seto, 2013; McDonald et al., 2014). The rapid rate of urbanization is claimed to expedite the closeness of urban and protected areas. Globally, more than 25% of terrestrial protected areas are within 50 km of urban areas (McDonald et al., 2009). This may have both positive and negative repercussions. The negative impacts may include change in water quality reaching to protected areas from urban wastewater (both residential and industrial) (Carpenter et al., 1998); illegal logging, poaching and harvesting; feral pets; human–wildlife conflict; anthropogenic garbage; light (Longcore and Rich, 2004), noise (Forman and Alexander, 1998) and air pollution; anthropogenic wildfires (Matos et al., 2002) and introduction of invasive exotic species (Alston and Richardson, 2006). However, the positive effects may be seen in increased recreational experiences, ecotourism and ecoeducation and increased environmental awareness among citizens and visitors. It is predicted that by 2030, the urban areas in proximity to protected areas across the world will increase substantially (McDonald et al., 2008; Güneralp and Seto, 2013). Most of the 35 biodiversity hotspots worldwide (Mittermeier et al., 2004; Myers et al., 2000) are facing challenges from rapid expansion of urban areas. It is further forecasted that five biodiversity hotspots will be having the largest percentages of their areas to become urban: the Guinean forests of West Africa (7%), Japan (6%), the Caribbean Islands (4%), the Philippines (4%), the Western Ghats and Sri Lanka (4%); however, there are other hotspots that will be largely unaffected by urban expansion (Seto et al., 2012). This will further escalate the concerns and complicate the urban and natural interactions thereby affecting biodiversity (Tilman et al., 1994; Güneralp and Seto, 2013).

Rapid urbanization is also predicted to put pressure on freshwater biodiversity across the world. The drinking water needs of the ever-increasing urban population will be a major challenge in the coming years. For example, it is projected that Western Ghats of India will be having 81 million people with acute shortage of water by 2050. According to McDonald et al. (2008), the Western Ghats possesses 293 fish species, 29% of which are endemic, and therefore, shortage of water availability can potentially lead to greater species extinction.

2.2 Impacts of urbanization on the ecosystem and its services

Rapidly increasing population across world and associated socioeconomic development has stimulated the growth of urban landscapes. Urban expansion profoundly affects abiotic and biotic properties of an ecosystem. According to Millennium Ecosystem Assessment (2005), worldwide 15 out of 24 major ecosystems are deteriorating, and approximately 67% of the ecosystems that provide important services have already been destroyed by humans. Globally, urbanization has altered various ecosystems, putting ecological balance at risk through various biotic and abiotic stressors and consequently leading to a substantial decline in ecosystem services. Ecosystem services can be broadly defined as the benefits people obtain from ecosystems (Pataki et al., 2006; Tratalos et al., 2007). According to Millennium Ecosystem Assessment (2005), ecosystem services can be grouped into four broad categories: (i) provisioning services (e.g., food, including seafood and game, crops, wild foods and spices, water, pharmaceuticals, biochemicals, industrial products, energy hydropower, biomass fuels); (ii) regulating services (e.g., carbon sequestration and climate regulation, waste decomposition and detoxification, purification of water and air, crop pollination, pest and disease control); (iii) supporting services (e.g., nutrient dispersal and cycling, seed dispersal, primary production) and (iv) cultural services (e.g., cultural, intellectual and spiritual inspiration, recreational experiences, including ecotourism, scientific discovery). As discussed in above section, urban expansion has both direct and indirect effect on ecosystem and its services, a number of aspects require comprehensive studies to understand ecological processes in urban ecosystems, these include understanding the mechanisms, characteristics, time–space evolution and factors affecting the ecosystem services within urban landscapes and thereby, exploring how urban buildups onto other land-use types can be carried out with managing the ecosystem services.

Urbanization affects a myriad of ecosystem services ranging from regional to global scale. The availability of fresh water is one of the most important ecosystem services on regional to global scale (McDonald et al., 2014). Various household and commercial activities of urban areas are dependent upon the availability of fresh water; however, the quality of the same is greatly influenced by them. Indiscriminate exploitation of water resources in urban settlements and rapidly changing climate has increased the uncertainty of water availability, particularly in arid and semiarid climates (Güneralp et al., 2015; McDonald et al., 2011). At present more than 150 million people live in cities with acute shortage of water, and it is forecasted that by 2050, this number will reach almost a billion people due to growth in urban population. Similarly, ecosystem services like air purification have been compromised due to the rapid urban expansion and associated economic and industrial activities, vehicular emissions and land cover changes.

3. Ecosystem services and their valuation in urban settlement

3.1 Urban ecosystem services

Urban ecosystem services are specifically defined as services that are being provided by urban ecosystems and their components (Gómez-Baggethun and Barton, 2013; see Fig. 6.1). Ecosystems and their services are critical for sustenance of life in urban settlements

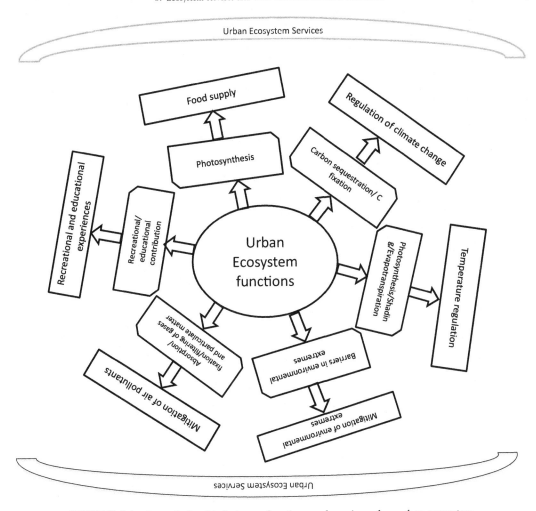

FIGURE 6.1 Interrelationship between functions and services of an urban ecosystem.

(Odum, 1989), maintenance of health (Tzoulas et al., 2007; Lovell and Taylor, 2013), amicable socioeconomical relationships (EEA, European Environmental Agency, 2011), social and food security (Costanza et al., 2006; Dixon and Richards., 2016; Clark and Nicholas, 2013) and overall human well-being (TEEB, The Economics of Ecosystems and Biodiversity, 2011). According to Pickett et al. (2001), urban ecosystems are those where people live in high densities or those where the built infrastructure covers a large proportion of the land. The infrastructures in these areas incorporate both planted vegetation such as urban forests, parks and gardens and water bodies like ponds, small lakes and wetlands. However, in a broader sense, areas around the cities, anyhow linked to or directly managed and affected by material and energy from core urban or suburban parts, are also components of urban systems (Pickett et al., 2001). There are various factors, such as population density, overall population, proximity or distance between urban dwellings and land-use patterns and employment outside primary sector, that determine the urban areas between countries or states (Gómez-Baggethun and Barton, 2013).

Recently, researches on ecosystem services have been focused on various dimensions such as economic (Jim and Chen, 2009; Sander et al., 2010), sociocultural (Barthel et al., 2010) and biophysical (Pataki et al., 2011), which have generated a comprehensive understanding. Further, various international initiatives like the Millennium Ecosystem Assessment (MEA, 2005) and the Economics of Ecosystems and Biodiversity (2011) have provided impetus to studies on urban ecosystem services. Though increasing attention is now being paid to urban ecosystems at the policy level, but as compared to natural ecosystems, they get modest interest. As discussed in Section 2 of this chapter that ecosystem services may be categorized broadly into four categories, many of which can be valuated at economic scales, however, others are of noneconomic value and largely remain unexplored (Chan et al., 2012). According to Millennium Ecosystem Assessment (2005), different ecosystem provides different types of ecosystem services, and therefore, general classification is needed to specific type of ecosystem. Various studies have been conducted to assess the ecosystem services rendered by an urban ecosystem. For instance, the urban ecosystem provides services with direct and immediate effects such as on health and security such as air purification, noise reduction, urban cooling and runoff mitigation (Bolund and Hunhammar, 1999; Villarreal and Bengtsson, 2005, Fig. 6.1), apart from others, e.g., food supply (Altieri et al., 1999; Andersson et al., 2007; Barthel and Isendahl, 2013), regulation of water flow (Bolund and Hunhammar, 1999; Pataki et al., 2011), thermal regulation (Chaparro and Terradas, 2009; Hardin and Jensen, 2007), noise reduction (Fang and Ling, 2003; Chaparro and Terradas, 2009), waste treatment (TEEB, 2011; Vauramo and Setälä, 2011), amelioration of environmental extremes (Costanza et al., 2006; Hardin and Jensen, 2007), air purifiers (Escobedo et al., 2008), pollination and seed dispersal (Andersson et al., 2007; McKinney, 2008; Müller et al., 2010), regulation of climate (Chaparro and Terradas, 2009; McPherson and Simpson, 1998; Nowak, 1994) and recreational and educational experiences (Andersson et al., 2007; Barthel et al., 2010; Tyrväinen et al., 2005).

3.2 Ecosystem disservices

Much less attention has been paid to urban 'EDSs' as compared to the concept of ecosystem services. EDS may be defined as 'the ecosystem-generated functions, processes and attributes that result in perceived or actual negative impacts on human well-being' (Shackleton et al., 2016). Where urban ecosystem services have positive contribution towards human life and health, the EDSs are economically and socially harmful, life-threatening or can endanger the health (Campagne et al., 2018). Recently, much debate has been generated on EDSs (TEEB, 2010; Lyytimäki, 2014; Barot et al., 2017). For instance, there are many plants that emit or secrete volatile organic compounds that substantially contribute to smog and ozone generation (Chaparro and Terradas, 2009; Gómez-Baggethun and Barton, 2013; Campagne et al., 2018), allergies caused by pollens (Gómez-Baggethun and Barton, 2013), decomposition of woods through microbial activities (Lyytimäki and Sipilä, 2009), etc. Therefore, development of urban green spaces without paying proper attention to the flora being planted may have undesired consequences in longer run (Lyytimäki, 2014; Barot et al., 2017).

3.3 Valuation of urban ecosystem services

3.3.1 Plurality in values

According to Farber et al. (2002), ecosystem services valuation is an assessment of trade-offs one has to go through to achieve their target. Further, the valuation of ecosystem services indicates the relative contribution of ecosystem in achieving the target. However, the approach to valuate ecosystem services varies depending upon axiological, epistemological and ontological positions adopted at the beginning of valuation process (Farber et al., 2002; TEEB, 2010; Gómez-Baggethun and Barton, 2013). Since the Ecological Economics is the amalgamation of two disciplines viz. Economics and Ecology, each of which failed to address all the issues pertaining to the preservation of environment- as the principles of economics or markets are omnipresent in every action of individuals and these actions are in turn affecting environment, thus, calling for synthesis of approaches of both. The valuation of ecosystem services is at the very core of ecological economics as the economic dominate the functioning of world, we need a value to understand its importance, and thus, creating awareness about the fears of technological pessimists. Ecological economics borrows apt tools from these two disciplines to get a value for ecosystem services.

There are multiple ways or approaches to valuate ecosystem services that may range from an individual's perceptions to the sustainable welfare of communities (Costanza, 2000). The latter approach is preferred over the former as the individualistic approach may be biased (Kahneman, 2011). Moreover, the idea of 'value pluralism' seems to be more inclusive when valuation process is linked to overall benefits of the communities. This idea considers that valuation methods in socioecological systems deal with multiple and generally nonconsensual languages, and therefore, values may be combined without giving preference to single method, to arrive at an informed decision (Chan et al., 2012; Gómez-Baggethun and Barton, 2013; Martín-López et al., 2014).

Due to multiplicity in the ways in which ecosystem benefits are perceived differently by different disciplines (Farber et al., 2002) and even within a discipline, there is no single value measure that is regarded superior in terms of comparability (Rees and Wackernagel, 1996; Martinez-Alier et al., 1998) or as a single measuring rod of measurement (Gómez-Baggethun and de Groot, 2010; Gómez-Baggethun and Martín-López, 2015) so as it dictates decisions relating to preservation of ecosystem and biodiversity. Thus, it can be said that ecological economics stands at the crossroads of estimating the value of ecosystem services to justify the policies aimed at its preservation.

3.3.1.1 Defining values

Since there are multiple aspects or benefits of ecosystem services and they are perceived differently, there are multiple values (benefits) attached to ecosystem service. First is 'intrinsic value', which is the value of nonhuman entities in themselves or 'as an end' (Daly, 1994; Davidson, 2013). However, in sharp contrast is 'instrumental value', which is the value of nonhuman entities as a means to desired ends of humans. But there is warning in using this value as using it may 'erode public attention for nature's intrinsic value' (Davidosn, 2013). At the same time, it focuses on using and manipulating nature as per human convenience, it being a 'means to the desired end'.

Krutilla (1967) divided monetary value into use value and nonuse value. The use value refers to the provisioning services such as food, water, timber and regulating services such as climate regulation, prevention of soil erosion and water purification. The nonuse value refers to the satisfaction derived from knowledge that biodiversity and ecosystem services will be available for future generations (bequest value or options value). Bequest value is one of the three types of warm glow values, the other two being satisfaction from knowledge that other people have access to nature (altruistic value) and nature will benefit other species (existence values) (Davidson, 2013; Gómez-Baggethun and Martín-López, 2015). The existence value had a different interpretation too and that is the 'satisfaction of knowing that nature exists but originating in altruism' (Davidson, 2013). The total economic value is sum of use values, nonuse values and intrinsic values (Davidson, 2013; Gómez-Baggethun and Martín-López, 2015). Another role of ecosystem is in increasing resilience to natural shocks, and the premium should be paid to cover oneself from loss resulting due to this shock. This premium can be referred to as 'Insurance value' (Gómez-Baggethun and Barton, 2013).

3.3.1.2 What is valuation?

3.3.1.2.1 Conventional economics Conventional economics did not take into consideration the value of ecosystem services. But the notion of value in conventional economics reminds us of the water–diamond paradox as posed by Smith, which was finally resolved by Schumpeter (1978), and the long debates about value as use and exchange value ended. Though water is a necessity for living (high use value), it has low exchange value, whereas not-so-useful diamonds have high exchange value. Ricardo (1817) measures value as the value of labour embodied in each product.

3.3.1.2.2 Environmental economics Environmental economists extended the classical economic framework and applied it to environmental goods. Environmental economists considered pollution and environmental degradation to be 'externalities' (Pigou, 1920). The value of an ecosystem is the additional benefit that can be reaped from saving it or additional loss if damage to the environment takes place. Environmental economics focuses only on monetary valuation of ecosystem services that is based on individual preferences (Costanza et al., 1991).

3.3.1.2.3 Conventional ecology The valuation in conventional ecology hinges mainly on the second law of thermodynamics (Farber et al., 2002), and the cost of depletion of any resource or species (and thus its value) is measured in entropy terms (Rees and Wackernagel, 1996). The other way is to understand the material and loss of energy flows, and thus, comparing inputs and output of energy in a closed system. It can be concluded that energy-based valuation is at the heart of ecology based on two laws of thermodynamics. The computation of energy lost (entropy), embodied energy, transformity ratios, emergy and exergy are certain ways of finding value of an ecosystem (Odum, 1996).

3.3.1.2.4 Ecological economics The valuation in ecological economics calls for the comprehensive measure for valuation and which is broader in the sense it includes all aspects, viz. ecological, economic and sociocultural (Costanza et al., 1991; Gómez-Baggethun and Martín-López, 2015).

3.4 Modelling urban ecosystem and valuation of its services

Though there are many techniques that are deployed in measuring the value of the ecosystem, but here we will discuss only those that are frequently used empirically. For a deeper understanding of urban ecology, we need to distinguish three terms (Pickett et al., 2016), viz. 'ecology in cities', 'ecology of cities' and 'ecology for cities'. The ecology in cities focuses on green patches, water bodies within the city. The 'ecology of city' includes not only biological components but also social and built components. The 'ecology for' calls for use of knowledge so generated in abovementioned two paradigms to be applied by researchers in urban planning with an aim of urban sustainability.

DeGroot et al. (2012) reviewed 320 publications to provide a comprehensive database of the valuation of ecosystem services. Their analysis is also helpful in understanding valuation methods employed in different locations, to different services and biomes. A majority (274 out of 665 values) of values focused on direct market valuation of provisioning services indicating the fact that majority of literature confines to mainstream economic paradigm, and only 51 (out of 665) values used other than market and preference-based values. The study also presented a value ranging between 491 int \$/year (int \$ = international dollar) for marine biome to 352,429 int\$/year for coral reef.

3.4.1 Ecological footprint

Human beings differ from other species as they have 'industrial metabolism' (Rees and Wackernagel, 1996; Fischer-Kowalski, 1998). To sustain the maximum load in a city, we need to assess an area-based estimate of resources required for them, which is known as the ecological footprint. In doing so, we need to compute annual per capita consumption of people in cities first and then we need to estimate the land area per capita required for this consumption and assimilate wastes from cities. If ecological footprint of a city is greater than consistent supply of productive land, then this difference is referred to as 'sustainability gap' or 'ecological deficit' (Rees, 1992). Rees and Wackernagel (1996) showed that 60% of resource depletion and pollution are due to cities, and 'cities have become black holes drawing in energy'. The ecological deficits as the ratio of ecological deficit to ecological productive land of select developed nations have been shown to be raging between 80% (United States) and 1400% (Belgium), which depicts the unsustainable nature of these countries. Though this method is conceptually simpler, it has several disadvantages as this model is static, whereas both nature and economy are dynamic systems. Since it considers consumption of certain specific goods, there are wastes other than CO_2 due to which value of ecological footprint is an underestimated one (Rees and Wackernagel, 1996).

3.4.2 Urban metabolism

The cities are assumed to be superorganisms that have metabolic requirements (Zhang et al., 2014; Wu et al., 2017) and 'material and energetic processes within the economy and society vis-à-vis various natural system' (Fischer-Kowalski, 1998). Urban metabolism could be energy-related and/or carbon metabolism.

3.4.2.1 Energy metabolism

The emergy-related metabolism of cities focuses on requirement of resources from outside measured in emergy terms, i.e., total energy required to produce a good or service.

Transformity is the ratio of energy of one type required to produce unit of energy of other types (Eq. 6.1). Thus,

$$T = \frac{I}{O} \qquad (6.1)$$

where T is the transformity ratio,

I *is the input of energy, i.e., emergy,*
O is the output of energy, i.e., embodied energy in product.

Greater the transformity, greater is the requirement of energy to produce a product.

Various techniques are employed, first being computation of various emergy related indices as proposed by Odum (1996). The emergy computations are useful, since to sustain urban ecological system cities require solar energy (or its equivalent). The energy flows of the urban system are represented by emergy value, which is the product of energy content of flow and transformity. The four emergy indices are emergy sources, emergy intensity, emergy structure and ecological economic interface (Huang, 1998; Brown and Ulgiati, 2010; Zhang et al., 2011; Wu et al., 2017), which together depicts energy flow in the city and also hints at the sustainability of cities. As economy progresses, there is a transition from rural and underdeveloped areas with energy self-sufficiency to high entropy (Rees and Wackernagel, 1996) and increased purchased emergy from outside cities like in Taiwan in 1970s (Huang, 1998). Wu et al. (2017) employed principal component analysis and date envelopment analysis to time series data of energy to capture flows of energy when technology is assumed to change every year. Taking into account, changes in technology capture urban metabolism better. Emergy analysis is the only measure of ecosystem service valuation since energy is a measure common to both economics and ecological value systems (Huang, 1998; Zhang et al., 2011).

3.4.2.2 Carbon metabolism

Another type of metabolism is carbon metabolism. There are carbon sinks such as forests, wetlands, oceans, fossil fuels, atmosphere and carbon sources such as combustion of fuels, fires, decomposition, respiration and digestion. The cities are considered to be carbon sources (Zhang et al., 2014). Zhang et al. (2014) provided value to coefficient of carbon sequestration to each type of land use. Similarly, they computed emissions as sum of carbon emission by different sources and concluded that in Bejing (a city in China) that carbon sequestration can only offset 2.4% of carbon emissions pointing to the imbalance of city's carbon metabolism.

Canu et al. (2015) provided estimates for the economic value of one of the ecosystem services, i.e., carbon sequestration in marine ecosystems. The computations pertain to the carbon sequestration services of the Mediterranean Sea. The biogeochemical model that has been used is OPATM-BFM (Ocean PArallelize Transport Model-Biological Flux Model). The OPA system has been used to 'assess the CO_2 fluxes at the air–ocean interface', and BFM captures biogeochemical dynamics by simulating chlorophyll synthesis and carbon dynamics based on plankton function type representation of food web. The economic model aimed at computing monetary value of ecosystem service of carbon sequestration is built by equating benefit of one unit of carbon sequestration to marginal damage of last carbon unit if not removed. The total value has been taken as the product of CO_2 flux and social cost of carbon (SCC, i.e., marginal damage). Thus, economic value obtained from carbon sequestration for area A for time period from t_0 to t_1 is as follows (Eq. 6.2):

$$\int_A \int_{t_0}^{t_1} \text{Flux}(x, y, t) \text{SCC}(t) dt \, dA \tag{6.2}$$

where (x,y) is the geographical coordinates,
   T represents time,
   Flux(x,y,t) is the CO_2 measured in ton/km^2/day (Eq. 6.3).

$$\text{SCC} = \int_0^T \frac{\partial D_t}{\partial E_t} (1+s)^{-t} dt \tag{6.3}$$

where $(1+s)^{-t}$ is the discounting factor and $\frac{\partial D_t}{\partial E_t}$ is incremental damage of CO_2 emissions.

They evaluated the carbon sequestration value to be ranging between 127 and 1722 million/year. According to this study, the Mediterranean Sea has been proven to be carbon sink, and at the same time, they concluded that the exclusive economic zones of six countries (Cyprus, Egypt, Lebanon, Israel, Syria and Turkey) are carbon sources and majority of carbon sequestration occurs over EEZ of four countries, viz. Spain, Algeria, Italy and Greece.

3.4.2.3 Material footprints

According to Brunner and Rechberger (2001), consider that 'modern cities are material hotspots more hazardous than landfills'. The metabolic processes can better be understood by analyzing all the flows of materials or substances in a well-defined system. It can be studied under material flow analysis, which is based on the law of matter, and flows are expressed in kg/year. For this purpose, cities' flows within and from outside are analyzed. At the input side is demand for resources by cities and at the output side is waste produced and greenhouse gases emitted by cities (Bai, 2009). Though there are advantages in terms of creating a comprehensive understanding of city's way of functioning of cities (Fischer-Kowalski, 1998; Daniels and Moore, 2001; Kennedy et al., 2007), which makes it vital tool in urban planning, it can also be helpful in detecting early warning signs for sustainability of cities and various other problems along with identification of source of imbalance in the flows in the city. But the biggest disadvantage of this method is that it requires huge set of data. Another way of analyzing material footprints is using these flows as social cost accounted for extended input–output (IO) model. Thus, all the inputs from ecosystem like any other economic input (like coal, iron, steel, etc.) and ecosystem output (and services) are considered in an integrated framework where an impact of benefits and impact on costs or inputs required by ecosystem can be examined (Giljum et al., 2009). Material footprints along with economic IO models assess the nexus between economy and environment. Such 'an integration of material flow and land-use accounts with monetary IO models is a relatively young field' (Giljum et al., 2009). Thus, there is greater scope for research in this assimilated field of ecological economics.

3.4.3 Monetary valuation

According to Farber et al. (2002), 'the exchange value of ecosystem services is the trading ratios for those services'. It explicitly indicates that the 'exchange value' is the directly traded monetary benefits received from services. On the one hand, where the exchange value can directly be measured through observable trade activity, the social value of services is difficult to valuate; therefore, they got attention of environmental and resource economists (Farber et al., 2002; Kopp and Smith, 2013; Costanza et al., 2014). The two major concepts dealing with the social value of ecosystem services focus upon the notion that (i) what society is able or willing to pay for services and (ii) what society is willing to accept if those services are no longer available; however, these concepts differ substantially (Farber et al., 2002). Thus, socially, the economic valuation of ecosystem services either directly deals with the payments for services or provides adequate compensation for the loss of services.

Urban ecosystem services may be valuated at the economic, sociocultural and insurance levels (Gómez-Baggethun and Barton, 2013). Economic valuation methodology includes provision for payment for ecosystem services or provides compensation if the services are lost. However, indirect valuation methods were developed for ecosystem services that are not explicitly tradable and are not 'private' in nature (Farber et al., 2002). These methods have their strengths and weakness, respectively. The application of 'willingness to pay' or 'willingness to accept' approach to valuation of ecosystem services has several disadvantages. Firstly, it rests merely on judgemental value of individuals and is not a scientific method. Secondly, since they are placing value on other species or ecosystem services and there is a difference between 'harm to nonhuman organisms and "similar" harm to ourselves' (Davidson, 2013), they might underestimate the value of it (Costanza and Waigner, 1991). Thirdly, the results also depend on how well-informed and well-aware people are about the value of ecosystem service. Fourthly, there is population density effect, i.e., if there is greater scarcity, then people give higher values to ecosystem services, leading to overvaluation (Gómez-Baggethun and Barton, 2013). Hedonic pricing of houses in different ecosystems should be used cautiously as people have preferences not only for ecosystem services but also for ethnic minorities' preference for urban ecosystem.

3.5 Challenges in valuation of urban ecosystem services

The first and foremost challenge in valuation is finding an integrated value and singular value language. Rather, the challenge is more difficult to find a comprehensive measure that best captures all values. The one valuation method that has cropped up is the 'monetary value of emergy' as a common unit in the economic and ecological valuations, and there is a greater scope for monetary valuation of material flows and social cost accounting extended IO model. The challenge pertaining to monetary valuation in specific is identifying the ecosystem services that are beneficial, the beneficiaries (further see Small et al., 2017) and the value of such perceived benefits/costs.

So far, in the studies ecosystem, disservices' value is not accounted for in any of the calculations pertaining to valuation. Some models like ecological footprints are static and values do not change when technology updates and the spatial pattern changes.

4. A way forward

There are advantages of urban areas since the concentration of population reduces cost of providing municipal services, energy consumption by motor vehicles and per capita land requirement (Rees and Wackernagel, 1996). Despite this, cities are unsustainable because of 'distancing effect of urbanization' that blinds individuals in cities to the fact that they are surviving on resources imported from a distance, and thus, make them believe that urban ecosystem can be sustained indefinitely (Rees, 1992).

Today, we are at a junction where certain steps are required to support fast-growing urbanization:

- There is a need of decoupling economic growth processes from materials (Giljum et al., 2009, 2014; Sanyé-Mengual et al., 2019) and carbon dioxide emissions (for more details, see Andreoni and Galmarini, 2012). Further, we require carbon budgets for each city based on carbon metabolism (Zhang et al., 2014).
- Urban planning, building up smart cities, construction of ecoindustrial parks and vertical gardens are other steps that government might take.
- Since it is the absence of markets that create a valuation problem, a market for ecosystem services should be made operational similar to the market for carbon where carbon credits were bought and sold or else payment should be made for wildlife conservation. Though there are issues associated with commodification of ecosystem services, it is believed that valuation is not commodification but an important step towards ecosystem preservation by highlighting its vital importance (Martín-López et al., 2009; Costanza et al., 2014). On the other hand, it is believed that there is threat that it will bring us back to the problem of cost—benefit analysis where if ecosystem services preservation has lower benefits or higher costs it may be overlooked.

Though authors have suggested various ways the application of each will depend on the nature of problem specific to an area and context in which these can be adopted. The optimal urban planning is inevitably the need of the hour.

References

Alston, K.P., Richardson, D.M., 2006. The roles of habitat features, disturbance, and distance from putative source populations in structuring alien plant invasions at the urban/wildland interface on the Cape Peninsula, South Africa. Biological Conservation 132 (2), 183—198.

Altieri, M.A., Companioni, N., Cañizares, K., Murphy, C., Rosset, P., Bourque, M., Nicholls, C.I., 1999. Greening of the 'barrios': urban agriculture for food security in Cuba. Agriculture and Human Values 16, 131—140.

Andersson, E., Barthel, S., Ahrné, K., 2007. Measuring social-ecological dynamics behind the generation of ecosystem services. Ecological Applications 17, 1267—1278.

Andreoni, V., Galmarini, S., 2012. Decoupling economic growth from carbon dioxide emissions: a decomposition analysis of Italian energy consumption. Energy 44 (1), 682—691.

Bai, X., 2009. Industrial ecology and the global impacts of cities. Journal of Industrial Ecology 11 (2), 1—5.

Barot, S., Yé, L., Abbadie, L., Blouin, M., Frascaria-Lacoste, N., 2017. Ecosystem services must tackle anthropized ecosystems and ecological engineering. Ecological Engineering 99, 486—495.

Barthel, S., Isendahl, C., 2013. Urban gardens, agriculture, and water management: sources of resilience for long-term food security in cities. Ecological Economics 86, 224—234.

Barthel, S., Folke, C., Colding, J., 2010. Social—ecological memory in urban gardens—retaining the capacity for management of ecosystem services. Global Environmental Change 20 (2), 255—265.

Bierwagen, B.G., 2007. Connectivity in urbanizing landscapes: the importance of habitat configuration, urban area size, and dispersal. Urban Ecosystems 10 (1), 29—42.

Bolund, P., Hunhammar, S., 1999. Ecosystem services in urban areas. Ecological Economics 29 (2), 293—301.

Brown, M.T., Ulgati, S., 2010. Updated evaluation of exergy and emergy driving the geobiosphere: a review and refinement of the emergy baseline. Ecological Modelling 221, 2501–2508.

Brunner, P.H., Rechberger, H., 2001. Anthropogenic metabolism and environmental legacies. In: Munn, T. (Ed.), Encyclopedia of Global Environmental Change, vol. 3. Wiley, Chichester, UK.

Bruntland Commission, 1987. Our Common Future (Report of the World Commission on Environment and Development).

Campagne, C.S., Roche, P.K., Salles, J.M., 2018. Looking into Pandora's Box: ecosystem disservices assessment and correlations with ecosystem services. Ecosystem Services 30, 126–136.

Canu, D.M., Gheramndi, A., Nunes, P.A.L.D., Lazzari, P., Cossarini, G., Silidiro, C., 2015. Estimating the value of carbon sequestration ecosystem services in the Mediterranean Sea: an ecological economic approach. Global Environmental Change 32, 87–95.

Carpenter, S.R., Caraco, N.F., Correll, D.L., Howarth, R.W., Sharpley, A.N., Smith, V.H., 1998. Nonpoint pollution of surface waters with phosphorus and nitrogen. Ecological Applications 8 (3), 559–568.

Chan, K.M.A., Satterfield, T., Goldstein, J., 2012. Rethinking ecosystem services to better address and navigate cultural values. Ecological Economics 74, 8–18.

Chaparro, L., Terradas, J., 2009. Ecological Services of Urban Forest in Barcelona. Institut Municipal de Parcs i Jardins Ajuntament de Barcelona, Àrea de Medi Ambient.

Clark, K.H., Nicholas, K.A., 2013. Introducing urban food forestry: a multifunctional approach to increase food security and provide ecosystem services. Landscape Ecology 28 (9), 1649–1669.

Concepción, E.D., Moretti, M., Altermatt, F., Nobis, M.P., Obrist, M.K., 2015. Impacts of urbanisation on biodiversity: the role of species mobility, degree of specialisation and spatial scale. Oikos 124 (12), 1571–1582.

Costanza, R., 1989. What is ecological economics. Ecological Economics 1, 1–7.

Costanza, R., 2000. Social goals and the valuation of ecosystem services. Ecosystems 4–10.

Costanza, R., Wainger, L., 1991. Ecological economics. Business Economics 26 (4), 45–48.

Costanza, R., Daly, H.E., Bartholomew, J.A., 1991. Goals, agenda, and policy recommendations for ecological economics. Ecological Economics: The Science and Management of Sustainability 1–20. Columbia University Press.

Costanza, R., Mitsch, W.J., Day Jr., J.W., 2006. A new vision for New Orleans and the Mississippi delta: applying ecological economics and ecological engineering. Frontiers in Ecology and the Environment 4 (9), 465–472.

Costanza, R., de Groot, R., Sutton, P., Van Der Ploge, S., et al., 2014. Changes in the global value of ecosystem services. Global Environmental Change 26, 152–158.

Daly, H.E., 1994. Operationalizing sustainable development by investing in natural capital. In: Jansson, A., Hammer, M., Folke, C., Costanza (Eds.), Investing in Natural Capital: The Ecological Economics Approach to Sustainability. Island Press, Washington (DC), pp. 22–37.

Daly, H.E., Townsend, K.N., 1993. Valuing the earth: economics, ecology, ethics. Sustainable Growth: An Impossibility Theorem 267.

Daniels, P.L., Moore, S., 2001. Approaches for quantifying the metabolism of physical economies: Part I: methodological overview. Journal of Industrial Ecology 5 (4), 69–93.

Davidson, M.D., 2013. On the relation between ecosystem services, intrinsic value, existence value and economic valuation. Ecological Economics 95 (C), 171–177. Elsevier.

DeGroot, R., Brander, L., van der Ploeg, S., et al., 2012. Global estimates of the value of ecosystems and their services in monetary units. Ecosystem Services 1, 50–61.

Dixon, J., Richards, C., 2016. On food security and alternative food networks: understanding and performing food security in the context of urban bias. Agriculture and Human Values 33 (1), 191–202.

EEA (European Environmental Agency), 2011. Green Infrastructure and Territorial Cohesion. EEA Technical report. In: The Concept of Green Infrastructure and its Integration into Policies Using Monitoring Systems, vol. 18. European Environment Agency.

Ehrlich, P.R., 1994. Ecological economics and the carrying capacity of the earth. In: Jansson, A., Hammer, M., Folke, C., Costanza (Eds.), Investing in Natural Capital: The Ecological Economics Approach to Sustainability. Island Press, Washington (DC), pp. 38–56.

Elmqvist, T., Fragkias, M., Goodness, J., Güneralp, B., Marcotullio, P., McDonald, R.I., Parnell, S., Schewenius, M., Sendstad, M., Seto, K.C., Wilkinson, C. (Eds.), 2013. Urbanization, Biodiversity and Ecosystem Services: Challenges and Opportunities: A Global Assessment. Springer, Dordrecht, The Netherlands.

Escobedo, F.J., Wagner, J.E., Nowak, D., De La Maza, C.L., Rodriguez, M., Crane, D.E., 2008. Analyzing the cost-effectiveness of Santiago, Chile's policy of using urban forests to improve air quality. Journal of Environmental Management 86, 148–157.

Fang, C.F., Ling, D.L., 2003. Investigation of the noise reduction provided by tree belts. Landscape and Urban Planning (63), 187–195.

Farber, S.C., Costanza, R., Wilson, M.A., 2002. Economic and ecological concepts for valuing ecosystem services. Ecological Economics 41, 375–392.

Fischer-Kowalski, M., 1998. Society's metabolism: the intellectual history of materials flow analysis, Part I, 1860–1970. Journal of Industrial Ecology 2 (1), 61–78.

Forman, R.T., Alexander, L.E., 1998. Roads and their major ecological effects. Annual Review of Ecology and Systematics 29 (1), 207–231.

Georgescu-Roegen, N., 1971. The Entropy Law and the Economic Process. Harvard University Press, MA.

Giljum, S., Hinterberger, F., Lutz, C., Meyer, B., 2009. Accounting and modelling global resource use. In: Handbook of Input-Output Economics in Industrial Ecology. Springer, Dordrecht, pp. 139–160.

Giljum, S., Dittrich, M., Lieber, M., Lutter, S., 2014. Global patterns of material flows and their socio-economic and environmental implications: a MFA study on all countries world-wide from 1980 to 2009. Resources 3 (1), 319–339.

Gómez -Baggethun, E., Martín-López, B., 2015. Ecological economics perspectives on ecosystem services valuation. In: Martinez-Alier, J., Muradian, R. (Eds.), Handbook of Ecological Economics. Edward Elgar Publishing, United Kingdom, pp. 260–282.

Gómez-Baggethun, E., Barton, D.N., 2013. Classifying and valuing ecosystem services for urban planning. Ecological Economics 86, 235–245.

Gómez-Baggethun, E., de Groot, R., 2010. Natural capital and ecosystem services: the ecological foundation of human society. In: Hester, R.E., Harrison, R.M. (Eds.), Ecosystem Services: Issues in Environmental Science and Technology. Royal Society of Chemistry, Cambridge, pp. 105–121.

Gowdy, J.M., 2000. Terms and concepts in ecological economics. Wildlife Society Bulletin 28 (1), 26–33.

Gowdy, J., Erickson, 2005. The approach of ecological economics. Cambridge Journal of Economics 29 (2), 207–222.

Grimm, N.B., Faeth, S.H., Golubiewski, N.E., Redman, C.L., Wu, J., Bai, X., Briggs, J.M., 2008. Global change and the ecology of cities. Science 319 (5864), 756–760.

Güneralp, B., Seto, K.C., 2013. Futures of global urban expansion: uncertainties and implications for biodiversity conservation. Environmental Research Letters 8 (1), 014–025.

Güneralp, B., Perlstein, A.S., Seto, K.C., 2015. Balancing urban growth and ecological conservation: a challenge for planning and governance in China. Ambio 44 (6), 532–543.

Hardin, P.J., Jensen, R.R., 2007. The effect of urban leaf area on summertime urban surface kinetic temperatures: a Terre Haute case study. Urban Forestry and Urban Greening 6, 63–72.

Hope, D., Gries, C., Zhu, W., Fagan, W.F., Redman, C.L., Grimm, N.B., et al., 2003. Socioeconomics drive urban plant diversity. Proceedings of the National Academy of Sciences 100 (15), 8788–8792.

Huang, S.-L., 1998. Urban ecosystems, energetic hierarchies, and ecological economics of Taipei metropolis. Journal of Environmental Management 52, 39–51.

Jim, C.Y., Chen, W.Y., 2009. Ecosystem services and valuation of urban forests in China. Cities 26 (4), 187–194.

Kahneman, D., 2011. Thinking, Fast and Slow. Doubleday Canada, Toronto, Ontario.

Kennedy, C., Cuddihy, J., Engel-Yan, J., 2007. The changing metabolism of cities. Journal of Industrial Ecology 11 (2), 43–59.

Kopp, R.J., Smith, V.K., 2013. Implementing natural resource damage assessments. In: Valuing Natural Assets. Rff Press, pp. 138–166.

Krutilla, J.V., 1967. Conservation reconsidered. The American Economic Review 57, 777–786.

Longcore, T., Rich, C., 2004. Ecological light pollution. Frontiers in Ecology and the Environment 2 (4), 191–198.

Lovell, S.T., Taylor, J.R., 2013. Supplying urban ecosystem services through multifunctional green infrastructure in the United States. Landscape Ecology 28 (8), 1447–1463.

Lovins, A.B., Lovins, H.L., 1991. Least-cost climatic stabilization. Annual Review of Energy and the Environment 16, 433–531.

Lyytimäki, J., 2014. Bad nature: newspaper representations of ecosystem disservices. Urban Forestry and Urban Greening 13 (3), 418–424.

Lyytimäki, J., Sipilä, M., 2009. Hopping on one leg—The challenge of ecosystem disservices for urban green management. Urban Forestry and Urban Greening 8 (4), 309–315.

Malthus, T.R., 1986. An Essay on the Principle of Population. Pickering & Chatto Publishers Ltd, London.

Martín-López, B., Gómez-Baggethun, E., González, J.A., Lomas, P.L., Montes, C., 2009. The assessment of ecosystem services provided by biodiversity: re-thinking concepts and research needs. Handbook of nature conservation: Global, Environmental and Economic Issues 261–282.

Martín-López, B., Gómez-Baggethun, E., García-Llorente, M., Montes, C., 2014. Trade-offs across value-domains in ecosystem services assessment. Ecological Indicators 37, 220–228.

Martinez-Alier, J., Munda, G., O'Neill, J., 1998. Weak comparability of values as a foundation for ecological economics. Ecological Economics 26 (3), 277–286.

Matos, D.S., Santos, C.J.F., Chevalier, D.D.R., 2002. Fire and restoration of the largest urban forest of the world in Rio de Janeiro City, Brazil. Urban Ecosystems 6 (3), 151–161.

McDonald, R.I., Kareiva, P., Forman, R.T., 2008. The implications of current and future urbanization for global protected areas and biodiversity conservation. Biological Conservation 141 (6), 1695–1703.

McDonald, R.I., Forman, R.T., Kareiva, P., Neugarten, R., Salzer, D., Fisher, J., 2009. Urban effects, distance, and protected areas in an urbanizing world. Landscape and Urban Planning 93 (1), 63–75.

McDonald, R.I., Green, P., Balk, D., Fekete, B.M., Revenga, C., Todd, M., Montgomery, M., 2011. Urban growth, climate change, and freshwater availability. Proceedings of the National Academy of Sciences 108 (15), 6312–6317.

McDonald, R.I., Weber, K., Padowski, J., Flörke, M., Schneider, C., Green, P.A., et al., 2014. Water on an urban planet: urbanization and the reach of urban water infrastructure. Global Environmental Change 27, 96–105.

McKinney, M.L., 2002. Urbanization, Biodiversity, and Conservation-The impacts of urbanization on native species are poorly studied, but educating a highly urbanized human population about these impacts can greatly improve species conservation in all ecosystems. BioScience 52 (10), 883–890.

McKinney, M.L., 2006. Urbanization as a major cause of biotic homogenization. Biological Conservation 127 (3), 247–260.

McKinney, M.L., 2008. Effects of urbanization on species richness: a review of plants and animals. Urban Ecosystems 11, 161–176.

McPherson, E.G., Simpson, J.R., 1998. Carbon Dioxide Reduction through Urban Forestry: Guidelines for Professional and Volunteer Tree Planters. Rep. PSW-171 Gen. Tech.,80. U.S. Department of Agriculture, Forest Service, Pacific Southwest Research Station, Albany CA, p. 237.

Meadows, D.H., Meadows, D.L., Randers, J., Behrens, W.W., 1972. The limits to growth, 102, p. 27. New York.

Millennium Ecosystem Assessment, 2005. Ecosystems and Human Well Being: Synthesis. Island Press, Washington DC.

Mittermeier, R.A., Gil, P.R., Hoffman, M., Pilgrim, J., Brooks, T., Mittermeier, C.G., et al., 2004. Hotspots Revisited: Earth's Biologically Richest and Most Endangered Terrestrial. Mexico, CEMEX SA, XI+ 392pp.

Müller, N., Werner, P., Kelcey, J.G., 2010. Urban Biodiversity and Design. Wiley-Blackwell.

Müller, N., Ignatieva, M., Nilon, C.H., Werner, P., Zipperer, W.C., 2013. Patterns and trends in urban biodiversity and landscape design. In: Urbanization, Biodiversity and Ecosystem Services: Challenges and Opportunities. Springer, Dordrecht, pp. 123–174.

Myers, N., Mittermeier, R.A., Mittermeier, C.G., Da Fonseca, G.A., Kent, J., 2000. Biodiversity hotspots for conservation priorities. Nature 403 (6772), 853.

Nobel Media AB, 2017. Richard H. Thaler — Facts.

Nowak, D.J., 1994. Atmospheric carbon dioxide reduction by Chicago's urban forest. In: McPherson, E.G., Nowak, D.J., Rowntree, R.A. (Eds.), Chicago's Urban Forest Ecosystem: Results of the Chicago Urban Forest Climate Project. Gen. Tech. Rep., NE-186. U.S. Department of Agriculture, Forest Service, Northeastern Forest Experiment Station, Radnor, PA, pp. 83–94.

Odum, E.P., 1989. Ecology and Our Endangered Life-Support Systems. Sinauer, Sunderland.

Odum, H.T., 1996. Environmental Accounting: Emergy and Decision Making. John Wiley, New York.

Pataki, D.E., Alig, R.J., Fung, A.S., Golubiewski, N.E., Kennedy, C.A., McPherson, E.G., et al., 2006. Urban ecosystems and the North American carbon cycle. Global Change Biology 12 (11), 2092–2102.

Pataki, D.E., Carreiro, M.M., Cherrier, J., Grulke, N.E., Jennings, V., Pincetl, S., Pouyat, R.V., Whitlow, T.H., Zipperer, W.C., 2011. Coupling biogeochemical cycles in urban environments: ecosystem services, green solutions, and misconceptions. Frontiers in Ecology and the Environment 9, 27–36.

Pauchard, A., Aguayo, M., Peña, E., Urrutia, R., 2006. Multiple effects of urbanization on the biodiversity of developing countries: the case of a fast-growing metropolitan area (Concepción, Chile). Biological Conservation 127 (3), 272–281.

Peters, G.P., Weber, C.L., Guan, D., et al., 2007. China's growing CO_2 emissions: a race between increasing consumption and efficiency gains. Environmental Science and Technology 41 (17), 5939–5944.

Pickett, S.T., Cadenasso, M.L., Grove, J.M., Nilon, C.H., Pouyat, R.V., Zipperer, W.C., Costanza, R., 2001. Urban ecological systems: linking terrestrial ecological, physical, and socioeconomic components of metropolitan areas. Annual Review of Ecology and Systematics 32 (1), 127–157.

Pickett, S., Cadenasso, M.L., Meiners, S.J., 2009. Ever since Clements: from succession to vegetation dynamics and understanding to intervention. Applied Vegetation Science 12 (1), 9–21.

Pickett, S.,T.A., Cadenasso, M.L., Childers, D.L., Mcdonnel, M.J., Zhou, W., 2016. Evolution and future of urban ecological science: ecology in, of, and for the city. Ecosystem Health and Sustainability 2 (7), 1–16.

Pigou, A.C., 1920. Economics of Welfare. Macmillan Press, London, 876 pp.

Raupach, M.R., Marland, G., Ciais, P., et al., 2007. Global and regional drivers of accelerating CO_2 emissions. Proceedings of the National Academy of Sciences of the United States of America 104 (24), 10288–10293.

Rees, W.E., 1992. Ecological footprints and appropriated carrying capacity: what urban economics leaves out. Environment and Urbanization 4 (2), 121–130.

Rees, W.E., Wackernagel, M., 1996. "Urban ecological footprints: why cities cannot be sustainable—and why they are a key to sustainability". Environmental Impact Assessment Review 16, 223–248.

Ricardo, D., 1951 [1817]. In: Sraffa, P. (Ed.), The Works and Correspondence of David Ricardo. Cambridge University Press for the Royal Economic Society, Cambridge (UK).

Ricketts, T.H., Daily, G.C., Ehrlich, P.R., Fay, J.P., 2001. Countryside biogeography of moths in a fragmented landscape: biodiversity in native and agricultural habitats. Conservation Biology 15 (2), 378–388.

Sagoff, M., 1995. Carrying capacity and ecological economics. BioScience 45 (9), 610–620.

Sander, H., Polasky, S., Haight, R.G., 2010. The value of urban tree cover: a hedonic property price model in Ramsey and Dakota Counties, Minnesota, USA. Ecological Economics 69 (8), 1646–1656.

Sanyé-Mengual, E., Secchi, M., Corrado, S., Beylot, A., Sala, S., 2019. Assessing the decoupling of economic growth from environmental impacts in the European Union: a consumption-based approach. Journal of Cleaner Production 236, 117535.

Schumpeter, J.A., 1978. History of Economic Analysis. Oxford University Press, New York.

Seto, K.C., Güneralp, B., Hutyra, L.R., 2012. Global forecasts of urban expansion to 2030 and direct impacts on biodiversity and carbon pools. Proceedings of the National Academy of Sciences 109 (40), 16083–16088.

Shackleton, C.M., Ruwanza, S., Sanni, G.S., Bennett, S., De Lacy, P., Modipa, R., et al., 2016. Unpacking Pandora's box: understanding and categorising ecosystem disservices for environmental management and human well-being. Ecosystems 19 (4), 587–600.

Small, N., Munday, M., Durence, I., 2017. The challenge of valuing of ecosystem services that have no material benefits". Global Environmental Change 44, 57–67.

TEEB (The Economics of Ecosystems and Biodiversity), 2010. The Economics of Ecosystems and Biodiversity: Ecological and Economic Foundations. Earthscan, London.

TEEB (The Economics of Ecosystems and Biodiversity), 2011. Manual for Cities: Ecosystem Services in Urban Management. UNEP and the European Commission.

Tilman, D., May, R.M., Lehman, C.L., Nowak, M.A., 1994. Habitat destruction and the extinction debt. Nature 371 (6492), 65.

Tratalos, J., Fuller, R.A., Warren, P.H., Davies, R.G., Gaston, K.J., 2007. Urban form, biodiversity potential and ecosystem services. Landscape and Urban Planning 83 (4), 308–317.

Tyrväinen, L., Pauleit, S., Seeland, K., de Vries, S., 2005. Benefits and uses of urban forests and trees. In: Konijnendijk, C., Nilsson, K., Randrup, T., Schipperijn, J. (Eds.), Urban Forests and Trees. Springer.

Tzoulas, K., Korpela, K., Venn, S., Yli-Pelkonen, V., Kaźmierczak, A., Niemela, J., James, P., 2007. Promoting ecosystem and human health in urban areas using Green Infrastructure: a literature review. Landscape and Urban Planning 81 (3), 167–178.

Vauramo, S., Setälä, H., 2011. Decomposition of labile and recalcitrant litter types under different plant communities in urban soils. Urban Ecosystems 14, 59–70.

Villarreal, E.L., Bengtsson, L., 2005. Response of a Sedum green-roof to individual rain vents. Ecological Engineering 25 (1), 1–7.

Wu, Y., Que, W., Liu, Y.-G., Li, J., et al., 2017. Efficiency estimation of urban metabolism via Emergy, DEA of time series. Ecological Indicators 20 (8), 276–284.

Yansui, L., Fang, F., Li, Y., 2014. Key issues of land use in China and implications for policy making. Land Use Policy 40, 6–12.

Zhang, Y., Yang, Z., Liu, G., Yu, X., 2011. Energy analysis of urban metabolism of Beijing. Ecological Modelling 222, 2377–2384.

Zhang, Y., Linlin, X., Weining, X., 2014. Analyzing spatial patterns of urban carbon metabolism: a case study in Beijing, China. Landscape and Urban Planning 130, 184–200.

CHAPTER

7

Urban green space, social equity and human wellbeing

Nuala Stewart[1]

[1]Master of Sustainable Development, Macquarie University, Sydney, NSW, Australia

1. Introduction

Definitions of greenspace vary tremendously. The definition utilized by Wolch et al. (2014) defines urban greenspace as parks, woodlands, green infrastructures such as green roofs and walls, riverside green areas and community gardens. This is the definition utilized for this paper. Greenspaces in cities are under pressure from urbanization (Taylor and Hochuli, 2017) because increasing urban density means competing demands for land in cities (Cox et al., 2018). This is particularly true for Greater Sydney, Australia, given the high rate of population growth (OECD, 2016) and the high level of urbanization (86%) (Australian Bureau of Statistics, 2016b). The 2016 national census states the population of Greater Sydney is 5.2 million, up 17% since 2006 (Australian Bureau of Statistics, 2018). This is consistent with World Bank (2019) data which reports the Australian population growth is 1.7% per annum, more than double that of the UK and the USA. (both 0.6%). Since the 1980s there has been a move away from single home dwellings that has seen the population density of Sydney increase (Lin et al., 2015), for example, in the city of Sydney (the central business district suburb called 'City of Sydney', not to be confused with Greater Sydney) 95% of residents live in medium- or high-density housing (Johnson, 2014). The New South Wales urban planning policy standard makes provision for 2.83 ha of urban greenspace for every 1000 of population (Hooper et al., 2018), implying a tension between urban densification and provision of urban greenspace. Given the paucity of environmental justice research in Australia, learnings in European and American research may inform Australian environmental and urban planning policy and standards. Assessing urban greenspace, primarily parks, and the needs of local communities on a case-by-case basis may offer significant savings of public sector funds in the form of an opportunity to optimize existing urban green space, or other underutilized open space, to meet the needs of the community. American research in this field of study may provide inspiration for creative urban greenspace projects in Greater Sydney.

2. Background Australia

Australia is the sixth largest country after Russia, Canada, China, the United States of America and Brazil. The area of Australia is 7,656,127 km^2 (Geoscience Australia, n.d.). The population in 2016, when the most recent census took place, was 23,401,892 (Australian Bureau of Statistics, 2016b). Eighty six percent of the population lives in an urban setting (United Nations Population Division, 2018). Population growth is concentrated in urban areas: the growth in population in Sydney (capital city of New South Wales), Melbourne (capital city of Victoria) and Brisbane (capital city of Queensland) constituted 65% of Australia's population growth in 2017–18. Australia's population density is a mere 3.1 people per square kilometre (Australian Bureau of Statistics, 2019a).

Table 7.1: Population and area (city and State) in 2017–18 (below) demonstrate that the population of the four biggest cities in Australia constitutes the majority of the population of each state.

Not only is the population density significantly higher in the cities, the rate of growth is also considerably higher than the rest of each state (Table 7.2).

TABLE 7.1 Population and area (city and state) in 2017–18.

City	City population	State population	City area (km^2)	State area (km^2)
Data from	Australian Bureau of Statistics (2019a)	Australian Bureau of Statistics (2019b)	Geoscience Australia (n.d.)	Geoscience Australia (n.d.)
Greater Sydney	5,029,711	7,480,228	12,000	801,137
Greater Melbourne	4,485,211	5,926,624	9,925	227,444
Greater Brisbane	2,270,000	4 ,703,193	6,000	1,729,742
Greater Perth	2,140,000	2,474,410	6,418	2,527,013

TABLE 7.2 Population increase in percentage – city versus state (2017–18)

City	Percentage (%)	Region	Percentage (%)
Greater Sydney	1.8	NSW State	1
Greater Melbourne	2.5	State of Victoria	1.3
Greater Brisbane	2.1	State of Queensland	1.3
Greater Perth	1.1	State of Western Australia	−0.1

Data from: Australian Bureau of Statistics, 2019a. 3218.0 – regional population growth, Australia, 2017 – 2018. Available at: https://www.abs.gov.au/ausstats/abs@.nsf/latestProducts/3218.0Media%20Release12017-18. (Accessed 5 September 2019).

3. Background: Greater Sydney

Greater Sydney covers 12,367.7 km^2 and is made up of 6 districts and 35 local councils (City of Sydney, 2018). In the 2016 census the population for Greater Sydney was 5,029,711. Based on a total site area of 12,367.7 km^2 (1,236,770 ha), the current population density of the Greater Sydney area is 407 persons per km^2 which includes surrounding national parks. A population density based on the built urban area of 4064 km^2 translates to a density of 1237 people per square kilometre (City of Sydney, 2018). In contrast, the population density for the state of New South Wales (in which Greater Sydney is situated) is a mere 9.7 people per square kilometre (Australian Bureau of Statistics, 2019a):

Across Greater Sydney the amount of public open space as a percentage of the six district areas varies considerably, due in part to the inclusion of large national parks in the district areas (Greater Sydney Commission, 2016) (Graph 7.1).

4. Background: Greater Sydney demographics

The population of Greater Sydney is richly diverse (see Graphs. 7.2 and 7.3). Of the Australian born population of Sydney, 70,135 (1.4%) identify as indigenous Aboriginal (Australian Bureau of Statistics, 2016b).

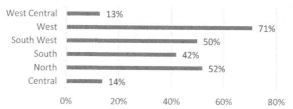

GRAPH 7.1 Public open space as a percentage of district area, Greater Sydney. *Data sourced from Greater Sydney Commission, 2016. Public Open Space Audit. Available at: https://gsc-public-1.s3.amazonaws.com/s3fs-public/Greater_Sydney_Open_Space_Audit.pdf. (Accessed on 31 August 2019).*

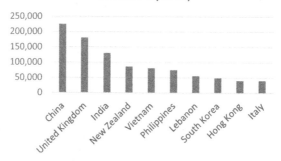

GRAPH 7.2 Birthplace of overseas born residents, Greater Sydney. *Source: Data sourced from Australian Bureau of Statistics, 2019a. 3218.0 − Regional Population Growth, Australia, 2017 − 2018. Available at: https://www.abs.gov.au/ausstats/abs@.nsf/latestProducts/3218.0Media%20Release12017-18. (Accessed 5 September 2019).*

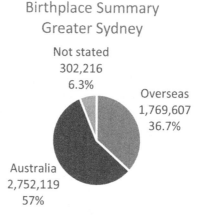

GRAPH 7.3 Birthplace summary, Greater Sydney. Source: Data sourced from Australian Bureau of Statistics, 2019a. *3218.0 − Regional Population Growth, Australia, 2017 − 2018. Available at: https://www.abs.gov.au/ausstats/abs@.nsf/latestProducts/3218.0Media%20Release12017-18.*

Greater Sydney employment

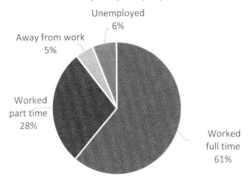

Source: Australian Bureau of Statistics, 2016a

GRAPH 7.4 Greater Sydney unemployment. *Source: Australian Bureau of Statistics, 2016a. Census Quickstats. Available at: https://quickstats.censusdata.abs.gov.au/census_services/getproduct/census/2016/quickstat/1030?opendocument. (Accessed 31 August 2019).*

5. Background: Greater Sydney employment

The unemployment rate is 6% in Greater Sydney, which is consistent with figures for the rest of the State of New South Wales and the country as a whole (State of New South Wales: 6%, Australia: 7%) (Australian Bureau of Statistics, 2016b, Graph 7.4).

6. Methodology

Web of Science was searched for (greenspace) AND (urban) AND (health) between 1 January 2017 and 1 March 2019, which returned 192 articles. To limit the scope of this research, only articles from Europe, United States of America, Australia and New Zealand were included. Similarly, articles which addressed specific health conditions were excluded: only those focussing on mental and physical wellbeing generally were included. Due to the broad range of disciplines this search returned, the relevant articles were identified manually, rather than compiling a list of NOT qualifiers. Abstracts and titles from the search were scanned to identify the articles which warranted review of the full text based on relevance. The bibliography of each article was reviewed to identify further relevant articles. Themes were identified and the articles were grouped according to the following categories (see Appendix 1: Peer-reviewed articles on urban greenspace and health): defining greenspace, urban planning, gender, social inequity/environmental justice, proximity, age, quality, mental health, physical health/activity, metastudy, shade, ethnicity. Many articles were relevant to more than one category. A table of the articles and categories was compiled (see Appendix 1). A total of 65 articles were reviewed in detail.

7. Results

In terms of the geographical spread of area of research (as opposed to the location of researchers), the search results are as follows: USA (n = 29), Europe (n = 19), Australia

APPENDIX 1   Peer-reviewed articles on urban greenspace and health

Author	Year	Title	Country	Topics	Defining GS	Urban planning	Gender	Env justice	proximity	Age	quality	m health	phys health	meta-study	Shade	ethnicity
Broomhall	1996	Study of the availability and environmental quality of urban open space used for ad australia	Australia	quality; mental health; quantity;							1	1				
Hooper, et al.	2018	Testing spatial measures of public open space planning standards with walking an Australia	Australia	quantity;		1			1				1			
Francis, et al.	2012	Quality or quantity	Australia	quality; mental health;							1	1				
Crawford, et al.	2008	Do features of public open spaces vary according to neighbourhood socio-economi Australia	Australia	quality; socio-ec;				1			1					
Byrne, et al.	2010	Green around the gills? The challenge of density for urban greenspace planning in Australia	Australia	urban planning; density		1										
Byrne, et al.	2013	Bordering on neglect: 'Environmental justice' in Australian planning	Australia	env justice				1								
Anderson, et al.	2014	Shade in urban playgrounds in availla lower socio-ec	Australia	env justice; shade				1							1	
Astell-Burt, et al.	2014	Do low-income neighbourhoods have the least green space? lower socio-ec	Australia	env justice				1								
Lily, et al.	2015	Understanding the potential loss and inequities of green space distribution with an Australia	Australia	social equity; socio-ec;				1								
Mavoa, et al.	2015	Area-Level Disparities of Public Open Space: A Geographic Information Systems A Australia	Australia	socio-ec				1								
Feng, et al.	2017	The relationship between neighbourhood green space and child mental wellbeing Australia	Australia	mental health; children						1		1				
Kimpton	2017	A spatial analytic approach for classifying and comparing greenspace - Australia	Australia	env justice; env justice; social equi				1		1	1					
Taylor & Hochuli	2017	Defining greenspace: Multiple uses across multiple disciplines	AUStralia	meta; def	1									1		
Wood, et al.	2017	Public green spaces and positive mental health – Investigating the relationship bet Australia	Australia	mental health; quality								1				
Cleary, et al.	2017	Predictors of Nature Connection Among Urban Residents: Assessing the Role of Ch Australia	Australia	youth						1						
Ekkel, et al.	2017	Nearby green space and human health: Evaluating accessibility metrics	Europe	quality					1		1					
Dzhambov, et al.	2018	Multiple pathways link urban green- and bluespace to mental/young adults	Europe	mental health youth						1		1				
Dzhambov, et al.	2018	Urban residential greenspace and mental health in youth: Dif young adults	Europe	mental health; youth	1					1		1				
Helbich, et al.	2018	More green space is related to less antidepressant prescription rates in the Nethe Europe	Europe	mental health								1				
Köngmaker	2018	Green space definition affects associations of green space with overweight and pł Europe	Europe	physical health; definition; quality	1						1		1			
Schneider, et al.	2019	Exposed children, protected parents; shade in playgrounds as a previously unstudie Europe	Europe	quality; shade; physical health						1	1		1		1	
Engemann, et al.	2019	Residential green space in childhood is associated with lower risk of psychiatric di Europe	Europe	mental health; children						1		1				
Elwood	2004	Who gets skin cancer: individual risk factors. In: Hill D et al, editor. Prevention of S Europe	Europe	physical health; env justice; age				1		1			1			
White, et al.	2018	Neighbourhood greenspace is related to physical activity in England, but only for d Europe, UK	Europe, UK	demographics; deg; physical health									1			
Mullin, et al.	2018	Natural capital and the poor in England: Towards an environmental justice analysi Europe, UK	Europe, UK	env justice; socio-ec;				1		1						
Cole, et al.	2019	Determining the health benefits of green space: does gentrification matter lower socio-ec	lower socio-ec	env justice				1		1						
Kabisch, et al.	2015	Human-environment interactions in urban green spaces - A systematic review of contemporary iss		mental health; physical health; meta								1		1		
Roe, et al.	2016	Understanding relationships between health, ethnicity, place and the role of urban green space in		env justice; ethnicity; low socio, health				1								1
Ferguson, et al.	2018	Contrasting distributions of urban green infrastructure across social and ethno-racio-socio-ec		ethni-socio-ec; ethnicity				1								
McEachan, et al.	2018	Availability, use of, and satisfaction with green space, and children's mental wellbeing at age 4 ye		mental health												
Twohig-Bennett, et al	2018	The health benefits of the great outdoors: A systematic review and meta-analysis of greenspace e		env justice										1		
Boyd, et al.	2018	Who doesn't visit natural environments for recreation and why: A population repr Europe, UK	Europe, UK	socio-ec; visitation	1			1								
Caldwell, et al.	2018	Visits to urban green-space and the countryside associate with different compone Europe, UK	Europe, UK	mental health								1				
Cox, et al.	2017	The impact of urbanisation on nature dose and the implications for human health Europe, UK	Europe, UK	mental health; physical health								1	1			
Gage, et al.	2018	Using Google Earth to Assess Shade for Sun Protection in Urban Recreation Space NZ	NZ	shade											1	
Gage, et al.	2018	Wellington Playgrounds Uncovered: An Examination of Solar Ultraviolet Radiation NZ	NZ	shade											1	
Gage, et al.	2018	Shade in playgrounds: findings from a nationwide survey and implications for urban USA	USA	shade											1	
Dunton	2009	Physical environmental correlates of childhood obesity	USA	health									1			
Joassart-Marcelli, et a	2011	Building the Healthy City: The Role of Nonprofits in Creating Active Urban Parks USA	USA	urban planning		1										
Newell, et al.	2013	Green Alley Programs: Planning for a sustainable urban infrastructure?		quality; env justice				1			1					
Wolch, et al.	2014	Urban green space, public health, and environmental justice: The challenge of mal USA	USA	quality; socio-ec; env justice				1			1					
Weinstein, et al.	2015	Seeing Community for the Trees: The Links among Contact with Natural Environm USA	USA	crime; safety; social cohesion;				1								
Anguelowski	2015	From Toxic Sites to Parks as (Green) LULUs? New Challenges of Inequity, Privilege USA	USA	low socio-ec; env justice	1			1								
Sandifer, et al.	2015	Exploring connections among nature, biodiversity, ecosystem services, and human USA	USA	mental health; physical health; biodiversity								1	1			
Rigdon	2016	A complex landscape of inequity in access to urban parks: A literature review USA	USA	env justice	1			1								
Cohen, et al.	2016	The Paradox of Parks in Low-Income Areas: Park Use and Perceived Threats USA	USA	safety; low socio-ec;				1		1						
Morgan Hughey, et al.	2016	Green and lean: Is neighborhood park and playground availability associated with USA	USA	youth; obesity; socio-ec; ethnicity				1		1						1
Rigolon	2017	Parks and young people: An environmental justice study of park proximity, acreage USA	USA	quality; socio-ec; env justice				1			1					
Whiting, et al.	2017	Outdoor recreation motivation and site preferences across diverse racial/ethnic gr USA	USA	socio-ec; ethnicity				1								1
Cronin, et al.	2017	Moving beyond the neighborhood: Daily exposure to nature and adolescents' mooc USA	USA	adolescents; age; mental health						1		1				
Holman, et al.	2018	Why neighborhood park proximity is not associated with total physical activity USA	USA	physical health; park proximity					1				1			
Mennis, et al.	2018	Shade as environmental justice design tool for skin cancer prevention USA	USA	env justice				1								
Zuk, et al.	2018	Gentrification, Displacement, and the Role of Public Investment	USA	gentrification	1											
Kondo, et al.	2018	Urban green space and it's impact on human health	USA	meta										1		
Gibson	2018	'Let's go to the park.' An investigation of older adults in Australia and their motiva USA	USA	mental health						1		1				
Derose, et al.	2018	Gender Disparities in Park Use and Physical Activity among Residents of High-Pov USA	USA	demographics; socio-ec; gender			1	1		1			1			
Douglas, et al.	2018	Social and environmental determinants of physical activity among ethnic minority children in low-i USA	USA	quality; socio-ec; demographics; age; ethnicity				1		1	1					1
Li, et al.	2019	Not a level playing field: A qualitative study exploring structur lower socio-ec USA	USA	env justice				1								
Stewart, et al.	2019	Park use preferences and physical activity among ethnic minority children in low-i USA	USA	low socio-ec; ethnicity				1		1			1			1
Neshit, et al.	2019	Who has access to urban vegetation? A spatial analysis of distributional green eq USA	USA	env justice				1	1							
Houston & Zuniga	2019	Put a park on it: How freeway caps are reconnecting and greening divided cities USA	USA	quality							1					
Karpeski	2008	Association of park size, distance, and features with physical activity in neighborho USA	USA	physical health; park proximity					1		1		1			
Reich	2016	The High Line and the ideal of democratic public space.	USA	env justice				1								
Total					**5**	**9**	**1**	**35**	**10**	**16**	**19**	**13**	**17**	**2**	**5**	**6**

Summary of eligible articles by geographic region

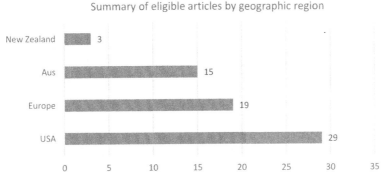

GRAPH 7.5 Summary of eligible articles by geographic region.

(n = 15) papers, NZ (n = 3). Of the 65 peer-reviewed articles included, more than half (35 out of 65) contained content relating to greenspace social inequity as an environmental justice issue (Graph 7.5).

More than half (35 out of 65) of the articles selected for inclusion in this paper tackled the topic of greenspace social inequity as an environmental justice issue. A large proportion of the Australian (11 out of 15) and USA articles (18 out of 29) focus of this topic, whereas the majority of articles from Europe and the UK tend to be focussed on the health aspects of access to urban greenspace. The three articles from New Zealand relate to the social inequity of shade as an environmental justice issue, suggesting that sun-related health impacts as a nascent area of research are regional. The literature reviewed can be divided into categories (see Appendix 1). Some articles address more than one category:

1. Urbanization as a health challenge
2. Defining greenspace
3. Mental and physical health benefits of urban greenspace
4. Social inequity in access to green space as an environmental justice issue
5. Quality of greenspace
6. Summary of Australian research on environmental justice
7. Greenspace as Locally Unwanted Land Use (LULUs)

8. Urbanization as a health challenge

The World Health Organization recognizes urbanization as a health challenge. Living in an urban environment is associated with less physical activity, and a higher incidence of noncommunicable diseases, such as cardiovascular diseases, diabetes, cancers and chronic respiratory diseases (WHO, 2015). Urbanization as a health challenge (WHO, 2015) necessitates a deeper understanding of the complexity of factors influencing the use of urban greenspace. Of the papers reviewed almost a half (29 out of 65) focussed on the health benefits of spending time in nature (see Appendix 1), and more focussed on physical health (17) than mental health (13), with seven referring to both mental and physical health.

Densification of urban areas in Greater Sydney is significant for residents because it is assumed that less time spent in nature (Cox et al., 2018) will have negative consequences for their health. Although it is not clear how contact with nature positively impacts on human health (Ekkel and de Vries, 2017), the myriad health benefits gained from spending time in nature are well documented (Boyd et al., 2018). Around the globe scholars have sought to explain the exact mechanism of action, i.e., how spending time in nature improves health, in the process contributing to a considerable body of work demonstrating correlation, not causation.

Urban greenspace is associated with physical activity which reduces the risk of chronic preventable diseases, such as diabetes (Wolch et al., 2014). Access to urban greenspace in childhood offers lifelong protection against psychiatric disorders (Engemann et al., 2019). Urban residents have less exposure to nature and worse health than rural residents (Cox et al., 2018), and there is social inequity in the distribution of urban greenspace (Cronin-de-Chavez et al., 2019) which has negative health implications. It therefore follows that the WHO (2020) names urbanization as one of the most significant health challenges of this century. This is significant because more than 55% of the world's population lives in an urban setting and is predicted to reach 68% by 2050 (United Nations, 2018).

9. Defining greenspace

The definition of greenspace is problematic in the literature reviewed. In the literature reviewed three widely cited papers concede that no standardized definition for greenspace exists (Taylor and Hochuli, 2017; Dzhambov et al., 2018a,b; Kondo et al., 2018). Gibson (2018) noted the use of terms equivalent to greenspace which are commonly used, such as a park or open space. This limits the usefulness of research in the broader context and for researchers attempting to undertake metaanalysis (Ekkel and de Vries, 2017; Taylor and Hochuli, 2017). Furthermore, inconsistency in defining greenspace may offer an explanation for inconsistent findings regarding the association of greenspace and human health (Klompmaker et al., 2018). Taylor and Hochuli (2017), in an analysis of 125 journal entries about greenspace, found that less than half of the studies reviewed defined their understanding of green space. It follows that differing definitions result in different findings in research (Klompmaker et al., 2018; Dzhambov et al., 2018a,b). The section that follows serves to highlight the breadth of definitions for greenspace (it is not intended to be exhaustive) and the number of variables under consideration for inclusion.

Definitions of greenspace vary tremendously. The definition utilized by Wolch et al. (2014) defines urban greenspace as parks, woodlands, green infrastructures such as green roofs and walls, riverside green areas and community gardens. This is the definition utilized for this paper. This definition is useful in its specificity and implication that it includes only green space accessible to the public. Lin et al. (2015) define greenspace as a parkland area and tree canopy cover, a far less comprehensive definition. Hooper et al. (2018) utilizes the term public open space rather than greenspace and focuses on the functionality in terms of suitability for leisure and recreational activities. Rather than clarify what constitutes greenspace, it is proposed that these definitions raise more questions, such as whether to include greenspace which is not public, for example, school grounds and golf courses. A recent study which did not define green space referred instead to nature-dose as a metric for consideration, irrespective of whether the dose of nature was achieved irrespective of location, limiting the usefulness of the research (Cox et al., 2018). Given that urban greenspace is an area of interest to governments, architects, ecologists, geographers and landscape designers, to name but a few, a single standardized definition may be useful in fostering cross-sectoral collaboration.

10. Mental and physical health benefits of urban greenspace

In addition to ecosystem services associated with urban greenspace (Dzhambov et al., 2018a,b) and enhancing social cohesion (Cox et al., 2018), access to and time spent in urban greenspace have a myriad of benefits for human health and have generated a considerable body of research. What follows constitutes a brief summary of the research reviewed.

10.1 Mental health

The 13 papers reviewed that focused on mental health and exposure to greenspace demonstrate the heterogeneity of this area of study. The majority of papers make reference to the incongruity of findings, find a positive or negative association, and offer a variety of explanations for the incongruence.

Evidence suggests that exposure to greenspace may enhance mental and physical wellbeing, evidenced in studies ranging from a few hundred participants up to nationwide on scale. One nationally representative Australian study of 5000 children found a positive correlation between urban greenspace quality and quantity and children's wellbeing (Feng and Astell-Burt, 2017). In a smaller study in Perth, Western Australia Wood et al. (2017) positive association was seen between number and quality of open greenspaces and positive mental health. Engemann et al. (2019) undertook a study of 943,027 Danish citizens and found that presence of large greenspaces in childhood offers lifelong protection from a wide range of psychiatric disorders. Two small American studies found a positive association between urban greenspace and adolescents' self-reported stress (Mennis and Ambrus, 2018) and mood (Li et al., 2018). White et al. (2018) found that 'walking the dog' is the most common greenspace activity in England and suggests that this may be the confounder in associating greenspace with positive health. This study utilized data from a nationwide survey of 280,790 people. Efforts to identify the missing link in associating greenspace with human wellbeing has elicited a swathe of research.

The literature reviewed included three metastudies. In a metastudy of 125 papers focussed on defining greenspace the authors noted that less than half included an explicit definition of greenspace (Taylor and Hochuli, 2017). Explicit definitions, when provided, fall into one of two categories: either as a synonym for nature or referring to cultivated urban greenspace, such as parks. The metastudy by Kabisch et al. (2015) reviewed 215 papers. The authors acknowledge the complex web of variables in this area of research and call for multidisciplinary collaboration in the context of increasing urbanization to facilitate transferability of knowledge between disciplines and to inform best practice. Incongruence in findings has spawned research efforts into objectively measureable health outcomes in relation to urban greenspace; however, the findings are inconsistent.

Twohig-Bennett and Jones (2018) in a metastudy of 143 academic papers found a positive association between urban greenspace and a long list of objectively measurable health outcomes, for example, lower blood pressure (associated with longevity). In contrast, Helbich et al. (2018) investigated the correlation of antidepressant prescription to greenspace availability, concluding that not all research confirms the positive impact of greenspace on mental health. Twohig-Bennett and Jones (2018) identify a gap in research into low socioeconomic status groups. All three meta-studies make reference to the heterogeneity of the papers reviewed and suggest that this makes meta-studies difficult because comparison of papers is problematic. Critique of methodology and study design is a recurring topic in the papers reviewed under the theme of mental health.

Dzhambov et al. (2018a,b) and Roe et al. (2016) suggest the perception of greenspace versus objectively measurable greenspace as a possible explanation for mixed research outcomes.

Helbich et al. (2018), Sandifer (2015) and Cox et al. (2018) offer the predominance of self-reported survey methodology as a potential stumbling block to establishing a positive association between human wellbeing and urban greenspace exposure. Similarly, Kabisch et al. (2015) state that the majority of the 215 papers reviewed utilize self-reporting surveys as a methodology. A multimethod approach may reduce subjectivity and result in more consistent research findings (Kabisch et al., 2015).

It is evident that to date there is no consensus among scholars regarding a positive association between urban greenspace and mental health. Progress towards a generally accepted definition for greenspace, or at least an awareness of the importance of explicit definition of greenspace, may facilitate robust metaanalysis and progress insight into the mechanism of action of greenspace exposure on mental health. Furthermore, collaboration between disparate disciplines engaged in this area of research may be enhanced, which is important given the public health implications.

10.2 Physical health and an active lifestyle

Of the literature reviewed, 25 papers researched the association between health and urban greenspace revealing that disparate subgroups are motivated to visit urban greenspace by a number of different factors (Rigolon, 2017). Many focused on one or more population subgroups; however, the overall findings emphasized the heterogeneity of urban residents. In a study of 474,000 American children, it was found that racial and ethnic differences account for significant variability in the association between park use and obesity (Morgan Hughey et al., 2017). Gender differences impact the use of urban green space such as parks (Astell-Burt et al., 2014; Derose et al., 2018). White et al. (2018) found that dog owners in the UK, irrespective of demographic differences, meet the recommended weekly physical activity, while non-dog owners do not. Age is also a determinant: older visitors to greenspace are concerned primarily with safety and infrastructure appropriate to their level of mobility, such as benches and level pathways (Gibson, 2018). Thus, differences in use of, and benefits derived from, greenspace vary according to culture, age, socioeconomic status and many other factors (Cronin-de-Chavez et al., 2019) pointing to the importance of understanding the heterogeneity of urban greenspace users and their motivations.

11. Social inequity in access to greenspace as an environmental justice issue

Thirty-three of the peer-reviewed articles selected referred to environmental justice in greenspace social equity. Evidence suggests that the benefits of green space are not equally distributed by all members of the population. US and European research focusses on social inequity of greenspace access: Areas with low-quality natural environments tend to be closest to lower socioeconomic neighbourhoods (Mullin et al., 2018). There is consensus around the fact that lower socioeconomic status is associated with less accessible urban greenspace (Astell-Burt et al., 2014), with fewer amenities or features to make them more appealing (Mavoa et al., 2015). A recent study by Anderson et al. (2014) found differences in access to shade in urban greenspace in Australia, suggesting that this may be an environmental injustice issue for the region: Gibson (2018) confirmed in his study of older Australians that among other features, shade is an important determinant of greenspace visitation. Lower socioeconomic areas tend to have less shade and children in playgrounds have less shade than their parents sitting and watching them (Anderson et al., 2014), indicating an additional aspect of social inequity. Further Australian research on shade may have implications for urban and public health planning.

12. Quality of urban green space

Quality of greenspace is emerging as an important metric which may explain inconsistent findings for a positive association between urban greenspace and health. A quarter of the academic papers reviewed (15 out of 62) focus on the quality of urban greenspace. Urban greenspace makes the journey to work longer for commuters (Klompmaker et al., 2018); therefore, greenspace should be of sufficient quality to fulfil its function. A tendency to value urban greenspace irrespective of the state of repair draws attention to the primacy of environmental ethics. Ekkel and de Vries (2017) take a more pragmatic approach, suggesting measuring proximity, size *and quality* of all green space within a certain distance into account in their search for a positive association between health and urban greenspace. Inclusion of qualities of greenspace then opens up questions regarding which characteristics to include. If physical features such as greenness, amenities (such as benches and play equipment) are under consideration, should experiential qualities such as overcrowding be included? Categorizing Australian urban greenspace, namely parks, by physical characteristics (Kimpton, 2017) revealed that affluent neighbourhoods have access to more amenity-rich greenspace than areas of relatively lower socioeconomic status. Gibson (2018) found that it is the amenities or features of a park rather than proximity or size which attract visits from older Australians. Thus, research of quantifiable qualities of urban greenspace offers an opportunity to better inform urban-planning decisions aimed at social equity in greenspace accessibility, and at the same time a new avenue for further research.

13. Summary of Australian research on environmental justice

Australian research on environmental justice is scant (Byrne and MacCallum, 2013). The Australian academic papers reviewed are predominantly focused on inequitable access to urban greenspace. Byrne and MacCallum (2013) attribute this to two factors: (i) the legacy of a political system in which environment and planning were separate and the absence of environmental justice topics in education for urban planners (Byrne and MacCallum, 2013). There is a consensus that planning policy may inadvertently exacerbate the social inequity in greenspace accessibility (Byrne et al., 2010; Mavoa et al., 2015; Kimpton, 2017) by failing to consider greenspace typology and suitability.

Evidence suggests greenspace social inequity in the Australian states of Victoria (Mavoa et al., 2015), Queensland (Byrne and MacCallum, 2013; Cleary et al., 2018), New South Wales (Lin et al., 2015) and Western Australia (Francis et al., 2012). Two of the Australian papers reviewed allude to the paradox of greenspace social equity and population densification as an urban planning strategy (Byrne et al., 2010; Lin et al., 2015).

Research undertaken by Francis et al. (2012) found a positive association between the quality, rather than the quantity, of urban greenspace and mental health, irrespective of whether the participants visited the urban greenspace or not. Furthermore, these findings suggest that tangible, objectively identifiable features of urban greenspace such as walking trails, water features and seating are more important for mental health than subjective features, such as safety.

14. Urban greenspace as locally unwanted land use (Lulu)

Environmental justice emerges strongly as a theme in the US research reviewed. Environmental justice is defined as equitable exposure to environmental good and harm (Wolch et al., 2014). Environmental justice mobilization in the USA. in the 1970s originated to address environmental pollution from locally unwanted land uses (LULUs) (such as contaminating factories) near low socioeconomic neighbourhoods. However, in the present day, Anguelovski (2016) suggests that parks and playgrounds are LULUs due to the land appreciation associated with urban greening that causes gentrification and displacement of the lower wage earners in those previously unappealing neighbourhoods. This suggests that urban greening policy intended to address inequity and the reality on the ground in terms of access to urban green space. Urban greening policy may inadvertently be an agent of further environmental injustice. This theme is present in several of the American research papers reviewed. Further themes in the international research reviewed are referenced in the discussion that follows for the sake of brevity.

15. Discussion

As the American environmental justice movement regarding greenspace social inequity is established, it may be that Greater Sydney can benefit from their experience rather than reinventing the wheel. Urban green space and human interaction with it is a nascent field of study internationally (Kabisch et al., 2015). Given that Australian environmental justice research is scant (Byrne and MacCallum, 2013), Australian scholars can look to Anglo-American research as a context for Australian research rather than direct application of international findings to the Australian context. The historical patterns of city development and urban planning in Australia, the USA. and the UK are similar ((Byrne and MacCallum, 2013). In this regard, aspects of the Anglo-American research may be applicable to Australia and/or inform new research directions.

15.1 Demographic considerations

In environmental justice research in the UK and the USA, the terms low-socioeconomic and ethnic minority are used interchangeably (Cronin-de-Chavez et al., 2019; Marquet et al., 2019). In Australia, however, migrants tend to earn above the median wage (Australian Bureau of Statistics, 2019a), indicating a significant departure from the Anglo-American experience and implying that a swathe of relevant research may not apply to the Australian context. Connolly (2018) is an advocate for the type of person-centred city Jane Jacobs (1961) describes in her book *The Death and Life of Great American Cities*. Jacobs values parks only for their potential as gathering places for social events and states that cities should be envisaged as part of nature. Planning should be motivated by protecting urban social diversity so that each city maintains its unique, distinct character. This concept of preservation of the uniqueness of the city may also translate to the urban greenspace-scale: seeking to identify discernible qualities in and around an urban green space and enhance them rather than a one-size-fits-all approach to facilitating the community's healthy lifestyle. If public funds can inadvertently cause harm, then surely with research the reverse can be true too.

15.2 Inclusive green of cities to avoid gentrification

Early scholarly research into renewal of urban areas and the subsequent displacement of lower-income residents underplayed the role of public investment; however Zuk et al. (2018) suggest that this is an area for future research, which may have implications for government departments on every scale across Greater Sydney, such as urban planning and public health. Critical consideration of greening initiatives with regard to potential social outcomes may allow for inclusive greening of cities (Haase et al., 2017) The increasing role of nongovernmental actors in provision and maintenance of public amenities in the United States may be a model for consideration in Greater Sydney when budgets are reined in.

15.3 Innovative approaches

Evidence suggests the urban planning policy should aim to balance both ecological and social considerations to facilitate the improvement of neighbourhoods to avoid causing gentrification on implementation (Checker, 2011). Checker (2011) proposes that only when all stakeholders are truly engaged in the urban planning process can this happen. Evidence suggests creative alternatives to traditional parks which create new urban greenspace. Newell et al. (2019) propose that greening of alleys in US cities, which initially served to address stormwater runoff and heat island effect, has proven to serve as parks with social and urban renewal cobenefits. Wolch et al. (2014) call this phenomenon 'just green enough' (p. 245). This type of unusual greening project benefits the neighbourhood without triggering gentrification and displacement of the most vulnerable residents, which is both green and just (Connolly,

2018). With the same goal in mind, Houston and Zuniga (2019) propose freeway cap parks as an innovative way to create new urban greenspace by utilizing airspace above freeways. This demonstrates creativity where open land is not available and maybe a solution for not only Greater Sydney, which is becoming increasingly densely populated, but any major city. Anglo-American research may offer Greater Sydney an alternative perspective from which to view urban areas.

15.4 Collaboration

Cronin-de-Chavez et al. (2019) and Joassart-Marcelli et al. (2011) state that nonprofits can play a role in supporting park use through maintenance and supervised activity programs. However, research suggests that the abundance and distribution of nonprofits follows existing stratifications along socioeconomic lines, possibly perpetuating social inequities rather than eliminating them. Furthermore, nonprofit activity is virtually nonexistent in the poorest cities (Joassart-Marcelli et al., 2011). Including a wider group of stakeholders, such as but not limited to nonprofits, in urban planning may facilitate a balance between economic, environmental and social equity (Connolly, 2018). Anglo-American scholarly research may be useful in the Australian context given that both the USA and Europe are home to multicultural, heterogeneous populations. However, it is important to keep in mind that migrants in Australia typically earn above the median wage.

Insight into culture-specific motivations for visiting urban greenspace is relevant for informing urban planning in multicultural, globalized cities (Kabisch et al., 2015) like Greater Sydney. Failure to consider all social outcomes of urban greening projects may threaten the Australian quality of life (Byrne et al., 2010). Urban greening is not benevolent: one cannot presume that urban greenspace provides benefits for all members of a community equally (Cole et al., 2019). Recent political developments in Australia may trigger urban planning policy review.

16. Conclusion

In answer to the research question: how does the availability and quality of urban greenspace affect the health of urban residents in Greater Sydney? The evidence suggests that although there is a consensus regarding the positive association between physical health and exposure to urban greenspace, further research is indicated to establish the causation. Greater Sydney, given the high rate of migration, population growth and urban population densification, may be an ideal site for further research. Research into the unique urban greenspace needs of individual communities in this richly multicultural city may serve as a microcosm for highly urbanized nations around the world. Reichl (2016) invites further exploration of diverse social mix as a consideration in designing urban greenspace when reminding us that no two parks are the same, stating that who visits them is a complex interplay of many disparate factors such as of culture and accessibility, some yet unknown.

Several options have become apparent in the literature reviewed in terms of resource-efficient and equitable provision of urban greenspace. City authorities engaging in nontraditional partnerships, such as collaboration with communities and nonprofits, may offer a resource-efficient pathway to optimize the health benefits of urban greenspace. Creative use of disused city infrastructure to create urban greenspace may of necessity become the norm in Greater Sydney, as infill development competes for urban land. Such a project is Paddington Reservoir Gardens, originally a reservoir which was decommissioned in 1899. Paddington Reservoir Gardens is now an attractive sunken garden with a pond and grassy areas covering 2350 m², popular as a venue for wedding photographs (City of Sydney, 2017).

Given the paucity of environmental justice research in Australia, the international research may inform environmental and urban planning policy and standards. The literature reviewed cautions against the primacy of environmental ethics. A balance of social, environmental and economic considerations in urban greening may facilitate a healthy and just Greater Sydney.

References

Anderson, C., et al., 2014. Shade in urban playgrounds in Sydney and inequities in availability for those living in lower socioeconomic areas. Australian and New Zealand Journal of Public Health 38 (1), 49–53. https://doi.org/10.1111/1753-6405.12130. John Wiley & Sons, Ltd (10.1111).

Anguelovski, I., 2016. From toxic sites to parks as (green) LULUs? New challenges of inequity, privilege, gentrification, and exclusion for urban environmental justice. Journal of Planning Literature 31 (1), 23–36.

Astell-Burt, T., et al., 2014. Do low-income neighbourhoods have the least green space? A cross-sectional study of Australia's most populous cities. BMC Public Health 14 (1), 19–21. https://doi.org/10.1186/1471-2458-14-292.

Australian Bureau of Statistics, 2016a. Census Quickstats. Available at: https://quickstats.censusdata.abs.gov.au/census_services/getproduct/census/2016/quickstat/1030?opendocument (Accessed 31 August 2019).

Australian Bureau of Statistics, 2016b. 3105.0.65.001 - Australian Historical Population Statistics. Available at: https://www.abs.gov.au/AUSSTATS/abs@.nsf/mf/3105.0.65.001 (Accessed 28 August 2019).

Australian Bureau of Statistics, 2018. 3222.0 - Population Projections, Australia, 2017 (base) − 2066. Available at: https://www.abs.gov.au/AUSSTATS/abs@.nsf/Latestproducts/3222.0Main%20Features62017%20(base)%20-%202066?opendocument&tabname=Summary&prodno=3222.0&issue=2017%20(base)%20-%202066&num=&view= (Accessed 1 September 2019).

Australian Bureau of Statistics, 2019a. 3218.0 − Regional Population Growth, Australia, 2017 − 2018. Available at: https://www.abs.gov.au/ausstats/abs@.nsf/latestProducts/3218.0Media%20Release12017-18 (Accessed 5 September 2019).

Australian Bureau of Statistics, March 2019b. 3101.0 - Australian Demographic Statistics. Available at: https://www.abs.gov.au/ausstats/abs@.nsf/mf/3101.0.

Boyd, F., et al., 2018. Who doesn't visit natural environments for recreation and why: a population representative analysis of spatial, individual and temporal factors among adults in England. Elsevier Landscape and Urban Planning 175, 102–113. https://doi.org/10.1016/j.landurbplan.2018.03.016. November 2017.

Byrne, J., MacCallum, D., 2013. 'Bordering on neglect: "Environmental justice" in Australian planning'. Australian Planner 50 (2), 164–173. https://doi.org/10.1080/07293682.2013.776984.

Byrne, J., Sipe, N., Searle, G., 2010. Green around the gills? The challenge of density for urban greenspace planning in SEQ. Australian Planner 47 (3), 162–177. https://doi.org/10.1080/07293682.2010.508204.

Checker, M., 2011. 'Wiped out by the "Greenwave": environmental gentrification and the paradoxical politics of urban sustainability'. City and Society 23 (2), 210–229. https://doi.org/10.1111/j.1548-744X.2011.01063.x.

City of Sydney, 2017. Paddington Resevoir Gardens. Available at: https://www.cityofsydney.nsw.gov.au/explore/facilities/parks/major-parks/paddington-reservoir-gardens (Accessed 2 Septmber 2019).

City of Sydney, 2018. Greater Sydney. Available at: https://www.cityofsydney.nsw.gov.au/learn/research-and-statistics/the-city-at-a-glance/greater-sydney (Accessed 31 August 2019).

Cleary, A., et al., 2018. 'Predictors of nature connection among urban residents: assessing the role of childhood and adult nature experiences'. Environment and Behavior. https://doi.org/10.1177/0013916518811431 xocs:firstpage xmlns:xocs=.

Cole, H.V.S., et al., 2019. Determining the health benefits of green space: does gentrification matter? Health and Place 57, 1−11. https://doi.org/10.1016/j.healthplace.2019.02.001.

Connolly, J., 2018. From Jacobs to the just city: a foundation for challenging the green planning orthodoxy. Elsevier Cities 1−7. https://doi.org/10.1016/j.cities.2018.05.011. September 2017.

Cox, D.T.C., et al., 2018. The impact of urbanisation on nature dose and the implications for human health. Elsevier Landscape and Urban Planning 179 (August), 72−80. https://doi.org/10.1016/j.landurbplan.2018.07.013.

Cronin-de-Chavez, A., Islam, S., McEachan, R.R.C., 2019. Not a level playing field: a qualitative study exploring structural, community and individual determinants of greenspace use amongst low-income multi-ethnic families. Elsevier Ltd Health and Place 56, 118−126. https://doi.org/10.1016/j.healthplace.2019.01.018. January.

Derose, et al., 2018. Gender disparities in park use and physical activity among residents of high-poverty neighborhoods in los angeles. Women's Health Issues 28 (1), 6−13.

Dzhambov, et al., 2018a. Multiple pathways link urban green- and bluespace to mental health in young adults. Environmental Research 166, 223−233.

Dzhambov, A., et al., 2018b. Urban residential greenspace and mental health in youth: different approaches to testing multiple pathways yield different conclusions. Environmental Research 160, 47−59. https://doi.org/10.1016/j.envres.2017.09.015. June 2017.

Ekkel, E.D., de Vries, S., 2017. Nearby green space and human health: evaluating accessibility metrics. Elsevier B.V. Landscape and Urban Planning 157, 214−220. https://doi.org/10.1016/j.landurbplan.2016.06.008.

Engemann, K., et al., 2019. Residential green space in childhood is associated with lower risk of psychiatric disorders from adolescence into adulthood. Proceedings of the National Academy of Sciences 116 (11), 5188−5193. https://doi.org/10.1073/pnas.1807504116.

Feng, X., Astell-Burt, T., 2017. The relationship between neighbourhood green space and child mental wellbeing depends upon whom you ask: multilevel evidence from 3083 children aged 12−13 years. International Journal of Environmental Research and Public Health 14 (3). https://doi.org/10.3390/ijerph14030235.

Francis, J., et al., 2012. Quality or quantity? Exploring the relationship between public open space attributes and mental health in Perth, Western Australia. Elsevier Ltd Social Science and Medicine 74 (10), 1570−1577. https://doi.org/10.1016/j.socscimed.2012.01.032.

Geoscience Australia, n.d. Area of Australia − states and territories. Available at: https://www.ga.gov.au/scientific-topics/national-location-information/dimensions/area-of-australia-states-and-territories. Accessed 2 September 2019.

Gibson, S.C., 2018. 'Let's go to the park.' an investigation of older adults in Australia and their motivations for park visitation'. Elsevier Landscape and Urban Planning 180, 234−246. https://doi.org/10.1016/j.landurbplan.2018.08.019. September.

Greater Sydney Commission, 2016. Public Open Space Audit. Available at: https://gsc-public-1.s3.amazonaws.com/s3fs-public/Greater_Sydney_Open_Space_Audit.pdf (Accessed 31 August 2019).

Haase, D., et al., 2017. Greening cities − to be socially inclusive? About the alleged paradox of society and ecology in cities. Habitat International 64, 41−48. https://doi.org/10.1016/j.habitatint.2017.04.005.

Helbich, et al., 2018. More green space is related to less antidepressant prescription rates in the Netherlands: a Bayesian geoadditive quantile regression approach. Environmental Research 166, 290−297.

Hooper, et al., 2018. Testing spatial measures of public open space planning standards with walking and physical activity health outcomes: findings from the Australian national liveability study. Landscape and Urban Planning 171, 57−67.

Houston, D., Zuniga, M.E., 2019. Put a park on it: how freeway caps are reconnecting and greening divided cities. Cities 85, 98−109.

Jacobs, J., 1961. The Death and Life of Great American Cities. Vintage 1993 Modern Library Edition.

Joassart-Marcelli, P., Wolch, J., Salim, Z., 2011. Building the healthy city: the role of nonprofits in creating active urban parks. Urban Geography 32 (5), 682−711. https://doi.org/10.2747/0272-3638.32.5.682.

Johnson, J., 2014. Sydney City's Green Vision, pp. 32−35.

Kabisch, N., Qureshi, S., Haase, D., 2015. Human-environment interactions in urban green spaces - a systematic review of contemporary issues and prospects for future research. Elsevier Inc., Environmental Impact Assessment Review 50, 25−34. https://doi.org/10.1016/j.eiar.2014.08.007.

Kimpton, A., 2017. A spatial analytic approach for classifying greenspace and comparing greenspace social equity. Elsevier Ltd Applied Geography 82, 129−142. https://doi.org/10.1016/j.apgeog.2017.03.016.

Klompmaker, J.O., et al., 2018. Green space definition affects associations of green space with overweight and physical activity. Elsevier Inc. Environmental Research 160, 531−540. https://doi.org/10.1016/j.envres.2017.10.027 (October 2017).

Kondo, M.C., et al., 2018. Urban green space and its impact on human health. International Journal of Environmental Research and Public Health 15 (3). https://doi.org/10.3390/ijerph15030445.

Li, et al., 2018. Moving beyond the neighborhood: daily exposure to nature and adolescents' mood. Landscape and Urban Planning 173, 33–43.

Lin, B., Meyers, J., Barnett, G., 2015. Understanding the potential loss and inequities of green space distribution with urban densification. Elsevier GmbH Urban Forestry and Urban Greening 14 (4), 952–958. https://doi.org/10.1016/j.ufug.2015.09.003.

Marquet, O., et al., 2019. Park use preferences and physical activity among ethnic minority children in low-income neighborhoods in New York City. Elsevier Urban Forestry and Urban Greening 38, 346–353. https://doi.org/10.1016/j.ufug.2019.01.018 (September 2018).

Mavoa, S., et al., 2015. Area-level disparities of public open space: a geographic information systems analysis in metropolitan Melbourne. Routledge Urban Policy and Research 33 (3), 306–323. https://doi.org/10.1080/08111146.2014.974747.

Mennis, M., Ambrus, 2018. Urban greenspace is associated with reduced psychological stress among adolescents: a Geographic Ecological Momentary Assessment (GEMA) analysis of activity space. Landscape and Urban Planning 174, 1–9.

Morgan Hughey, S., et al., 2017. Green and lean: is neighborhood park and playground availability associated with youth obesity? Variations by gender, socioeconomic status, and race/ethnicity. Elsevier Inc. Preventive Medicine 95, S101–S108. https://doi.org/10.1016/j.ypmed.2016.11.024.

Mullin, K., et al., 2018. Natural capital and the poor in England: towards an environmental justice analysis of ecosystem services in a high income country. Elsevier Landscape and Urban Planning 176, 10–21. https://doi.org/10.1016/j.landurbplan.2018.03.022. March.

Newell, et al., 2013. Green alley programs: planning for a sustainable urban infrastructure? Cities 31, 144–155.

OECD, 2016. Population growth rates: annual growth in percentage. In: Population and Migration. OECD Publishing, Paris. https://doi.org/10.1787/factbook-2015-graph2-en.

Reichl, A.J., 2016. The high line and the ideal of democratic public space. Urban Geography 37 (6), 904–925. https://doi.org/10.1080/02723638.2016.1152843. ISSN 0272-3638. (Accessed 1 June 2019).

Rigolon, A., 2017. Parks and young people: an environmental justice study of park, proximity, acreage, and quality in Denver, Colorado. Elsevier Landscape and Urban Planning 165, 73–83. https://doi.org/10.1016/j.landurbplan.2017.05.007 (November 2016).

Roe, J., Aspinall, P.A., Ward Thompson, C., 05 July 2016. Understanding relationships between health, ethnicity, place and the role of urban green space in deprived urban communities. International Journal of Environmental Research and Public Health 13 (7).

Sandifer, P., et al., 2015. Exploring connections among nature, biodiversity, ecosystem services, and human health and well-being: opportunities to enhance health and biodiversity conservation. Ecosystem Services 12, 1–15.

Taylor, L., Hochuli, D.F., 2017. Defining greenspace: multiple uses across multiple disciplines. Elsevier B.V. Landscape and Urban Planning 158, 25–38. https://doi.org/10.1016/j.landurbplan.2016.09.024.

The World Bank Group, 2019. Population growth (% annual). Available at: https://data.worldbank.org/indicator/SP.POP.GROW. Accessed on 2 May 2020.

Twohig-Bennett, Jones, 2018. The health benefits of the great outdoors: a systematic review and meta-analysis of greenspace exposure and health outcomes. Environmental Research 166, 628–637.

United Nations, 2018. World Urbanization Prospects: 2018 Revision. Available at: https://data.worldbank.org/indicator/SP.URB.TOTL.IN.ZS?locations=AU. Accessed 2 September 2019.

WHO, 2020. Urban Health. Available at: https://www.who.int/health-topics/urban-health. Accessed 24 June 2020.

White, et al., 2018. Neighbourhood greenspace is related to physical activity in England, but only for dog owners. Landscape and Urban Planning 174, 18–23.

Wolch, J., Byrne, J., Newell, J., 2014. Urban green space, public health, and environmental justice: the challenge of making cities "just green enough". Elsevier B.V. Landscape and Urban Planning 125, 234–244. https://doi.org/10.1016/j.landurbplan.2014.01.017.

Wood, et al., 2017. Public green spaces and positive mental health — investigating the relationship between access, quantity and types of parks and mental wellbeing. Health and Place 48, 63–71.

Zuk, M., et al., 2018. Gentrification, displacement, and the role of public investment. Journal of Planning Literature 33 (1), 31–44.

Urban environment

Urbanization, urban agriculture and food security

Antonia D. Bousbaine[1], Christopher Bryant[2]

[1]Département de Géographie, Laboratoire LAPLEC, Université de Liège, Liège, Belgium;
[2]Géographie, Université de Montréal, Canada & Adjunct Professor, School of Environmental Design and Rural Development, University of Guelph, Montréal, Québec, Canada

1. Introduction

1.1 Urban ecology and the roles of urban agriculture

Urban Ecology is generally accepted as a field of study related to how living organisms (and we include human beings) create and interact with each other in urban environments. Immediately, we have to recognize that these living organisms are not all alike (including human beings of course), and their presence can vary substantially even within a particular city environment and also within the units in a given type of urban agriculture. In addition,

TABLE 8.1   The roles of urban ecology and urban agriculture.

Roles of urban ecology	Roles of urban agriculture and links to urban ecology
- Protection of wild animals, birds, and environments for these as well as valued insects	- Food production, healthy food production
- Green spaces (grasslands and forests)	- Sustainable agricultures (which should generate minimal pollution of water resources and air pollution)
- Green spaces produced by human beings (lawns, orchards), which can be privately owned or public spaces; these green spaces can also support multiple functions such as areas in which citizens can exercise, appreciate the landscape qualities and take photos	- Potential for food supplies with limited costs for populations in need (e.g., Santropol Rouland, near the centre of Montréal) and that are essentially healthy or sustainable foodstuffs
- Protection of water resources with limited or no water pollution	- Putting in place projects producing healthy food (e.g., food land belts, multiple individual healthy food projects — even involving projects that sell baked farm produce)
- Contributions to tourism activities, recreational activities (e.g., footpaths, spectacular views)	- Protection of certain humanized environmental spaces (e.g., grasslands)

the environment in which these living organisms exist can vary tremendously in terms of the openness of the environment, the presence or not of water bodies and streams, rivers and other water flows and the presence of different types of vegetation as well as wild animals and insects. As noted in Table 8.1, some of these ecological roles can also be valued substantially by certain segments of the human population, e.g., the conservation of certain types of landscapes which can also frequently incorporate the conservation of agricultural activities (Bryant et al., 2019).

The roles of urban agriculture (Table 8.1) can also vary substantially between different cities and their surrounding territories. These roles can include: a. supporting certain types of landscapes that can attract both human residents as well as tourists; b. contributing to the maintenance of the water quality of lakes and rivers (e.g., through the development of sustainable agricultures (Bryant et al., 2018)); and c. supporting the production of healthy foodstuff and selling (either directly or indirectly) this healthy foodstuff to citizens. This is also part of the multifunctionality of agricultural production, which contributes to the perceived value of Urban Agriculture.

1.2 Productivist agriculture

In urban agriculture that is located in urban agglomerations and periurban areas, it is important to understand that in some of the geographical components of urban agriculture we can also find examples of productivist agriculture with its negative effects on the environment, and when we go further from the urban agglomerations and their periurban zones, we find significant territories where productivist agriculture is still dominant. Thus, before we focus on urban agriculture, it is appropriate for us to review the relationships among productivist agriculture, the environment and human health.

This productivist system has been increasingly singled out, since the late 1970s, and protest movements have been heard, especially in the Anglo-Saxon world including the United States and Australia (Larrère and Larrère, 1997). This agricultural system affected European countries in the 1980s with the overproduction of food and nitrate pollution problems with its climax reached during the health crises that affected some European Union countries in the late 1980s and early 1990s (Goulet, 2008). Faced with these situations, scientists, politicians, environmental organizations and representatives of civil society came forward to denounce the environmental degradation caused by this dominant industrial agricultural system whose consequences go even beyond the environment because they can also have an impact on the climate (Mzoughi and Napoléone, 2013).

Very quickly the 'ecological limits of this conventional model' became known (Caplat, 2014: 23), calling into question this system which admittedly had made it possible to reduce hunger in the world but with so many negative repercussions. The environmental consequences can be summarized as follows:

- The massive use of pesticides (herbicides, fungicides and insecticides) has contaminated soil as well as groundwater. The first victims of this massive use of pesticides are the farmers themselves, and these pesticides have an impact on the air given their volatility.
- Concentration of high doses of nitrate, a problem both for human health and ecosystems.
- Artificialization of natural environments and destruction of biodiversity through deforestation (in 2010 the forest area in the world covered 4 billion hectares or 31% of the land area, while in 1970 two-thirds of the land was forested).
- Unprecedented decline in natural habitats correlated with the disappearance of certain animal species as a result of the change in land use revealed as the root cause of the 'biodiversity crisis' (Groombridge, 1992; Watson and Vitousek, 2012).
- Perturbation of biogeochemical cycles, atmospheric chemistry and biodiversity as a whole (Fresco et al., 1993; Watson and Vitousek, 2012; Chapin et al., 2000; Lambin et al., 2003).
- Deterioration of soil quality and erosion of soils.
- Destructuring traditional agrarian landscapes (e.g., passage from farmland to open-field).
- Emission of greenhouse gases caused by intensive farming above ground and participation in global warming. Agriculture accounts for 23% of direct emissions and 0.87% of indirect emissions (IPPC, 2014).
- The scarcity of species of birds, mammals and plants in Europe (Krebs et al., 1999; Donald et al., 2002; Kleijn and Sutherland, 2003; Green et al., 2005), right up to the extinction of some of them (Robinson and Sutherland, 2002).
- The progressive disappearance of pollinizer insects.
- High dependence on fossil energies.

These different environmental consequences appear to have led to substantial efforts internationally to measure the real impacts, in particular in relation to biodiversity and ecosystem services (MEA, 2005; and TEEB, 2009). These impacts are not the only ones that have been highlighted, since at the social level this type of agriculture also appears to have had significant negative consequences.

IV. Urban environment

1.3 Urban agriculture

Over the last 10–15 years, the term 'Urban Agriculture' has increasingly referred to agricultural activities located directly in the urban environment of cities including their agglomerations **and** the periurban areas around cities (e.g., Lohrburg et al., 2016); (Charvet and Laureau, 2018). The project which led to the publication of the book by Lohrburg et al., in 2016, was the result of the largest research project at that time ever undertaken and supported financially based on the broader definition of urban agriculture; it was primarily European with many European Union countries participating, as well as EU Universities and their researchers, but it also included a researcher from Canada (C.R. Bryant), another from Cuba, another from Japan and another from Africa. Furthermore, another book has been produced in which various chapters have dealt with urban agriculture (e.g., Bousbaine et al., 2019).

Thus, it is not surprising that there is now a fairly broad recognition that there is a wide range of types of urban agriculture (e.g., small-scale fermettes, 'normal' farms, part-time farms, farm types differentiated by the nature of their produce, rooftop farms and gardens (e.g., the Fermes Lufa in Montreal)), linkages between farm production/food production and consumption/consumers via different forms of food projects, e.g., regional projects including Food Land Belts such as around the city of Liège, Belgium (Bousbaineet al., 2019); organizations that link consumers to farmers, such as the AMAP (*Associations pour le maintien d'une agriculture paysanne,*or in English, Associations for Maintaining Small Scale Family Farming) in France (in and around cities of different sizes) as well as the tremendous number of types of food projects that do not integrate producers and consumers in any formal manner. The structure of food projects can thus vary substantially reflecting, among other factors, different cultural values on the part of consumers, as well as of producers. As already mentioned, the human values of the farmers as well as consumers can vary substantially even within the same city-periurban environment. Furthermore, in the context of the urban environment, in many cities there are community gardens of different types. Of course, these gardens usually involve the production of food produce. However, there are other functions potentially such as providing opportunities for immigrants to become involved actively in these community gardens which can help some migrants integrate more effectively into the broader urban community.

And while generally the focus is on the production (comestible but also horticultural, e.g., flowers, bushes), some types of farms also have informal or sometimes even formal characteristics that attract urban citizens to visit the production units. This can include for instance being able to communicate directly with the farmers and other producers, as well as having a bakery on the farm which produces a variety of cooked food produce for sale, based on produce grown on the farm.

1.4 Urban agriculture and food security

Interest in food security is a relatively recent phenomenon in many countries, and it has emerged especially in many of the poorer neighbourhoods of major cities. It has also been recognized in many rural communities. Food security is a complex concept, based on four main components (Bryant and Bousbaine, 2019):

A. The volume of food production in relation to the demand of the population, especially in large cities and metropoles. The demand for food is becoming more and more oriented towards 'local' products, often because many people associate such local food produce with being synonymous with a so-called healthy diet, although that is not necessarily the case! Faced with these growing demands, the preservation of 'nurturing' farmland must be taken into consideration. An example of this is the small group of farmers in Senneville, at the western end of the Island of Montréal (see the location map of Fig. 8.1). The farmers there are principally people who had no family connections with farming before they became farmers. Interestingly enough, these farmers decided that they needed to take action to conserve their agricultural land as well to become more sustainable (see Figs 8.2 and 8.3 that show photos of farm houses and properties in the agricultural area of Senneville). They initiated this by asking Bryant to accompany them in their development which became in effect an Action Research project (Bousbaine and Bryant, 2015, 2016; Bryant et al., 2009). This case study is presented and discussed later in this chapter.

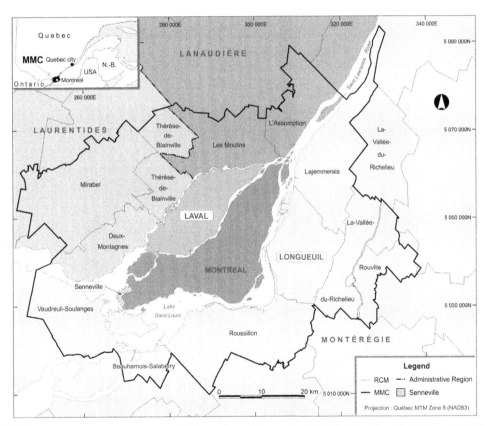

FIGURE 8.1   Senneville in the Montreal Metropolitan Community (MMC). *From Bryant, C. R., and Chahine, G. (2016). Action research and reducing the vulnerability of peri-urban agriculture: a case study from the M ontreal Region. Geographical Research 54(2), 165–175, Conception and Creation: Marc Girard, cartographer for the Department of Geography, University de Montréal.*

FIGURE 8.2   Farmland and a Greenhouse in Senneville. *From Bryant, C. R., and Chahine, G. 2016. Action research and reducing the vulnerability of peri-urban agriculture: a case study from the M ontreal Region. Geographical Research, 54(2) 165—175.*

FIGURE 8.3   Farm Buildings in Senneville. *From Bryant, C. R., and Chahine, G. 2016. Action research and reducing the vulnerability of peri-urban agriculture: a case study from the M ontreal Region. Geographical Research 54(2), 165—175.*

B. A diet that is convergent with human health. While such a diet can be quite variable, the key is that on the one hand, the agricultural production processes have limited negative effects upon the environment such as water resources (i.e., they are sustainable agricultures) as well as focusing on producing food produce that can contribute to human health. This requires substantial confidence on the part of consumers in the farmers' production processes. This certainly involves being able to visit a farm and exchange with the farmer and his or her employees which is essential to contribute to the building of confidence; as the consumer visits the farm or other production units, this confidence can become reinforced.

C. Accessibility to food, which is determined by various parameters such as actual food prices partly linked to production costs and sales costs, as well as the 'costs' of access for consumers (e.g., food prices in shops and the various types of short (food) circuits and the costs of accessibility to visit the food producers). Accessibility can also be linked to production structures that can be very different for the same food and the same territory ,e.g., productivist farms, family farms, agricultural cooperatives, community gardens, and formal systems versus informal production and sales systems. The issues involved in this multiple set of farm types have been recognized both in developed and developing countries (Bousbaine et al., 2019), and of course the variability can be even greater as soon as we take account of human values and priorities.

D. The existence of certain sources of stress that may affect food production and food security (e.g., climate change and variability, different forms of competition, new food production technologies). Focusing on the first stressor, climate change and variability, has become increasingly recognized as being real! But the effects of climate change and variability can vary substantially between different territories. Climate change and variability can also certainly affect urban agriculture (as well as, of course, the different dimensions of urban ecology). Climate change and variability can also affect competition since it can affect different agricultural territories to different degrees, even around the periurban areas of large urban agglomerations (Bryant et al., 2016a,b). Climate change and variability is finally being recognized as reality as citizens — including farmers — have become increasingly convinced of the reality of climate change and variability, particularly as many countries both in North America and Western Europe have had to face enormous hot spells in the last 2 years; and for some of the farming communities, e.g., in the province of Québec, farmers in many territories have had to deal with climate change and variability since the mid-1990s. Bryant undertook many years of research into how agriculture could manage to cope with climate change and variability particularly from 1990 onwards when he became a professor at the Université de Montréal.

One of the significant events that helped Bryant's research teams undertake research that ended up convincing many farmers of the reality of climate change and variability was when the FADQ (Financière Agricole du Québec — the Financial Agricultural Corporation of Québec) called Bryant at the Université de Montréal and offered to contribute to the research that had been underway since close to the beginning of the 1990s. The FADQ asked whether the research project needed financial support, but this was not an issue as the research team had research grants to support their research. The issue that was presented to the FADQ was the lack of adequate data on the history of farmers' experiences with coping with climate change and variability; it was known by the research teams that the FADQ had been managing crop insurance programmes for a long time. Thus, the FADQ was asked if it was possible for the FADQ to provide detailed data over a long period of time for the agricultural territories in which the research team was working. The FADQ agreed to do this for the regions where Bryant's research team was interviewing farmers and managing focus group discussions with the farmers. Then, after consecrating an important team of its members to fulfil this request, the FADQ provided detailed data over periods of 25—30 years concerning the requests of farmers to benefit from the FADQ agricultural insurance programme (Bryant et al., 2005). These data included the nature of the farmers' problems (e.g., lack of significant water resources, poor climatic conditions for crop growth, damage by insects), the effect of these problems on the farmer's crop production and the dates. These data were then analyzed by the research team, and the results in the form of temporal graphs were presented to groups of farmers in each region being studied … who finally saw what had been happening in their territory and they became convinced that they had to become more constructive in developing adaptation strategies to climate change and variability (Akkari and Bryant, 2016). The only requirement of the FADQ was that they asked to be informed about any presentations made by the research team or workshops managed for farmers to discuss adaptation possibilities, as well as any presentations made in conferences (and not only in the province of Québec). The FADQ thus accompanied the researchers when they were managing workshops for the farmers as well as when more academic presentations were made in conferences and colloquia.

This experience has also proven useful to farmers and food producers in urban agriculture in the broader Montreal region. This simply underscores the notion that although a particular stressor has long been present, pertinent information often needs to be analyzed and presented appropriately to reinforce the need for farmers (or other segments of the population) to decide they need to adapt. Some of the decisions taken by groups of farmers were quite remarkable; for instance, the farmers in one group began looking at the future patterns of climate in their territory that another research team had been working on and made available. They then looked at what their territory's climate would look like in 30 years' time. And then they looked across the USA to find territories where farming had been underway in a current climate that was very similar to what the group of Québec farmers had been informed that their own climate would look like in 30 years' time. They then sent a group of farmers from their territory in Québec to the US territory the current climate of which was similar to what their own territory's climate would be like in 30 years' time and exchanged talks with American farmers there to find out how the American farmers were coping with these different climatic conditions. Simple but original! And it just shows that people (farmers in this instance) who are not researchers can understand what is happening and identify appropriate adaptation strategies.

2. Sustainability of agricultural produce and food produce and the linkages between producers and consumers

Sustainable agricultures can be seen as agricultures that do not produce negative impacts of agricultural production technology (including fertilizers and other chemical inputs) on food production and on the environment (e.g., water quality). Around many cities, there have been major negative effects of productivist agriculture on water quality.

Not only must sustainable agricultures produce in such a way that reduces such negative impacts to a minimum but, in addition, when producing food produce, the processes used to create sustainable agricultures must ensure that this food produce is healthy. These projects can be either individual projects or collective projects. The development of sustainable agricultures can be planned, even to the point of using some form of strategic development planning to produce appropriate sustainable agricultural development (Bryant and Loudiyi, 2017).

A form of strategic development planning for agricultural spaces can be seen in Québec, one of only two Canadian provinces that initiated legislation to conserve agricultural land

(British Columbia in 1972, and Québec in 1978). In Québec, agricultural reserves were created. However, in several instances agricultural land was removed from some agricultural reserves by the provincial government or by certain municipalities (with the provincial government's approval). The Québec legislation was modified in the late 1990s to include the conservation of farm activities as well as agricultural land, but there was still pressure to remove agricultural land from some agricultural reserves. The situation began to change dramatically early in the new century when the Provincial government recommended that the Regional Municipal Counties (RCM) put together PDZAs (Plans de Développement des Zones Agricoles- or Development Plans for the Agricultural Zones), which were in effect similar to strategic development plans for agricultural development (Bryant and Loudiyi, 2017). The RCM were encouraged to involve a range of legitimate actors in developing the plans, including not only farmers and their associations, but also, for example, tourist agencies and different segments of the population. In most cases, these PDZA were successful and the involvement of a range of actors more or less ensured that the agricultural reserves would be left intact! Around Montreal there are also RCMs in which there are agricultural reserves with PDZAs.

Individual Projects generally involve one producer (and frequently his or her family members) and a set of consumers either formally where the consumers become members of a formal network (such as the AMAP in France) or informally where the consumers simply go to the store on the farm and purchase the produce they want. Sometimes individual farm projects can be quite complex. This is the case of the Laureau Farm west of the Chateau de Versailles, near Paris, France. Apart from having several farms across France, the one near Paris has for long been based on organically produced food. The farm also has a supermarket in which farm produce from other farmers practising sustainable agricultural production in the region is sold to consumers. Finally, the family leader (Mr. Xavier Laureau) has had his staff involved in managing workshops for primary school children on 'la bonne bouffe' (healthy foodstuff). Obviously, this could be seen as a method of marketing! But it is much more than that and it is also a way of educating youngsters to what healthy food produce consists of. Mr. Xavier Laureau himself is also involved in communicating with the different municipalities in the region as well as with organizing meetings with municipal officials and with ministry representatives to discuss issues in the region relating to agriculture, sustainable agricultures and healthy foodstuffs, and the maintenance of highly valued historical landscapes (Charvet, 2018). So once more, human values and priorities mean that not all farmers in the territory in which M. Laureau works and lives have necessarily the same values and priorities as he does.

Collective Projects have become increasingly important in some countries and around their cities. For instance, Food Land Belts with their multitude of collective and individual projects (e.g., the Ceinture Alimentaire de Liège (The Liège Food Land Belt) (CATL) (Bousbaine et al., 2019) and the Système Alimentaire Montréalais (the Montreal Food System) (SAM) which also includes a large number of individual agricultural and food projects). For these to function well, it is necessary that the colleagues managing the various individual projects are capable of working together and at least communicating effectively with each other when necessary. One of the additional constructive aspects of the food land belts is that when successful (Bousbaine and Bryant, 2018) they attract many consumers, i.e., nonfarm citizens for the most part. When these collective projects end up attracting large numbers of nonfarm citizens, they become also supporters of the maintenance of the agricultural lands which support such a project. And when this occurs, it becomes less and less likely that a municipality or even an upper level of government will be prepared to remove the land from its agricultural land-use category or agricultural reserve in order to support an industrial park or a new subdivision development.

3. The role of human values

Human Values (Table 8.2) are critical for understanding the development of sustainable agricultures and healthy food produce both on the part of producers (farmers …) and of consumers, as well as other actors, including local municipal officials, representatives of ministries of agriculture, researchers and many others (each geographical context can implicate some similar actors as well as some very different ones). It is important to remember that each actor (government officials and representatives, local farmers, researchers, among many other categories of actors …) has his or her own priorities and values. Often, this characteristic has been overlooked, even though some actors have had negative effects on particular projects which somehow did not appear to fit in with their personal values and priorities! In some instances, two cities in Belgium both with the aim to develop some form of agricultural and food produce ended up having enormous differences. On the one hand, the CATL has been remarkably successful with considerable support from the city and other municipalities, as well as from large numbers of citizens. On the other hand, the city of Charleroi where there was an attempt to develop some a similar sort of food production project has not been very successful, although there was an orientation towards the production of organic food produce. The problem appears to have been related to the history and culture of this city and its surrounding area. It had been a major industrial city and much of the employment in the city was based on industrial activities. But then these industries became subjected to significant competition and an important part of the employment base was lost. Despite

TABLE 8.2   Human values pertinent for urban agriculture from the perspective of urban ecology.

The value of locally and sustainably produced healthy food produce and the maintenance of agricultural spaces

- The values associated with being able to know the farm producers and their personnel and understand the values of these producers in relation to healthy and sustainably produced food; these values lie at the base of success of these projects

- Being able to understand the production processes and their effects on the quality of the food produced and the environment

- The capacity of certain agricultural spaces to offer physical exercises for nonfarmers and their dogs (e.g., on some of the Senneville farms at the western end of the Island of Montreal, where the footpaths near agricultural fields and the farmers themselves actually encourage this type of use close to their farmland)

- The conservation of agricultural spaces with the capacity to conserve environments with appreciated landscapes and wildlife notably birds, and in some contexts opportunities for fishing this, many people there still had 'dreams' of the possibility of the industrial employment base becoming renewed again. Such values and preoccupations have made it difficult for successful agricultural and food production projects to be established there.

4. The role of action research in developing sustainable agriculture and healthy food produce

Action Research has been directed frequently at different territories involving urban agriculture in urban environments in the last 25 years (i.e., urban agglomerations and periurban territories) (e.g., in the Montréal region (Bousbaine and Bryant, 2015; 2016; Bryant and Chahine, 2015; Bryant et al., 2009)), and at a more general level INRA (Institut National de la Recherche International, 2019).

Action research is a very different type of research compared to the 'traditional' type of research that has been practised for a long period of time; Action research necessitates the researcher taking on a role that traditional researchers have frequently not been comfortable with. This is partly because a researcher taking on an action research role must not take over what is being investigated and tell farmers what to do or how to improve things. The researcher following an action research process must never dictate to the farmers what they should do nor how they should construct their own sustainable agricultural production system or food project. The researcher must accompany the farmers and respond to their questions when the farmers pose them. This is not always easy for some researchers who have been used to being regarded as the 'expert', but it is up to the farmers to ask the pertinent questions and take control. For instance, Bryant received a visitor at his office at the Université de Montréal who asked him if a group of farmers in Senneville could come to his office at the Université de Montréal to ask him some questions about what they would like to achieve with their farms. After a few seconds of thinking, Bryant said 'No! ... I would prefer to go and visit them and their farms in Senneville and discuss what they would like to achieve'. Bryant had previously met and discussed with one of the farmers who had been present at a colloquium on agriculture at the University of Guelph which Bryant also attended and made a presentation.

Thus, Bryant went to Senneville and visited the farmers who told him they would like to be able to ensure that their agricultural lands would not be taken from them by the municipality (and if so, with the approval of the provincial government!), because they believed that their farming activities were an important source of foodstuff to many citizens, and they honestly believed that they should do all they could to ensure that they could continue to produce healthy foodstuff.

The farmers asked whether it would be useful to start the process by organizing a colloquium with a few presenters who had experience in the domain and if it would be possible to invite a few dozen participants to attend the colloquium and also participate in workshops. It was an excellent idea and so this was followed up by Bryant and the two interns who at the time were in his laboratory. The results regarding the colloquium were amazing. Initially it was expected that together (through the farmers' contacts and through our own contacts) we could invite about 40 participants and hold the colloquium in the nearby Ecomuseum.

But as the time passed, we ended up with over 100 people who really wanted to attend! The Ecomuseum was not large enough, so the Dean of the Faculty of Agriculture at that part of McGill University that was based in St. Anne de Bellevue (also at the west end of the Island of Montreal) was contacted and he made available a large amphitheater for the colloquium as well as several smaller rooms for the workshops that took place in the afternoon.

The results were quite constructive. The farmers of Senneville suggested we could organize some committees involving them and other actors with an interest in what they were trying to do. This was undertaken and several meetings were held over the next year or so. The results were quite amazing, particularly when presentations were prepared and delivered by one or two of the farmers to the municipality and to the organization interested in a Green Belt for the Montréal region! This was so unusual and unexpected but the farmers were remarkably successful. And the farmers of Senneville carried on farming; they are also involved with Santropol Rouland which is based near the centre of the City of Montreal; this is a social organization with access to a small amount of farmland in Senneville, and is helped by the farmers at Senneville. This organization produces foodstuff for its members; and those who are well-off pay a higher price for the foodstuff while members who are 'in need' pay much less, i.e., one group of members subsidizes the others — a real contribution to food security; this social organization has also received grants from the provincial government.

5. Conclusions

Urban agriculture has become a major phenomenon in many urban cities and their surrounding periurban territories. In fact, for many people urban and periurban territories are both part of urban (see the work of Lohrburg et al. in the production of the book *Urban Agriculture Europe* published in 2016). At the time, the project upon which this book was based was undoubtedly the largest urban agriculture project ever conceived, with 26 countries involved, an even larger number of universities and their researchers, and with a small group of researchers who were specialists in this field, from Canada, Japan, Cuba Africa. Nowadays, Urban agriculture is increasingly recognized as a significant recognition of an enormous reality which is highly varied in terms of its structure and orientation, e.g., to sustainable agricultures and/or to healthy produce, or to using farm practices for the education of young children (e.g., the Laureau Farm just west of the Chateau de Versailles, near Paris, France). Urban Agriculture has become a fundamental part of urban ecology as well and it contributes to many of the roles and functions of urban ecology. Perhaps one day the two will merge together! One dimension that appears more and more important is the human values associated with urban agriculture, or perhaps we should say urban agricultures. These human values are partly the values associated with the producers (e.g., farmers, operators of community gardens, rooftop garden managers or owners), as well as people who become involved in the sale of farm produce especially directly to consumers. And of course, the consumers occupy a critical role because they are able to choose which farmers or producers they buy food produce from. Producers can clearly influence consumers. Families with children can be encouraged to visit some farms and purchase food produce, but they can also be easily encouraged to visit certain farms and farm shops when, for instance, the farm has an orchard near the farm store large enough for the farmer to organize visits for the parents and their children to visit the orchard in a wagon pulled by a horse, and/or when the farmer has organized a small fenced in 'zoo' with small farm animals that attract children to come over and look at them (e.g., just to the west of the Island of Montreal). Furthermore, there are also urban agglomerations in which there are rooftop gardens and businesses, such as in New York City … and even in the Montréal urban agglomeration, where there are currently now several rooftop gardens and businesses. The first one also delivers baskets of food produce (and flowers and other horticultural products) to clients in the agglomeration. It is almost as if one needs to keep an eye on all the neighbourhoods in the urban agglomeration and the surrounding periurban areas because potentially there are so many dimensions of the farms and food production processes that can be subject to transformation or replacement.

References

Akkari, C., Bryant, C.R., 2016. The co-construction approach as approach to developing adaptation strategies in the face of climate change and variability: a conceptual framework. Agricultural Research. https://doi.org/10.1007/s40003-016-0208-88, 2016, March online, 1-12. ISSN 2249-720X.

Bousbaine, A.D., Bryant, C.R., 2015. The integration of action research and traditional field research to provide sustainable solutions to maintaining periurban agriculture. Geographical Research (for a Special Issue: Rural Action Research), 2015. https://doi.org/10.1111/1745-5871.12134. On-line, Wiley and Sons). ISSN: 1745-5871.

Bousbaine, A.D., Bryant, C.R., 2016. Action research: an essential approach for constructing the development of sustainable urban agricultural systems. Challenges in Sustainability 4 (1), 20−27. https://doi.org/10.12924/cis2016.04010020, 2016. ISSN 2297-6477. Available freely on line: http://www.librelloph.com/challengesinsustainability/article/view/cis-4.1.20.

Bousbaine, A.D., Bryant, C.R., 2018. Agri-food projects in food land Belts: conditions for success (editorial), current Investigations in Agriculture and current research. Current Investigation in Agriculture and Current Research 2 (1). https://doi.org/10.32474/CIACR.2018.02.000126, 1/3. 2018. CIACR.

Bousbaine, A.D., Akkari, C.J., Bryant, C.R., 2017. Strategic development planning for agricultural development and the integration of other domains important for the territory. International Journal of Avian & Wildlife Biology 2, 6−2017.

Bousbaine, A., Nguendo-Yongsi, B., Bryant, C., 2019. Urban agriculture in and around cities in developed and developing countries: a conceptualization of urban agriculture dynamics and challenges. In: Thornton, A. (Ed.), Urban Food Democracy and Governance in North and South. 2019, International Political Economy Series. Palgrave MacMillan, 10.1007/978-3-030-17187-2.

Bryant, C.R., Bousbaine, A.D., 2019. La sécurité alimentaire et l'agriculture urbaine au Québec : leurs composantes et variabilités. In: Book : Pour la sécurisation alimentaire au Québec : perspective territoriale, Presses de l'Université du Québec, M. Doyon et J.-L. Klein, pp. 77−86.

Bryant, C.R., Chahine, G., 2015. Action research and reducing the vulnerability of peri-urban agriculture: a case study from the Montreal Region. Geographical Research 1−11. https://doi.org/10.1111/1745-5871.12119. On-line Wiley and Sons). ISSN: 1745-5871.

Bryant, C.R., Loudiyi, S. (Eds.), 2017. Des espaces agricoles dans la metropolisation. Perspectives franco-quebecoises ? L'Harmattan, p. 330.

Bryant, C.R., Singh, B., DesRoches, S., Thomassin, P., Baker, L., Madramootoo, C., Délusca, K., Savoie, M., 2005. Climate variability in Quebec: lessons for farm adaptation from an analysis of the temporal and spatial patterns of crop insurance claims in Quebec. National Conférence on: Adapting to Climate Change in Canada 2005: Understanding Risks and Building Capacity, Natural Resources Canada, Montréal. May 4-7, 2005.

Bryant, C.R., Chahine, G., Saymard, È., Poulot, M., Charvet, J.-P., Fleury, A., Vidal, R., Loudiyi, S., 2009. The Direct Contribution of Research to Modifying Spatial Patterns of Local Development: Action Research to Reduce Vulnerabilities and Re-build Agricultural Activity in the Urban Fringes of Montreal and Paris. In: Proceedings of the 40th Conference of the Mid-continent Regional Science Association and the 31st Annual Conference of the Canadian Association of Regional Science, pp. 67−78. Milwaukie, 28 to 30 May 2009), 2009.

Bryant, C.R., Diaz, J.P., Karaita, B., Lohgberg, F., Yokohari, M., 2016a. Urban agriculture from a global perspective (2016). In: Lohrberg, F., Lica, L., Scazzosi, L., Timpe, A. (Eds.), Urban Agriculture Europe, Chapter 1.3, pp. 30–37 (Jovis).

Bryant, C.R., Bousbaine, A.D., Akkari, C.J., Daouda, O., Delusca, K., Épule, T.E., Sarr, M.A., Drouin-Lavigne, C., 2016b. Climate Change and Global Food Security in the Face of Other Stressors: The Challenges for Agricultural Transformation, Adaptation and Conservation. Presentation Made at the World Conference on Climate Change, 24-26 October 2016, Valencia.

Bryant, C.R., Bousbaine, A.D., Akkari, C.J., 2018. The complexities of sustainable agricultures in and around urban agglomerations. Nutri Diet Probiotics 2018 1 (2), 5, 180010.

Bryant, C.R., Bousbaine, A.D., Akkari, C.J., 2019. The conservation of agricultural land and farm activities in Canada. *ASAG* (Acta Scientifique Agriculture), ASAG-19-CON-089, 3 (4), 162.

Caplat, J., 2014. Changeons d'agriculture - Réussir la transition. Actes Sud, Paris, p. 160.

Chapin III, F.S.», Zavaleta, E.S., Eviner, V.T., Naylor, R.L., Vitousek, P.M., Reynolds, H.L., Hooper, D.U., Lavorel, O.E., Hobbie, S.E., Mack, M.C., Diaz, S., 2000. Consequences of changing biodiversity. Nature 405, 234–242.

Charvet, J.-P., Laureau, X., 2018. Révolution des Agricultures Urbaines, des Utopies à la Réalité – Vers des Métropoles Agri-Urbaines. Éditions France Agricoles, p. 202.

Charvet, J.-P., 2018. Atlas de l'agriculture : Mieux nourrir le monde.

Donald, P.F., Sanderson, F.J., Ian, J., Burfield, I.J., van Rommel, F.P.J., 2002. Further Evidence of Continent-wide Impacts of Agricultural Intensification on European Farmland Birds, 1990–2000, vol. 116. Agriculture, Ecosystems & Environment. Issues 3–4, September 2006. 189-19.

Fresco, Smaling, E.M.A., Fresco, L.O., 1993. A decision - support model for monitoring nutrient balances under agricultural land use NUTMON. Geoderma 60 (1-4-2017), 235–256.

Goulet, F., 2008. De tensions épistémiques et professionnelles en agriculture. Revue d'anthropologie des connaissances. 2008/2 2, 291–310 n° 2).

Green, E.M., Antczak, A.J., Bailey, A.O., Franco, A.A., Wu, K.J., Yates, J.R., Kaufman, P.D., 2005. Replication-independent histone deposition by the HIR complex and Asf1. Current Biology 15 (22), 2044–2049.

Groombridge, B. (Ed.), 1992. Global Biodiversity. Status of the Earth's Living Resources. A Report Compiled by the World Conservation Monitoring Centre. Chapman & Hall. Science, xx + London, Glasgow, New York, Tokyo, Melbourne, Madras, p. 585.

INRA (Institut National de la Recherche International), 2019. A Series of Publications on Participatory Research (Research Action) (consulted again on July 21st. https://sciencesparticipatives.inra.fr/publications/.

(IPCC) Intergovernmental Panel on Climate Change, 2014. Working Group 11, AR5 Climate Change 2014: Impacts, Adaptation, and Vulnerability.

Kleijn, D., Sutherland, W.J., 2003. How effective are European agri-environment schemes in conserving and promoting biodiversity? Journal of Applied Ecology 40 (6), 947–969.

Krebs, J.R., Wilson, J.D., Bradbury, R.B., Siriwardena, G.M., 1999. The second silent spring? Nature 400, 611–612.

Lambin, E.F., Geist, H.J., Lepers, E., 2003. Dynamics of land-use and land-cover change in tropical regions. Annual Review of Environment and Resources 28, 205–241.

Larrère, C., Larrère, R., 1997. La Crise Environnementale. Éditions de l'INRA, Paris, p. 302, 1997.

Lohrberg, F., Lica, L., Scazzosi, L., Timpe, A. (Eds.), 2016. Urban Agriculture Europe, Jovis, p. 231.

F. Lufa. Montréal (web site visited on July 21st, 2019). https://montreal.lufa.com/fr/.

MEA: Millenium Ecosystem Assessment, 2005. Montréal community gardens (visited July 21st, 2019). http://ville.montreal.qc.ca/portal/page?_pageid=5977,68887600&_dad=portal&_schema=PORTAL.

Mzoughi, N., Napoléone, C., 2013. L'écologisation, une voie pour reconditionner les modèles agricoles et dépasser leur simple évolution incrémentale. Introduction. Article in: Natures Sciences Sociétés · April 21, 161–165, 2013.

Robinson, R.A., Sutherland, W.J., 2002. Post-war changes in arable farming and biodiversity in Great Britain. Journal of Applied Ecology 39 (1), 157–176.

Santropol Rouland (Montréal). (consulted on July 21st, 2019). https://santropolroulant.org/en/.

TEEB, 2009. The Economics of Ecosystems and Biodiversity.

Watson, P.A., Vitousek, P.M., 2012. Principles of Terrestrial Ecosystem Ecology. Book. Springer (Second Edition): ISBN 978-1-1007/978-1-4419-9504, 271.

Carbon reduction strategies for the built environment in a tropical city

Dilawar Husain[1], Ravi Prakash[2]

[1]Department of Mechanical Engineering, School of Engineering and Technology, Sandip University, Nashik, India; [2]Department of Mechanical Engineering, Motilal Nehru National Institute of Technology, Allahabad, Uttar Pradesh, India

OUTLINE

Urban Ecology
https://doi.org/10.1016/B978-0-12-820730-7.00009-4

1. Introduction

Urbanization is the most influencing factor responsible for huge energy and natural resource consumption along with waste generation, greenhouse gas (GHG) emissions etc (Mahapatra et al., 2017). According to the Paris Agreement (UNFCCC, 2015), all member countries have committed to reducing GHG emissions in order to limit the global average temperature increase to below 2°C. Sustainable urban development can help to meet this goal because the building sector is responsible for about 40% (Becqué et al., 2016) of the materials consumed globally; it is also related with about 32% of the total global energy consumption and 19% of energy-related GHG emissions (Lucon and Ürge-Vorsatz, 2014). The construction industry in India is expected to grow annually at the rate of 5.6% during the period 2016–20, and it may further grow up to 7.1% per year by 2025 (Invest India, 2018). In India, one-quarter primary energy and one-third of electrical energy are consumed in buildings (Wagner et al., 2007; Ramesh et al., 2013); with commensurate GHG emissions. Therefore, energy efficiency is now considered an essential requirement for all new and existing buildings.

Some previous studies suggested that the selection of low environmental impact building materials has significant potential to reduce CO_2 emissions during the construction phase of the building (González and García Navarro, 2006; Husain and Prakash, 2019a). Many studies have emphasized the dominance of the operation stage in life cycle carbon emissions assessment irrespective of a country's climatic conditions (Ramesh et al., 2010; Varun et al., 2012). Azzouz et al. (2017) investigated that operational carbon and energy consumption of a building is 6.8 and 10.5 times higher than their embodied carbon and energy, respectively. Suresh et al. (2017) investigated the annusal emission (operational) of TERI University, Delhi, and found that there is nearly 0.72 tCO_{2e} emission per capita in the university campus.

Various techniques may be utilized to reduce the environmental impacts of buildings. Omar and Mahmoud (2018) have studied the technical performance and economic feasibility of grid-connected rooftop solar PV systems in Palestine and have reported that the system has a payback of 4.9 years and an annual yield of 1756 kWh/kWp. Earth air tunnel (EAT) system is a prominent passive technique used for heating and cooling applications of the buildings as well as it reduces the building-related GHGs emission (Singh et al., 2018). Some studies suggest that 20%–50% of the cooling/heating energy consumed in the building is due to the built envelope (Yu et al., 2015; Bano and Kamal, 2016). The operational environmental impact reduction potential per unit area of a green roof compared to a conventional roof is in the range of 62%–67% (Husain and Prakash, 2019b).

This study focuses on some carbon reduction strategies for the built envelope of a tropical urban area by examining four representative buildings in the city of Prayagraj (Allahabad), Uttar Pradesh, India. According to the Global Footprint Network (GFN, 2016), Indian resource demands have already surpassed the available biocapacity of the country. The building sector in India has significant potential to reduce its resource consumption by adopting sustainable systems.

2. Case studies

This study focuses on some carbon reduction strategies for the built envelope of a tropical urban area by examining four representative buildings in India. Through simulation models using eQuest software (TQEST, 2016) and RETScreen software (Natural Resources Canda

NRC, 2016), the impact of renewable energy and energy conservation on four different types of buildings has been examined. The technical feasibility, economic viability, and carbon reduction potential of solar PV systems, wind turbines, solar hot water, efficient lighting (LED bulbs), and EAT for the city buildings have been evaluated and presented as follows:

2.1 Building type: school (primary education)

2.1.1 Site details

The first case study has been performed for Bal Bharti Vidyalaya Junior High School building, located at Motilal Nehru National Institute of Technology (MNNIT), Allahabad, Prayagraj (U.P.) India. The geographical location of the school campus is 25.45°N latitude, 81.73°E longitude. The school building consists of 18 classrooms, 1 office room, 1 staff room, 1 corridor and 1 dining room. The working schedule of the classrooms is 6 days/week with 6 h/day from 7:00 a.m. to 1:00 p.m. in summer season and 8:00 a.m. to 2:00 p.m. in the winter season. Bal Bharti Vidyalaya Junior High School building and 3D simulation model of school building are shown in Fig. 9.1A and Fig. 9.1B, respectively.

The internal loads of the building are interior lighting load and miscellaneous loads. For interior lighting, there are 2 bulbs of 100 W in each room and 5 bulbs in the corridor. For equipment loads, 3 fans of 60 W have been used in each room. The total floor area of the school building is about 2500 ft^2. The loads are reduced by using efficient lighting in place of conventional bulbs. This reduced load is the input for the modelling of renewable energy systems. The detailed methodology is depicted in Fig. 9.2. For the proposed model analysis, 6 kW off-grid rooftop solar PV system has been proposed to meet the required energy consumption for the school building.

2.1.1.1 Climatic conditions

For techno-economic feasibility of the proposed system, accurate climate data of Prayagraj (Allahabad), India, is required. The solar radiation and wind speed (at 10 m height) are 5.79 kWh/m^2/d and 2.5 m/s for Prayagraj (Allahabad) city. In summer, ambient air temperature reaches up to 50°C during peak hours, while in winter, it drops to freezing point. In the year of 2018, the annual recorded precipitation/rainfall has been 980 mm. The relative humidity varies between 29% and 81%.

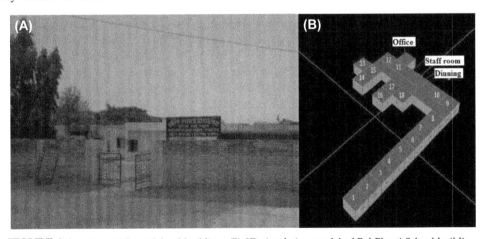

FIGURE 9.1   (A) Bal Bharti School building, (B) 3D simulation model of Bal Bharti School building.

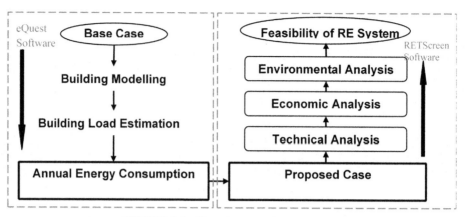

FIGURE 9.2   Flow diagram of methodology.

2.1.2 Results

The total annual electricity consumption in lighting and miscellaneous loads for the school campus is 11.63 MWh. To save the electrical energy, 100 W bulbs are replaced by 10 W LEDs, to save about 90% of the required lighting load. After proposing the efficient lighting, the load becomes 9.40 MWh which is about 19.2% reduction in base model load. Thirty-one units of solar PV panels (195 W_P capacity each) have been proposed for meeting the required load. The technical details of the proposed solar PV power generation system are given in Table 9.1. This system annually generates 11.3 MWh electricity, in which 9.40 MWh is utilized for building consumption and excess energy of 1.9 MWh can be useful for cooking applications for a mid-day meal or local area lighting during night time.

The use of renewable energy for producing electricity is advantageous in the reduction of CO_2 emission. In India, the amount of CO_2 emission is 0.91 tonne per MWh grid electricity generation. Also, there are transmission and distribution losses which are about 23.04% in India (Ministry of power, 2016); hence the total CO_2 emission becomes 1.1 tonnes per MWh electricity. For this case study, the annual GHG emission is 13.5 tCO_2 and it is reduced by renewable energy systems. The annual saving includes the cost of electricity produced and the excess electricity produced by renewable energy systems. The total initial cost of the system is about Rs. 0.604 million with annual savings of Rs. 74,080. A project is feasible when the Net Present Value (NPV) is positive. The NPV value of the system is about Rs. 0.312 million, which is an indication of the feasibility of the proposed system. The payback period of the system is 8.5 years, with the system's life of 25 years as shown in Fig. 9.3.

2.2 Building type: temple complex

2.2.1 Site details

The second case study has been performed for ISKCON Temple, Prayagraj (Allahabad), India. The geographical location of the complex is 25.43°N latitude, 81.84°E longitude. The total floor area of the temple complex is 5145 ft^2. It consists of the main temple building block (400 ft^2), office building block (1500 ft^2), kitchen (200 ft^2), and restroom building block

TABLE 9.1 The detailed technical specifications of the solar PV system.

S.N.	Power systems	Variables	Specification
1.	Solar PV panel	Manufacturer	Trina solar
		PV module type	Mono-Si
		Module number	TSM-195DC/DA01A
		Power capacity	195 W
		Efficiency	15.3%
		Frame area	1.28 m^2
		Life	25 years
2.	Battery	Manufacturer	Exide
		Model	12 V, 150 Ah
3.	Charge controller	Manufacturer	Sungrow
		Model	GSB-B-16kw220v
4.	Inverter	Capacity	6 kVA
		Efficiency	97%
		Losses	3%

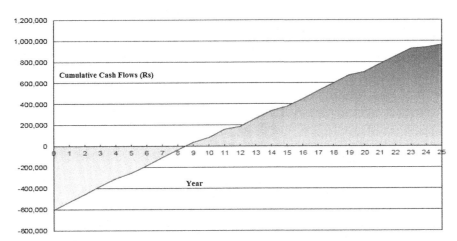

FIGURE 9.3 Payback period of the system using Solar PV system.

(4120 ft^2). The ISKCON temple consists of one single storey main temple building block (20 × 20 × 11 ft^3 in dimension) attached with kitchen (20 × 10 × 11 ft^3 in dimension), restroom of 22 rooms (each room of 12 × 12 × 10 ft^3 in dimension) with two-storey building block, and one single-storey office building block of five rooms (each of 10 × 10 × 10 ft^3 in dimension).

The total floor area of the built environment is 5145 ft^2 with a capacity of 108 persons during peak occupancy. Common brick and cement mortar are used for wall construction of the building. The roof is constructed by 4 inch concrete with cement mortar plaster. The floor is in earth contact with 4 inches thick concrete and ceramic tile. The exterior doors are wooden door with wooden frame in which single core wooden block has been used. The constructional details of the building envelope are given in Table 9.2 and photographs of the ISKCON temple are shown in Fig. 9.4A.

Internal loads like occupancy, lighting power density and equipment power density are also important parameters that affect a building's energy consumption. The data is mostly collected via surveys. Human occupancy can be assessed based on experimental information set by zone type: The maximum design occupancy of a restroom is 100 sq. ft/person, the main temple building is 8 sq. ft/person, the kitchen is 50 sq. ft/person and office building is 200 sq. ft/person. Lighting power density is set by zone type: 0.04 W/ft^2 in the main temple building, 0.5 W/ft^2 in the office building, 0.3 W/ft^2 in the restroom building and 0.1 W/ft^2 in the kitchen. Equipment power density is set by zone type: 0.17 W/ft^2 in the main temple building, 0.5 W/ft^2 in the office building, 0.3 W/ft^2 in the restroom building and 0.1 W/ft^2 in the kitchen. In the proposed case system, earth air tunnels are used to provide thermal comfort in the built environment. In addition, 100 W bulbs and 27 W CFL bulbs are proposed to be replaced by 10 W LED bulbs, which minimize lighting load. A centrifugal blower of 1 hp power is used in the earth air tunnel. It is important to note that all appliances are assumed to be used at 30%–100% activities of the occupants throughout the day.

TABLE 9.2   Constructional details of building envelope.

S.N	Variables	Materials	U-value (W/m^{2}K) (US Department Of Energy, 2016)
1.	Wall	1 inch cement plaster, 9 inch brick, 1 inch cement plaster	1.86
2.	Roof	1 inch cement plaster, 4 inch concrete, 1 inch cement plaster	2.035
3.	Floor	4 inch concrete, interior finish with ceramic/stone tiles	6.91
4.	Exterior doors	Wooden doors with 2.5 inch frame width	6.6
5.	Exterior windows	Single clear (1001) glass type windows with 2.0 inch frame width	2.95

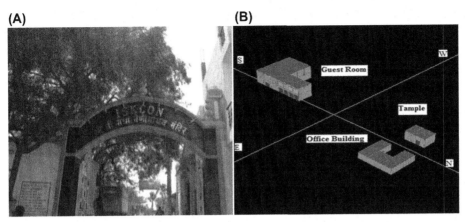

FIGURE 9.4 (A) Image of ISKCON temple entrance, (B) 3D model of the building (north-east side view).

The 3D view of the ISKCON temple building generated by eQuest simulation software is shown in Fig. 9.4B. Fig. 9.5 shows the schematic of the hybrid power system (HPS) considered for the proposed case. Solar and wind power systems are combined together to supply clean energy and at the same time to compensate for the irregular variation in sunlight or wind speed. For modelling of the HPS, the following input data were used:

(i) climatic data of Prayagraj (Allahabad) city in order to predict energy outputs of the HPS
(ii) the annual load profile of ISKCON temple is used for an adequate size of the HPS
(iii) the component cost of HPS to meet the proposed load

The proposed HPS of total capacity 16 kW (solar PV on the temple and office roofs with wind turbines on double-storey restroom block) can successfully meet all loads throughout the year such as lights, fans, air blower, water cooler, etc. The required wind speed

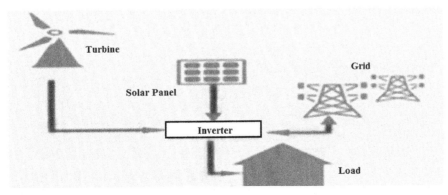

FIGURE 9.5 Proposed hybrid power system.

(2.5 m/s) for the selected wind turbines is available at a height of 10 m above the ground. Hence, wind turbines are proposed to be installed at the second storey of the temple complex. Under normal mode of operation, the solar PV-Wind system will supply electricity to the load simultaneously. Any excess energy generated by the hybrid system will be transferred to the grid that provides additional savings. If there is insufficient power supply from the hybrid power system, the grid will automatically supply the deficit power to the load. The details of all the components of the proposed case systems are shown in Table 9.3.

Three modes of HPS are considered in this case study, which can generate sufficient power throughout the year for the proposed case. In the first mode (mode 1), solar PV-Wind HPS is connected in the ratio of 75%:25% of total power capacity. Solar PV of 12 kW (185 W_p peak

TABLE 9.3 Component specifications of proposed system.

S.N.	Power systems	Variables	Specification
1.	Solar PV panel[a]	Manufacturer	Trina solar
		PV module type	Mono-Si
		Module number	TSM-195DC/DA01A
		Efficiency	15.3%
		Frame area	1.28 m^2
		Peak power capacity	185 W
		Life	25 years
2.	Wind turbine[a]	Manufacturer	ReDriven
		Model	2 kW
		Rated output voltage	240 V, 60 hz, 1 phase
		Start-up wind speed	2.5 m/s
		Cut out speed	18 m/s
		Survival speed	40 m/s
		Rated wind speed	10 m/s
		Hub height	30 m/s
		No of blades	3
3.	Inverter[a]	Manufacturer	Su-kam
		Capacity	16 kVA
		Efficiency	90%
		Losses	5%
4.	Earth air tunnel[a]	Blower	1 hp, 4000 cfm (6796 m^3/h), (Rs. 18,000)
		PVC pipe	0.152 m diameter, 20 m long, (Rs.500/piece)

[a]*Cost has been taken from local market.*

power capacity of each solar PV panel) and two wind turbines of 4 kW (2 kW power capacity of each wind turbine) are connected in series for supplying the required power to the load. In the second mode (mode II), solar PV-Wind HPS is connected in the ratio of 50%:50% of total power capacity. Four wind turbines of 8 kW and solar PV panels of 8 kW are connected in series for supplying the required power to the load. In the third mode (model III), solar PV-Wind HPS is connected in the ratio of 25%:75% of total power capacity. Solar PV of 4 kW and six wind turbines of 12 kW are connected in series for supplying required power to the load.

2.2.2 Results

The energy simulation of the base case and proposed case systems by eQuest provides annual electricity consumption as 23.49 MWh and 21.12 MWh, respectively. The monthly electricity consumption is shown in Fig. 9.6. The difference in annual electricity consumption in proposed case system is 10.09%. However, it will increase the thermal comfort of the built environment due to the use of earth air tunnel (Bansal et al., 2009). In this case study, mode I, mode II and mode III HPS can annually generate 25.08 MWh, 22.73 MWh, and 21.12 MWh electrical energy, respectively. The mode I and mode II annually generate excess electrical energy of 3968 kWh and 1615 kWh, which can be supplied to the grid. Mode III is just sufficient to meet the required load. This indicated that mode I HPS gives better result based on electricity generation.

A significant amount of CO_2 emissions are reduced by installing different modes of hybrid power systems in place of conventional grid electricity. The potential reduction in annual CO_2 emissions of proposed case system with mode I, mode II and mode III (HPS) are evaluated as 36.2 tCO_2, 31.2 tCO_2 and 29.8 tCO_2, respectively. Therefore, proposed case system with mode I (HPS) offers the best alternative for CO_2 reduction. For the proposed case system, initial cost, annual savings and the payback period for the three HPS modes are shown in Fig. 9.7. The results indicate that all proposed HPS systems are feasible for 25 years lifetime. The mode I HPS gives better result based on initial cost and annual saving.

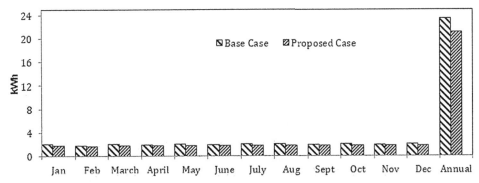

FIGURE 9.6   Annual and monthly electricity consumption of simulation models for ISKCON temple.

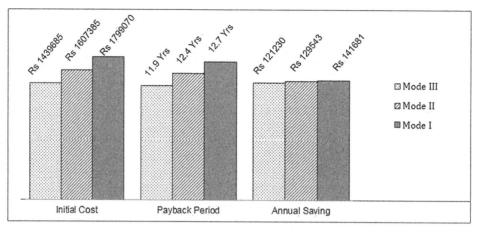

FIGURE 9.7   Economic analysis data of proposed systems.

2.3 Building type: college (secondary education)

2.3.1 Site details

Building taken for this case study is Vishnu Bhagwan Public School building that is located at Jhalwa, Prayagraj (Allahabad) (U.P.), India. The geographical coordinates of the building are 25.45°N latitude, 81.73°E longitude. The school building has two blocks in which one four-storey building block is used for studies and a second three-storey building block is used for residential purposes. The building orientation is south facing and the total floor area is18,500 m^2. The school building image is shown in Fig. 9.8A. The detailed information about internal loads of both building blocks is depicted in Table 9.4.

The parameter settings involved were south-facing envelope model, common brick walls used for building wall construction, reinforced cement concrete (RCC) M20 used for roof construction and building envelope of wooden-based doors and simple glass windows. The simulated 3D envelope model of the school building blocks is shown in Fig. 9.8B. After

FIGURE 9.8   (A) Vishnu Bhagwan public school building; (B) 3D model of building envelope.

TABLE 9.4 Internal load of building blocks.

Attributes	Unit	Block 1 (Classroom)	Block 2 (Hostel)
Lighting load	W/sq. ft	2.93	2.5
Hot water system	kWh/sq. ft	0	1.41
Water cooler	W/sq. ft	3.59	40.76
Miscellaneous load	W/sq. ft	4.13	6.52
Water pump	W/sq. ft	6.52	4.13

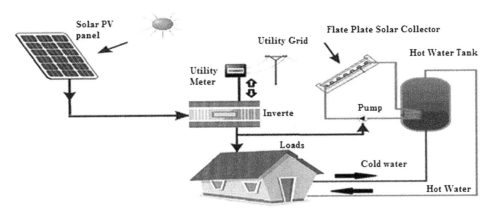

FIGURE 9.9 Schematic diagram of hybrid system.

estimation of the electricity and hot water demand of the school building, the proposed grid-connected solar PV with solar hot water hybrid system (Fig. 9.9) assessment has been done to meet the building demand. The technical details of integrated solar PV system with solar hot water hybrid system are given in Table 9.5.

2.3.2 Results

The electricity consumption of existing building has been estimated; then the feasibility of the proposed hybrid system is analyzed on the basis of energy, economic, and environmental assessment. The annual electricity consumption in lighting and water heating are 37 MWh and 31 MWh, respectively. The proposed solar PV power system can generate 37.4 MWh electricity annually, while 0.4 MWh electricity is required for water circulation in the solar hot water system. Due to reduced grid electricity consumption for the proposed hybrid system, the carbon emissions may be reduced by 86.6 tCO$_2$ annually. The total cost of the proposed hybrid system is nearly Rs. 4 million, while annual savings are 1.05 million due to reduced consumption of grid electricity (purchased at the rate of Rs 7.30/kWh). Therefore, the simple payback period of the proposed hybrid system is nearly 4 years.

TABLE 9.5 Component specifications of hybrid system.

S.N.	Power systems	Variables	Specification
1.	Solar PV panel	Manufacturer	Trina solar
		PV module type	Mono- Si
		Module number	TSM-195DC/DA01A
		Efficiency	15.3%
		Frame area	1.28 m^2
		Peak power capacity	185 W
		Life	25 years
2.	Solar hot water System	Manufacturer	ARINNA
		System capacity	7000 LPD
		System output temperature	55–60°C avg.
		Solar collector area on terrace	140 m^2 (approx.)
		Collector type	Flat plate parallel flow type
		Gross area	2.01 m^2
		Absorber material	Copper (absorptive >90%; Emissivity <12%)
		Transparent cover (glass)	Tempered low iron glass 4 mm (transmissivity >88%)
		Insulation material	Polyurethane + glass wool 32 kg/m^3
3.	Inverter	Manufacturer	Su-kam
		Capacity	16 kVA
		Efficiency	95%
		Losses	5%

Cost of the components have been taken from local market.

2.4 Building type: public auditorium

2.4.1 Site details

This case study examines a public auditorium located at M.N.N.I.T, Prayagraj (Allahabad) (U.P.), India, which is used from Monday to Friday from 10:00 a.m. to 09:00 p.m. throughout the year. Energy consuming equipment is powered on and off according to the workday. The building has a single storey, east-facing orientation, pitched roof (angle 8 degrees) and has a floor area of 11,775 ft^2 with 100% air-conditioning using conventional systems. The building image is shown in Fig. 9.10A. The HVAC (heating, ventilation and air conditioning) system in this building consists of seven packaged air conditioners (working refrigerant R22 ($CHClF_2$), 25 TR cooling capacity each); which provide conditioned air through the insulated ducts to the ceiling of the hall. These packaged AC units are only meant for cooling and cannot provide thermal comfort during winters.

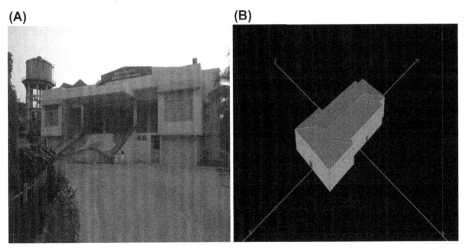

FIGURE 9.10   (A) Multipurpose hall building, (B) 3D model of multi-purpose hall building.

The software parameter settings involved an east-facing building envelope model, common brick walls used for building wall construction, asbestos cement board used for roof construction, gypsum board used for ceiling construction and building envelope consists of six wooden-based doors facing one east, one north, two west and two south with no windows. Full details are shown in Table 9.6 and a 3D simulation model of the building is shown in Fig. 9.10B. The building HVAC system is integrated with packaged 175 TR cooling capacity air conditioners (25 TR cooling capacity, 9.1 EER (COP = 2.66), 10,000 CFM air circulation each), and cold air supply to the built environment is through the ducts. The HVAC settings in eQuest include the type of air conditioner, cooling capacity, temperature control, which are similar to actual HVAC system.

TABLE 9.6   Details of the building envelope.

Variables	Description	Overall U-value (W/K.m²) (US Department Of Energy, 2016)
External wall	Brick wall. Thickness: 0.23 m, 12 mm plywood insulation with 9 inch air cavity	0.717
Roof	Asbestos cement board. 12 mm thick, roof pitch angle 8 degree.	0.163
Ceiling	Gypsum board. Thickness: 12 mm with metal frame	0.218
Floor	4 inch concrete, interior finish with ceramic/stone tiles	6.91
Exterior door	Wooden doors with 2.5 inch frame width	6.6

The assessment simulation tool RETScreen software which is used for preliminary evaluation of the energy feasibility and financial viability of potential offgrid connected wind systems is used. Hub height of 10 m is suggested for all turbines because wind speed tends to increase with hub height and is associated with higher wind power output. Monthly wind speed data is used in the RETScreen simulations to capture variability in electricity production at the building location. For energy calculations, airfoil losses and miscellaneous losses are taken as 2% and 5%, respectively. For the proposed model, electricity would be generated by seven ReDriven wind turbines of 2 kW capacity, which operate throughout the year. The reason for choosing the specific wind turbine is its availability and suitable performance characteristics (like power curve data) such as 2.7 m/s cut-in speed.

2.4.2 Results

The comparison between base case (HVAC system, lighting, miscellaneous loads) and proposed case (earth air tunnel, LED lighting, miscellaneous loads) has been done and the monthly electricity consumption obtained by energy simulation modelling. The annual electricity consumptions are 217.6 MWh (as shown in Fig. 9.11) for the base model while 9.06 MWh for proposed model, respectively. The drastic difference in annual electricity consumption in proposed model comes due to earth air tunnel used in place of conventional air conditioner.

In this case study, the annual electricity generated by wind turbines is 9.28 MWh, which is sufficient for the proposed annual electricity demand. Earth air tunnel consists of ninety 0.15 m inner diameter PVC pipe (23.42 m) and three blowers of 1300 W (4000 cfm, 1.3 kW, 12,000 rpm) each (Bansal et al., 2009). Seven (7) wind turbines of 2 kW rated power are selected for proposed model, which are based on the wind characteristics for the Prayagraj location.

The annual reduction potential of carbon emission is nearly 338 tCO_2 through the proposed system. The NPV value of the proposed system is Rs 2.446 million, which indicates the feasibility of the proposed system. The initial cost of the proposed system is Rs 1.883 million, and total annual saving is Rs 1.318 million, hence a payback period of 1.3 years. The details of the cost of various items involved in the proposed system are shown in Table 9.7.

Electric Consumption (kWh x000)

	Jan	Feb	Mar	Apr	May	Jun	Jul	Aug	Sep	Oct	Nov	Dec	Total
Space Cool	3.27	8.20	15.93	27.29	16.50	-	18.34	30.42	28.78	22.63	11.99	5.56	188.93
Heat Reject.	-	-	-	-	-	-	-	-	-	-	-	-	-
Refrigeration	-	-	-	-	-	-	-	-	-	-	-	-	-
Space Heat	-	-	-	-	-	-	-	-	-	-	-	-	-
HP Supp.	-	-	-	-	-	-	-	-	-	-	-	-	-
Hot Water	-	-	-	-	-	-	-	-	-	-	-	-	-
Vent. Fans	1.63	1.55	1.71	1.71	0.90	-	1.06	1.71	1.71	1.71	1.55	1.79	17.02
Pumps & Aux.	0.01	0.00	-	-	-	-	-	-	-	-	-	0.00	0.01
Ext. Usage	-	-	-	-	-	-	-	-	-	-	-	-	-
Misc. Equip.	0.11	0.10	0.11	0.11	0.06	-	0.07	0.11	0.11	0.11	0.10	0.12	1.11
Task Lights	1.02	0.97	1.07	1.07	0.56	-	0.66	1.07	1.07	1.07	0.97	1.12	10.65
Area Lights	-	-	-	-	-	-	-	-	-	-	-	-	-
Total	6.04	10.82	18.82	30.19	18.02	-	20.13	33.31	31.67	25.52	14.61	8.59	217.71

FIGURE 9.11   eQuest Simulation result for base case.

TABLE 9.7 Description of various equipment in proposed model.

Items	Specifications	Units	Total cost in Rs. (million)
Turbine	2 kW, ReDriven Ltd	7	1.092
Battery	12 V,150 Ah	24	0.288
Inverter	9 kW capacity, 48 V	1	0.070
PVC pipe	0.15 m inner diameter, 20 feet	90	0.225
Blower	4000 CFM,1.3 kW, single phase	3	0.054

3. Discussion

This chapter presents four different case studies for a tropical city for assessing the potential of carbon reduction through retrofitting of some sustainable energy systems. The selection of sustainable systems is dependent on the design of the built envelope, operational energy requirements, location and geography of the buildings within the urban area, etc. For the temple building, the use of wind turbines was proposed, due to its proximity to the river Yamuna, which provided higher wind speeds. Similarly, for the public auditorium, sufficient height was available to provide the required wind speeds. The use of EATs was also only feasible for these two buildings in view of sufficient land availability there. The solar hot water system was only meant for the college building due to its need for the hostel inmates. However, solar PV and efficient LED lighting were proposed for all building types due to sufficient roof area availability. Thus any generalized recommendations for all building types were not feasible. Hence, this study demonstrates that proper site survey and data collection of urban buildings is a prerequisite for suitable recommendations to achieve carbon reduction.

The energy, environmental, and economic aspects of the proposed sustainable systems for the four case studies are as follows:

4. Energy savings

With efficient lighting for the school building, the electrical load reduces by 19.2%; this reduced load can be easily met through the proposed rooftop solar PV system. For the ISKCON temple complex, the electrical energy consumption reduces by approximately 10% even with the use of EAT for better thermal comfort. For the college building, the annual electricity consumption of 31 MWh for water heating is avoided by the proposed solar hot water system, integrated with rooftop solar PV system. For the public auditorium, a drastic reduction in electricity consumption (more than 90%) can be achieved by using EAT in place of mechanical air-conditioning. Hence all the proposed energy systems are not only energy savers but also facilitate in providing better thermal comfort.

5. Carbon reduction

For the school building, the annual GHG emissions of 13.5 tCO_2 can be avoided by proposed energy systems. For the ISKCON temple complex, the potential reduction in annual CO_2 emissions can be as high as 36.2 tCO_2 annually. For the college building, the carbon emissions may be reduced by 86.6 tCO_2 annually. The maximum potential reduction in carbon emissions of 338 tCO_2 can be achieved for the public auditorium. Hence, all the proposed energy systems can lead to significant carbon reduction if adopted for the urban built environment.

6. Economic feasibility

For the school building, the payback period of the solar PV system with battery storage is 8.5 years with the system's lifespan of 25 years. For the ISKCON temple, the hybrid solar PV-wind energy systems have a payback period of nearly 12 years, with annual savings of Rs. 0.12–0.14 million. For the college building, the proposed solar thermal and PV systems require an investment of Rs 4 million with a payback period of only 4 years. The use of EAT in place of mechanical air-conditioning for the public auditorium was found to be most attractive in economic terms with a payback period of only 1.3 years. Thus all the proposed energy systems are found to be economically viable.

The significance of the studies presented can also be seen in light of the recent report of the World Meteorological Organization (World Meteorological Organisation, 2019). As per the report, the global CO_2 emissions growth rates are nearly 20% higher than the previous 5 years, and the global average temperature has now increased by 1.1°C since the preindustrial period. Since building operation typically consumes 80%–90% of its life cycle energy with commensurate carbon emissions (Ramesh et al., 2010), there is an urgent need to adopt strategies for carbon reduction in the built environment.

7. Conclusions

The case studies presented in this chapter assess the feasibility of carbon reduction from the built environment in the tropical climatic zone of India. They suggest some sustainable design features based on the simulation results that can be helpful to reduce the environmental impact of the building envelope. The feasibility of renewable energy systems (solar PV and wind turbines) based on technical, economic and environmental aspects have been presented for the selected buildings. The simulation results indicate that the electricity supply for the buildings can be generated most economically and ecologically by renewable systems. They significantly reduce CO_2 emissions, which helps the country to meet the emission's goal according to the Paris Climate Agreement. Further carbon reduction is feasible through solar hot water systems as a substitute for electric geysers. Efficient lighting using LED bulbs has been proposed for all types of buildings to reduce electricity consumption. Passive design techniques such as EAT may provide better thermal comfort, as a substitute for conventional HVAC systems.

Environmental impact reduction of tropical urban buildings may also be examined on the basis of other solar passive techniques such as natural insulation, green roofs, reflective roofs with cool paint, low embodied energy construction materials, e.g., fly ash bricks and aerated concrete blocks, as well as smart building operations.

References

Azzouz, A., Borchers, M., Moreira, J., Mavrogianni, A., 2017. Life cycle assessment of energy conservation measures during early stage office building design: a case study in London, UK. Energy and Buildings 139, 547–568.

Bano, F., Kamal, M.A., 2016. Examining the role of building envelope for energy efficiency in office buildings in India. Architecture Research 6 (5), 107–115. https://doi.org/10.5923/j.arch.20160605.01.

Bansal, V., Misra, R., Agrawal, G.D., Mathur, J., 2009. Performance analysis of earth–pipe–air heat exchanger for winter heating. Energy and Buildings 41, 1151–1154.

Becqué, R., Mackres, E., Layke, J., Aden, N., Liu, S., Managan, K., Graham, P., 2016. Accelerating Building Efficiency: Eight Actions for Urban Leaders. World Resources Institute (US), Washington, DC. https://www.wri.org/publication/accelerating-building-efficiency-actions-city-leaders.

Global Footprint Network, 2016. GFN. http://data.footprintnetwork.org/analyzeTrends.html?cn=100&type=EFCtot. (Accessed Nov 2017).

González, M.J., García Navarro, J., 2006. Assessment of the decrease of CO_2 emissions in the construction field through the selection of materials: practical case study of three houses of low environmental impact. Building and Environment 41 (7), 902–909.

Husain, D., Prakash, R., 2019a. Ecological footprint reduction of built envelope in India. Journal of Building Engineering 21, 278–286.

Husain, D., Prakash, R., 2019b. Ecological footprint reduction of building envelope in a tropical climate. Journal of The Institution of Engineers (India): Series A 100 (1), 41–48. https://doi.org/10.1007/s40030-018-0333-4.

Invest India, 2018. Natinal Investment Promotion and Facilitation Agency. https://www.investindia.gov.in/sector/construction. Accessed on August, 2018.

Lucon, O., Ürge-Vorsatz, D., 2014. Buildings "Climate Change 2014: Mitigation of Climate Change. Contribution of Working Group III to the Fifth Assessment Report of the Intergovernmental Panel on Climate Change". Cambridge University Press, Cambridge, United Kingdom and New York, NY, USA. https://www.ipcc.ch/pdf/assessment-report/ar5/wg3/ipcc_wg3_ar5_chapter9.pdf.

Mahapatra, R., Jeevan, S.S., Das, S., 2017. Environment Reader. For University, ISBN 978-81-86906-03-3. http://www.downtoearth.org.in/reviews/environment-reader-for-universities-57295.

Ministry of Power Central Electricity Authority (MPCEA), "Government of India, 2016. CO2 Baseline Database for the Indian Power Sector, User Guide, 2016. http://www.cea.nic.in/reports/others/thermal/tpece/cdm_co2/user_guide_ver11.pdf. Accessed on December 2018.

Natural Resources Canada, NRC, 2016. RETScreen International. http://www.nrcan.gc.ca/energy/software-tools/7465.

Omar, M.A., Mahmoud, M.M., 2018. Grid connected PV- home systems in Palestine: a review on technical performance, effects and economic feasibility. Renewable and Sustainable Energy Reviews 82 (3), 2490–24971.

Ramesh, T., Prakash, R., Shukla, K.K., 2013. "Life cycle energy analysis of a multifamily residential house: a case study in Indian context" Open. Journal of Energy Efficiency 2, 34–41.

Ramesh, T., Prakash, R., Shukla, K.K., 2010. Life cycle analysis of buildings: an overview. Energy and Buildings 42 (10), 1592–1600.

Singh, R., Sawhney, R.L., Lazarus, I.J., Kishore, V.V.N., 2018. Recent advancements in earth air tunnel heat exchanger (EATHE) system for indoor thermal comfort application. A review" Renewable and Sustainable Energy Reviews82 3, 2162–2185.

Suresh, J., Agarwal, A., Jani, V., Singhal, S., Sharma, P., Jalan, R., 2017. Assessment of carbon neutrality and sustainability in educational campuses (CaNSEC): a general framework. Ecological Indicators 76, 131–143.

The Quick Energy Simulation Tool (TQEST, 2016). http://www.doe2.com/equest/.

U.S. Department of Energy, 2016. (USDOE, 2016). http://www.doe2.com. Accessed on May 23, 2019.

United Nations Framework Convention on Climate Change (UNFCCC, 2015. Adoption of the Paris Agreement Report No. FCCC/CP/2015/L.9/Rev.1, p. 2. https://unfccc.int/resource/docs/2015/cop21/eng/l09r01.pdf. Article 2.

Varun, A.S., Shree, V., Nautiyal, H., 2012. Life cycle environmental assessment of an educational building in Northern India: a case study. Sustainable Cities and Society 4, 22–28.

Wagner, H.J., Mathur, J., Bansal, N.K., 2007. Energy Security Climate Change and Sustainable Development. Anamaya Publishers, New Delhi, 9788188342815, 8188342815.

World Meteorological Organization (WMO), 2019. The global climate 2015–2019. https://public.wmo.int/en/media/press-release/global-climate-2015-2019-climate-change-accelerates.

Yu, J., Tian, L., Xu, X., Wang, J., 2015. Evaluation on energy and thermal performance for office building envelope in different climate zones of China. Energy and Buildings 86, 626–639.

Trends in active and sustainable mobility: experiences from emerging cycling territories of Dhaka and Innsbruck

Rumana Islam Sarker[1], Golam Morshed[1],
Sujit kumar Sikder[2], Fariya Sharmeen[3,4]

[1]Department of Infrastructure Engineering, University of Innsbruck, Innsbruck, Austria;
[2]Leibniz Institute of Ecological Urban and Regional Development (IOER), Dresden, Germany;
[3]Institute for Management Research, Radboud University, Nijmegen, the Netherlands; [4]Faculty
of Civil Engineering and Geosciences, Delft University of Technology, Delft, the
Netherlands e-mail address: rumana.sarker@uibk.ac.at

1. Background

Promoting green mobility modes is widely accepted for securing the pathway to sustainable transition in response to global environmental challenges (e.g., climate change) by many urban/metropolitan scholarships. Sustainable mobility often refers to green and active modes - such as walking, cycling and public transport (Pucher and Dijkstra, 2003; Tolley, 2016; Sarker et al., 2019). More recently, the micromobility modes (e-scooter) are branded as a green urban mobility solution, but argued to taint the status-quo of urban mobility practice mostly in the largest cities in United States and Europe with the blessing of crowd-source-based digital innovations (McKenzie, 2019). Some critics have already started to ask if this trend may adversely affect human health due to a reduction in the active cycling and walking trends. Nevertheless cycling remains the foremost form of green and sustainable mobility and is already adopted in many global countries, particularly in the global north as one of the effective urban mobility mode in consideration of its environmental and health-related benefits. However, there is not much visible progress in case of global south, which may be linked to the unfavorable social construction, infrastructure, lifestyle and environmental awareness. Several premises on cycling have been studied in different contexts: factors affecting use and barriers to cycling (Iwińska et al., 2018), gender-sensitive cycling infrastructure (Garrard et al., 2008), difference in cycling perception (Haustein et al., 2019) and cycling as a social practice (Spotswood et al., 2015). The evidence-base is nevertheless marginal, particularly in case of fast-growing cities in economically less developed countries.

Visions for cycling are often homogeneous and convergent to the Western world, ignoring the socioeconomic, cultural and spatial contexts of urban mobility in developing countries. Here we question the effectiveness of smart visions for sustainable mobility in a non-Western megacity context. In order to explore sustainability of smart cycling, it is important to understand the prevailing institutional and sociotechnical practices. Institutional practices incorporate policy, governance mechanisms, industry structure, and strategy while sociotechnical practices include interaction of users, suppliers of operators with technology. For effective stabilization of sustainable trends, a harmony between these two practices must be established.

In case of Bangladesh, studies on urban cycling reported ineffectiveness of both the aforementioned practices, viz., the lack of cycling-friendly urban road infrastructure (Rana et al., 2016) and deficient user motivation for cycling (Nawaz, 2015). Empirical evidences show that the basic requirements of promoting cycling as an urban transportation are often absent. Moreover, the availability of alternative cheaper modes such as rickshaws play a major alternative to active cycling. Other broader issues like negative social perception, road safety and women empowerment have also been known to contribute to the whole status-quo (Hossain and Susilo, 2011). Despite these issues cycling is becoming a visible urban mode of transportation in the nation's capital — Dhaka. In fact, the major adopters are young professionals, students. A similar rise in cycling has also been evident in Innsbruck in Austria in recent years. It is common practice to engage pioneer bicycling cities in terms of setting development goals for contemporary cities. However, this chapter takes the opportunity to offer a counterperspective by studying the case of emerging cities in the cycling landscape from both ends of the developing and developed world — Innsbruck and Dhaka. Innsbruck is an emerging cycling city in Europe with substantial bicycle use despite its topographical barrier and adverse weather conditions. While the widely famous *Copenhaganize Index*[1] in its recent ranking recognizes Austria's capital Vienna within the global top ten, Innsbruck has great potential to become a leading cycling city according to the *Global Bicycling Cities Index*[2] combining scores in investment, infrastructure quality and safety provision for cyclists. Dhaka is a rising metropolitan city with one of the highest population densities in the world, with low cycling share but a high walking share. The context in Dhaka differs considerably from Innsbruck, but the learning elements from overcoming barriers could be a valuable addition to drive sustainable transport policy.

Such perspectives are valuable for several reasons — first the journey offers more insights and detailed nuances on overcoming barriers and utilizing opportunities than the peak. Particularly when details of primary survey data can be analyzed at the emerging stages, which is the case here. 'Formal approaches of bicycling knowledge creation and application only serve to verify general insights of little practical use. Understanding and advocating bicycling warrants knowledge of how cycling evolves as part of life trajectories, social practices, and entrepreneurial initiatives' (Krizek et al., 2018). Second, cycling is often put outside the mainframe planning and policy agenda even in high cycling countries, for instance, the Netherlands (Sharmeen et al., 2016). Often local initiatives, grassroot movements and social entrepreneurships pave the way to steer policy debates and bring about change. Documenting innovative initiatives in a comparative context would be valuable to understand process dynamics towards establishing sustainable practices.

And finally, while the contexts may differ inspirations often resonate beyond geography and polity. Previous research in the area of green mobility shows interesting findings by comparing case studies in diverse regions, relating socioeconomic factors with individual travel decisions and their impact on developing policy measures (Wang, 2015; Brussel and Zuidgeest, 2012). As there is no one-size-fits-all approach in terms of global policy implications, this cross-cultural study will bring insights on motivation of cycling in contrasting socioeconomic and geographical settings, addressing the impact of nonconventional initiatives to promote cycling at the community level to overcome the challenges in targeted policy design.

In this chapter, the findings present the attitudes towards cycling and the underlying effects on the frequency of cycling in case of Dhaka (Bangladesh) and Innsbruck (Austria). The empirical evidence for Dhaka was collected by administrating a web-based cycling survey and the data for Innsbruck was analyzed based on the last mobility survey in 2011. With an in-depth analysis of the available empirical data of Innsbruck and Dhaka this study will explore — what are the potential policy inputs necessary — in order to support positive attitude to the urban cycling adoption? The results may help to formulate better strategic planning policy to trigger active and nonmotorized mobility innovation in the context of fast growing cities.

[1]https://copenhagenizeindex.eu/.

[2]https://www.coya.com/bike/index-2019.

2. Urban cycling: rising trend in Dhaka

2.1 Modal shift in urban transport

With a growing population of about 25 million, Dhaka is already known as one of the world's most densely populated cities (JICA and DTCA, 2016). Undoubtedly, it has an impact on the overall traffic situation in Dhaka. Due to the unplanned urbanization and poor traffic system, the hourly traffic speed has slowed down from 21 km to 7 km within a decade, wasting 3.2 million daily working hours in traffic congestion (Nabi, 2018). Moreover, insufficient supply of public transport with poor vehicular condition and no price regulation has made travelling a dreadful experience for the mass population. Amidst this turmoil, recently cycling has gained popularity as a 'faster' urban transport mode on the congested roads of Dhaka. Until now bicycle has been mostly used by the rural and working-class people in Bangladesh (Hammadi, 2013). However, in Dhaka mostly young professionals and students embraced cycle as their main mode of transport to save travel time. This newest cycling trend was primarily initiated in 2011 by a Dhaka-based Facebook cyclist group, called 'BDcyclists'. Since then with its 1,24,170 members (numbers are growing), this group regularly organizes critical mass to create awareness about cycling, which is a popular monthly cycling event around the world (Furness, 2007). In 2017 this group created a Guinness World Record for the longest single line of moving bicycles with 1186 cyclists (Chowdhury, 2016). This widespread popularity of a single Facebook group motivated many other cycling enthusiasts to form similar groups, not only in Dhaka but also in other cities to encourage cycling within their communities. Primarily, this organization intended to persuade at least 5% (ca. 35,000 people) car-commuters to use bicycle for commute by 2014 (Hammadi, 2013). In the above-mentioned background, to explore changing travel behaviour of the aspiring cyclists, a tailor-made online survey was conducted by the authors in 2015 among various Facebook cyclists groups, collecting a representative sample of 326 cyclists. This survey comprises mainly students (74%) and professionals from private sectors (21%) in their mid-twenties (73%) and early thirties (22%). The average daily cycling distance covers 16–20 km for more than half of the respondents (49%) and bicycle is frequently used for commuting to work or educational purpose (58%). Our online survey in 2015 illustrates recreational cycling as one of the frequent activities performed by the respondents (68%). It can also be associated with the commonly practised activities of weekly leisure rides (typically early morning rides with breakfast or evening rides at the outskirts of Dhaka) among different cyclist groups in Dhaka (Chowdhury, 2019). These coordinated group rides not only facilitate physical activity, socializing and exploring new places but also can significantly induce behavioral change in the long term. Although there is lack of data on community rides and their effect on cycling awareness in the context of Bangladesh, corporate cycling campaigns are popular promotional measures among the best and emerging cycling cities of the developed countries. Studies show that these programmes deliver sustainable change in travel behaviour rather than just raising awareness. From a web-based survey on the participants of 2004 ride-to-work-day event in Victoria (Australia), positive influence of the event has been observed and one-quarter of the first-time riders continued to commute by bicycle afterwards (Rose and Marfurt, 2007). Piatkowski et al. (2014) analyzed similar type of event in Denver (USA) showing that participants reported increased bicycle use for both work and nonwork trips after the event. Inclusion of gamification also plays an important role in promotion of cycling. Apart from substantial infrastructural improvement, social dynamics and emotional engagement (e.g., enjoyment from rides) of such measures have influential effect on individuals (Millionig et al., 2016). During the attempt to break the Guinness World Record for longest single line of bicycles in 2016, the 'BDcyclists' group has gained media attention recognizing their valuable contribution in changing the way people travel. Cyclists from various parts of Bangladesh (from 26 districts out of 64 districts) took part in this cycling extravaganza (Chowdhury, 2016). This publicity added another 54,000 members in the 'BDcyclists' group and inspired other cyclists groups as well. In this case, positive word-of-mouth and a successful event organization motivated more cyclists in Dhaka to take part in further community activities and sporting cycle as their main travel mode. The social impact of such events is not empirically examined yet. Nonetheless in an informal exchange with one of the authors during the 2015 online survey, a quote from an active group member and endurance cyclist displays a glimpse of how one individual attempt can influence others.

Initially I was the only person in my office to travel with bicycle, but now I can see some of my colleagues are also commuting on bicycle. Surprisingly, they are even investing in expensive gears and accessories to make daily commute more comfortable. (Private Job Holder, 25 Years Old)

In our online survey, the main influential factors to use bicycle instead of other modes were its convenience and flexibility of route choice in terms of avoiding traffic congestion and uncertainty with timely arrival of public transport (54%) (Fig. 10.1) This is not surprising

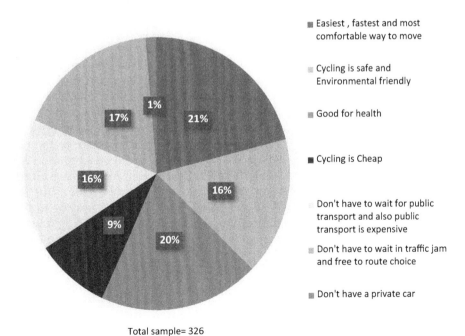

Total sample= 326

FIGURE 10.1   Influencing factors to use bicycle in Dhaka. *Credit: Authors own survey in 2015.*

due to Dhaka's dysfunctional traffic system, where timing one's trip is a challenging task. According to Bangladesh Road Transport Authority, the current total number of registered vehicles in Dhaka is 1.1 Million (BRTA, 2018). The city has only 9% roadways which is far apart from the minimum road network requirement of a megacity (25%) (Mahmud et al., 2008). Overflowing number of motorized vehicles as well as constant violation of traffic regulations creates day-long congestion, which makes travelling by cars and buses immensely difficult. On the contrary, bicycle takes up less road space and facilitates spontaneous manoeuvring, making it a better travel mode choice during gridlocks.

Dhaka is not only suffering from disruptive traffic situation and overwhelming load of population but also lacks in providing recreational facilities (e.g., park, playgrounds) to foster physical and emotional wellbeing of the citizens. The disproportionate ratio of the recreational areas to the total area of Dhaka (1:0.04) is one of the lowest in the world, affecting the overall health of children and young adults (Ahmed and Sohail, 2008). Cycling has numerous health benefits and can aid daily recommended physical activity without the need of sports facilities; therefore, respondents in our survey also considered physical fitness (20%) as a motivating factor to cycle regularly. One-third of the respondents also considered cycling as a safe and environmental friendly way to travel. When considering either car or bicycle use, only 1% of the total sample relied on cycling because of not owning private car. As the survey was administered before the ride-sharing giant Uber and other local companies emerged as a substitute of existing public transport, this number might not represent current situation.

2.2 Barriers in cycling: incompetent planning, social hindrance and negative reaction from other modes

Cycling is the most sustainable urban transport mode facilitating both short and medium-distance trips (Pucher and Buehler, 2017). In Dhaka 76% of the trips are short distance (within 3 km) and nonmotorized transport (e.g., walking, rickshaw) still remains the most dominant mode (40%) considering the higher motorization rate in the city (Bhuiyan, 2007; DTCA, 2010; Gallagher, 2016; JICA and DTCA, 2016; Sikder et al., 2018). Despite its social, economic and environmental benefits, cycling is somewhat neglected in the transport planning procedure of Dhaka. In a middle-income country like Bangladesh, a significant share of the population owns a bicycle but reliable statistical evidence is nonexistent. Based on a report in 2013 about promoting clean transport in Asia, a similar trend is observed in India where number of bicycles per household (45%) is larger in comparison with car ownership, but is scarcely acknowledged in mode share data (Clean Air Asia Center, 2013). According to the revised strategic transport plan for Dhaka, significant increase in the share of public transport is expected with the implementation of mass rapid transit by 2035; however, improving infrastructure to attract more utilitarian cycling for an affordable transport system has not been given appropriate attention (JICA and DCTA, 2016). Based on the responses from our 2015 online survey, inadequacy and insecurity with parking facilities (60%) and absence of dedicated lanes (52%) were perceived as the main infrastructural barriers for cycling in Dhaka. Notably, these are perceived as barriers by both frequent and occasional cyclists in the survey (Fig. 10.2). Nevertheless, poor road condition was less likely to prevent frequent cyclists from using their bicycle (30%) compared to the occasional riders (54%). In line with previous

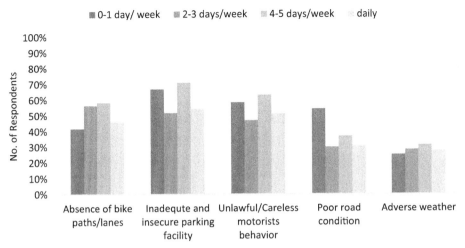

FIGURE 10.2   Barriers in cycling with trip frequency in Dhaka. *Credit: Authors own survey in 2015.*

studies, it can be explained as regular bicycle trips can induce positivity in surroundings whereas occasional riders perceive it as a barrier due to lack of familiarity and confidence (Forsyth and Oakes, 2015).

While evaluating the types of roads generally used by the cyclists in our survey, mostly inner city trips along the primary roads (85%) in mixed traffic was observed. As the participants could select multiple routes, significant use of outer city ring roads (64%), roads in the neighborhood (62%) as well as regional highways (43%) for cycling have been identified. However, cycling routes along the national highway (27%) were taken less frequently, interpreting the fear of collision with speedy motorized vehicles and dangerous overtaking. Hoque and Salehin (2013) in their study on vulnerable road users' safety assessed two of the busiest national highways connecting Dhaka to northern parts of Bangladesh with high level of accident risks for nonmotorized transport. Poor road design and high overtaking tendency with inadequate median separation are the main causes to identify these highways as 'fatal' for pedestrians and cyclists. On the other hand, bicycle and rickshaw are equally vulnerable on urban mixed traffic conditions and account for 12% of the urban road casualties, with an average of 30 bicycle deaths per year reported by the police (Hoque and Salehin, 2013). When asked about the accidents in the past 3 years while cycling, 60% of the respondents had major accidents and 8% had minor accidents in our online survey, whereas most of the respondents maintained safety precautions while travelling (Figs 10.3 and 10.4).

With an open-ended question, some of the risk factors prompting cycling accidents on the busy streets of Dhaka were reported by the respondents in our survey. Underscoring factors were namely complicated intersections, conflicting road use with high volume of autorickshaw, speedy motorbike and human hauler, difficulties in U-turn, noncompliant traffic rules, conflict with motorized vehicle at the merging point of flyovers, etc. In agreement with the factors reported by the respondents, the report of accident database in 2008 confirms rear end collision (58%) and fatalities on the midsection of the road (70%) among cyclists (Hoque and Salehin, 2013; MAAP5, 2008). It also states that children (below 16 years) and

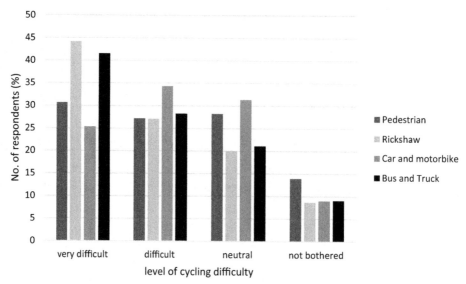

FIGURE 10.3   Interaction with other modes on the road in Dhaka. *Credit: Authors own survey in 2015.*

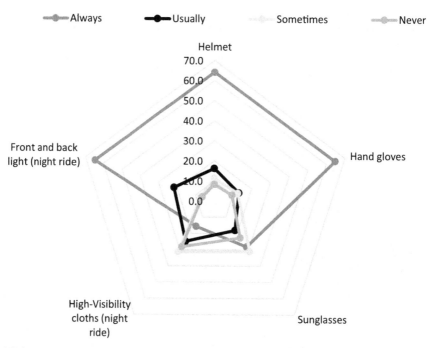

FIGURE 10.4   Safety measures taken by the respondents while cycling in Dhaka. *Credit: Authors own survey in 2015.*

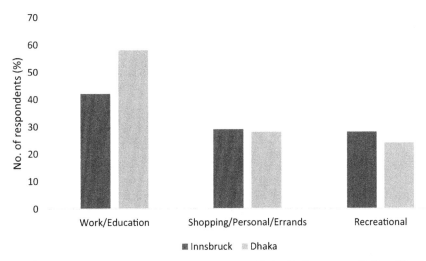

FIGURE 10.5 Bicycle trip purpose of survey participants in Dhaka (Online survey 2015) and Innsbruck (Mobility survey 2011). *Credit: Authors own analysis with mobility survey data.*

young people have higher rate of fatalities in cycling accidents, stressing the vulnerability of current cycling demographics to accidents. This indicates the need of traffic segregation for the safety of current and potential bicyclists, which was one of the highly anticipated infrastructural improvements (74%) by the respondents of our 2015 survey (Fig. 10.8).

In addition to the infrastructural barrier, cyclists often face hostility and inconsiderate driving behaviour from other road users. Threads of tweets and posts regarding cyclist-motorist interaction as well as previous researches on this topic shows severity of such

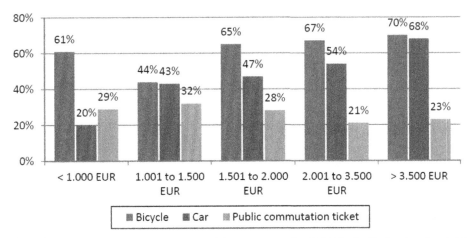

FIGURE 10.6 Bicycle use with income from Mobility survey in Tyrol (2011). *Credit: Based on Pospischil, F. and Mailer, M., 2014. The potential of cycling for sustainable mobility in metropolitan regions—the facts behind the success story of Innsbruck. Transportation Research Procedia, 4, 80-89. doi:10.1016/j.trpro.2014.11.007.*

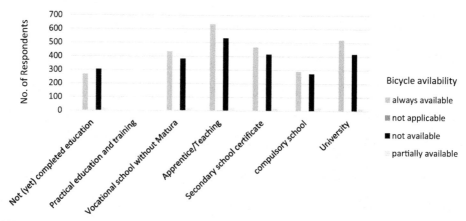

FIGURE 10.7 Level of education with bicycle availability in Tyrol Mobility Survey 2011. *Credit: Authors own analysis from the Mobility survey data.*

problems around the world (Fruhen and Flin, 2015; Joshi et al., 2001). Dhaka is no exception and most of the respondents (70%) in our survey found it difficult to cycle in mixed-mode traffic (Fig. 10.3). During a focus group interview in preparation of our survey, cyclists complained about recurrent verbal harassment from motorists and also experienced chasing behaviour from bus drivers in extreme cases. Fruhen and Flin (2015) explained such behaviours as a result of overall social norms about aggressive driving towards cyclists and negative attitudes towards cycling from noncyclist drivers. Cyclists' appearance also influences harassment and usually experienced, young to middle aged adults, cycling at relatively high speed become the targets of the motorists (Heesch et al., 2012). Often the consequence of conflicts in road-sharing results in abandoning cycling or habitual change, i.e., altering routes or become occasional riders (Kaplan and Prato, 2016). In order to promote safe cycling and encourage more cyclists on the road, awareness campaigns should focus on psychological attributes of the motorists in addition to improved infrastructure and strengthening the traffic rules of Dhaka.

Suggested Improvement need by the respondents	Would influence more cycling trip				
	Very likely	Likely	Unlikely	Very unlikely	Missing Value
Separate lanes for bikes on road	74,2	17,2	3,7	1,5	3,4
Bike route wayfinding signs and bike maps	53,7	27,9	8,0	1,8	8,6
Shared roadways designed to give priority to cycling traffic	57,7	29,8	2,8	,9	8,9
Reduced speed for motorized traffic	31,6	31,6	19,6	5,5	11,7
Increased enforcement and education of traffic laws	31,6	31,6	19,6	5,5	11,7
More secure bike parking/storage	87,1	6,7	1,5	,3	4,3

FIGURE 10.8 Suggested improvement by the respondents to influence more cycling trips in Dhaka. *Credit: Authors own survey in 2015.*

Apart from barriers, studies have repeatedly found gender participation gap in cycling, where significant differences in cycling pattern (e.g., frequency, purpose, length, road type) between men and women have been observed (Judge, 2011; Heesch, 2012). Our survey sample only comprises 1% female respondents, following the notion of female cyclists in most of the researches and representing the perception of overall religious and cultural prohibition. Generally middle and lower class male-dominated families are prominent in Bangladeshi social structure, where seclusion of women with specific attires and keeping them within boundaries through religious custom is commonly practised (LBGI and BCL, 2005). Additional factors to aggravate female cycling comprises safety concern, complex trip-chain behaviour due to differing roles in household and work, especially with childcare (Mcgukin and Murakami, 1999; Bianco and Lawson, 1996; Prati, 2018), as well as discomfort of cycling in traditional clothes (saree, salwar kameez).

Based on United Nations Development Programme (UNDP) report, Bangladesh also had a lower gender-related development ranking in 2000 (LBGI and BCL, 2005). However, women empowerment is on the rise in Bangladesh since past decades with substantial growth in female workers, as entrepreneurs and participants in economic growth (Schwab et al., 2017). Public and private interventions in the field of female education, family planning, access to information technology are also being undertaken. Despite social and cultural boundaries, there are extraordinary examples of 'Infoladies', where young, educated women in rural areas are cycling to remote villages providing internet to the people in need and also advocating them on health, harvesting, government entitlements, etc. (Hilliar, 2014). Encouragement of female cycling in urban areas is also noticeable, where more female cyclists are visible than before. Especially, cycling training programmes for women organized by the cyclists group are encouraging potential female cyclists to overcome their fear and hesitation in Dhaka (Chowdhury, 2019). Women participation in the Guinness Record also demonstrates the interest and courage of female cyclists to change the existing situation. Perhaps not to a greater extent but with due institutional, cultural and infrastructural change, number of female cyclists can also increase in Dhaka.

To summarize, absence of infrastructure, failure to impose traffic regulations, safety concern in sharing road, gender issue and lack of social acceptance are not the only barriers of cycling in Dhaka. Unavailability of reliable data on cycling behaviour as well as not integrating cycling in principle planning strategies may deviate the current positive trend of cycling. Unlike successful Dutch, Danish and German cycling cities, where cycling has been extensively promoted with appropriate infrastructure and policies (Pucher and Boehler, 2017), reluctance on acquiring the data to understand mass acceptability of cycling in Dhaka and lack of interventions may slow down the comprehensive development of overall transport system of the city.

3. Story of transition in Innsbruck: the cycling capital of Austria

Whether people cycle frequently or occasionally, the practice of cycling is global (Horton et al., 2007). Certainly favorable environment and ease of cycling varies in time and space. Northern European cities are renowned as the frontiers of cycling and often their policies and strategies are adopted in other countries as the best practices in cycling (Pucher and

Buehler, 2007; 2016; Pucher et al., 2011a,b). On a mission of reducing car dependency and promoting active mobility, unlike other high-income European cities, Austrian cities didn't gather as much attention. Over the past two decades, the reduced percentage of car use in Vienna is notably higher than other major western European cities (Buehler et al., 2017). Innsbruck, the fifth largest city in Austria located in the Alpine region of Tyrol, also witnessed major changes in their transport preferences. Remarkable rise in bicycle use (+10%) and significant decrease in car use have been observed in the last mobility survey in 2011, compared to the previous survey in 2002 (IMAD, 2011). Innsbruck also accumulates the highest percentage of bicycle mode share (23%) in Austria (VCÖ, 2012). While our previous discussion comprised the phenomenal re-embracement of cycling in Dhaka despite the discrepancies in overall transport system, at the other end of the world with notable socioeconomic and geographical differences similar trends have been observed in Innsbruck.

3.1 Key drivers in the last decades

Infrastructure provision, mobility culture, urban density, as well as the sociodemographics, affect the level of cycling in cities (Haustein et al., 2019). With a dense urban structure, 82% of all the trips in Innsbruck are within 5 km, favouring suitable cycling distance (Civitas Wiki, 2016). Hence 94% of the bicycle trips in Innsbruck lies within this radius (Pospischil and Mailer, 2014; Neyer, 2013). Geographically located at the western part of Austria, embedded in the Inn Valley with a population of 132,493 inhabitants, Innsbruck is known as a student-friendly city and 21% of its population are students (University of Innsbruck, 2019). As a young and vibrant city with plenty of winter sport activities, it is appealing for mountain-sport-enthusiasts to live and study here, contributing mostly young cyclists (80% in their twenties) in the city (Pospischil and Mailer, 2014). The Tyrolian Mobility survey in 2011 with 5093 observations shows that Innsbruck resembles Dhaka in cycling population and trip purposes (Fig. 10.5).

Regardless of different spatiotemporal location and variance in level of cycling, various travel surveys also represent young adults as the core adaptors of cycle as an everyday mode for all of their trips compared to their peers (DTU Transport, 2013; Sener et al., 2009). Proenvironmental behaviour of the younger generation could be the plausible reason for the higher propensity of young cyclists in the Mobility surveys (Wray-Lake et al., 2010). Studies from six industrialized countries in Europe, Asia and USA also confirmed declining car ownership and licensing rate among younger generation with increased use of environmental friendly transport (Kuhnimhof, 2012). In the case of Austria, increased fuel price (+70% between 2002 and 2011) could also contribute in changed travel preferences in Innsbruck (Neyer, 2013). Travel mode choice is also often influenced by the household income and vehicle ownership (Meead et al., 2009; Madhuwanti et al., 2015). Results from the 2011 Mobility survey in Innsbruck show steady increase in car ownership with growing income. Interestingly, the rate of bicycle availability descended up to a certain growth in income level, however afterwards with the growing income bicycle availability peaked and continued to increase (Fig. 10.6). This can be explained with the higher level of education nurturing environmental consciousness of the young demographics (Poscpischil, 2014) (Fig. 10.7).

Besides the contribution of young educated adults to make Innsbruck the 'Cycling Capital of Austria' with the highest bicycle mode share, significant infrastructural improvements in past decades featuring segregated cycle lanes and better accessibility to the city center also made an impact to bring more people into cycling. Furthermore, safe and asphalted cycling path along the river 'Inn' throughout the city as well as establishing cycling bridges and underpasses connecting the city districts offered generous route choices to cater cyclists' need. Innsbruck, with its current 130 km cycling path and 827.2 km-long mountain bike tracks, accommodates not only daily city trips but also leisure bike tours for the relatively young and sporty population (BMLFUW, 2015; Lang, 2013). Additionally, expansion of parking facilities in the inner city and around the central station with bike and ride facility contribute in more cycling trips.

In addition to the provision of cycling lanes, shared space and implementing one way cycling routes, integration of bicycle with public transport was a great way to extend the attractiveness of the public transport and promote green mobility in Innsbruck (Kager and Hams, 2017; Buehler and Pucher, 2012). Evidence showed that cycle-access-based public transport doubles the weekly public transport trips, resulting in reduced car use, congestion, and profiting the overall urban transport economy (Kager and Harms, 2017, Veryard and Perkins, 2018). The bike-sharing system in Innsbruck 'Stadtrad' in cooperation with the local transport authority IVB (Innsbrucker Verkehrsverbund) started its operation in 2014. Complementary discount (60%) on annual bike-sharing fee with IVB annual public transport ticket makes it an attractive possibility for the public transport users to cycle. Reportedly, there are 8500 daily users of 310 cycles stored at 37 docking stations around the city (IVB, 2017; Stadt Innsbruck, 2018). Additionally, city buses, trams, as well as regional and long distance trains, are well-equipped with bike-on-board facilities, allowing cyclists to optimize their travel time with access to a wide range of stations (BMLFUW, 2015).

With the increased number of cyclists in Europe, cyclists' share of total road deaths is also rising slowly (8% in 2016) (European Commission, 2018). However, the cycling accident data of 12 years (1990–2002) in Innsbruck in comparison with the population growth shows a declining trend. This could be another reason to increase the bike share by 10% in 2011. Weather also has a substantial effect on travel mode choice. A 2015 case study in Rotterdam (the Netherlands) shows that cycling is directly affected with precipitation and air temperature; however, wind speed and snow cover have limited to no effect (Böcker et al., 2016). Innsbruck has a moderate climate comprising of fairly warm summer, heavy snowfall in winter and average rainfall from June till September with precipitation fluctuating between 110 and 135 mm (Tirol Info, 2019). Usually the cyclists in Innsbruck are mostly weather resistant, cycling throughout the year (Pospischil and Mailer, 2014). High-quality road maintenance by the city could be potentially the reason to such traits as poor road maintenance sometimes averts even loyal cyclists from cycling in winter season (Iwinska et al., 2018).

In terms of bike availability, the Mobility survey in 2011 shows no difference in female and male respondents (26%) of Innsbruck and the overall sample is also equally distributed (IMAD, 2011). Nevertheless, an attitudinal barometer towards mobility shows that in Europe females outnumber males as noncyclists (European Commission, 2013). Although higher female participation is observed in cities with well-developed cycling infrastructure and high level of cycling (Pucher and Boehler, 2007; Garrard et al., 2008), women's participation in cycling is underrepresented in most travel surveys. Notably, perceived barriers in terms of safety and attitudinal difference towards segregated versus nonsegregated infrastructure exists more in women compared to men (Krizek et al., 2010; Aldred et al., 2016; Akar et al., 2013; Heesch et al., 2012; Garrard et al., 2012). Another important factor is different time distribution in household responsibilities as well as trip characteristics between male and female (Prati, 2018). Women tend to make more escort trips while carrying heavy belongings; especially with the birth of a child suitability of cycling decreases (Spotswood et al., 2015; Scheiner, 2014; McCarthy et al., 2017). Study on care responsibilities in Austrian rural Alpine region also confirmed that women (74%) were solely in charge of childcare comprising higher pickup/drop-off trips compared to men with only 4% women cycling to work (Unbehaun et al., 2014). The mobility survey in 2011 did not assess change in mobility pattern with child birth in Tyrol, but mothers carrying their children on a bicycle seat or in a bike trailer are a common scenario in Innsbruck. Although there is limitation in empirical evidence on gender-based mobility characteristics, it can be assumed that there are a considerable number of female cyclists in Innsbruck with the higher percentage being female students (53%) at the University, in relation with the higher number of young cyclists in the city.

While the bicycle mode share in Innsbruck is the highest in Austria with immense infra-structural development, there is still room for improvement to retain this positivity. Topolog-ically Innsbruck is 574 m above sea-level and surrounding areas are another 500 m higher (Tirol Info, 2019). Although Austrian and German literature shows limited to no consensus on the effect of spatial structure on mobility behaviour, a 2009 study in Zürich (Switzerland) shows the negative impact of topographical condition on the cycling utility (Blanc and Figliozzi, 2016). Therefore, interventions to provide e-bike sharing system in addition to the conventional bike-sharing system can attract more utilitarian trips, resulting in reduced car use and improved air quality.

4. Policy recommendation

While there is a hope for Dhaka to become an exemplary cycling city in Bangladesh with appropriate policy interventions and social campaigns, bicycle share in Tyrolian capital Innsbruck is already on the rise among other regions in Austria. Nevertheless, 43% of energy consumption in Tyrol is accounted by transport (Amt der Tiroler Landesregierung, 2017). So, part of the 'Tyrol on the bike' Programme, Innsbruck aims to increase its share from 23% to 25% by 2020 (BMLFUW, 2015). Several infrastructural and awareness-raising interventions have been made such as organization of UCI Road world championship for cyclists in 2018, attracting cyclists from all over the world. Community activities such as 'The Whole of Tyrol is Cycling'[3] competition, 'BIKEline'[4] school campaigns as well as training courses for migrants and senior citizens have been also organized to motivate cycling beyond age, gender and ethnicity (BMLFUW, 2015). Although conventional city bikes are available for rent, due to topographic conditions, establishing electric bicycle hire system could encourage more up-hill cycling trips regardless of different cycling expertise. In 2014, sale of electric bicycles corresponded to the market share of 12% (VSSÖ, 2014). So, the need of electric bicycle and charging facilities is evident not only for daily use but also for cycling tourism. For instance, workshops and investment have been made to promote Austrian cities as a tourist

---

[3]https://tirol.radelt.at/.

[4]http://www.schulenmobil.at/start.asp?ID=348&b=65.

destination for cycling, where 41% of the cycling-related jobs account for tourism. Cities like Eisenstadt and Klagenfurt already invested in electric bike-sharing system and charging stations, respectively (BMLFUW, 2015). Following the footsteps of other cities, Innsbruck could also further invest in electric cycling facilities to retain its title as 'Cycling Capital of Austria'.

Concerning Dhaka's future in embracing bicycle as a mass mode of transport, in addition to privately organized initiatives, use of smartphone-based application such as Crowd mapping route data could be an effective tool compared to the traditional approaches. Real-time information about traffic volume, topography and safety considerations of a particular road could be collected by the participation of Facebook cyclists group (e.g., BDCyclist group). This measure could also be integrated as the part of government-led initiative 'Digital Bangladesh'[5] envisioning improved quality of life with technology-oriented government service. Furthermore, assisting the transport authorities in a shared platform to understand the need of bicycle amenities on the streets of Dhaka will also work as an incentive for these cyclists. With the help of advanced technology and greater intentions of cyclists groups to make the change, bicycle crowd maps could help to create future cycling network of Dhaka. 'Dynamic Connections' in Berlin as well as experiments in Bangalore already show the usefulness of such projects to improve cyclist environment (Joshi et al., 2014). Besides, safe roads and safer intersections as well as promotional activities led by public and private organizations could encourage more cyclists on the road.

5. Discussion and conclusion

'Dhaka's transport system suffers for lack of vision, not money — Experts', The Daily Star, Dhaka, 1 September (UNB, 2018).

The revised strategy for Dhaka's transport (JICA and DTCA, 2016) also echoes the above statement. The strategic plan includes 4-phase developmental plans between 2016 and 2035. The urban development budget for greater Dhaka in the phase 1 (2016–20) takes 25% of the total transport sector budget, promising to develop roadways, MRT and BRT lines in Dhaka costing nearly 6000 Million USD in total (JICA and DTCA, 2016). Indeed, these are significant measures which should be given attention; however cycling infrastructure is almost absent when developing new roadways in the city. As argued in the introduction, cycling is outside to the transportation regime and has been freelancing its way through the local enterprises. The cycling mode share of Dhaka is about 2%, similar to the cycling share in London (Gallagher, 2016). Whereas, London's future transport strategies envision 5% increase in cycling mode share by 2026 with high performance cycling network and safe road for cyclists along with public transport improvement (Greater London Authority, 2013), the potential of cycling in Dhaka's transport system is merely overlooked by the policymakers. Cycling should be integrated and promoted as a complementary mode of public transport rather than a competing mode in policy implications (Braun et al., 2016). For instance, the encouragement from local public transport operators of Innsbruck through discounts on bike-sharing fee for public transport users and in-vehicle facilities for travelling with bicycle certainly promoted cycling in Innsbruck.

[5]https://a2i.gov.bd/innovation-culture/#1509863576593-89df8881-c977.

Apart from the altruistic nature of the educated young adults in using environmental-friendly transport as well as the effect of income level in their travel pattern, substantial infrastructural improvement, namely separate lanes and extended parking facilities, led to more cyclists in Innsbruck. Our survey participants in 2015 resonate the need of cycling infrastructure in Dhaka to attract more cycling trips with better parking facilities (87%) and separate lanes (74%) (Fig. 10.8). Despite the topographical barrier to establish well-accessed bicycle lanes and facilities, the city of Innsbruck managed to increase accessibility to the central districts by constructing bridges and underpasses to provide ample cycling routes. On the other hand, Dhaka characterizes flat land enabling relatively convenient infrastructure provision. Therefore, partial investment in cycling facilities along the development in public transport infrastructure could increase the bicycle share in Dhaka (Gallagher, 2016). Previous researchers also show that presence of bicycling amenities may affect positively on utilitarian cycling; especially cycling to work, which could considerably minimize traffic congestion at peak hours in Dhaka (Sener et al., 2009; Sacks, 1994; Guttenplan and Pattern, 1995; Hunt and Abraham, 2006).

Innsbruck's success as 'cycling capital of Austria' is an outcome of both traditional and innovative government-led initiatives and engaging community activities for cyclists of all ages. Although public initiatives are merely visible in cycling promotion, privately organized social media campaigns of BDcyclists group voiced the need of cyclists in Dhaka and inspired other private interventions to promote would-be cyclists. Since 2018 'JoBike'[6] is the only private bicycle rental service in Bangladesh focusing mainly on University students in Dhaka and other parts of the country with more than 200 bicycles (Islam, 2018). This is an app-based service with a minimal cost of three taka per 5 min, offering a convenient alternative for short and medium distance trips. The lessons from Innsbruck on cycling promotion could be taken by the public authorities in Dhaka and they could be benefitted from these existing small-scale cycling initiatives, by incorporating them into traditional promotional measures. For instance, inauguration of car-free day movement at the two busy roads of Dhaka in 2016, initiated by government and nongovernment organizations was acknowledged by the transport ministry by including a third road to be part of the event (Jahan and Aziz, 2018). Continuation of such campaigns along with significant policies and infrastructural provision could retain the existing positivity on cycling, but only prioritizing traditional goals rather than gender equity could affect females taking up cycling (Polk, 2008). Gender gap is often not incremental in proportion to the overall cycling level (Dalton, 2016; Aldred et al., 2016). Therefore, a more supportive cycling culture like Innsbruck as well as responding to the infrastructural preferences for women could increase female cyclists in Dhaka. Moreover, the way traditionally women dress in Bangladesh, gender-specific cycle production with better functionality and convenience can play a pivotal role (Cheung, 2014).

Innsbruck is part of the 'Cycling Competence Austria'[7] which aims to form a bicycle trade industry with best possible cycling solutions combining manufacturers, research institutes, municipalities and companies. The city also contributes in bicycle tourism, which is the main provider (42%) of the jobs related to cycling in Austria (BMLFUW, 2015). As the cyclists in Dhaka are relatively younger population, there is also a demand for trendy bicycles with

---

[6]https://play.google.com/store/apps/details?id=rider.bike.jo&hl=en.

[7]https://radkompetenz.at/category/mitglieder/.

fashionable accessories (Nawaz, 2015). Increase in the production and marketing of high-quality bicycles, accessories, as well as cleaning and repair services, could influence the economic growth as well as increase the number of cyclists.

In this chapter, we discuss the similarity in the rising cycling modal share in two cities at the contrasting ends of the geographical and social spectrum, one megacity of global south and one 'cycling capital' of global north. While some demographics of the cyclists (age) bare resemblance others (gender) lie further apart. However, we note that the harmony of infrastructural and social practices is a major contributor to sustainable transitions. As also evident, in both cases the role of social media, bottom-up initiatives and motivational campaigns have contributed significantly to the rising share of cycling. Users, therefore, are central in driving transitions and success is achieved when effectively supported by policy and infrastructural initiatives to overcome barriers — natural as well as man-made.

References

Ahmed, K., sohail, M., 2008. Child's play and recreation in Dhaka city, Bangladesh. Municipal Engineer: Proceedings of the Institute of Civil Engineers 161 (4), 263—270.

Akar, G., Fischer, N., Namgung, M., 2013. Bicycling choice and gender case study: the Ohio State University. International Journal of Sustainable Transportation 7 (5), 347—365. https://doi.org/10.1080/15568318.2012.673694.

Aldred, R., Woodcock, J., Goodman, A., 2016. Does more cycling mean more diversity in cycling? Transport Reviews 36 (1), 28—44. https://doi.org/10.1080/01441647.2015.1014451.

Amt der Tiroler Landesregierung, 2017. Aktionsprogramme-mobilität (ONLINE) Available at: https://www.tirol2050.at/uploads/tx_bh/aktionsprogramm_e_mob.pdf. (Accessed, 28 March, 2019).

Bangladesh Road Transport Authority (BRTA), 2018. Year wise number of registered vehicle in Dhaka Metro (ONLINE) Available at: https://brta.portal.gov.bd/sites/default/files/files/brta.portal.gov.bd/monthly_report/c6b00557_49aa_412d_b2e3_5b1409db3152/Dhaka%20Metro.pdf (Accessed 2 September 2019).

Bhuiyan, A.A., 2007. Final Report, Study on Bus Operation in Dhaka City, Air Quality Management Project (AQMP), Department of Environment (DoE), Ministry of Environment and Forest (MoEF).

Bianco, M., Lawson, C., 1996. Trip Chaining, Childcare and Personal Safety: Critical Issues in Women's Travel Behavior, Proceedings from the Second National Conference on Women's Travel Issues. US Department of Transportation, Federal Highway Administration, Washington DC.

Blanc, B., Figliozzi, M., 2016. Modeling the impacts of facility type, trip characteristics, and trip stressors on cyclists' comfort levels utilizing crowdsourced data. Transportation Research Record: Journal of the Transportation Research Board (2587), 100—108.

Böcker, L., Dijst, M., Faber, J., 2016. Weather, transport mode choices and emotional travel experiences. Transportation Research Part A: Policy and Practice 94, 360—373.

Braun, L.M., Rodriguez, D.A., Cole-Hunter, T., Ambros, A., Donaire-Gonzalez, D., Jerrett, M., Mendez, M.A., Nieuwenhuijsen, M.J., de Nazelle, A., 2016. Short-term planning and policy interventions to promote cycling in urban centers: findings from a commute mode choice analysis in Barcelona, Spain. Transportation Research Part A: Policy and Practice 89, 164—183.

Brussel, M., Zuidgeest, M., 2012. Chapter 8 Cycling in Developing Countries: Context, Challenges and Policy Relevant Research. Cycling and Sustainability. Emerald Group Publishing Limited, pp. 181—216.

Buehler, R., Pucher, J., 2012. Cycling to work in 90 large American cities: new evidence on the role of bike paths and lanes. Transportation 39 (2), 409—432.

Buehler, R., Pucher, J., Altshuler, A., 2017. Vienna's path to sustainable transport. International Journal of Sustainable Transportation 11 (4), 257—271.

Bundesministerium für Land- und Forstwirtschaft, 2015. Umwelt und Wasserwirtschaft (BMLFUW). In: Cycling Master Plan 2015—2025, Vienna, Austria.

Cheung, K., 2014. Bicylcle Planning, Gender & Marketing: Creation of Publications for Women, Senior Project, City and Regional Planning Department, California Polytechnic State University. San Luis Obispo.

Chowdhury, F., 2016. BDCyclists (ONLINE) Available at: http://bdcyclists.com/gwrbdc2016/index.html (Accessed 2 September 2019).

Chowdhury, F., 2019. BDCyclists (ONLINE) Available at: http://bdcyclists.com (Accessed 2 September 2019).

Clean Air Asia Center, 2013. Promoting Non-motorized Transport in Asian Cities: Policymakers' Toolbox, pp. 1–63. Pasig City, Philippines.

Dalton, A., 2016. Cycling Experiences: Exploring Social Influence and Gender Perspectives. Doctoral dissertation, University of the West of England.

Dhaka Transport Co-Ordination Authority (DTCA), 2010. Dhaka Urban Transport Development Study (DHUTS), Final Report. Bangladesh University of Engineering and Technology (BUET) and Japan International Cooperation Agency (JICA) Study Team.

European Commission, 2013. Attitudes of Europeans towards Urban Mobility. Special Eurobarometer 406/Wave EB79.4-TNS Opinion & Social.

European Commission, 2018. Traffic Safety Basic Facts on Cyclists. Directorate General for Transport, June.

Forsyth, A., Oakes, J.M., 2015. Cycling, the built environment, and health: results of a midwestern study. International Journal of Sustainable Transportation 9 (1), 49–58. https://doi.org/10.1080/15568318.2012.725801.

Fruhen, R., Flin, R., 2015. Car driver attitudes, perceptions of social norms and aggressive driving behaviour towards cyclists. Accident Analysis and Prevention 83, 162–170.

Furness, Z., 2007. Critical mass, urban space and Vélomobility. Mobilities 2 (2), 299–319. https://doi.org/10.1080/17450100701381607.

Gallagher, R., 2016. Cost-Benefit Analysis: Dhaka's Future Urban Transport, Bangladesh Priorities. Copenhagen Consensus Center.

Garrard, J., Rose, G., Lo, S.K., 2008. Promoting transportation cycling for women: the role of bicycle infrastructure. Preventive Medicine 46 (1), 55–59.

Garrard, J., Handy, S., Dill, J., 2012. Women and cycling. In: Pucher, J., Buehler, R. (Eds.), City Cycling. MIT Press, Cambridge, USA.

Greater London Authority, 2013. The Mayor's Vision for Cycling in London – an Olympic Legacy for All Londoners. Mayor of London/Transport for London.

Guttenplan, M., Patten, R., 1995. Off-road but on track. Transportation Research News 178 (3), 7–11.

Hammadi, S., 2013. 'The Humble Cycle Making a Comeback on Dhaka Streets,' the Guardian, Dhaka, 20 June.

Haustein, S., Koglin, T., Nielsen, T.A.S., Svensson, Å., 2019. A comparison of cycling cultures in Stockholm and Copenhagen. International Journal of Sustainable Transportation 1–14.

Heesch, K.C., Sahlqvist, S., Garrard, J., 2012. Gender differences in recreational and transport cycling: a cross-sectional mixed-methods comparison of cycling patterns, motivators, and constraints. International Journal of Behavioral Nutrition and Physical Activity 9 (1), 1–12. https://doi.org/10.1186/1479-5868-9-106.

Hilliar, A., 2014. 'Meet the 'Infoladies': The Women Who Get Bangladesh's Villages Connected', France 24, France, 8 June.

Hoque, M., Salehin, F., 2013. Vulnerable road users (VRUS) safety in Bangladesh. In: 16th Road Safety on Four Continents Conference, 15–17 May, Beijing, China.

Horton, D., Rosen, P., Cox, P., 2007. Cycling and Society. Ashgate publishing limited, Hampshire, England, ISBN 9780754648444.

Hossain, M., Susilo, Y.O., 2011. Rickshaw use and social impacts in Dhaka, Bangladesh. Transportation Research Record 2239 (1), 74–83.

Hunt, J.D., Abraham, J.E., 2006. Influences on bicycle use. Transportation 34 (4), 453–470.

IMAD, 2011. Mobilitätsstudie, Amt der Tiroler Landesregierung, Stadt Innsbruck. Innsbrucker Verkehrsbetriebe und Stubaitalbahn GmbH (IVB), 2017. Bericht über die Prüfung des fahrradverleihsystems, stadtrad innsbruck.

Info, T., 2019. Weather in Innsbruck, 574 m. Innsbruck. Available at: https://www.tyrol.com/good-to-know/weather-innsbruck (Accessed 9 September, 2019).

Innsbruck, S., 2018. Statistik rund um das Fahrrad in Innsbruck, Redaktion "Innsbruck informiert". Innsbruck.

Innsbrucker Verkehrsbetriebe und Stubaitalbahn GmbH (IVB), 2017. Bericht über die Prüfung des fahrradverleihsystems stadtrad innsbruck).

Islam, Z.M., 2018. 'Bicycle-rental startup JoBike plans big,' the Daily Star, Dhaka, 8 September.

Iwińska, K., Blicharska, M., Pierotti, L., Tainio, M., de Nazelle, A., 2018. Cycling in Warsaw, Poland—Perceived enablers and barriers according to cyclists and non-cyclists. Transportation Research Part A: Policy and Practice 113, 291—301.

Jahan, M., Aziz, T., 2018. 'Dhaka introduced a "car free day". Dhaka, Bangladesh'. City-Exploration-Network. April.

Japan International Cooperation Agency (JICA) and Dhaka Transport Coordination Authority DTCA, 2016. The Project on the Revision and Updating of the Strategic Transport Plan for Dhaka, pp. 1—84. Dhaka.

Joshi, N., Helmut, I., 2014. Opportunities and challenges of employing crowdmapping in bicycle mobility projects. In: International Scientific Conference on Mobility and Transport, 19—20 May, Munich, Germany.

Joshi, M., Senior, V., Smith, G., 2001. A diary study of the risk perceptions of road users, Health. Risk and Society 3 (3), 261—279.

Judge, H., 2011. Designing more inclusive streets: the bicycle, gender, and infrastructure, geography honors Projects.-Paper29. Available at: http://digitalcommons.macalester.edu/geography_honors/29 (Accessed 2 September 2019).

Kager, R., Harms, L., 2017. Synergies from improved cycling-transit integration: towards an integrated urban mobility system. ITF Discussion Paper (23).

Kaplan, S., Prato, G., 2016. "Them or Us": perceptions, cognitions, emotions, and overt behavior associated with cyclists and motorists sharing the road. International Journal of Sustainable Transportation 10 (3), 193—200. https://doi.org/10.1080/15568318.2014.885621.

Krizek, K.J., Johnson, P.J., Tilahun, N., 2010. Gender Differences in Bicycling Behavior and Facility Preferences. In: Rosenbloom, S. (Ed.), Paper Presented at the Research on Women's Issues in Transportation. Transportation Research Board, Chicago Illinois, United States.

Krizek, K., Sharmeen, F., Martens, K., 2018. Bicycling in changing urban regions. Journal of Transport and Land Use 11 (1), 805—810.

Kuhnimhof, T., Armoogum, J., Buehler, R., Dargay, J., Denstadli, J., Yamamoto, T., 2012. Men shape a downward trend in car use among young adults — evidence from six industrialized countries. Transport Reviews 32 (6), 761—779.

Lang, E., 2013. Mountainbiken in und um Innsbruck. Innsbruck, Institute for Infrastructure Engineering, Unit for Intelligent Transport Systems, University of Innsbruck.

Luis Berger Group Incorporated (LBGI) and Bangladesh Consultant limited (BCL), 2005. Strategic Transport Plan. Ministry of Communications, Dhaka Transport Coordination Board, pp. 1—84.

MAAP5, 2008. Accident Database. Maintained by Accident Research Institute, Bangladesh University of Engineering and Technology, Dhaka, Bangladesh.

Madhuwanthi, M., Marasinghe, A., Dharmawansa, A., Janaka, R., Nomura, S., 2015. Factors influencing to travel behavior on transport mode choice. International Journal of Affective Engineering. https://doi.org/10.5057/ijae.IJAE-D-15-00044.

Mahmud, S., Haque, S., Bashir, M., 2008. Deficiencies of existing mass transit system in metropolitan Dhaka and improvement options. In: The 13 CODATU Conference of Urban Transport (CODATU XIII), 12—14 November, Ho Chi Minh City, Vietnam.

McCarthy, L., Delbosc, A., Currie, G., Molloy, A., 2017. Factors influencing travel mode choice among families with young children (aged 0—4): a review of the literature. Transport Reviews 37 (6), 767—781. https://doi.org/10.1080/01441647.2017.1354942.

McGuckin, N., Murakami, E., 1999. Examining trip-chaining behavior: comparison of travel by men and women. Transportation Research Record 1693 (1), 79—85.

McKenzie, G., 2019. Spatiotemporal comparative analysis of scooter-share and bike-share usage patterns in Washington, DC. Journal of Transport Geography 78, 19—28.

Meead, S.K., Mohammad, R.R., Mohammad, R.A., Gholam, A.S., 2009. Evaluating the Factors Affecting Student Travel Mode Choice (No. 1429-2016-118652).

Millonig, A., Wunsch, M., Stibe, A., Seer, S., Dai, C., Schechtner, K., Chin, R.C., 2016. Gamification and social dynamics behind corporate cycling campaigns. Transportation research procedia 19, 33—39.

Nabi, S., 2018. Dhaka loses 3.2m working hours to traffic congestion daily. In: Dhaka Tribune, Dhaka, 5 July.

Nawaz, T., 2015. Motivation behind bicycling habit in Sylhet city of Bangladesh. Asian Business Review 5 (1), 38—42.

Neyer, T., 2013. Fahrradverkehr in Innsbruck, Masterarbeit. Institute for Infrastructure Engineering, Unit for Intelligent Transport Systems, University of Innsbruck.

Piatkowski, D., Bronson, R., Marshall, W., Krizek, K.J., 2014. Measuring the impacts of bike-to-work day events and identifying barriers to increased commuter cycling. Journal of Urban Planning and Development 141 (4). https://doi.org/10.1061/(ASCE)UP.1943-5444.0000239.

Polk, M., 2008. Gender mainstreaming in Swedish transport policy. In: Uteng, T.P., Cresswell, T. (Eds.), Gendered Mobilities. Ashgate, Aldershot, pp. 229–243.

Pospischil, F., Mailer, M., 2014. The potential of cycling for sustainable mobility in metropolitan regions—the facts behind the success story of Innsbruck. Transportation Research Procedia 4, 80–89.

Prati, G., 2018. Gender equality and women's participation in transport cycling. Journal of Transport Geography 66, 369–375.

Pucher, J., Buehler, R., 2007. At the frontiers of cycling: policy innovations in the Netherlands, Denmark, and Germany. World Transport Policy and Practice 13 (3), 8–57.

Pucher, J., Buehler, R., 2016. Safer cycling through improved infrastructure. American Journal of Public Health 106 (12), 2089–2091.

Pucher, J., Buehler, R., 2017. Cycling towards a more sustainable transport future. Transport Reviews 37 (6), 689–694. https://doi.org/10.1080/01441647.2017.1340234.

Pucher, J., Dijkstra, L., 2003. Promoting safe walking and cycling to improve public health: lessons from The Netherlands and Germany. American Journal of Public Health 93 (9), 1509–1516.

Pucher, J., Buehler, R., Seinen, M., 2011a. Bicycling renaissance in North America? An update and re-appraisal of cycling trends and policies. Transportation Research Part A: Policy and Practice 45 (6), 451–475.

Pucher, J., Garrard, J., Greaves, S., 2011b. Cycling down under: a comparative analysis of bicycling trends and policies in Sydney and Melbourne. Journal of Transport Geography 19 (2), 332–345.

Rana, M.S., Uddin, M.S., Al Azad, M.A.S., 2016. Evaluation of bicycling environment for urban mobility: a case of selected roads in Khulna Metropolitan City. Social Sciences 5 (6), 77–85.

Rose, G., Marfurt, H., 2007. Travel behaviour change impacts of a major ride to work day event. Transportation Research Part A: Policy and Practice 41 (4), 351–364.

Sacks, D.W., 1994. Greenways as Alternative Transportation Routes: A Case Study of Selected Greenways in the Baltimore, Washington Area. MS thesis. Towson State University.

Sarker, R., Mailer, M., Sikder, S., 2019. Walking to a public transport station: Empirical evidence on willingness and acceptance in Munich, Germany. Smart and Sustainable Built Environment 9 (1), 38–53. https://doi.org/10.1108/SASBE-07-2017-0031.

Scheiner, J., 2014. Gendered key events in the life course: effects on changes in travel mode choice over time. Journal of Transport Geography 37, 47–60. https://doi.org/10.1016/j.jtrangeo.2014.04.007.

Schwab, K., Samans, R., Zahidi, S., Leopold, T.A., Ratcheva, V., Hausmann, R., 2017. The Global Gender Gap Report. World Economic Forum, Geneva, ISBN 978-1-944835-12-5.

Sener, I., Eluru, N., Chandra, R.B., 2009. Who are bicyclists? Why and how much are they bicycling? Transportation research record: Journal of the Transportation Research Board, 2134. In: Transportation Research Board of the National Academies, Washington, D.C., pp. 63–72. https://doi.org/10.3141/2134-08.

Sharmeen, F., Martens, K., Lagendijk, A., 2016. Cycling Innovations towards Urban Transitions in Energy, Policy and New Modalities. Cycling and Society Annual Symposium, Lancaster, UK.

Sikder, S.K., Nagarajan, M., Kar, S., Koetter, T., 2018. A geospatial approach of downscaling urban energy consumption density in mega-city Dhaka, Bangladesh. Urban Climate 26, 10–30.

Spotswood, F., Chatterton, T., Tapp, A., Williams, D., 2015. Analysing cycling as a social practice: an empirical grounding for behaviour change. Transportation Research Part F: Traffic Psychology and Behaviour 29, 22–33.

Tolley, R., 2016. Supporting walking in cities. In: Paper Presented at the Walk the City International Conference.

Transport, D.T.U., 2013. Facts Sheet about Danish National Travel Survey, Data- and Model Center. Department of Transport, Technical University of Denmark (ONLINE) Available at: http://www.cycling-embassy.dk/wp-content/uploads/2015/01/2011-2013-Fact-sheet-cycling-in-DK-1.pdf (Accessed 2 September 2019).

Unbehaun, W., Uhlmann, T., Hössinger, R., Leisch, F., Gerike, R., 2014. Women and men with care responsibilities in the Austrian Alps: activity and mobility patterns of a diverse group. Mountain Research and Development 34 (3), 276–291.

University of Innsbruck, 2019. Detaillierte Zahlen von Studium (ONLINE) Available at: https://www.uibk.ac.at/universitaet/profil/dokumente/uni-in-zahlen-2019.pdf (Accessed 2 September 2019).

Verkehrsclub Österreich, 2012. Verkehrsclub Österreich. Available at: www.vcoe.at.

Veryard, D., Perkins, S., 2018. Integrating urban public transport systems and cycling. In: Summary and Conclusions of the ITF Roundtable on Integrated and Sustainable Urban Transport, 24–25 April 2017, Tokyo, Japan.

VSSÖ, 2014. Verband der Sportartikelerzeuger und Sportausrüster Österreichs – ARGE Radverkehr.

Wang, S., 2015. The function of individual factors on travel behaviour. In: Comparative Studies on Perth and Shanghai. State of Australian Cities Conference 2015: Refereed Proceedings.

Wiki, C., 2016. Smart Choices for Cities Cycling in the City, Policy Note, pp. 4–25.

Wray-Lake, L., Flanagan, C.A., Osgood, D.W., 2010. Examining trends in adolescent environmental attitudes, beliefs, and behaviors across three decades. Environment and Behavior 42 (1), 61–85.

CHAPTER

11

Air quality and its impact on urban environment

Sushil Kumar, Pyarimohan Maharana

School of Environmental Sciences, Jawaharlal Nehru University, New Delhi, India

OUTLINE

1. Introduction

The rapid growth of industries, vehicular load, urbanization, construction activity and population growth has resulted in the degradation of air quality in the urban environment in recent times. The shifting of air quality index (AQI) from good to hazardous affects the atmospheric chemistry, climate, nutrient dynamics and human health. Serious concerns regarding ambient air pollution started during the mid-20th century on account of major air pollution episodes in Europe and the United States. Among the several air pollution episodes, the London smog and Los Angeles episodes during the industrial revolution period are very well known. In comparison with the urban areas around the world, the AQI over cities of South Asian countries (China, India, Pakistan, Afghanistan) is in poor to hazardous category (Pant et al., 2019; Wang et al., 2019a,b). Air pollution is one of the major threats to human health, which is evident from 4.2 million reported deaths worldwide in 2015 due to exposure to outdoor air pollution in the urban environment (Cohen et al., 2017; Salmond et al., 2018). The European countries are also experiencing daily severe poor air quality and associated health problems. The recent report of annual premature death of around 790,000 in 28 countries of European Union on account of exposure to air pollution indicates the impact of air pollution (Goodkind et al., 2019; Lelieveld et al., 2019). Further, the global burden study carried out in 2015 over key cities in China reported 1.1 million annual deaths and 21.8 million lifelong disabilities on account of air pollution exposure (Huang et al., 2018).

This chapter discusses various aspects such as the global, regional and local aspects of the air quality and its impact on the environment and also provides a detailed case study of air pollutants over the Indian capital, New Delhi. The following section deals with the urban pollutant emission flux and their fate in the major cities around the world. Section 3 focuses on the impact of air quality on human health. Section 4 deals with the impact of air pollution on urban vegetation, while the increasing temperature in the urban microclimate is discussed in Section 5. The atmospheric pollutants removal ability and the influence of urban infrastructure on the ventilation coefficient (VC) are illustrated in Sections 6 and 7, respectively. The seasonal variation of pollutants and air quality over New Delhi is provided as a case study in Section 8, and the concluding remarks of the study are provided in Section 9.

2. Urban pollutant emission flux and their fate in the major cities around the world

The ideal ambient atmosphere is a mixture of several gases such as N_2, O_2, trace gases and particulate matters. The urban atmosphere is a complex mixture of some additional toxic gases in addition to the nitrogen, oxygen and carbon dioxide. The presence of volatile organic compounds (VOCs) and ground-level ozone (tropospheric ozone) also affects air quality in the urban environment. The rapid growth of vehicular transport, automobiles, industries, construction and resuspension activities alters the composition of the ambient atmosphere. The increasing concentration of NO_x and SO_x is alarming in the urban areas of the megacities in India, China and European countries. The biomass burning, vehicular and industrial emission are rapidly increasing the carbon monoxide (CO) and Pb concentrations in urban areas. The particulate matters present in the ambient atmosphere are a major concern for human health due to its wide range of particle size (penetration to the human organs) and complex chemical composition. The particles present in the atmosphere from size range 0.1 nm to 100 μm are termed as suspended particulate matter. The size and chemical composition of the particles is governed by source, its strength and other meteorological factors.

New Delhi, the capital of India, the most talked-about city in recent times, due to its air quality, is influenced by multiple sources of particulate matter such as anthropogenic (15%), primary crustal sources (59%) and fine dust (26%) (Yadav et al., 2016). The fine particles ($PM_{2.5}$) can enter the respiratory and blood vessels and become the main cause of cardiovascular and several respiratory diseases. Kumar et al. (2018) have reported 50%–60% mass contribution of particulate matter by $PM_{2.5}$ to PM_{10} over New Delhi, India, and attributed to multiple sources of emission such as industrial, resuspension of road dust, biomass burning and vehicular emission. Panko et al. (2019) have suggested the $PM_{2.5}$ contributed by exhaust (fossils fuel burning) and nonexhaust emission (resuspension of road dust, tier wear and brake wear abrasion) in the urban environment over London, Los Angeles and Tokyo. Xiang et al. (2019) reported that Beijing is severely affected by $PM_{2.5}$ particularly during winter, which is influenced by the shallow atmospheric boundary layer along with the other sources such as the crop residue burning, industrial emission and vehicular emission. It is interesting to mention that the coal burning contributed around 26%–45% of polyaromatic compounds (PAH) over Dalian, China. Similarly, the meteorological condition played a dominant part in influencing the $PM_{2.5}$ and ozone concentration over California (Zhu et al., 2019).

Major gaseous pollutants (SO_X, NO_X, O_3, CO) emitted from multiple sources in urban areas affect the plant and human health. SO_X is mainly emitted from combustion activities such as coal burning, fuel burning in several industries and biomass burning (Rastogi et al., 2016). In addition, domestic burning of wood, cow dung and sulphur ore smelting contribute a significant fraction in the atmospheric budget of the urban environment. The oxides of sulphur in the atmosphere on reaction with water vapour form acidic components and play a major role in acid rain. The gas-to-particle conversion (GPC) of oxides of sulphur significantly contributes a fine fraction of atmospheric aerosols, especially in the $PM_{2.5}$ size range. High-temperature burning, such as vehicular and industrial emission, is the major contributor of NO_X in the atmosphere. A recent study showed the emission of 2.35 and 1.03 million metric tons of SO_X and NO_X, respectively, in 2015 only from New York, United States (Blanchard et al., 2019). In the urban environment of a highly industrialized area, the biogenic and transport sector contributes around 20%–25% of total ozone emission, and transportation activity contributes around 10%–15% of the total emission (Wang et al., 2019a,b).

3. Air quality and human health

Air quality and human health are directly linked to each other, which also affect the urban ecosystem. Ambient air pollution is the main cause of respiratory diseases, breathing problems, cardiovascular diseases and premature death. Urban air pollution is one of the important risk factors for premature death in the world. Recent researches suggest that annually 7 million premature deaths are reported on account of air pollution in the world (Karambelas et al., 2018). The reports from the World Health Organization showed that 4 million annual deaths took place due to exposure of fine particulate matters ($PM_{2.5}$) (Li et al., 2019). In addition, 6 lakh premature death of children were reported worldwide due to exposure to air pollutants. Among the air pollutants, fine particulate matter ($PM_{2.5}$) has been identified as the major culprit for premature deaths in the urban environment.

The health effects associated with fine particles mainly depend on the nature and chemical composition of particles (Li et al., 2019). The mixing of natural particles with anthropogenic emitted particles and airborne pathogens intensified the toxicity of these particles. Recent reports suggest that mental health is directly linked with the air quality status of the urban environment. The person suffering from respiratory and cardiovascular diseases are at a greater risk of health problem. Neophytou et al. (2013) reported that 10 μgm^{-3} mass concentration of PM_{10} in dusty days increases the respiratory and cardiovascular diseases by 2.43%, while nondusty days have a limited effect on health. Similarly, the premature death count sharply decreases with a decrease in the mass concentration of $PM_{2.5}$ (Li et al., 2019). This has further been confirmed from the decrease in the number of premature death from 1,078,800 in 2014 to 962,900 in 2017 over Henan Province, China. Another study illustrated that air pollution was the fifth major important risk factor for premature death in 2015. Their results also show that the premature death associated with exposure to $PM_{2.5}$ is around 3.5 million in 1994 to 4.5 million in 2015.

4. Air pollution and urban vegetation

The vegetation and plant cover present in and around the urban environment, directly and indirectly, influences the air quality and atmospheric chemistry.

The four main processes by which trees elevate human health and air quality are

- The atmospheric deposition and absorption of particulate matter and gaseous pollutant.
- Modification of circulation pattern of ambient air or dispersion of pollutant through trees.
- Formation of ground-level ozone through the biogenic VOCs.
- Synergetic effect of pollen grains with the anthropogenic pollutants.

5. Temperature and urban microclimate

The ambient atmospheric temperature of the urban environment is slightly higher than the surrounding area and is considered as urban heat island (Duncan et al., 2019). The surface area of plant leaves and plant canopies affects earth radiation budget through alteration of albedo. The wind speed, humidity and mixing layer height are influenced by the plants in urban environment. Therefore, the tree-covered area shows slightly minimum ambient temperature compared to bare area on account of restricted vertical mixing of air. Duncan et al. (2019) have performed a case study over Perth and the western part of Australia and reported the temperature variation is slightly high over the tree and shrub-dominated area compared to grassland.

6. Atmospheric pollutants removal

Several studies have been carried out to identify the pollutant removal ability through modelling studies and primary data analysis. The studies are based on air pollutant deposition and temperature variation. The formation of ground-level ozone is directly linked with the biogenic volatile organic emission and NOx concentration in the ambient atmosphere. The naturally emitted VOCs such as monoterpenes, isoprene, etc., react with NOx to form O_3. The rate of emission of VOCs depends upon the plant species from which they are formed. Kumar and Yadav (2016) reported mineral dust—mediated photochemical reactions in the atmosphere of Delhi over northwestern India. Another study in Berlin reported an increase of 60% of the O_3 level on account of biogenic emission of VOCs from plants (Churkina et al., 2017). Ren et al. (2017) found a significant effect of biogenic emission of the VOCs and their conversion into O_3 and particulate matter over Beijing, China. The topography and vegetation greatly affect the ambient temperature and pollution sources. The barren lands over arid regions are an important source of atmospheric aerosols. The earth's surface covered by green vegetation restricts the erosion and hence minimizes the transport of surface dust to the atmosphere. The green plants also have the ability to absorb the gaseous pollutants and act as a natural sink for particulate matter. The rate of removal of atmospheric pollutants is directly proportional to the biomass of the tree (Nowak, 2019). Usually, healthy plants with 70—80 cm stem diameter absorb 60—70 times more pollutants as compared to the plants with 5—10 cm stem diameter (Nowak, 2019). Green plants absorb gaseous pollutants through their stomata and leaf surface. In addition, some gaseous substances pass into the intercellular space and are stored for a longer duration. The absorbed pollutants on the leaf surface and other parts are removed through washout and rainout processes. The absorption efficiency of pollutant is specific to plant species with respect to leaf structure, texture, morphology and length. The leaves having broad and rough surface store pollutants for a long duration.

7. Urban infrastructure and ventilation coefficient

Urban air quality over megacities is deteriorating due to the rapid growth of industries, urban infrastructure and economic development. The dispersion of atmospheric pollutants depends on VC of the ambient atmosphere. VC is the product of wind speed and atmospheric boundary layer (mixing height) over the region (Chen et al., 2019; Saha et al., 2019). The mixing height of the atmosphere is linked with atmospheric temperature and seasonal variation. The high VC suggests that the dispersion of pollutants is high, while low VC over a region restricts the dispersion of pollutants in the ambient atmosphere.

$$\text{Ventilation Coefficient(VC)} = \text{Mixing height} \times \text{wind speed}$$

8. Case study of urban air quality of Delhi

Delhi is located in semiarid climate in the upper Indo-Gangetic plain (IGP) of the northwestern part of India. The present case study utilizes MERRA-2 (Gelaro et al., 2017) datasets to study the behaviour of different types of aerosols over the Indian region with a particular focus on the Delhi region. The black carbon (BC), SO_4, total dust column density (TDCD) and total aerosol (TA) have been considered. The dataset has been examined for the period ranging from 1980 to 2016 and extracted over the longitude of Delhi for the analysis of mean annual cycle and interannual variability.

The mean annual cycle is computed by taking the climatological monthly average of each component for the entire period. It reflects the temporal evolution of each variable within the year and helps to find the period where they show peaks and lows. The annual time series is utilized to compute the mean value. This mean value is deducted from each value of the time series to give rise to an anomaly time series for each variable. Afterwards, the standard deviation of the annual time series is computed, and the anomaly time series is divided by the standard deviation to get a standardized anomaly time series. The standardized anomaly value greater (smaller) than $+1$ (-1) represents the years with excess (deficit) dust loading in the atmosphere. The years with values between $+1$ and -1 represents the normal years in terms of atmospheric dust loading.

8.1 Temporal evolution and interannual variability of anthropogenic and natural aerosols over Delhi

The present work considers the study of different natural and anthropogenic aerosols, their mean evolution pattern and interannual variability over Delhi. BC in the ambient atmosphere is the second most light-absorbing component after CO_2 and plays an important role in positive radiative forcing (Jacobson, 2001). The temporal evolution of the BC extinction coefficient (Fig.11.1A) shows that the highest value is found during the winter (October–January) as compared to other periods (February–September). The biomass, crop residue burning, vehicular emission, favourable wind direction (northwesterly) and wind speed over North India particularly Punjab and Haryana increase the BC content of the atmosphere after September (Rastogi et al., 2016; Kumar et al., 2018). This leads to a gradual increase of BC in the atmosphere, and hence the extinction value starts to increase. The highest concentration of BC was found during winter due to the shallow mixing height and relatively stable atmospheric condition. In addition, to get rid of the lowering of temperature, the burning activity of coal, crop residue and solid waste increase the concentration of BC. The GPC of atmospheric SO_2 to SO_4 in the presence of water content in the ambient atmosphere depends on the sources, their strength and meteorology. Interestingly, the SO_4 concentration (a direct measure of scattering) is found to be lowest during the premonsoon months (March–May), while its concentration is higher during the rest of the month with the highest peak during peak monsoon months (July–August) (Fig. 11.1B). Although the SO_4 emissions from the industry and automobiles constantly emit SO_4 into the atmosphere, a higher temperature during premonsoon allows for more vertical mixing and hence its concentration is relatively low. The dust column mass density is highest during premonsoon and peaks around June over Delhi (Fig. 11.1C). The premonsoon period over the Indian subcontinent is generally hot,

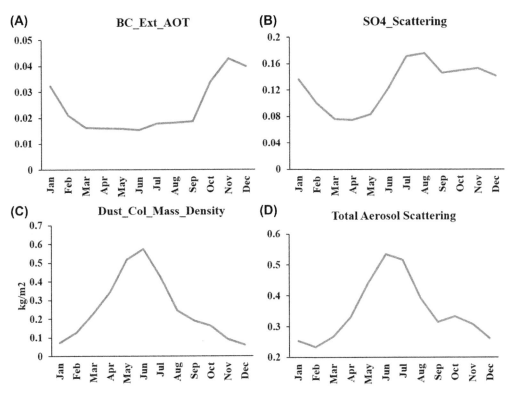

FIGURE 11.1 The mean annual cycle (temporal evolution) of (A) BC extinction coefficient, (B) SO$_4$ scattering, (C) dust column mass density and (D) total aerosol scattering over Delhi.

dry and devoid of rainfall. Therefore, the atmospheric dust load starts to increase over Delhi and North India (Lau et al., 2006; Kim et al., 2016; Maharana et al., 2019). Delhi region receives the monsoonal rain around the last week of June, and hence the dust load peaks during June over Delhi. With the arrival of monsoon rain, the dust particles in the atmosphere get washed away due to the wet deposition and its concentration starts to decline. However, the dust concentration is higher during summer months (June–September) due to the constant transport of dust from the Thar Desert region towards Delhi through the low-level westerlies (Lau and Kim, 2006, 2010). The reversal of southwest monsoon after September results in less dust transport towards Delhi, and hence the dust column density declines gradually. The wet deposition during winter months (December–February) further declines the concentration leading to the lowest value during the year. The TA scattering reflects the aerosol load on the atmospheric column. More the concentration of aerosol, more the scattering. It reflects the cumulative aerosol burden (both anthropogenic and natural aerosols) in the atmospheres, and its behaviour is determined by each of the individual components. The temporal evolution mostly follows the pattern of the natural dust aerosol, it starts to peak during the premonsoon period and reaches its peak in June (Fig. 11.1D). The onset of monsoon settles the aerosol load gradually due to wet scavenging. A second peak around postmonsoon (October–November) is mainly contributed by the anthropogenic aerosol around Delhi.

The annual-averaged aerosol concentration is considered for the analysis of the interannual variability of aerosol over Delhi. The variability of natural and anthropogenic aerosols reflects an interesting behaviour over Delhi. Based on the behaviour, the entire study period can be divided into two major subperiods. The initial years (1980–1999) show a relatively lesser concentration of the natural as well as anthropogenic aerosols as compared to the recent years (2000–2015) with the exception of SO$_4$ during 1990 and 1991 (Fig. 11.2). The gradual increase of the aerosols is evident from the initial years until the end of the study period. The negative values of the standardized anomaly (-1) in the initial years illustrate the lower atmospheric aerosol loading, which gradually increases to positive value after 2000. Afterwards, all the years show excess aerosol loading. It is interesting to observe that although the natural dust load is gradually rising, the major increase in the anthropogenic aerosols (BC and SO$_4$) is apparently evident from a higher frequency of crossing the standardized anomaly of $+1$. The rising atmospheric load of anthropogenic aerosols may be attributed to a gradual increase in the motor vehicular emission, industrial emission and burning of crop residue during winter over Delhi.

8.2 Spatial distribution of different aerosols component during different seasons

The spatial variation of seasonal climatology of the BC extinction coefficient, SO$_4$ scattering, dust column density and total scattering due to the aerosols are described in the following subsections.

8.2.1 BC extinction coefficient

The BC extinct coefficient is a measure of the solar radiation attenuated due to the presence of BC in the atmospheric column and hence represents its atmospheric concentration. The seasonal distribution of BC is over North India and IGPs with a spehcial focus on Delhi marked in *dark black dot*; (Fig. 11.3). During DJF (winter), the BC concentration is found to

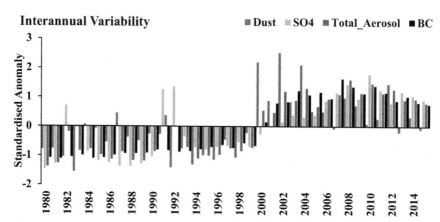

FIGURE 11.2 The interannual variability of (A) BC extinction coefficient, (B) SO$_4$ scattering, (C) dust column mass density and (D) total aerosol scattering over Delhi.

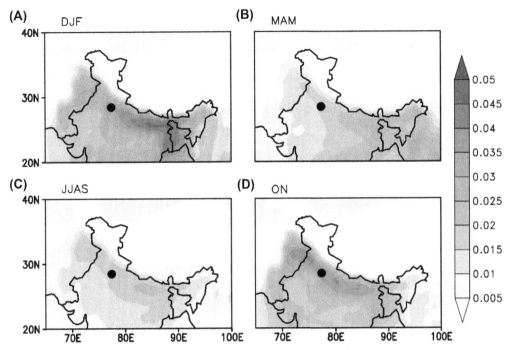

FIGURE 11.3 The spatial distribution of BC extinction coefficient (shaded) for (A) winter (DJF), (B) premonsoon (MAM), (C) monsoon (JJAS) and (D) postmonsoon (ON). The *black dot* represents the location of Delhi.

be highest over Delhi and IGP (Fig. 11.3A). The major reason for the higher concentration is the biomass burning over adjoining states (Haryana and Punjab), which gets transported towards Delhi. In addition, the favourable wind direction (northwesterly) and wind speed enhance the BC concentration over Delhi. In addition, the low-temperature over North India and IGP resulted in a very shallow mixing depth. Therefore, pollutants are concentrated at the lower part of the atmosphere and hence show a higher concentration of BC. Another source of the BC in IGP is the burning of wood and cow dung cakes as fuel for cooking. During premonsoon, the temperature gradually starts to rise over the IGP and Northern parts of India (Fig. 11.3B). This rising temperature increases the mixing height, and hence the vertical mixing of the pollutant takes place, which results in a decreased concentration of BC. The onset of rainfall during the summer monsoon period greatly reduces the BC concentration over Delhi due to wet scavenging, and hence the BC concentration is least during summer (Fig. 11.3C). The initial of crop residue burning starts to increase its concentration gradually over Delhi (Fig. 11.3D).

8.2.2 SO_4 scattering

The SO_4 in the ambient atmosphere is mainly contributed by crustal and GPC, and its concentration depends on source and meteorology (Xiao et al., 1998; Wang et al., 2005). The seasonal variation of SO_4 shows different behaviour than BC. Although both of them are of anthropogenic origin, the variation may be attributed to the source of origin of these aerosols

(Ramanathan et al., 2001). The major sources of SO_4 are the burning of coal and fossil fuel in the industries. There are many coal-based operational power plants around the national capital region. The emitted SO_2 through coal-based thermal power plant and biomass burning activity gets converted into SO_4 in the presence of water content through secondary reaction. The spatial pattern of the distribution of SO_4 is similar to the BC distribution. The shallow mixing depth due to a lower temperature during winter again traps the SO_4 close to the surface (Fig. 11.4A). Similarly, the gradually rising temperature during premonsoon allows the vertical mixing of SO_4, and hence the concentration gradually decreases (Fig. 11.4B). Interestingly, the concentration of SO_4 further increases during summer although the precipitation is highest during summer. It is also important to remember that the mean rainfall over Delhi is not very high, and the rainfall is scattered and scanty. Further, the use of fossil fuel and the burning of coal are continuous due to the constantly increasing power requirement. All these lead to a higher concentration of SO_4 during summer (Fig. 11.4C). During postmonsoon, the SO_4 concentration further increases over Delhi and IGP as compared to Delhi (Fig. 11.4D).

8.2.3 Dust column density

The seasonal dust column density over Delhi and IGP is represented along with the 10m wind pattern (Fig. 11.5A−D). The western disturbances along the westerlies bring in moisture

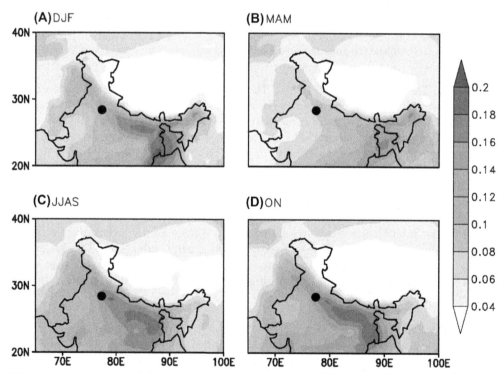

FIGURE 11.4   The spatial distribution of SO_4 scattering (shaded) for (A) winter (DJF), (B) premonsoon (MAM), (C) monsoon (JJAS) and (D) postmonsoon (ON). The *black dot* represents the location of Delhi.

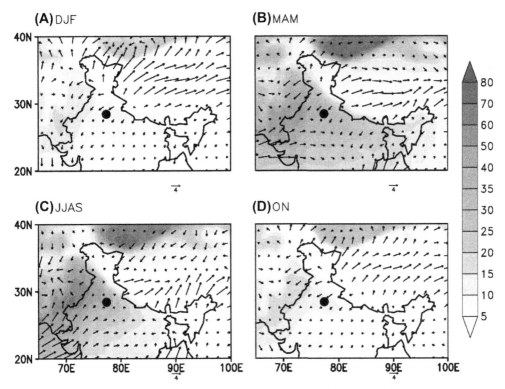

FIGURE 11.5 The spatial distribution of dust column mass density (shaded) and 10m wind (m/s; vector) for (A) winter (DJF), (B) premonsoon (MAM), (C) monsoon (JJAS) and (D) postmonsoon (ON). The *black dot* represents the location of Delhi.

from the Mediterranean Sea and cause wintertime rainfall over North India and Delhi. The rainfall settles the natural dust over this region, and hence the dust column density is very low (Fig. 11.5A). The major source of dust over the Indian region is the Thar Desert region in the northwest part of India. The dust aerosols get transported towards the east through the southwest. After the winter, North India becomes relatively dry and the dust transport from the Thar region towards Delhi takes place. The maximum dust concentration is found during May (Maharana et al., 2019), and the dust penetration towards the east of India is highest during premonsoon (Fig. 11.5B). The arrival of monsoonal rain washes out the atmospheric dust, and the spatial structure of the dust column burden during summer shows a drastic reduction in dust concentration over the eastern part of India while the dust concentration remains higher over the dust source region (western part of India) (Fig. 11.5C). The higher rainfall in the eastern Indian region reduces the atmospheric dust concentration. But the southwestern parts constantly transport the dust from the Thar region towards Delhi throughout the monsoon period, hence maintaining the dust concentration over Delhi (Yadav and Rajamani, 2004). The wind pattern reverses during postmonsoon, the southwesterly (SW) winds change to northwesterly (NW), and hence the dust transport significantly decreases over Delhi.

8.2.4 Total scattering

The seasonal variation of total scattering is dependent upon the various natural and anthropogenic aerosols and their sources (Fig. 11.6). The TA concentration (scattering) during winter is found to be less over Delhi, but relatively higher over eastern India. The major contribution comes from the BC and SO_4, which emerges in the atmosphere due to the crop residue, coal and biomass burning and use of fossil fuel. The aerosol scattering over the North India and IGP is quite uniform. The premonsoon dust concentration is normally very high and gets transported to the eastern part of India; also the burning of firewood in the east leads to a higher concentration. The spatial pattern of total scattering follows the TDCD, which shows a decreasing scattering (concentration) from western to eastern parts of India. As discussed earlier, even though the rainfall and hence the wet scavenging is highest, the constant dust transport by the southwesterly towards Delhi never interrupts and the concentration is high over the western parts of India. Further, the dust concentration is very low during postmonsoon, and the major contributor of TA is mostly from the anthropogenic origin (SO_4 and BC). The total scattering is slightly higher than in wintertime.

As a result, air quality over Delhi region is influenced by multiple sources and meteorological factors. In addition, Delhi itself has multiple sources of pollution including anthropogenic and natural or both. The air quality over Delhi is the worst hit during November, December

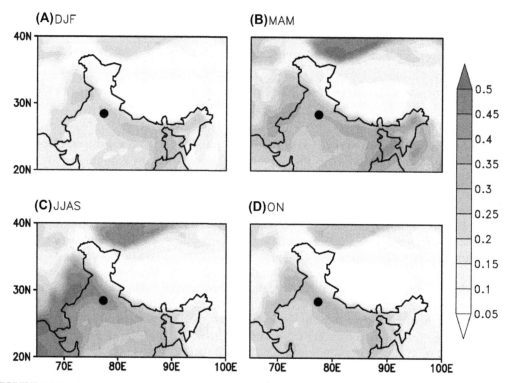

FIGURE 11.6 The spatial distribution of total aerosol scattering (shaded) for (A) winter (DJF), (B) premonsoon (MAM), (C) monsoon (JJAS) and (D) postmonsoon (ON). The *black dot* represents the location of Delhi.

and January in winter season, while April, May and June on account of long-range transport of particle along both wind corridor (Table 11.1). The air quality during summer season influences by on account of transported Thar Desert particle, while winter season is influenced by on account of biomass burning. The mass concentration of particulate matter and other gaseous pollutants are depending on the source, their strength and meteorological factors. Wintertime (October, November, December and January) is characterized by low wind speed, lowering mixing height, low temperature and closed atmosphere (Table 11.1). The low wind speed and mixing height favoured high mass concentration of particulate matter and SO_X, NO_X, CO and O_3 over Delhi (Table 11.1). The mass concentration of particulate

TABLE 11.1 Average mass concentration ($\mu g/m^3$) of criteria pollutant, total days (month wise) of air quality index (AQI) and meteorological parameter over New Delhi.

2018	PM$_{10}$	PM$_{2.5}$	SOx	NOx	O$_3$	CO
January	207 (135−362)	367 (227−619)	28 (5−63)	77 (38−99)	35 (11−105)	2.8 (1.2−4.8)
February	281 (149−435)	152 (69−357)	31 (11−63)	73 (26−122)	50 (27−77)	2.1 (0.3−3.1)
March	225 (148−345)	102 (67−133)	40 (19−66)	75 (50−100)	54 (36−78)	1.5 (0.9−3.0)
April	324 (141−459)	76 (36−114)	11 (2−34)	37 (24−58)	NA	0.5 (0.3−0.9)
May	277 (127−407)	101 (49−238)	11 (5−29)	33 (22−50)	64 (27−99)	0.4 (0.3−0.5)
June	289 (71−981)	72 (30−285)	6 (2−12)	25 (12−43)	56 (21−101)	0.4 (0.2−1.4)

2018	AQI	Good	Satisfactory	Moderate	Poor	Very poor	Severe
January	328 (244−408)	−	−	−	10	17	4
February	243 (140−320)	−	−	6	18	4	−
March	203 (150−290)	−	−	18	13	−	−
April	222 (99−356)	−	1	7	18	4	−
May	217 (125−303)	−	−	12	17	2	−
June	202 (72−447)	−	4	17	5	1	3

Time	Mixing height (m)	Wind speed (m/s)	Temperature (°C)	Relative humidity (%)
January	582 (400−766)	4.38 (1.5−4.4)	15.3 (11.1−18.8)	64.3 (49.7−85.2)
February	685 (450−1009)	4.9 (1.4−4.9)	21.5 (15.9−25.0)	52.9 (78.6−76.5)
March	790 (510−449)	3.8 (0.6−3.8)	26.1 (22.5−30.6)	42.2 (30.5−55.3)
April	723 (524−973)	4.6 (1.9−4.4)	30.8 (25.8−35.7)	35.0 (18.0−54.0)
May	718 (579−877)	3.8 (1.5−4.8)	34.0 (27.9−38.2)	34.8 (16.5−52.7)
June	785 (398−942)	3.5 (1.7−4.5)	33.0 (29.3−37.9)	55.1 (31.9−76.6)

Source: Central Pollution Control Board (CPCB) New Delhi, India.

matter and other gaseous pollutants over Delhi is low. In addition, summertime is character-ized by high wind speed, high ambient temperature, high mixing height and open atmosphere—favoured dilution of atmospheric pollutants. The AQI over New Delhi remained poor on maximum days during winter season, while in summer, the number of poor or very poor days is less (Table 11.1). The winter season is characterized by high biomass burning and coal burning activity to get rid of lowering of temperature and mini-mum dispersion. The AQI over Delhi become poor to severe and hazardous due to use of firecracker and favourable meteorological condition. In addition, dominant biomass burning activity over neighboring states release a huge amount of particulate matter and its transpor-tation towards Delhi via prevailing NW wind deteriorates the AQI of Delhi.

9. Conclusion

The air quality is an important aspect for determining human health, particularly in the urban areas of the world. It is mainly defined by ambient atmospheric particulate matters as well as gaseous components. A higher concentration of $PM_{2.5}$ and PM_{10} along with the gaseous pollutants (O_3, SO_x, NO_x, CO) and Pb affects human health, plants and animals pre-dominantly in the urban environment. Recently, the rapid increase in industrialization, vehic-ular emission, resuspension of road dust, unplanned construction activities, deforestation and the exponential population rise are deteriorating the ambient air quality of the cities. The ambient air pollution has become a serious concern after the Industrial Revolution, and the major events that draw the attention of the world are the London smog and Los Angeles ep-isodes. Recently, the AQI over the south Asian region is the worst in the world, which in-cludes major cities of China, India, Pakistan and Afghanistan, which are deteriorating day by day. The decline in AQI severely affects human health and results in loss of human life. This chapter highlights the temporal evolution, interannual variability and seasonal spatial distribution of few natural and anthropogenic aerosols with special reference to Delhi, along with its recent AQI.

The temporal evolution reflects that the BC extinct coefficient (concentration) is highest during winter, where the crop burning over the North Indian region is a common practice. This is also further supported by the relatively low atmospheric condition and shallow mix-ing layer leading to low VC. The wet scavenging during summer reduces the concentration of BC over Delhi. Similarly, SO_4 mostly generated from the burning of fossil fuel and coal shows its peak during monsoon, while the higher temperature during premonsoon leads to higher mixing depth and hence the lowest concentration is found during this period. However, the SO_4 concentration in the atmosphere does not vary much due to the constant burning of fossil fuel to meet the energy demand of the city and public transport. The hot, rainless and dry condition during the premonsoon supports the increase of the dust load which peaks June. The wet deposition during summer slightly reduces the dust load from the atmosphere, but the constant supply of dust from Thar region by low-level westerly still increase the dust concentration of dust over Delhi. However, it is lowest during the winter under the in-fluence of rainfall due to western disturbances. The behaviour of TA scattering (cumulative aerosol burden) is mostly determined by the behaviour of dust load (peaks during June) in the atmosphere. However, the second peak during postmonsoon to winter is mainly contributed by the anthropogenic aerosols. The interannual variability of these aerosols shows a constant increase of aerosol load over Delhi from 1080 to 2015. But the rapid rise in all aerosol concentration is observed after 2000, and the surging vehicular and industrial emissions along with crop residue burning are the major culprits.

Overall, the urban air quality degraded on account of the rapid growth of industries, construction, vehicular emission and population. The shifting of AQI from good to poor and hazardous results in several human health disorder. The increased level of SOx, NO_X and ground-level ozone becomes the main cause of respiratory diseases. The high mass concentration of particulate matter becomes the major risk factor of premature death in urban environment. However, the planned construction and infrastructure development enhance the VC to reduce the atmospheric pollutant load. The increasing the surface area of urban forest favoured in improving AQI on account of absorption and adsorption of particulate matter and toxic gaseous pollutant.

References

Blanchard, C.L., Shaw, S.L., Edgerton, E.S., Schwab, J.J., 2019. Emission influences on air pollutant concentrations in New York state: I. Ozone. Atmospheric Environment 3, 100033.

Chen, J., Brager, G.S., Augenbroe, G., Song, X., 2019. Impact of outdoor air quality on the natural ventilation usage of commercial buildings in the US. Applied Energy 235, 673–684.

Churkina, G., Kuik, F., Bonn, B., Lauer, A., Grote, R., Tomiak, K., Butler, T.M., 2017. Effect of VOC emissions from vegetation on air quality in Berlin during a heatwave. Environmental Science and Technology 51, 6120–6130.

Cohen, A.J., Brauer, M., Burnett, R., Anderson, H.R., Frostad, J., Estep, K., Balakrishnan, K., Brunekreef, B., Dandona, L., Dandona, R., Feigin, V., 2017. Estimates and 25-year trends of the global burden of disease attributable to ambient air pollution: an analysis of data from the Global Burden of Diseases Study 2015. The Lancet 389, 1907–1918.

Duncan, J.M.A., Boruff, B., Saunders, A., Sun, Q., Hurley, J., Amati, M., 2019. Turning down the heat: an enhanced understanding of the relationship between urban vegetation and surface temperature at the city scale. Science of the Total Environment 656, 118–128.

Gelaro, R., McCarty, W., Suárez, M.J., Todling, R., Molod, A., Takacs, L., Randles, C.A., Darmenov, A., Bosilovich, M.G., Reichle, R., Wargan, K., 2017. The modern-era retrospective analysis for research and applications, version 2 (MERRA-2). Journal of Climate 30, 5419–5454.

Goodkind, A.L., Tessum, C.W., Coggins, J.S., Hill, J.D., Marshall, J.D., 2019. Fine-scale damage estimates of particulate matter air pollution reveal opportunities for location-specific mitigation of emissions. Proceedings of the National Academy of Sciences of the United States of America 116, 8775–8780.

Huang, J., Pan, X., Guo, X., Li, G., 2018. Health impact of China's Air Pollution Prevention and Control Action Plan: an analysis of national air quality monitoring and mortality data. The Lancet Planetary Health 2, 313–323.

Jacobson, M.Z., 2001. Strong radiative heating due to the mixing state of black carbon in atmospheric aerosols. Nature 409, 695–697.

Karambelas, A., Holloway, T., Kinney, P.L., Fiore, A.M., DeFries, R., Kiesewetter, G., Heyes, C., 2018. Urban versus rural health impacts attributable to $PM_{2.5}$ and O_3 in northern India. Environmental Research Letters 13, 064010.

Kim, M.K., Lau, W.K., Kim, K.M., Sang, J., Kim, Y.H., Lee, W.S., 2016. Amplification of ENSO effects on Indian summer monsoon by absorbing aerosols. Climate Dynamics 46, 2657–2671.

Kumar, P., Yadav, S., 2016. Seasonal variations in water soluble inorganic ions, OC and EC in PM_{10} and PM>10 aerosols over Delhi: influence of sources and meteorological factors. Aerosol and Air Quality Research 16, 1165–1178.

Kumar, S., Nath, S., Bhatti, M.S., Yadav, S., 2018. Chemical characteristics of fine and coarse particles during wintertime over two urban cities in north India. Aerosol and Air Quality Research 18, 1573–1590.

Lau, K.M., Kim, M.K., Kim, K.M., 2006. Asian summer monsoon anomalies induced by aerosol direct forcing: the role of the Tibetan Plateau. Climate Dynamics 26, 855–864.

Lau, K.M., Kim, K.M., 2006. Observational relationships between aerosol and Asian monsoon rainfall, and circulation. Geophysical Research Letters 33, 1–5.

Lau, W.K., Kim, K.M., 2010. Fingerprinting the impacts of aerosols on long-term trends of the Indian summer monsoon regional rainfall. Geophysical Research Letters 37 (16), 1–5.

Lelieveld, J., Klingmüller, K., Pozzer, A., Pöschl, U., Fnais, M., Daiber, A., Münzel, T., 2019. Cardiovascular disease burden from ambient air pollution in Europe reassessed using novel hazard ratio functions. European Heart Journal 40, 1590–1596.

Li, X., Jin, L., Kan, H., 2019. Air Pollution: A Global Problem Needs Local Fixes.

Maharana, P., Dimri, A.P., Choudhary, A., 2019. Redistribution of Indian summer monsoon by dust aerosol forcing. Meteorological Applications 26 (4), 584–596.

Neophytou, A.M., Yiallouros, P., Coull, B.A., Kleanthous, S., Pavlou, P., Pashiardis, S., Dockery, D.W., Koutrakis, P., Laden, F., 2013. Particulate matter concentrations during desert dust outbreaks and daily mortality in Nicosia, Cyprus. Journal of Exposure Science and Environmental Epidemiology 23, 275.

Nowak, D.J., 2019. The atmospheric system: air quality and greenhouse gases. In: Understanding Urban Ecology, pp. 175–199.

Panko, J.M., Hitchcock, K.M., Fuller, G.W., Green, D., 2019. Evaluation of tire wear contribution to $PM_{2.5}$ in urban environments. Atmosphere 10, 99.

Pant, P., Lal, R.M., Guttikunda, S.K., Russell, A.G., Nagpure, A.S., Ramaswami, A., Peltier, R.E., 2019. Monitoring particulate matter in India: recent trends and future outlook. Air Quality, Atmosphere and Health 12, 45–58.

Ramanathan, V., Crutzen, P.J., Kiehl, J.T., Rosenfeld, D., 2001. Aerosols, climate, and the hydrological cycle. Science 294, 119–2124.

Rastogi, N., Singh, A., Sarin, M.M., Singh, D., 2016. Temporal variability of primary and secondary aerosols over northern India: impact of biomass burning emissions. Atmospheric Environment 125, 396–403.

Ren, Y., Qu, Z., Du, Y., Xu, R., Ma, D., Yang, G., Shi, Y., Fan, X., Tani, A., Guo, P., Ge, Y., 2017. Air quality and health effects of biogenic volatile organic compounds emissions from urban green spaces and the mitigation strategies. Environmental Pollution 230, 849–861.

Saha, D., Soni, K., Mohanan, M.N., Singh, M., 2019. Long-term trend of ventilation coefficient over Delhi and its potential impacts on air quality. Remote Sensing Applications: Society and Environment 15, 100234.

Salmond, J., Sabel, C., Vardoulakis, S., 2018. Towards the Integrated Study of Urban Climate, Air Pollution, and Public Health.

Wang, K., Yin, H., Chen, Y., 2019a. The effect of environmental regulation on air quality: a study of new ambient air quality standards in China. Journal of Cleaner Production 215, 268–279.

Wang, P., Chen, Y., Hu, J., Zhang, H., Ying, Q., 2019b. Source apportionment of summertime ozone in China using a source-oriented chemical transport model. Atmospheric Environment 211, 79–90.

Wang, Y., Zhuang, G., Sun, Y., An, Z., 2005. Water-soluble part of the aerosol in the dust storm season—evidence of the mixing between mineral and pollution aerosols. Atmospheric Environment 39, 7020–7029.

Xiang, Y., Zhang, T., Liu, J., Lv, L., Dong, Y., Chen, Z., 2019. Atmosphere boundary layer height and its effect on air pollutants in Beijing during winter heavy pollution. Atmospheric Research 215, 305–316.

Xiao, H., Carmichael, G.R., Zhang, Y., 1998. A modeling evaluation of the impact of mineral aerosols on the particulate sulfate formation in East Asia. Atmospheric Science 22, 343–353.

Yadav, S., Rajamani, V., 2004. Geochemistry of aerosols of northwestern part of India adjoining the Thar desert. Geochimica et Cosmochimica Acta 68, 1975–1988.

Yadav, S., Tandon, A., Tripathi, J.K., Yadav, S., Attri, A.K., 2016. Statistical assessment of respirable and coarser size ambient aerosol sources and their timeline trend profile determination: a four year study from Delhi. Atmospheric Pollution Research 7, 190–200.

Zhu, S., Horne, J.R., Mac Kinnon, M., Samuelsen, G.S., Dabdub, D., 2019. Comprehensively assessing the drivers of future air quality in California. Environment International 125, 386–398.

CHAPTER

12

Sustainable water management in megacities of the future

D. Pavlova, Y. Milshina

National Research University Higher School of Economics, Moscow, Russia

OUTLINE

1. Introduction

Nowadays more than half of the world's people — nearly 4 billion — lives in cities. Urban areas are expected to host 2.5 billion more dwellers by 2050, which is almost 70% of global population (UNESCO, 2019). The major contribution to the growth of urban citizens (90%) is projected to be made by Asia and Africa (UN, 2014).

The growth is anticipated to be concentrated in urban centers, particularly — megacities — metropolitan areas of more than 10 million inhabitants (UN, 2014). For the several years, from 2014 to 2018, the share of urban dwellers, who live in megacities, increased almost twice — from 12% to 21% (UNESCO, 2019). Their number, in turn, increased from approximately 450 million to 530 million people for the same period of time (UNESCO, 2019). The speed of growth of megacities' inhabitants is extremely fast considering that in 1990 this number was only about 150 million people (UNESCO, 2016).

From a retrospective point of view, this phenomenon is relatively new since in 1970 there were only three megacities in the world — Tokyo, London and New York (RGS, 2016). Due to the rapid urbanization, in 1990, their number was already 10 (UNESCO, 2016). In the next century the trend has strengthened and in 2014 there were 28 megacities worldwide, located in Asia (16 cities), Latin America (4 cities), Africa (3 cities), Europe (3 cities) and Northern America (2 cities) (UN, 2014). Currently there are already 33 megacities on all continents except Oceania (UN, 2018). In the future, 43 megacities are projected to exist in 2030, most of which will be located in the southern part of the world (UNESCO, 2019).

Megacities concentrate enormous amounts of people in relatively small areas, which usually keep on expanding. This imposes both positive and negative effects on the global socio-economic structure (Euromonitor, 2018). On the one hand, these urban areas are the centers of wealth, social diversity, economic growth and innovation. In this way, they attract talents and investments. On the other hand, citizens of megacities face numerous challenges such as overcrowding, traffic congestion, air pollution and natural resources depletion.

Despite the fact that megacities are generally located close to rivers, lakes or seas, water scarcity represents one of the key vulnerabilities for their population at the moment (UNESCO, 2019). According to the UNESCO (2019), if this issue is ignored or left unaddressed, it can destroy the natural water cycle, cause long-term water shortage, lead to the degradation of the quality of other natural resources and loss of many iconic landscapes and ecosystems.

The megacities of the future are more vulnerable to this danger in comparison to the old megacities (UNESCO, 2019). One of the reasons for it is that most of them are expected to be located in Africa and Southeast Asia, where the climate is naturally more hot and weather conditions more extreme: floods, droughts, tsunami, etc., are more prevalent in these regions in contrast to the Northern part of the world. Another reason, which might be even more important, is that the emerging megacities are developing extremely quickly and, therefore, experience swift population growth and rise in demand for natural resources, while the old ones have been developing gradually and had some time to adopt to new conditions. Thus, new megacities definitely lack time, tools and competencies to establish adequate urban services, including water-related ones.

UNESCO (2019) identifies several critical factors, which should be considered in order to establish a sustainable water management system in a megacity of the future:

- *Population* (controlling urban growth and access to basic services; managing drinking water services and meeting the demand for fresh water; providing access to water sanitation and satisfying the need for wastewater treatment; ensuring the protection of water resources and improving the management of runoff water)
- *Climate* (reducing the carbon footprint and recovering resources such as energy, gas, nitrogen and phosphorous; adapting to the risks of flood, drought, rising sea levels, erosion and salination, changes in temperature)
- *Governance* (constructing regulatory and legal framework; employing service operators; mobilizing civil society)

This study as a descriptive one is aimed to deepen and expand the scientific knowledge base about the sustainable water management systems in megacities of the future by identifying the global trends in water management.

To achieve this goal, in the next section we discuss the results of the previous studies on this topic, particularly the studies dedicated to the investigation of urban water management issues and sustainable water management problems in megacities. After that, we conduct a text-mining analysis across the leading academic and patent databases, professional media and analytical market reports, in order to distinguish the prevailing and changing agendas in global water management, identify the semantic units and their clusters, detect their temporal spread over the observation period (2010−19) and relationships between each other. Finally, we summarize the main points of our study and discuss the compatibility of the identified water-related issues and detected global trends with the critical factors of sustainability highlighted by UNESCO.

2. Literature review

In this section we analyze a variety of studies dedicated to the investigation of urban water management in general, sustainability issues of such a kind of management in particular and megacities' solutions related to these issues. To clarify the logic of our analysis, we provide a step-by-step description of it below.

a. We choose a reliable source of information, which, in our case, is 'Web of Science' (further − WoS) − online platform that provides subscription-based access to numerous databases of scientific publications.
b. We construct relevant search queries using the relevant keywords ('urban', 'water', 'management', 'sustainability', 'megacity') and logic operators ('NEAR', 'OR').
c. We set a relevant time period (1900−2019) and field (topic) for the search.
d. We describe the dynamics of publication activity and results of the previous publications on these topics.

2.1 Urban water management − historical perspective

For this subsection we constructed the following search query: (water* NEAR manage* NEAR (city OR cities OR urban)). On this request WoS provided 4639 publications dating back from 1961 to present − 2019 (Fig. 12.1).

As it follows from Fig. 12.1, the topic of urban water management has become of crucial importance in recent years since the number of publications has risen significantly and reached 479 in the previous year (this year − 2019 − still represents the incomplete statistics), whereas it was only 153 ten years ago. Overall, there were two peaks at which the number of publications increased more than twice: between 1992 and 1993, when this number grew from 9 to 19, and between 2004 and 2005, when it rose from 64 to 128. While trying to explain it, a time lag between writing and publishing a paper (3−4 years) should be considered. In this case, such changes might be explained by the global shift from the economic competition, growth, productivity and efficiency to environmental concerns and sustainability, which started in 1990s (Martin, 2016).

We applied a filter 'highly cited publications' to the results of the search and obtained 33 articles on urban water management published in 2009−19.

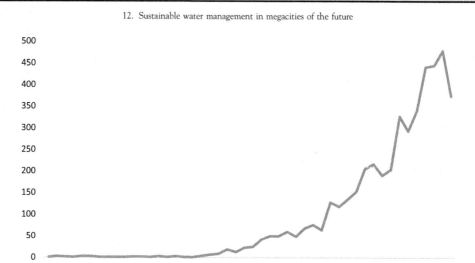

FIGURE 12.1   Publication activity on urban water management. *Source: WoS.*

Many scholars investigated water-related issues connected with rapid global urbanization such as an increasing number of storms and floods and degree of water stress. For this reason, they suggested to employ *management tools* to address these problems. Weather and Evans (2009), for example, explored the linkages between land use and water management in rural and urban areas and demonstrated the opportunities to minimize the flood risks by conducting future studies, particularly Foresight Future Flooding project. García-Ruiz et al. (2011) constructed several future scenarios for water resources development in the Mediterranean region in the context of demographic and climatic global changes such as decline in the average streamflow, transformation of river regimes characteristics and alterations in reservoir outputs and management. Zhou et al. (2013), for example, explored the hydrological response of urbanization, particularly land-use and land cover changes, in the Yangtze River Delta region and argued that this phenomenon will have a slight impact on annual water yield, but a profound impact on surface runoff, peak discharge and flood volume, especially in suburban areas. In this way, they highlighted the importance of flood mitigation and water resources management systems in urban areas. McDonald et al. (2014), while investigating the global hydrologic models and urban water infrastructure, figured out that one in four cities in world with economic activity at 4.8 ± 0.7 trillion dollars stay water stressed because of geographical and financial barriers, which should be addressed by tools of strategic management. Booth et al. (2016) investigated the 'urban stream syndrome', which is characterized by degraded physical, chemical and biological conditions, and emphasized the need of urban stormwater management improvement in order to protect downstream watercourses and construct watershed to tackle this issue. Yang et al. (2019) examined the impact of urban expansion on watershed in the Dianchi Lake region and emphasized a significant role of urban water management for protecting the natural hydrological cycle, surface energy balance and biodiversity.

Other researchers suggested the use of *highly technological solutions* for overcoming the urban water challenges. Willis et al. (2013), for instance, while analyzing the end use water consumption and water savings in households, figured out that sociodemographic clusters have higher water consumption across end uses and quantified the payback period for some water efficient devices. Fletcher et al. (2013), considering the stormwater as an important resource, discuss the recent technological advances in urban rainfall prediction (e.g., radar, microwave networks), which help to restore the natural water balance and remove the pollutants, identified the need for further research into the spationatural dynamics of urban rainfalls in order to enhance the short-term prediction. Michael-Kordatou et al. (2015), in turn, considering the wastewater reuse as a key element of sustainable water management system, investigated the performance of different advanced treatment techniques that remove wastewater effluents (e.g., membrane filtration and separation, activated carbon adsorption, ion-exchange resin, chemical oxidation), the efficiency of which occurred to be dependent on the type and amount of organic compounds in the aqueous matrix as well as operational parameters. Yaseen et al. (2016) demonstrated the application of an innovative data-driven approach — the extreme learning machine method — to forecasting monthly stream-flow discharge rates and approve its efficiency while studying the case of the Tigris River in Iraq. Sun et al. (2018) emphasized an important role of water yield and water purification in urban areas that should be enhanced in order to protect the ecosystem services seriously damaged by urban expansion. Dillon et al. (2019) concentrated on the problem of increasing groundwater extraction and overdraft and suggested to consider the managed aquifer recharge, or intentional groundwater replenishment, as an efficient strategy of water management in urban environment, which can secure the stressed groundwater systems and improve water quality.

Some researchers, in turn, suggested to use *'green' solutions* to tackle the global urban water-related issues. Berndtsson (2010), for instance, examined the performance of green roofs as urban drainage systems and highlighted their high potential for maintaining the quality of runoff water. Fioretti et al. (2010) also figured out that vegetated roofs are highly efficient in mitigating storm water runoff generation. Berardi et al. (2014) examined the benefits of green roofs such as reduced building energy consumption, minimization of urban heat island effect and air pollution, increased sound insulation and ecological preservation, and demonstrated their economic feasibility in terms of the life-cycle cost and provided several examples of policies that promote their usage worldwide. Barthel and Isendahl (2013), while analyzing the urban water management practices of historic and prehistoric cities (e.g., the Classic Maya civilization, Byzantine Constantinople), emphasized an important role of urban gardens, agriculture and water management in the long-term food security. Livesley et al. (2016) highlighted the importance of urban trees, or urban forest, for water management in cities, particularly for the prevention of natural hazards such as storms and floods, and identified their positive effects (air cooling and pollutants removing, releasing harmful organic compounds and allergenic particulates). Berland et al. (2017) argued that urban trees play a crucial role for the urban stormwater management as a part of the green infrastructure, which nowadays includes mainly infiltration technologies such as rain gardens, bioswales and permeable pavements, but ignores the arboriculture, particularly wooden plants that remove water from the soil, enhance infiltration and performance of other green technologies.

The complex of these solutions was considered as 'green' infrastructure by some researchers and tightly connected with governance issues. Xia et al. (2017) analyzed such issue as waterlogging in Chinese cities and highlight the importance of both gray infrastructures (e.g., piped, conveyances, drainage network, water tank) and green ones (e.g., river, lake) for improving urban water management, which should be also supported by municipal engineering, urban hydrology, environmental and social sciences for the integral and systematic view. Liu and Jensen (2018) showed best practices of green infrastructure establishment in sustainable urban water management, which are aimed at tackling such issues as environment protection, water supply and flood risk mitigation, and identified the key problems for their implementation: space and cost constraints, low level of intersectional and stakeholder collaboration. In order to solve these challenges, the scholars suggested to find a balance between the top-down and bottom-up governance approaches, develop citywide strategies and pilot project programmes, create guidelines and regulation programmes. Chan et al. (2018) focused on the issue of surface water flooding caused by land-use change, particularly rapid urbanization and socioeconomic development. To address this issue, the authors suggest to use 'blue' and 'green' spaces for the urban stormwater management, or nature-based solutions, in addition to grey engineer-based infrastructure, and increase the effectiveness of municipal governance by the implementation of land-use guidance and assessment tools (e.g., flood risk assessment, climate projection methods, drainage guidance).

The *governance solutions* based on well-established communication between different institutions were separately highlighted by some scholars. Qadir et al. (2010), for example, investigated wastewater irrigation issues in developing countries, which harm human health and environment and could be addressed by the implementation of the comprehensive wastewater treatment programmes including improved policies (e.g., effluent standards), institutional dialogues between water delivery and sanitation sectors, and financial incentives. Ernstson (2013), in turn, studied environmental justice in the context of social and political processes (e.g., urban development practices, day-to-day management of urban ecosystems) and constructed a framework for maintaining the ecosystem services at the city-wide and local levels. Bach et al. (2014) reviewed a broad scope of research studies in urban drainage modelling, classified the existing models by key considerations (data, structure, computing and integration), as well as common methodology, or systems approach, and point out the need to address interplay between social and economical, biophysical and technical issues. Fletcher et al. (2015) analyzed the history, scope and application of urban water drainage practices aimed at urban stormwater management and clarified the underlying principles of communication terms used in this field. Hommes and Boelens (2018) analyzed the 'hydrosocial' linkage between the city of Lima and the Rimac watershed and emphasized the need to consider not only engineering technical solutions, but also political agendas and visions of citizens in urban water management. Alda-Vidal et al. (2018), while examining the issue of urban water inequalities in African cities, demonstrated that centralized water supply network, which focuses mostly on formal operators rather than actual water users, worsens the existing water stress in poor areas and should be addressed by a more progressive water politics. Nguyen et al. (2019) demonstrated a comprehensive and balanced approach to urban water management of a Sponge City that includes urban water resourcing, green infrastructures, ecological water management and urban permeable pavement. At the same time, the authors emphasized the need to establish an intelligent decisionmaking mechanism and close cooperation between different government agencies in order to overcome the existing barriers such as uncertainties in urban design and planning and financial insufficiencies.

Finally, some scholars emphasized the *sustainability issues* in urban water management. Farrelly and Brown (2011), for instance, while analyzing the difficulties of transition to more sustainable practices in the Australian urban water sector identified a need to construct the complex governance framework, which combine experimentation, or learning-by-doing and organizational cultures. Brown et al. (2013) conducted a longitudinal case study of Melbourne's stormwater management regime and identified the need to establish the networked bridging institutions, including government, private, community and scientific sectors, in order to improve its sustainability. At the same time, Larsen et al. (2016) studied numerous innovative approaches to sustainable urban water management such as stormwater drainage, wastewater treatment, increased water productivity, institutional and organization reforms, and pointed out the need to aggregate the transdisciplinary efforts of the research of public and private sectors. Olawumi and Chan (2018), while conducting a scientometric review of global trends in sustainability research, identified the sustainable urban water management as an emerging trend in sustainable development studies.

2.2 Sustainable water management in megacities — new agenda

For this subsection we constructed the following search query: ((water* NEAR sustainab* NEAR megacit*)). On this request WoS provided 22 publications dating back from 2013 to present — 2019 (Fig. 12.2).

Fig. 12.2 illustrates the publication activity on sustainable water management in megacities. In comparison with the previous search query, which statistical analysis of results is demonstrated on Fig. 12.1, this query shows much fewer publications. On the one hand, this could be explained by the fact that such topics as urban water management are broader

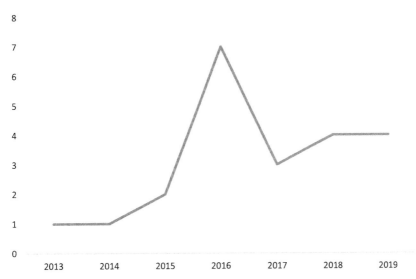

FIGURE 12.2 Publication activity on sustainable water management in megacities. *Source: WoS.*

that sustainable water management in megacities and partially include the sustainability issues, which were already discussed in the previous subchapter. On the other hand, the temporal dynamics of these publications make us assume that this is a rather new agenda for the scientific community, since the first publications on this topic date back to 2013, while the previous to 1961. Moreover, the number of publications on sustainable water management in megacities has dramatically increased in previous years. This means that this topic is gaining popularity among the scholars. In this regard, in this subchapter we will discuss the results of the previous studies on this topic.

To start with, the researchers demonstrated a variety of novel *managerial approaches* to the investigation of water sustainability challenges of megacities. Yang et al. (2016) applied an integrated framework of 'telecoupling', which considers socioeconomic and environmental interactions across telecoupled human and natural systems, to the analysis of water sustainability challenges of Beijing. As a result, the scholars highlighted the significance of considering both socioeconomic and environmental consequences, including feedback effect, for the development of sustainable water management in water-receiving (like Beijing) as well as water-sending megacities. Li et al. (2016) developed a novel methodological approach to characterization of karst aquifers in Beijing's outlying districts based on high-frequency datasets as well as routine manual flow measurements combined with traditional geochemical sampling techniques. Baskaya and Tekeli (2016) studied one of the major drinking water basins of Istanbul — Buyukcekmece — which is seriously affected by the expanding megacity. The authors emphasized the importance of establishing multiscale landscape strategies based on the evaluation of two key parameters such as 'green infrastructure capacity' and 'openness to intervention' for the environmental protection of urban streams and drinking water basins. Esculier et al. (2019) explored the water-agro-food system of Paris Megacity and suggested to optimize management of nitrogen and phosphorus in the three subsystems (agriculture, waste management and sanitation) by means of a special approach that links agriculture and urban residues.

Moreover, while investigating the sustainability aspects of water-related issues of megacities, the scholars employed and suggested for further application and testing a number of innovative *methodological techniques*. To start with, Kaya et al. (2014) analyzed the impact of urbanization, particularly rapid population growth, on two drinking water reservoirs of Istanbul (Omerli and Buyukcekmece). For this purpose, the scholars investigated three types of satellite images (vegetation layer — forests, parks and other green areas; impervious layer — settlements, transportation networks; soil layer) and identified an urgent need the sustainable urban water management through the development of sensitive watershed in order to satisfy the future water demand of the city. Ma et al. (2015), in turn, explored water footprint as an innovative approach to urban water management in contrast to the traditional measurement of water withdrawal, which is rather limited, in a megacity in North China — Beijing. The case study based on real and virtual water consumption showed that water footprint of citizens is decreasing, whereas water import dependency is increasing. In this regard, the scholars concluded that water utilization in this megacity remains unsustainable and water shortageis its key challenge since the water footprint per capita is 10 times higher than the amount of the existing water resources. To solve this problem, the authors suggested to encompass water saving.

Furthermore, some researchers suggested for use and exploration of sustainable water management in megacities *simulation and modelling methods*. Shi et al. (2013) examined an issue of increased urban water consumption in megacities of developing countries caused by booming urbanization. On the basis of multiple regression forecasting model and a back-propagation artificial neuron network the scholars simulated the water consumption changes for the rapidly developing megacity — Shenzhen — in South China and constructed a number of socioeconomic and demographic scenarios for 2011−20, which predicted slower rates of increase of water consumption over the decade. Meshgi et al. (2015) investigated the changes of hydrological cycle in tropical urban environment. By means of genetic programming, the authors developed a modular model to simulate and predict hydrograph flow components. As a result, they suggested to integrate water-sensitive urban infrastructure for sustainable water management in tropical megacities. Banihabib et al. (2016) examined the water demand management in megacities of Iran in order to optimize water allocation. For this purpose, the scholars constructed an innovative hierarchical optimization model that improves water economic efficiency and maintains agricultural production in deficit-irrigated regions. Yang et al. (2016) analyzed the issue of increasing water consumption in Chinese megacities. According to the authors, the major challenge for urban water simulation nowadays is data insufficiency. To solve this problem, the scholars created a comprehensive framework for probabilistic prediction of water consumption in an incompletely informational environment that could be used for water resources planning and management in megacities. Kumar (2019) conducted a comparative study of water quality in eight different cities of South and Southeast Asia and provided two scenarios of their future development based on a numerical simulation tool. The results of the study demonstrated the potential water quality deterioration which will not be suitable for fishing in future as it is expected by local governments. To solve this problem, the author suggested upgrading wastewater handling infrastructure. Hyndman et al. (2017) developed a novel modelling framework which helps to quantify the impact of changes in land use, crop growth and urbanization on groundwater storage and tested it for the case of Beijing. According to the researches, this model is highly efficient for the improvement of mitigation and adaptation strategies aimed at addressing water challenges of megacities.

Finally, the scholars, while analyzing the opportunities for increasing the sustainability of water management in megacities, paid attention to the *governance issues*. Fonseca-Salazar et al. (2016), for example, examined the issues of water quality and sanitation in one of the megacities of Latin America — The Mexico City Metropolitan Area. On the basis of the microbiological analysis of water, particularly the presence of bacteria and pathogenic protozoa in it, the authors provided a current diagnosis of its quality and highlighted the need for transition to wastewater treatment from a local to regional level. Chen et al. (2017) conducted a case study of a Chinese megacity, Tianjin, which showed high efficiency of large-scale centralized water reuse project, which is based on a multiple barrier treatment approach and hierarchical distribution structure. The scholars suggested to employ such an approach for safe, reliable and economically feasible long-term water management in megacities. Joo et al. (2017) analyzed the Singapore's global hydrohub strategy and emphasized the importance of collaborative city branding for the development of sustainable water solutions on the basis of the Anholt's three-pronged conceptual framework, which includes substance, strategy and symbolism. Kim et al. (2018) investigated the urban water cycle and water governance regime of Seoul by using the City Blueprint Approach, which includes Trends and Pressures Framework, City Blueprint Framework and Governance Capacity Framework. As a result, the scholars identified a number of priority areas for this megacity such as nutrient recovery from wastewater, stormwater separation and cost recovery of water sanitation services, as well as the key barriers and opportunities for the improvement of government capacity. Van den Brandeler et al. (2019), while analyzing the literature on water sustainability issues of megacities, uncovered the fact that despite the principles of integrated water resource management are widely adopted nowadays, the urban water management still ignores the river basin issues. According to the scholars, such a mismatch between river basin management and megacity water governance should be eliminated in order to create a sustainable integrated water management system. Bédécarrats et al. (2019) conducted a comprehensive assessment of the efficiency and sustainability of decentralized water systems in the African megacity — Kinshasa. The author concluded that the fulfilment of some criteria remains uncertain and identified the key challenges.

3. Methodology

As it follows from the literature review of the previous studies on urban water management and sustainable water management in megacities, the scholars in this field mostly used either theoretical governance frameworks to examine the cases of particular cities, their major sustainability challenges and opportunities for improvement or quantified the influence of urbanization and land use on water demand and supply or changes in it by means of statistical analysis and special software tools.

In this study a novel methodology for the investigations of sustainable water management is proposed, which is based on the intellectual analysis of big data, or text mining that is conducted with the use of a special system — iFORA (Intelligent Foresight analytics) (Kuzminov et al., 2018). This system automates strategic analytics in foresight and long-term science and technology forecasting, which is used for the development of strategies, technological roadmaps and investment programme for public and private entities. Currently it analyzes more than 350 million documents in English and other Latin languages, as well as Russian and other Cyrillic languages, which include:

- paper conferences (more than 100 thousand documents);
- grants (more than 1 million documents);
- reports on research projects (more than 300 thousand documents);
- scientific articles (more than 250 million documents);
- patents (more than 100 million documents);
- educational programmes;
- reports released by consulting companies and international organizations (more than 1 million documents);
- market analysis reports and professional media (more than 18 million documents);
- legislation (more than 2 million documents);
- job offers (more than 2 million documents).

The main technology of iFORA is text mining, or semantic analysis of text documents. This is a highly specialized method of extraction of valuable information from the big volumes of unstructured data by means of computer linguistics, particularly machine learning algorithms and natural language processing techniques. The semantic analysis is aimed at the identification and assessment of different semantic units in text documents such as trends, technologies, markets, products, agendas, etc. (Kuzminov et al., 2018). The main functions of iFORA, therefore, are trends and markets analysis, benchmarking and forecasting, analysis of policies and corporate strategies, network analysis and project management support (Kuzminov et al., 2018).

The iFORA methodology combines two main algorithms: identification of compatibility of terms (topics) in sentences as well as vector representation of the identified terms, which is the assignment to each term of a unique numerical age obtained by analyzing the context of texts. In this way, terms used in similar context have similar numerical vectors and closely related to each other (Kuzminov et al., 2018). The reliability of the iFORA methodology was approved when it was tested at the international research workshop 'Semantic analysis for the purposes of innovation policy' organized by Organization for Economic Co-operation and Development (Kuzminov et al., 2018).

Regarding the automated trends analysis, which is the central point of our study, iFORA allows to conduct the topical and comparative analysis of trends, patent landscapes, structural changes in priority science and technology areas; assess their significance and dynamics; and explore the technological life cycle. The key tool of processing of big volumes of unstructured data that is integrated into iFORA is a multiplier of search terms. It analyzes the input search term and provides the derived terms from market analytics, research papers, patent and grants (Kuzminov et al., 2018). The application of this tool will be illustrated in the next chapter.

4. Results

Firstly, a number of word clouds were constructed by means of terms multiplier, which is integrated into iFORA and was discussed in the previous chapter.

As it follows from Fig. 12.3, some of the topics that appeared on the word clouds presented above have already been mentioned by the authors of the previous studies. Among these topics are 'watershed function', 'water storage', 'freshwater quality', 'biophysical impact', 'global wetland', 'farming landscape', 'land cover', 'land use', 'tropical forest vegetation', 'high pollutant concentration', 'water scarcity', 'tropical dryland', 'soil and water conservation', 'local regional national and global scale', 'hydrological regime', 'water quality enhancement', 'watershed health', 'change adaptation', 'change mitigation', 'water planning', 'water demand management', 'water management policy', 'sustainable irrigation', 'climate change adaptation', 'land and water management', 'water and food security', 'water footprint', 'water conservation', 'urban water blueprint', 'clean water', 'waste water system', 'Beijing water supply', 'sprawl impact', 'sewage system', 'sanitation system', 'leaky infrastructure', 'water decision allocation', 'clean and safe drinking water'.

Still, there are some topics that have not been mentioned by the scholars: 'transboundary river basin', 'ocean base industry', 'wetland manager', 'geomorphic change', 'tropical forest logging', 'streamflow alteration', 'CO_2 emission source', 'scientific and local knowledge', 'water accounting', 'water use datum', 'sustainable groundwater', 'drought preparedness', 'bioenergy policy', 'climate proofing', 'planet water resource', 'new water technology', 'water

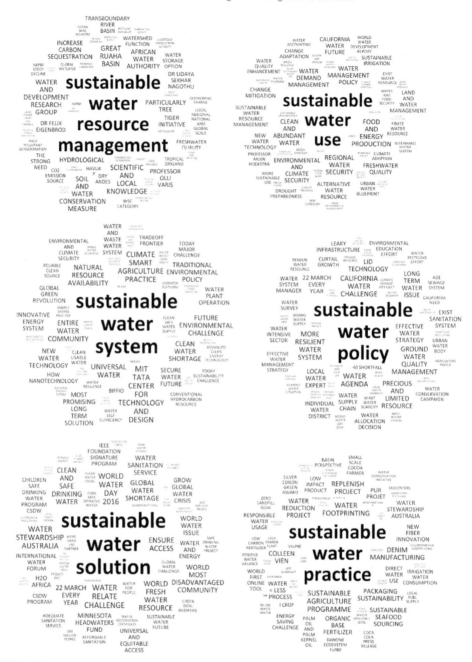

FIGURE 12.3   Word clouds for sustainable urban water management. *Source: National Research University Higher School of Economics's Text Mining System.*

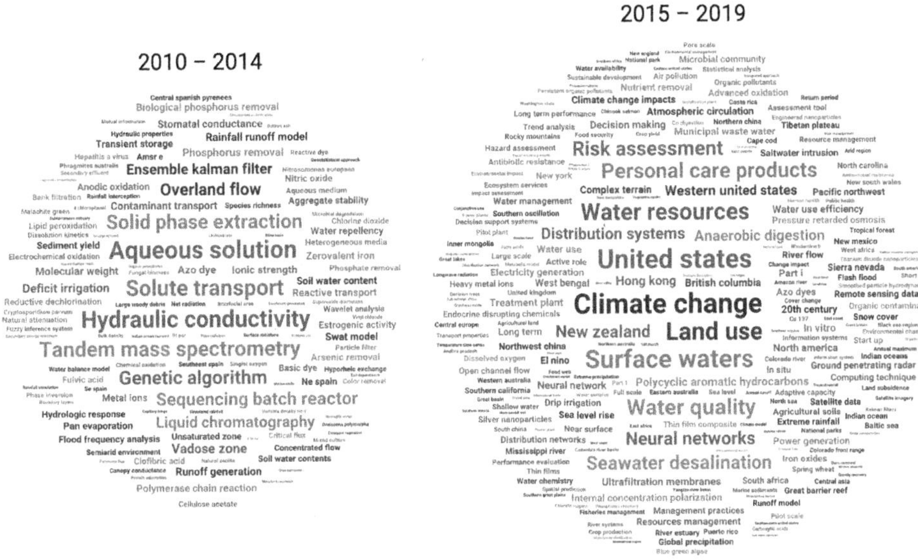

FIGURE 12.4   Changing agendas in global water management. *Source: National Research University Higher School of Economics's Text Mining System.*

resilience', 'climate smart agriculture practice', 'global green revolution', 'innovative energy system', 'entire water community', 'universal water', 'nanotechnology', 'green building', 'advanced clean energy technology', 'water plant operation', water self-sufficiency', 'environmental education effort', 'marine resource', 'LID technology', 'water recycling', 'post fire flooding', 'water intensive sector', 'water supply chain', 'individual water district', 'deforestation area', 'disadvantaged community', 'affordable sanitation', 'responsible water usage' 'low carbon fertilizer', 'positive water balance', 'new fiber innovation'.

To demonstrate the dynamics of the changing agendas in the global water management, we constructed two other word clouds for 2010−14 and 2015−19 years (Fig. 12.4). The colours on this Fig. 12.4 and other figures based on iFORA data (Figs. 12.5-12.8) indicate the presence of close relationships between the semantic units that establish separate clusters of topics.

As it follows from Fig. 12.4, the number of topics related to urban water management has increased considerably over the decade. In 2010−14 the most discussed topics in this sphere were 'aqueous solution', 'solute transport', 'hydraulic conductivity, 'tandem mass spectrometry' and 'genetic algorithm'. In 2015−19, in turn, the prevailing agendas have transformed significantly towards 'risk assessment', 'personal care products', 'water resources', 'United States' 'climate change', 'land use', 'surface waters', 'water quality', 'neural networks' and 'seawater desalination'. This means that the issue of urban water management now is considered in a more global and comprehensive way.

To illustrate the relationships existing between these topic and clusters that they establish, we represent a semantic map at Fig. 12.5.

According to Fig. 12.5, the most trending clusters of topic in global water management nowadays are 'aqueous solutions', 'sediment transport', 'time series', 'drinking water', 'waste

FIGURE 12.5 Semantic map and clusters of topics in global water management. *Source: National Research University Higher School of Economics's Text Mining System.*

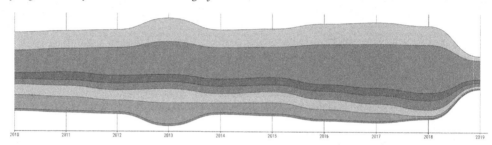

FIGURE 12.6 Clusters streamgraph in global water management, 2010—19. *Source: National Research University Higher School of Economics's Text Mining System.*

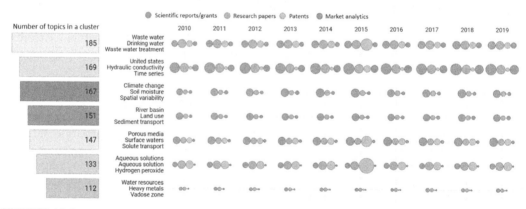

FIGURE 12.7 Life-cycle chart of clusters in global water management, 2010—2019. *Source: National Research University Higher School of Economics's Text Mining System.*

water', 'spatial variability' and 'surface waters'. To show the dynamic of the discussion of these clusters of topic in global water management, we provide the clusters streamgraph for 2010—19 at Fig. 12.6.

According to Fig. 12.6, the most discussed clusters of topic for the whole period of observation are 'waste water, drinking water and waste water treatment', 'United States, hydraulic conductivity and time series' and 'aqueous solution, aqueous solutions and hydrogen

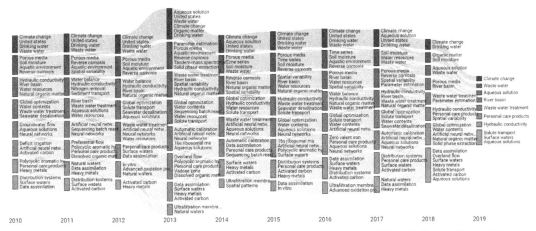

FIGURE 12.8 Structure dynamics of clusters in global water management, 2010–19. *Source: National Research University Higher School of Economics's Text Mining System.*

peroxide'. Such clusters of topics as 'climate change, soil moisture and spatial variability', 'river basin, land use and sediment transport' and 'porous media, surface waters and solute transport' were less discussed in the database of text documents examined for this period. A peak in the discussion of all the topics in 2013 should also be noticed.

To analyze the life cycles of the clusters mentioned above in more detail, we constructed a life-cycle chart of the topics in global water management for 2010–2019, which is presented at Fig. 12.7.

As it follows from Fig. 12.7, the most discussed cluster of topics in the scientific literature for the whole period of observation is 'United states, hydraulic conductivity and time series', whereas such clusters as 'waste water, drinking water and waste water treatment' and 'aqueous solution, aqueous solutions and hydrogen peroxide' showed a rapid increase in the number of patents in 2015.

To demonstrate the relationships existing between these clusters of topics, we represent a diagram of structural dynamics in global water management in 2010–19.

As it follows from Fig. 12.8, the topics from the clusters 'waste water, drinking water and waste water treatment' and 'aqueous solution, aqueous solutions and hydrogen peroxide', 'climate change, soil moisture and spatial variability' and 'river basin, land use and sediment transport' are tightly interconnected with each other.

5. Discussion and conclusion

To sum up, in this chapter the water-related sustainability issues in urban, especially megacities, environments were widely discussed.

Firstly, the scientific literature on these topics was analyzed using the citation database — WoS. With its use we conducted a brief bibliometric analysis, which showed that such themes as 'urban water management' has been widely discussed since 1961 and the number of publications has increased dramatically in recent years. At the same time, the number of publications on 'sustainable water management in megacities' is much less discussed in the academic literature since the first publications date back only to 2013. This means, that this topic is relatively new for the scientific community and, apparently, rapidly gaining popularity.

Secondly, we analyzed the major issues and solutions that were previously highlighted by the scholars in this field. In particular, most of the researchers identified the following problems existing in urban water management:

- natural hazards such as floods, storms, etc.;
- lack of clean drinking water;
- imperfection of drainage infrastructure;
- lack of 'green' infrastructure;
- imperfection of wastewater treatment system;
- need to move towards sustainability.

To solve these issues, the scholars suggested a number of tools that can be grouped into four categories:

- management tools (strategic management tools, foresight projects, scenarios building, flood mitigation and water resources management systems, urban stormwater management system);
- technological tools (water efficient devices, radars, microwave networks, membrane filtration and separation, activated carbon adsorption, ion-exchange resin, chemical oxidation, extreme learning machine method, water purification, intentional groundwater replenishment);
- 'green' tools (green roofs, urban gardens, urban trees, urban forest, 'green' infrastructure);
- governance tools (comprehensive wastewater treatment programmes including improved policies (e.g., effluent standards), citywide and local frameworks for maintaining the ecosystem services, urban drainage guidance, 'hydrosocial' linkages, unified terms for communication, political agendas and visions of citizens, intelligent decision-making mechanisms).

Thirdly, we analyzed a broad scope of research papers on sustainable water management in megacities and discovered that previously the scholars, while investigated this topic, established a number of new managerial approaches to addressing the challenges described above. Among these approaches are integrated framework of 'telecoupling', which considers socio-economic and environmental interactions across telecoupled human and natural systems, a novel methodological approach to characterization of karst aquifers based on high-frequency datasets, as well as routine manual flow measurements combined with traditional geochemical sampling techniques, multiscale landscape strategy development, linking agriculture and urban residues to optimize management of nitrogen and phosphorus. Moreover, we discovered a number of new methodological techniques in this literature such as satellite images analysis by several layers and water footprint exploration. In addition, we observed a wide use of simulation and modelling methods to investigate urban water consumption and demand, the hydrological cycle, the impact of changes in land use, crop growth and urbanization on groundwater storage, etc.

Finally, we conducted intellectual big data analysis by means of a specialized system — iFORA. On this basis, we identified a number of topics that were previously ignored by the researchers. These topics mostly include either highly innovative technological solutions, which might be underinvestigated at the moment, or global issues that attracted less attention from the researchers since most of the studies were dedicated to particular cities or mega-cities. Moreover, we detected the changing character of the agendas in global water management, the number of which is rapidly increasing and has become of a more comprehensive character. Finally, we provided a number of prevailing topics connected with global water issues and their clusters ('waste water, drinking water and waste water treatment', 'United States, hydraulic conductivity and time series', 'aqueous solution, aqueous solutions and hydrogen peroxide', 'climate change, soil moisture and spatial variability', 'river basin, land use and sediment transport', 'porous media, surface waters and solute transport') as well as showed their structural dynamics over the observation period (2010—19).

Overall, these findings show a relatively high degree of compatibility with the recommendations for sustainable water management and key critical factors represented by UNESCO. These results are highly suggested to be used for the establishment of sustainable water management strategies by policy actors in megacities of the future. Possibly, these trends could be implemented by national authorities into the roadmaps construction oriented towards the establishment of sustainable urban water management system in megacities or scenarios development.

Acknowledgment

The book chapter was prepared within the framework of the Basic Research Program at the National Research University Higher School of Economics.

References

Alda-Vidal, C., Kooy, M., Rusca, M., 2018. Mapping operation and maintenance: an everyday urbanism analysis of inequalities within piped water supply in Lilongwe. Malawi Urban Geography 39 (1), 104—121.

Bach, P.M., Rauch, W., Mikkelsen, P.S., Mccarthy, D.T., Deletic, A., 2014. A critical review of integrated urban water modelling—Urban drainage and beyond. Environmental Modelling and Software 54, 88—107.

Banihabib, M.E., Hosseinzadeh, M., Peralta, R.C., 2016. Optimization of inter-sectorial water reallocation for arid-zone megacity-dominated area. Urban Water Journal 13 (8), 852—860.

Baskaya, F.T., Tekeli, E., 2016. Revealing landscape strategies for the water basins—case of buyukcekmece lake, Istanbul. Journal of Environmental Protection and Ecology 17 (2), 703.

Barthel, S., Isendahl, C., 2013. Urban gardens, agriculture, and water management: sources of resilience for long-term food security in cities. Ecological Economics 86, 224—234.

Bédécarrats, F., Lafuente-Sampietro, O., Leménager, M., Sowa, D.L., 2019. Building commons to cope with chaotic urbanization? Performance and sustainability of decentralized water services in the outskirts of Kinshasa. Journal of Hydrology 573, 1096—1108.

Berndtsson, J.C., 2010. Green roof performance towards management of runoff water quantity and quality: a review. Ecological Engineering 36 (4), 351—360.

Booth, D.B., Roy, A.H., Smith, B., Capps, K.A., 2016. Global perspectives on the urban stream syndrome. Freshwater Science 35 (1), 412—420.

Berardi, U., GhaffarianHoseini, A., GhaffarianHoseini, A., 2014. State-of-the-art analysis of the environmental benefits of green roofs. Applied Energy 115, 411—428.

Berland, A., Shiflett, S.A., Shuster, W.D., Garmestani, A.S., Goddard, H.C., Herrmann, D.L., Hopton, M.E., 2017. The role of trees in urban stormwater management. Landscape and Urban Planning 162, 167—177.

Brown, R.R., Farrelly, M.A., Loorbach, D.A., 2013. Actors working the institutions in sustainability transitions: the case of Melbourne's stormwater management. Global Environmental Change 23 (4), 701–718.

Chan, F.K.S., Griffiths, J.A., Higgitt, D., Xu, S., Zhu, F., Tang, Y.T., et al., 2018. "Sponge City" in China—a breakthrough of planning and flood risk management in the urban context. Land Use Policy 76, 772–778.

Chen, Z., Wu, Q., Wu, G., Hu, H.Y., 2017. Centralized water reuse system with multiple applications in urban areas: lessons from China's experience. Resources, Conservation and Recycling 117, 125–136.

Dillon, P., Stuyfzand, P., Grischek, T., Lluria, M., Pyne, R.D.G., Jain, R.C., et al., 2019. Sixty years of global progress in managed aquifer recharge. Hydrogeology Journal 27 (1), 1–30.

Esculier, F., Le Noë, J., Barles, S., Billen, G., Créno, B., Garnier, J., et al., 2019. The biogeochemical imprint of human metabolism in Paris Megacity: a regionalized analysis of a water-agro-food system. Journal of Hydrology 573, 1028–1045.

Ernstson, H., 2013. The social production of ecosystem services: a framework for studying environmental justice and ecological complexity in urbanized landscapes. Landscape and Urban Planning 109 (1), 7–17.

Euromonitor, 2018. Megacities: Developing Country Domination. URL: https://go.euromonitor.com/strategy-briefing-cities-2018-megacities.html.

Farrelly, M., Brown, R., 2011. Rethinking urban water management: experimentation as a way forward? Global Environmental Change 21 (2), 721–732.

Fioretti, R., Palla, A., Lanza, L.G., Principi, P., 2010. Green roof energy and water related performance in the Mediterranean climate. Building and Environment 45 (8), 1890–1904.

Fletcher, T.D., Shuster, W., Hunt, W.F., Ashley, R., Butler, D., Arthur, S., et al., 2015. SUDS, LID, BMPs, WSUD and more—The evolution and application of terminology surrounding urban drainage. Urban Water Journal 12 (7), 525–542.

Fonseca-Salazar, M.A., Díaz-Ávalos, C., Castañón-Martínez, M.T., Tapia-Palacios, M.A., Mazari-Hiriart, M., 2016. Microbial indicators, opportunistic bacteria, and pathogenic protozoa for monitoring urban wastewater reused for irrigation in the proximity of a megacity. EcoHealth 13 (4), 672–686.

Fletcher, T.D., Andrieu, H., Hamel, P., 2013. Understanding, management and modelling of urban hydrology and its consequences for receiving waters: a state of the art. Advances in Water Resources 51, 261–279.

García-Ruiz, J.M., López-Moreno, J.I., Vicente-Serrano, S.M., Lasanta–Martínez, T., Beguería, S., 2011. Mediterranean water resources in a global change scenario. Earth-Science Reviews 105 (3–4), 121–139.

Hommes, L., Boelens, R., 2018. From natural flow to 'working river': hydropower development, modernity and socio-territorial transformations in Lima's Rímac watershed. Journal of Historical Geography 62, 85–95.

Hyndman, D.W., Xu, T., Deines, J.M., Cao, G., Nagelkirk, R., Viña, A., et al., 2017. Quantifying changes in water use and groundwater availability in a megacity using novel integrated systems modeling. Geophysical Research Letters 44 (16), 8359–8368.

Joo, Y.M., Heng, Y.K., 2017. Turning on the taps: Singapore's new branding as a global hydrohub. International Development Planning Review 39 (2), 209–227.

Kim, H., Son, J., Lee, S., Koop, S., Van Leeuwen, K., Choi, Y., Park, J., 2018. Assessing urban water management sustainability of a megacity: case study of Seoul, South Korea. Water 10 (6), 682.

Kaya, S., Seker, D.Z., Tanik, A., 2014. Temporal impact of urbanization on the protection zones of two drinking water reservoirs in Istanbul. Fresenius Environmental Bulletin 23 (12), 2984–2989.

Kumar, P., 2019. Numerical quantification of current status quo and future prediction of water quality in eight Asian megacities: challenges and opportunities for sustainable water management. Environmental Monitoring and Assessment 191 (6), 319.

Kuzminov, I., Bakhtin, P., Khabirova, E., Loginova, I.V., 2018. Detecting and validating global technology trends using quantitative and expert-based foresight techniques. Higher School of Economics Research Paper No. WP BRP 82.

Larsen, T.A., Hoffmann, S., Lüthi, C., Truffer, B., Maurer, M., 2016. Emerging solutions to the water challenges of an urbanizing world. Science 352 (6288), 928–933.

Li, J., Qi, Y., Zhong, Y., Yang, L., Xu, Y., Lin, P., et al., 2016. Karst aquifer characterization using storm event analysis for Black Dragon springshed, Beijing, China. Catena 145, 30–38.

Livesley, S.J., McPherson, E.G., Calfapietra, C., 2016. The urban forest and ecosystem services: impacts on urban water, heat, and pollution cycles at the tree, street, and city scale. Journal of Environmental Quality 45 (1), 119–124.

Liu, L., Jensen, M.B., 2018. Green infrastructure for sustainable urban water management: practices of five forerunner cities. Cities 74, 126–133.

Ma, D., Xian, C., Zhang, J., Zhang, R., Ouyang, Z., 2015. The evaluation of water footprints and sustainable water utilization in Beijing. Sustainability 7 (10), 13206–13221.

Martin, B.R., 2016. Twenty challenges for innovation studies. Science and Public Policy 43 (3), 432–450.

McDonald, R.I., Weber, K., Padowski, J., Flörke, M., Schneider, C., Green, P.A., et al., 2014. Water on an urban planet: urbanization and the reach of urban water infrastructure. Global Environmental Change 27, 96–105.

Meshgi, A., Schmitter, P., Chui, T.F.M., Babovic, V., 2015. Development of a modular streamflow model to quantify runoff contributions from different land uses in tropical urban environments using Genetic Programming. Journal of Hydrology 525, 711–723.

Michael-Kordatou, I., Michael, C., Duan, X., He, X., Dionysiou, D.D., Mills, M.A., Fatta-Kassinos, D., 2015. Dissolved effluent organic matter: characteristics and potential implications in wastewater treatment and reuse applications. Water Research 77, 213–248.

Nguyen, T.T., Ngo, H.H., Guo, W., Wang, X.C., Ren, N., Li, G., et al., 2019. Implementation of a specific urban water management-Sponge City. The Science of the Total Environment 652, 147–162.

Olawumi, T.O., Chan, D.W., 2018. A scientometric review of global research on sustainability and sustainable development. Journal of Cleaner Production 183, 231–250.

Qadir, M., Wichelns, D., Raschid-Sally, L., McCornick, P.G., Drechsel, P., Bahri, A., Minhas, P.S., 2010. The challenges of wastewater irrigation in developing countries. Agricultural Water Management 97 (4), 561–568.

RGS, 2016. Discovering Megacities. URL: https://www.rgs.org/schools/teaching-resources/discovering-megacities/.

Shi, P., Yang, T., Chen, X., Yu, Z., Acharyad, K., Xu, C., 2013. Urban water consumption in a rapidly developing flagship megacity of South China: prospective scenarios and implications. Stochastic Environmental Research and Risk Assessment 27 (6), 1359–1370.

Sun, X., Crittenden, J.C., Li, F., Lu, Z., Dou, X., 2018. Urban expansion simulation and the spatio-temporal changes of ecosystem services, a case study in Atlanta Metropolitan area, USA. The Science of the Total Environment 622, 974–987.

UN, 2018. The World's Cities in 2018. URL: https://www.un.org/en/events/citiesday/assets/pdf/the_worlds_cities_in_2018_data_booklet.pdf.

UN, 2014. A World of Cities. URL: https://www.un.org/en/development/desa/population/publications/pdf/popfacts/PopFacts_2014-2.pdf.

UNESCO, 2019. Water, Megacities and Global Change: Portraits of 16 Emblematic Cities of the World. URL: https://unesdoc.unesco.org/ark:/48223/pf0000367866?posInSet&=3&queryId&=ddec32dc-6796-443a-ab82-a1db8eb697ed.

UNESCO, 2016. Water, megacities and global change. Portraits of 15 Emblematic Cities of the World (URL: https://unesdoc.unesco.org/ark:/48223/pf0000245419.).

Van den Brandeler, F., Gupta, J., Hordijk, M., 2019. Megacities and rivers: scalar mismatches between urban water management and river basin management. Journal of Hydrology 573, 1067–1074.

Wheater, H., Evans, E., 2009. Land use, water management and future flood risk. Land Use Policy 26, S251–S264.

Willis, R.M., Stewart, R.A., Giurco, D.P., Talebpour, M.R., Mousavinejad, A., 2013. End use water consumption in households: impact of socio-demographic factors and efficient devices. Journal of Cleaner Production 60, 107–115.

Xia, J., Zhang, Y., Xiong, L., He, S., Wang, L., Yu, Z., 2017. Opportunities and challenges of the Sponge City construction related to urban water issues in China. Science China Earth Sciences 60 (4), 652–658.

Yang, K., Pan, M., Luo, Y., Chen, K., Zhao, Y., Zhou, X., 2019. A time-series analysis of urbanization-induced impervious surface area extent in the Dianchi Lake watershed from 1988–2017. International Journal of Remote Sensing 40 (2), 573–592.

Yang, T., Shi, P., Yu, Z., Li, Z., Wang, X., Zhou, X., 2016a. Probabilistic modeling and uncertainty estimation of urban water consumption under an incompletely informational circumstance. Stochastic Environmental Research and Risk Assessment 30 (2), 725–736.

Yang, W., Hyndman, D., Winkler, J., Viña, A., Deines, J., Lupi, F., et al., 2016b. Urban water sustainability: framework and application. Ecology and Society 21 (4).

Yaseen, Z.M., Jaafar, O., Deo, R.C., Kisi, O., Adamowski, J., Quilty, J., El-Shafie, A., 2016. Stream-flow forecasting using extreme learning machines: a case study in a semi-arid region in Iraq. Journal of Hydrology 542, 603–614.

Zhou, F., Xu, Y., Chen, Y., Xu, C.Y., Gao, Y., Du, J., 2013. Hydrological response to urbanization at different spatio-temporal scales simulated by coupling of CLUE-S and the SWAT model in the Yangtze River Delta region. Journal of Hydrology 485, 113–125.

13

Comparing invasive alien plant community composition between urban, peri-urban and rural areas; the city of Cape Town as a case study

Luca Afonso[1], Karen J. Esler[2], Mirijam Gaertner[3], Sjirk Geerts[4]

[1]Centre for Invasion Biology, Department of Botany and Zoology, Stellenbosch University, Matieland, South Africa; [2]Department of Conservation Ecology and Entomology and Centre for Invasion Biology, Stellenbosch University, Matieland, South Africa; [3]Nürtingen-Geislingen University of Applied Sciences (HFWU), Schelmenwasen 4-8, Nürtingen, Germany and Centre for Invasion Biology, Department of Botany and Zoology, Stellenbosch University, Matieland, South Africa; [4]Department of Conservation and Marine Sciences, Cape Peninsula University of Technology, Cape Town, South Africa

Urban Ecology
https://doi.org/10.1016/B978-0-12-820730-7.00013-6

1. Introduction

Urbanization influences biodiversity in many different ways (Collins et al., 2000; Pickett et al., 2001; Sukopp and Werner, 1983), e.g., by altering quality of air, water and soil (Gill, 1996; Willig and Walker, 1999); temperature regime; rainfall patterns (Landsberg, 1981; Sheperd et al., 2002), fragmenting and disturbing remaining habitat (Kowarik, 1990). Urban areas — defined as areas surrounded by city development and characterized by high-density human infrastructure such as residential areas, commercial buildings and transport infrastructure (National Geographic, 2016a) — are also hotspots for invasions and act as key points of entry for introduced invasive alien plants (IAPs) (Kowarik, 2011). Trade and horticulture are the most prolific pathways of human-mediated dispersal (Dehnen-Schmutz et al., 2007; Dehnen-Schmutz, 2011; Mack et al., 2000; Pyšek, 1998; Reichard and Hamilton, 1997; Vitousek, 1997). Consequently, large urban areas often contain a greater proportion of IAPs compared to surrounding rural areas (Khun and Klotz, 2006; Pyšek, 1998). The latter is defined as open swathes of land which have a low population density with agriculture being the primary economic activity (National Geographic, 2016b).

Disturbances associated with pulses of available nutrients (Davis and Thompson, 2000) and IAP propagule pressure (Lonsdale, 1999) have been put forward as two of the most significant factors which influence plant invasion in urban areas (Lososova et al., 2012). With increasing levels of human-mediated habitat disturbance, propagule pressure increases and so do the number of nonnative plant species (Gaertner et al., 2016; Kowarik, 2011; Lososova et al., 2012). These invasions are often facilitated through disturbances which are associated with roads, railways and riparian areas. Furthermore, in the urban/wildland interface humans are more likely to disturb natural habitats (natural areas or public open space which is undeveloped and sustains biodiversity) through trail construction, poor vegetation management, vegetation trampling or refuse dumping (Sullivan et al., 2005) and plant collecting (Alston and Richardson, 2006). These areas at the urban/wildland interface are also referred to as peri-urban areas and are at the interface or transition zone between urban and rural areas (Thapa and Murayama, 2008). Typically they contain both urban and rural features (Allen, 2003).

Natural environments at the urban fringe also experience increased levels of air and water pollution Struglia (Struglia and Winter, 2016) and erosion or sedimentation through water runoff from the hard surfaces characteristic of urban areas (Whitford et al., 2001). Many of these factors interact and have a synergistic effect on one another. For example, disturbance leads to increased solar radiation on soils, higher temperatures, altered soil conditions, and may result in increased IAPs and a decrease in native plant diversity (Drayton and Primack, 1996; Parendes and Jones, 2000). Furthermore, parks and gardens act as an important source of propagules for IAPs at the urban fringe (Alston and Richardson, 2006; Anderson et al., 2014). High propagule pressure increases the chance of establishment, naturalization and invasion of IAPs (Lonsdale, 1999; Rouget and Richardson, 2003).

Since urban areas can also act as sources of IAP invasions into natural areas that surround them (Pyšek, 1998), natural areas at the urban/wildland fringe are vulnerable to invasion because of their proximity to gardens (Marco et al., 2010; Bell et al., 2003). The abundance and richness of IAPs are known to be linked to the distance to the nearest urban area (Sullivan et al., 2005). Alston and Richardson (2006) found that the distance from putative source populations and human-mediated disturbance influence IAP richness in Newlands forest, South Africa. In contrast, rural areas contain few typical urban features and are strongly related to agricultural land uses or large open swaths of land (National Geographic, 2016a). Cilliers et al. (2008) discovered that IAP invasions penetrate deeper into native urban grasslands when compared to native rural grasslands.

Since urban areas are characterized by high levels of stress and disturbance compared to peri-urban and rural areas, one might expect that IAPs in urban areas may have different plant adaptive strategies to those found in peri-urban and rural areas. According to Grime (1977), stress (conditions which restrict production) and disturbance (the destruction of plant biomass) are two external limiting factors which affect plant biomass in any environment. As such, three plant adaptive strategies may occur, i.e., competitive plants (low stress and low disturbance); stress-tolerant plants (low disturbance and high stress); and ruderal plants (high disturbance, low stress). The difference in habitat condition between urban areas and rural areas means that plant adaptive strategies and growth forms should differ between these environments. For example, due to the higher levels of nutrients and high levels of human-mediated disturbance in urban areas, one would expect ruderal species to be more prevalent in those areas.

Despite the plethora of research on IAPs, distributions of IAPs within urban areas and surrounding areas are relatively unknown (Gulezian and Nyberg, 2010; Lososova et al., 2012). By investigating the difference in invasive species composition between urban, peri-urban and rural areas, we aimed to explore the mechanisms which promote IAP establishment in urban and surrounding areas and to formulate best-practice management suggestions. We further explore the origin of IAPs (whether horticultural, agricultural or forestry) and their adaptive strategies (Grime, 1977) within urban and surrounding areas.

Our study area is the City of Cape Town. Cape Town is located within the Cape Floristic Region (CFR) of South Africa and is considered a biodiversity hotspot since it has a high native species richness and high levels of endemism (Holmes et al., 2012; Myers et al., 2000). The CFR is also a recognized UNESCO world heritage site, being the smallest but most biologically diverse compared to the other plant kingdoms (Davison and Marshak, 2012). The region has a long history of human activity with resultant disturbance from farming practices, urban development and considerable recreational usage.

Despite the fact that South Africa has some of the most recognizable and unique plant communities in the world, it is also one of the most invaded and threatened (Bromilow, 2010). Many of the prominent IAPs found in Cape Town (acacias and pines) were introduced for multiple reasons. *Acacia saligna* and *A. cyclops* were introduced in the mid-19th century to stabilize sand dunes (Davison and Marshak, 2012). Pine species were introduced for the timber industry, remnants of which are still visible in Cape Town today (Anderson and O' Farrell, 2012). A number of these species have significant impacts on ecosystem services in natural, agricultural and transformed areas (Mangachena and Geerts, 2017; van Wilgen et al., 2008). For hundreds of years many agricultural IAPs were accidentally introduced with the sowing of crop seeds and these now persist as weeds on agricultural and natural land (Bromilow, 2010). Plants are also often intentionally introduced as ornamentals because of their flowers, high growth and germination rate and tolerance to a wide range of environmental conditions, which often leads to a conflict of interests between conservationists, land owners and plant nursery owners (Foxcroft et al., 2008; Geerts et al., 2017; Novoa et al., 2018).

To determine how natural areas within Cape Town are affected by IAPs, we selected sample sites within the Cape Town Biological Network (BioNet). The Cape Town BioNet consists of a series of interconnected protected areas known as Critical Biodiversity Areas (CBAs) (ranging from pristine habitats to more degraded, but highly threatened ecosystems) and Critical Ecological Support Areas (CESAs) (Davison and Marshak, 2012). The BioNet was developed by the City of Cape Town to identify vegetation remnants of high importance to avoid the loss of endangered and threatened flora to urban development. The Cape Town BioNet is, therefore, an intrinsic feature to the urban fabric of Cape Town. As such the BioNet offers a unique opportunity to study the abundance and composition of IAPs that invade natural areas from urban, peri-urban and rural land uses in the city of Cape Town municipal boundaries.

We aimed to investigate the composition (abundance and richness) of IAPs in urban, peri-urban and rural areas in the City of Cape Town, South Africa. More specifically we determined (1) if there is a difference between IAP species abundance and richness between urban, peri-urban and rural areas; (2) if habitat condition is responsible for promoting the establishment and facilitating invasion of IAPs into natural areas; (3) the reason for introduction and (4) the most prevalent plant adaptive strategies of IAPs in urban environments.

2. Methods

2.1 Study area

The dominant vegetation type in our study area is Swartland Shale Renosterveld which can be found growing on fertile fine-grained soils in the Swartland and Boland areas of the Western Cape (Walton, 2006). Approximately 95% of this vegetation type has been transformed due to agricultural activities (i.e., vineyards and orchids) with only 0.5% of its original extent being formally protected (Cowen, 2013; Walton, 2006). Swartland Shale Renosterveld is therefore classified as critically endangered (Cowen, 2013). It contains low to tall narrow-leaved shrubland growing in more arid areas.

Our study area is located in the greater Tygerberg area of the City of Cape Town where urban, peri-urban and rural sites meet. The area has a Mediterranean climate characterized by winter rainfall and summer drought. The annual rainfall ranges between 100 and 400 mm per annum (Harris et al., 2010). All study sites lie within the BioNet (Davison and Marshak, 2012). All the sites that were selected occur within naturally vegetated areas adjacent to urban, peri-urban or rural habitat types. The urban sites were located within Tygerberg Nature Reserve, Durbanville Nature Reserve and Uitkamp Nature Reserve, which are all urban nature reserves. All sites at all three habitat types are regularly cleared of conspic-uous IAPs by both the reserve management as well as the City of Cape Town Invasive Species Management Unit. Moreover, nature reserves are popular recreational areas in an urban matrix of wealthy suburbs with large and diverse gardens.

2.2 Site selection

For the selection of sites, different layers supplied by the South African National Biodiver-sity Institute (SANBI) Biodiversity GIS website (http://bgis.sanbi.org/[Biodiversity GIS, hereafter]) were used for analysis in Arc GIS 10.3.1. The City of Cape Town BioNet layer, as well as the Biodiversity GIS Vegetation map of 2012, was used in selecting sampling sites. The sites were selected by extracting the Swartland Shale Renosterveld (SSR) vegetation cate-gory from the VegMap 2012 (selected by attributes and exported as a layer) and selecting sampling sites (by location) for all BioNet sites (polygons) that intersected and overlapped with SSR vegetation type layer (Fig. 13.1). Sampling was done in one vegetation type (Swart-land Shale Renosteveld) in a localized area in order to limit environmental variability.

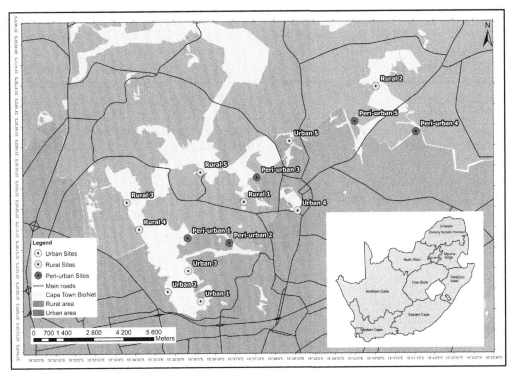

FIGURE 13.1 Study sites used to identify differences in invasive alien species composition between urban, peri-urban and rural areas within the greater Tygerberg area of the City of Cape Town. Swartland Shale Renosterveld within the Cape Town Bionet (highlighted in pink: light gray in printed version) is shown in relation to urban land cover (red: dark gray in printed version) and rural land cover (green: gray in printed version). The peri-urban zone is present at the interface of the urban and rural zones. The study sites selected across the urban-rural gradient are illustrated on the map as yellow (light gray in printed version) (urban cites), blue (dark gray in printed version) (peri-urban) and green (gray in printed version) (rural sites). The scale of this map is 1:80 000.

Urban sites are defined as areas which are in close proximity (within 100 m) to only residential, commercial or industrial land-use types, which represent typical urban land-use types. Peri-urban sites are defined as areas which are in close proximity (within 100 m) to residential, industrial or commercial land-use types and open public space or agricultural land-use types. Rural sites are those which are in close proximity (within 100 m) to agricultural, public open space and rural land-use types. The land-use types were verified with the use of the City of Cape Town Integrated Zoning Land Parcel layer (City of Cape Town, 2016).

2.3 Field sampling and experimental design

We sampled five replicates of each habitat type (urban, peri-urban or rural). Sites were sampled during the summer months of November and December 2016. Within each of these replicates, six 25 m^2 plots were sampled (n = 90 plots) to determine IAP species richness and abundance. The number of IAP species as well as their percentage cover was recorded for each plot. Voucher specimens of all unknown species were collected, pressed and identified to species level.

To identify factors promoting IAPs in urban areas as compared to peri-urban and rural areas we recorded habitat condition for each plot using a scoring system (developed by the City of Cape Town Invasive Species Management Unit) (Table 13.1). Each criterion was scored along a scale of 1–5. The criteria classes are as follows: 1 = good/natural; 2 = good/near-natural; 3 = fair/degraded; 4 = modified/very degraded; 5 = irreversibly degraded. The mean score for each plot was then calculated. According to their habitat condition, plots were classified as containing pristine vegetation (good/natural/near-natural vegetation) or degraded vegetation (fair/degraded/modified/very degraded/irreversibly degraded or modified vegetation). The environmental variables collected for each plot were: percentage exposed rock, percentage bare ground and degree of erosion as well as environmental disturbances.

TABLE 13.1   The criteria used to determine whether habitat condition is a mechanism for IAP species establishment in urban, peri-urban and rural areas

Criterion	Subcriterion
Physical disturbance	Top soil lost (e.g., quarry; road)
	Ploughed with or without fertilizing
	Altered hydrology
Biotic impacts	Overgrazing/brush cutting/over harvesting
Indigenous vegetation cover	Vegetation % cover
	Vegetation postfire age (year)
	Median height (m)
	Strata and spatial patterning of dominant growth forms
Composition	Indigenous species richness

Criteria classes: 1, good/natural; 2, good/near-natural; 3, fair/degraded; 4, modified/very degraded; 5, irreversibly degraded.

To determine the origin of the IAPs found in this study, they were classified on the basis of the likely reason for their introduction and use in the area, i.e., ornamental horticulture, forestry, agriculture or indigenous. Species origins were obtained from Bromilow (2010) and Fish et al. (2015). In addition, species were classified according to the C-S-R plant functional types (Grime, 1977), i.e., competitors, stress-tolerators and ruderal species. With the use of field guides, the invasion status (whether invasive, alien, naturalized or extralimital native species) of each species was determined, as was growth form type.

2.4 Statistical analyses

IAP richness and abundance were analyzed using variance estimation, precision and comparison (VEPAC Module of Statistica 13.5) analyses. VEPAC was used to (1) determine the variance and differences between habitat types for both IAP species richness and IAP species abundance and (2) to determine the difference in IAP species abundance for five different growth forms (grasses, herbs, shrubs, trees and climbers). The VEPAC was a mixed model ANOVA with habitat type as a fixed effect and area as random effect. This was followed by a Fishers Least Significant Difference (LSD) posthoc test (LSD). A mean habitat score was determined for each plot by averaging the score given for each of the subcriteria shown in Table 13.1.

A principal component analysis (done using the R package 'Vegan') was used to determine whether species composition is different between habitat types. In addition, multiple regression analyses were used to determine the relationship between habitat condition, IAP species abundance and richness. All analyses were conducted in STATISTICA 13 (Statistica, 2019) and R (R development core team, 2013).

3. Results

In total, 79 alien plant species were recorded in the greater Tygerberg area across all habitat types. Urban, peri-urban and rural habitat types had proportionately more invasive plant species than alien and naturalized plant species (Table 13.2). In urban areas, there were equal proportions of competitors (42%) and ruderal species (46%) with fewer stress-tolerators (12%). However, there was a higher proportion of ruderal species (54%) in both the peri-urban and rural habitat types compared to competitors (37%) and stress-tolerators (9%). IAPs from an agricultural origin were more common in all three habitat types than species from a horticultural or forestry origin (Table 13.2).

The two main components of the principal component analysis describe 14% (PC1 — x-axis) and 13% (PC2 — Y-axis) of the variation. The species composition for urban and rural habitat types was similar and tightly grouped together (Fig. 13.2). The grouping of peri-urban species was more widely scattered and the species were grouped adjacent to the urban and rural showing that the results for peri-urban sites differ from those of the urban and rural sites. *Bromus hordeaceus, B. diandrus, Lolium multiflorum x Liliom perene, Aevena barabarta* and *Cynodon dactylon* were the five major outliers from the PCA and may explain why the two principal components account for relatively little of the variation.

TABLE 13.2 The number (and percentage) of plant species across the three different habitat types for invasion status, plant adaptive strategies and the potential origins of introductions for all species sampled across the City of Cape Town. Percentages were derived using the total number of species found in each area.

		Urban (%)	Peri-urban (%)	Rural (%)
Invasion status	Invasive	31(56)	26(59)	18(53)
	Alien	8(15)	7(16)	4(12)
	Naturalized	11(20)	10(23)	11(32)
	Indigenous	5(9)	1(2)	1(3)
Plant adaptive strategies	Competitor	24(42)	17(37)	13(37)
	Ruderal	26(46)	25(54)	19(54)
	Stress-tolerator	7(12)	4(9)	3(9)
IAP origins	Forestry	3(5)	1(2)	1(3)
	Agricultural	33(58)	28(61)	23(66)
	Horticultural	7(12)	7(15)	6(17)
	Unknown	10(18)	8(17)	3(9)
	Indigenous	4(7)	2(4)	2(6)

IAP, Invasive Alien Plant

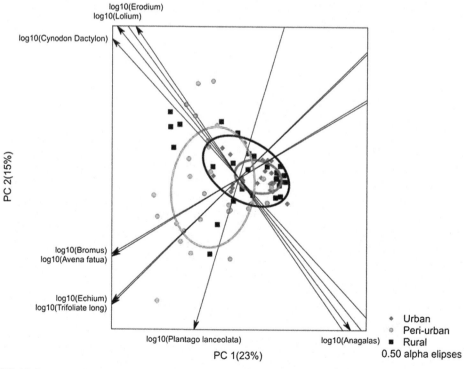

FIGURE 13.2 Principle component analysis showing the groupings of invasive alien species in urban (red: dark gray in printed version), peri-urban (blue: gray in printed version), and rural (black).

TABLE 13.3 Table showing the means and *P*-values for the variance estimation and precision and comparison undertaken for invasive alien plant abundance, invasive alien plant richness as well as the different growth forms sampled across all three habitat types. Abundance is the number of individuals per species in a community and richness is the number of species within a community. *P* values less than 0.05 are significant.

	Urban	Peri-urban	Rural	*P*-value
IAP species abundance	24[b]	57[a]	18[b]	0.000746
IAP species richness	8	9.2	6.2	0.10
Grasses	0.7[b]	1.4[a]	0.7[b]	0.01426
Herbs	1[ab]	1.15[a]	0.75[b]	0.02347
Trees	0.15	0.06	0.04	0.37130
Shrubs	0.056	0.054	0.002	0.41657
Creepers	0	0.12	0	0.39657

Mean values with the same letter(s) are not significantly different.

There was no significant difference in IAP abundance between urban and rural habitat types (Table 13.3). The peri-urban habitat type had a significantly higher abundance of IAPs compared to urban and rural habitat types (Table 13.3). There was no significant difference in IAP species richness between the different habitat types (Table 13.3).

There was no significant difference in tree, shrub or creeper abundances between the different habitat types (Table 13.3). However, grasses were significantly more abundant in peri-urban habitat compared to the urban and rural habitat types, but there was no significant difference between grasses in the urban and rural habitat types. Herbs found in peri-urban sites were significantly higher in abundance than herbs in the rural habitat type (Table 13.3).

The abundance of IAPs was positively correlated to habitat condition ($P = .00$, Fig. 13.3). Furthermore, the richness of IAPs was positively correlated to habitat condition ($P = .0008$, Fig. 13.3). Therefore, as the level of degradation increases so does the number and cover of IAPs.

4. Discussion

In determining the composition of IAPs across the urban-rural gradient, we show that IAP abundance and richness were marginally higher in urban areas compared to rural areas. However, the abundance and richness of IAPs in peri-urban areas were significantly higher than in both the urban and rural areas. These broad patterns may be associated with the types of IAP source populations in the different areas and/or habitat conditions.

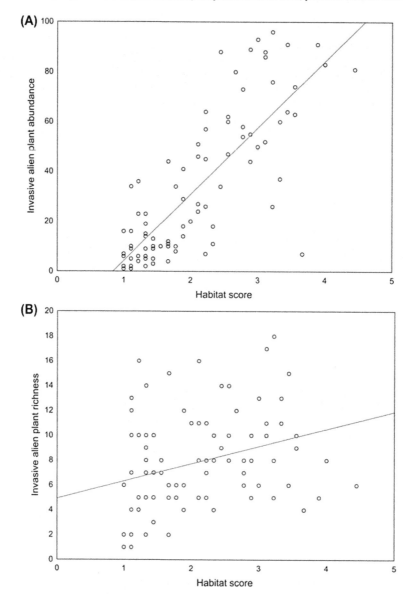

FIGURE 13.3   Regression analysis of the effect of habitat type on (A) IAP abundance [$P = .000$; $r^2 = 0.637$] and (B) richness [$P = .003$; $r^2 = 0.096$] in urban, peri-urban and rural habitat types. A higher habitat score indicates a higher level of degradation.

4.1 The difference in IAP species abundance and richness between urban, peri-urban and rural areas

Proximity to large propagule sources is an important determinant of successful invasion (Heinrichs and Pauchard, 2015; Richardson and Pyšek, 2006). The larger the number of propagules, the higher the chance for establishment, persistence and naturalization of IAPs (Rouget and Richardson, 2003). Because the initial invasion of an IAP into natural areas is often limited by the availability of propagules, in some cases areas with high propagule pressure (for example, from suburban gardens through ornamental horticulture) can be a better predictor of IAP cover than environmental factors (Von Holle and Simberloff, 2005).

Ornamental horticulture is one of the most important pathways for IAP introductions worldwide and spread of introduced IAPs often begins in urban areas (Mayer et al., 2017). However, this is not the case in our study. Only 12%, 15% and 17% of the IAPs found within urban, peri-urban and rural habitat types, respectively, were of horticultural origin. It is key to note that our study area is not within the core of the City of Cape Town but generally has a much higher proportion of gardens compared to the more densely developed inner city. The low proportion of horticultural IAPs found across all three sites indicates that the successful invasion of horticultural IAPs into natural areas may be dependent on other factors (e.g., elevation, aspect or disturbance). The majority (more than 50%) of the IAPs found in all three habitat types are of agricultural origin, possibly a legacy of agricultural land uses that occurred in the area prior to development. Therefore, a potential explanation for the highest abundance of IAPs in peri-urban areas is that these are receiving environments for both urban/suburban (horticultural) IAPs and those of agricultural origin.

The distance from IAP source populations was found to be an important factor in structuring IAP richness in Cape Town forests (Alston and Richardson, 2006). Similarly, the abundance of IAPs in New Zealand native forest fragments correlates with the distance to the nearest large town (Sullivan et al., 2005). In Southern Wisconsin (USA) rural housing variables and the distance to forest edge had the strongest association to the richness and abundance of IAPs in native forests (Gavier-Pizzaro et al., 2010). However, some studies contradict these general findings; for example, Fornwalt et al. (2003) concluded that anthropogenic disturbances (logging, grazing and other similar disturbances) have less of an impact on IAP establishment than topographic positioning.

4.2 Habitat condition is responsible for promoting the establishment and facilitating the invasion of IAPs into natural areas

Historic and current anthropogenic activities are important influences on the degree of invasion in natural areas (inside urban areas and beyond) and may actually diminish the influence of proximity to the urban edge. The City of Cape Town has a long history of human influence, including the presence of permanent settlements, contour farming and recreational usage; these disturbances are likely to have had a profound effect in increasing IAPs in the area.

Where intact native vegetation occurs adjacent to urban centres, competition among species may hamper the invasion of IAPs into these areas (Stajerova et al., 2017; Wania et al., 2006). Swartland Shale Renosterveld demonstrates some resistance to invasion (biotic resistance) by horticultural IAPs since the plots in the urban habitat types which were classified as good/natural or near-natural had lower abundances and richness of IAPs when compared to plots which were classified as modified/fairly degraded or irreversibly degraded. Similarly, biotic resistance, likely due to the saturation of the natural vegetation patch by a very competitive native species (Funk et al., 2008), was responsible for hindering invasion of IAPs in native grassland remnants in the North West, South Africa (Cilliers et al., 2008).

Our results show that the more degraded the habitat condition of the sites, the more likely IAPs occur in higher abundances and richness. As such, the higher abundances and richness of IAPs in peri-urban areas is most likely because these peri-urban areas are the most disturbed habitats in the sample area. Previous studies support our findings. Alston and Richardson (2006) found evidence that disturbance expedites IAP invasions while investigating the roles of habitat features, distance from alien source populations and disturbance structure IAP invasions at the urban/wildland interface. Invasive alien plant invasions were highest in areas of anthropogenic activity and near vegetation fragment edges (Aiko et al., 2012) as well as roadside and sparse habitats (Alston and Richardson, 2006; Catford et al., 2011; Geerts et al., 2016; Johnston and Johnston, 2004). Sites without disturbance did not support invasive alien plants in urban bushland, while sites which were disturbed by trampling did support invasive alien plants (Lake and Lieshman, 2004). Furthermore, Buonopane et al. (2013) found that noxious and exotic IAP frequency and cover significantly decreased with the distance from roads.

Roads and trails aid in increasing a site's invasibility through the creation of open spaces, which effectively fragment natural areas and change local ecosystem microclimates (Delgado et al., 2007; Gelbard and Belnap, 2003). Through the alteration of light availability, disturbance adds to the competitive advantage of invasive species by opening up spaces in the canopy for IAPs to potentially grow and become established among native plant species (Yates et al., 2004). Natural areas and surrounding areas in the City of Cape Town are very popular recreational areas creating disturbances and facilitating dispersal of IAPs. Furthermore, the firebreaks found on all of the study sites provide a sanctuary for a cocktail of IAPs which have the potential to spread further into the natural areas.

4.3 Practical implications for management

The accelerated rural land transformation brought on by the process of urbanization means that managers urgently need to appropriately manage and conserve natural and semi-natural habitat remnants (Kowarik, 2011). It is crucial to protect these remnants against further degradation and IAP invasion.

While horticultural IAPs were not the most prevalent in this study, a critical step in preventing the spread of IAPs across the urban/wildland interface would be to remove species that are known or suspected to be invasive from suburban gardens (Alston and Richardson, 2006). Furthermore, further housing developments adjacent to natural areas of high conservation importance should be discouraged from using IAPs in landscaping.

Since habitat condition was a key predictor of IAP abundance and richness, it follows that the susceptibility of invaded ecosystems can be managed to minimise factors that facilitate alien plant invasion such as reducing anthropogenic disturbance. Firebreaks, or any source of potential disturbance (roads, paths and rivers) must receive attention to ensure that IAPs are not allowed to disseminate into natural areas. This would demand a considerable investment of resources, but is indicative of the level of management that is required if small remnants of natural or seminatural vegetation are to be conserved.

Natural areas that are already heavily invaded must be monitored and buffer zones should be established in order to protect nature reserves (Heinrich and Pauchard, 2015). Furthermore, when planning and identifying BioNet areas which will be affected by the process of urbanization, long-term goals must be set in order to maintain already critically endangered vegetation types from becoming infested by IAPs. Habitat condition standards should be forecast and IAP control programmes implemented to maintain the ecological integrity of potential future conservation sites.

Swartland Shale Renosterveld and the Cape Town BioNet are under considerable threat from IAP invasion and from anthropogenic activities and immediate management intervention is required to maintain ecosystem integrity throughout the mosaic of natural areas found within urban areas. Most importantly, management should be prioritized for areas at the urban-wildland interface and beyond to prevent the degradation of potential future conservation sites as the city expands. Further research on the intrinsic characteristics of peri-urban areas may provide a better understanding of invasion success and spread of ornamental IAPs and biotic resistance.

Acknowledgements

The City of Cape Town is thanked for permission to work in the Cape Town biological network in the greater Tygerberg area. Dr Patricia Holmes is thanked for site-selection assistance, Jan-Hendrik Keet and Suzaan Kritzinger-Klopper are thanked for assistance with fieldwork and the identification of plant species. Prof. Martin Kidd provided statistical advice and assisted with analysis. Funding for this work was provided by the DST-NRF Centre of Excellence for Invasion Biology and Working for Water Program through their collaborative research project on 'Integrated Management of invasive alien species in South Africa'.

References

Aikio, S., Duncan, R.P., Hulme, P.E., 2012. The vulnerability of habitats to plant invasion: disentangling the roles of propagule pressure, time and sampling effort. Global Ecology and Biogeography 21 (8), 778–786.

Allen, A.E., 2003. Environmental planning and management of the peri-urban interface: perspectives on an emerging field. Environmental Planning and Management 5, 135–147.

Alston, K.P., Richardson, D.M., 2006. The roles of habitat features, disturbance, and distance from putative source populations in structuring alien plant invasions at the urban/wildland interface on the Cape Peninsula, South Africa. Biological Conservation 2, 183–198.

Anderson, P.M.L., O'Farrell, P.J., 2012. An ecological view of the history of the city of Cape town. Ecology and Society 17 (3).

Anderson, P.M.L., Avlonitis, G., Ernstson, H., 2014. Ecological outcomes of civic and expert-led urban greening projects using indigenous plant species in Cape Town, South Africa. Landscape and Urban Planning 127, 104–113.

Bell, C.E., Wilen, C.A., Stanton, A.E., 2003. Invasive plants of horticultural origin. Horticultural Science (Calcutta) 38 (1), 14–16.

Bromilow, C., 2010. Problem Plants and Alien Weeds of South Africa, third ed. Briza publications, Pretoria.

Buonopane, M., Snider, G., Kerns, B.K., Doescher, P.S., 2013. Forest ecology and management complex restoration challenges: weeds, seeds, and roads in a forested wildland urban interface. Forest Ecology and Management 295, 87–96.

Catford, J.A., Vesk, P.A., White, M.D., Wintle, B.A., 2011. Hotspots of plant invasion predicted by propagule pressure and ecosystem characteristics. Conservation Biogeography 17 (6), 1099–1110.

Cilliers, S.S., Williams, N.S.G., Barnard, F.J., 2008. Patterns of exotic plant invasions in fragmented urban and rural grasslands across continents. Landscape Ecology 23 (10), 1243–1256.

City of Cape Town, 2016. Open Portal Data. https://web1.capetown.gov.za/web1/OpenDataPortal/.

Collins, J., Kinzig, A., Grimm, N.B., Fagan, W.F., Hope, D., Wu, J., Borer, E.T., 2000. A New Urban Ecology: modeling human communities as integral parts of ecosystems poses special problems for the development and testing of ecological theory. American Scientist 88 (5), 416–425.

Cowan, O.S., 2013. The peninsula Shale Renosterveld of devil's peak: phytosociology, system drivers and restoration potential. University of Cape Town 1–141.

Davis, M.A., Thompson, K., 2000. Eight ways to Be a colonizer; two ways to Be an invader. A Proposed Nomenclature Scheme for Invasion Ecology 81 (3), 226–230.

Davison, A., Marshak, M., 2012. State of Environment Report (City of Cape Town. Cape Town).

Dehnen-Schmutz, K., 2011. Determining non-invasiveness in ornamental plants to build green lists. Journal of Applied Ecology 48 (6), 1374–1380.

Dehnen-Schmutz, K., Touza, J., Perrings, C., Williamson, M., 2007. A century of the ornamental plant trade and its impact on invasion success. Diversity and Distributions 13 (5), 527–534.

Delgado, J.D., Arroyo, N.L., Fern, M., 2007. Edge effects of roads on temperature, light, canopy cover, and canopy height in laurel and pine forests (Tenerife, Canary Islands). Landscape and Urban Planning 81, 328–340.

Drayton, B., Primack, R.B., 1996. Plant species lost in an isolated conservation area in Metropolitan Boston from 1894 to 1993. Conservation Biology 10 (1), 30–39.

Fish, L., Mashau, A.C., Moeaha, M.J., Nembudani, M.T., 2015. Identification Guide to South African Grasses: An Identification Manual with Keys, Descriptions and Distributions. South African National Biodiversity Institute, Pretoria.

Fornwalt, P.J., Kaufmann, M.R., Huckaby, L.S., Stoker, J.M., Stohlgren, T.J., 2003. Non-native plant invasions in managed and protected ponderosa pine/Douglas- fir forests of the Colorado Front Range. Forest Ecology and Management 177, 515–527.

Foxcroft, L.C., Richardson, D.M., Wilson, J.R.U., 2008. Ornamental plants as invasive aliens: problems and solutions in kruger National park, South Africa. Environmental Management 41, 32–51.

Funk, J.L., Cleland, E.E., Suding, K.N., Zavaleta, E.S., 2008. Restoration through reassembly: plant traits and invasion resistance. Trends in Ecology and Evolution 23 (12), 695–703.

Gaertner, M., Larson, B.M.H., Irlich, U.M., Holmes, P.M., Stafford, L., van Wilgen, B.W., Richardson, D.M., 2016. Landscape and urban planning managing invasive species in cities: a framework from Cape town, South Africa. Landscape and Urban Planning 151, 1–9.

Gavier-Pizarro, G.I., Radeloff, V.C., Stewart, S.I., Huebner, C.D., Keuler, N.S., 2010. Rural housing is related to plant invasions in forests of southern Wisconsin, USA. Landscape Ecology 25, 1505–1518.

Geerts, S., Mashele, B.V., Visser, V., Wilson, J.R.U., 2016. Lack of human assisted dispersal means *Pueraria montana* var. *lobata* (kudzu vine) could still be eradicated from South Africa. Biological Invasions 18 (11), 3119–3126.

Geerts, S., Rossenrode, T., Irlich, U.M., Visser, V., 2017. Emerging ornamental plant invaders in urban areas; *Centranthus ruber* in Cape Town, South Africa as a case study. Invasive Plant Science and Management 10 (4), 322–331.

Gelbard, J.L., Belnap, J., 2003. Roads as conduits for exotic plant invasions in a semiarid landscape. Conservation Biology 17 (2), 420–432.

Geographic, N., 2016a. http://nationalgeographic.org/encyclopedia/urban-area/.

Geographic, N., 2016b. http://nationalgeographic.org/encyclopedia/rural-area/.

Gill, T.E., 1996. Eolian sediments generated by anthropogenic disturbance of playas: human impacts on the geomorphic system and geomorphic impacts on the human system. Geomorphology 17, 207–228.

Grime, T.P., 1977. Evidence for the existence of three primary strategies in plants and its relevance to ecological and evolutionary theory. The American Naturalist 111 (982), 1169–1194.

Gulezian, P.Z., Nyberg, D.W., 2010. Distribution of invasive plants in a spatially structured urban landscape. Landscape and Urban Planning 95 (4), 161–168.

Harris, C., Burgers, C., Miller, J., Rawfoot, F., 2010. O-and H-isotope record of Cape Town rainfall from 1996 to 2008, and its application to recharde studies of Table Mountain ground water, South Africa. South African Journal of Geology 113, 33–56.

Heinrichs, S., Pauchard, A., 2015. Struggling to maintain native plant diversity in a peri-urban reserve surrounded by a highly anthropogenic matrix. Biodiversity and Conservation 24 (11), 2769–2788.

Holmes, P.M., Rebelo, A.G., Dorse, C., Wood, J., 2012. Can Cape Town's unique biodiversity be saved? Balancing conservation imperatives and development needs. Ecology and Society 17 (2), 1–13.

Johnston, F.M., Johnston, S.W., 2004. Impacts of road disturbance on soil properties and on exotic plant occurrence in subalpine areas. Arctic, Antartic, and Alpine Research 36 (2), 201–207.

Kowarik, I., January 1990. Ecological consequences of the introduction and dissemination of New plant species: an analogy with the release of genetically engineered organisms. In: European Workshop on Law and Genetic Engineering in Berlin, pp. 67–71.

Kowarik, I., 2011. Novel urban ecosystems, biodiversity, and conservation. Environmental Pollution 159 (8–9), 1974–1983.

Kühn, I., Klotz, S., 2006. Urbanization and homogenization - comparing the floras of urban and rural areas in Germany. Biological Conservation 127 (3), 292–300.

Lake, J.C., Leishman, M.R., 2004. Invasion success of exotic plants in natural ecosystems: the role of disturbance, plant attributes and freedom from herbivores. Biological Conservation 117, 215–226.

Landsberg, H.E., 1981. The Urban Climate. Academic press, inc. London LTD, London.

Lonsdale, W.M., 1999. Global patterns of plant invasions and the concept of invasibility. Ecology 80 (5), 1522–1536.

Lososova, Z., Chytry, M., Tichy, L., Danihelka, J., Fajmon, K., Hajek, O., Rehorek, V., 2012. Native and alien floras in urban habitats: a comparison across 32 cities of central Europe. Global Ecology and Biogeography 21 (5), 545–555.

Mack, R.N., Simberloff, D., Lonsdale, W.M., Evans, H., Clout, M., Bazzaz, F.A., 2000. Biotic Invasions: causes, epidemiology, global consequences, and control. Ecological Applications 86 (4), 689–710.

Mangachena, J.R., Geerts, S., 2017. Invasive alien trees reduce bird species richness and abundance of mutualistic frugivores and nectarivores; a bird's eye view on a conflict of interest species in riparian habitats. Ecological Research 32 (5), 667–676.

Marco, A., Lavergne, S., Dutoit, T., Montes, V.B., 2010. From the backyard to the backcountry: how ecological and biological traits explain the escape of garden plants into Mediterranean old fields. Biological Invasions 12 (4), 761–779.

Mayer, K., Haeuser, E., Dawson, W., Essl, F., Marten, W., Kreft, H., Pys, P., 2017. Naturalization of ornamental plant species in public green spaces and private gardens. Biological Invasions 3613–3627.

Myers, N., Mittermeier, R.A., Mittermeier, C.G., Fonseca, G.A.B., Kent, J., 2000. Biodiversity hotspots for conservation priorities. Nature 403, 853–858.

Novoa, A., Shackleton, R., Canavan, S., Cybèle, C., Davies, S., Dehnen-Schmutz, K., Fried, F., Gaertner, M., Geerts, S., Griffiths, C., Kaplan, H., Kumschick, S., Le Maitre, D., Measey, J., Nunes, A.L., Richardson, D.M., Robinson, T.B., Touza, J., Wilson, J.R.U., 2018. A framework for engaging stakeholders on the management of alien species. Journal of Environmental Management 205 (1), 286–297.

Parendes, L.A., Jones, J.A., 2000. Role of light availability and dispersal in exotic plant invasion along roads and streams in the H.J. Andrews experimental forest, Oregon. Conservation Biology 14 (1), 64–75.

Pickett, S.T.A., Cadenasso, M.L., Grove, J.M., Nilon, C.H., Pouyat, R.V., Zipperer, W.C., Costanza, R., 2001. Urban ecological sysytems: linking terrestrial ecological, physical, and socioeconomic of metropolitan areas. Annual Review of Ecology and Systematics 32, 127–157, 2001.

Pyšek, P., 1998. Alien and native species in Central European urban floras: a quantitative comparison. Journal of Biogeography 25 (1), 155–163.

R Core Team, 2013. R: A Language and Environment for Statistical Computing. R Foundation for Statistical Computing, Vienna, Austria. http://www.R-project.org/.

Reichard, S.H., Hamilton, C.W., 1997. Predicting invasions of woody plants introduced into North America. Conservation Biology 11 (1), 193–203.

Richardson, D.M., Pyšek, P., 2006. Plant invasions: merging the concepts of species invasiveness and community invasibility. Progress in Physical Geography 30 (3), 409–431.

Rouget, M., Richardson, D.M., 2003. Current patterns of habitat transformation and future threats to biodiversity in terrestrial ecosystems of the Cape Floristic Region, South Africa. The American Naturalist 162 (6), 713–724.

Sheperd, M.J., Pierce, H., Negri, A.J., 2002. Rainfall modification by major urban areas: observations from spaceborne rain radar on the TRMM satellite. Journal of Applied Meteorology 41, 689–701.

Stajerova, K., Smilauer, P., Bruna, J., Pyšek, P., 2017. Distribution of invasive plants in urban environment is strongly spatially structured. Landscape Ecology 32 (3), 681–692.

Statistica, 2019. Retrieved November. Statsoft.support.software.dell.com.

Struglia, R., Winter, P.L., 2016. The role of population projections in environmental management. Environmental Management 30 (1), 13–23.

Sukopp, H., Werner, P., 1983. Urban Environments and Vegetation. Junk Publishers, Netherlands.

Sullivan, J.J., Timmins, S.M., Williams, P.A., 2005. Movement of exotic plants into coastal native forests from gardens in northern New Zealand. New Zealand Journal of Ecology 29 (1), 1–10.

Thapa, R.B., Maruyama, Y., 2008. Land Evaluation for Peri-Urban Agriculture Using Analytical Hierarchical Process and Geographic Information System Techniques: A Case Study of Hanoi. University of Tsukuba. Japan.

van Wilgen, B.W., Reyers, B., Le Maitre, D.C., Richardson, D.M., Schonegevel, L., 2008. A biome-scale assessment of the impact of invasive alien plants on ecosystem services in South Africa. Journal of Environmental Management 89, 336–349.

Vitousek, P.M., D'Antonio, C.M., Loope, L.L., Rejmánek, M., Westbrooks, R., 1997. Introduced species: a significant component of human-caused glocal change. New Zealand Journal of Ecology 21 (1), 1–16.

Von Holle, B., Simberloff, D., 2005. Ecological resistance to biological invasion overwhelmed by propagule pressure. Ecology 86 (12), 3212–3218.

Walton, B.A., 2006. Vegetation Patterns and Dynamics of Renosterveld at Agter-Groeneberg. MSc thesis. Stellenbosch University, Stellenbosch.

Wania, A., Ingolf, K., Klotz, S., 2006. Plant richness patterns in agricultural and urban landscapes in Central Germany — spatial gradients of species richness. Landscape and Urban Planning 75, 97–110.

Whitford, V., Ennos, R., Handley, J.F., 2001. "City form and natural process" - indicators for the ecological performance of urban areas and their application to Merseyside, UK. Landscape and Urban Planning 57, 91–103.

Willig, M.R., Walker, L., 1999. Disturbance in terrestrial ecosystems: salient themes, synthesis, and future directions. Ecosystems of the World 747–768.

Yates, E.D., Levia, D.F., Williams, C.L., 2004. Recruitment of three non-native invasive plants into a fragmented forest in southern Illinois. Forest Ecology and Management 190, 119–130.

Urban material balance

Types, sources and management of urban wastes

Tahir Noor[1], Arshad Javid[1], Ali Hussain[1], Syed Mohsin Bukhari[1], Waqas Ali[1], Muhammad Akmal[2], Syed Makhdoom Hussain[3]

[1]Applied and Environmental Microbiology Laboratory, Department of Wildlife and Ecology, University of Veterinary and Animal Sciences, Lahore, Punjab, Pakistan; [2]Water Research Laboratory, Department of Fisheries and Aquaculture, University of Veterinary and Animal Sciences, Lahore, Punjab, Pakistan; [3]Aquaculture Research Laboratory, Department of Zoology, Government College University, Faisalabad, Punjab, Pakistan

OUTLINE

Urban Ecology
https://doi.org/10.1016/B978-0-12-820730-7.00014-8

1. Introduction

Increase in urbanization has resulted in enormous amount of waste generation. More than half percent of all world population lives in urban areas. The rate of urbanization is expected to increase by 1.5 times by 2045. The waste generation from urban areas is increasing at a pace double than the rate of urbanization itself. The waste generation will increase approximately 2 times by 2045. The chapter include a core analysis on the types of waste (liquid waste, solid waste, plastic waste, paper waste, tins and metals, ceramics and glass, organic waste, recyclable waste, nondegradable waste, hazardous waste, nonhazardous waste, radioactive waste, sanitary waste, construction and demolition waste), generation of waste (residential waste, industrial waste, hospital wastes, municipal solid waste, agricultural waste, biomedical wastes, e-waste), effects of waste on life, approaches to waste management (unregulated disposal, recycling, recovery, incineration and burial) and common methods of urban waste disposal (submergence, trade, attenuation, isolation, reduce, reuse and recycle). The waste generated from urban areas has very lethal effects on urban ecology. As the liquid wastes generated from industrial sectors badly affect the aquatic life which further threatens human life as we consume them. Similarly, the smoke that comes from the chimneys cause air pollution resulting in lungs disorder as we inhaled in polluted air. The exposure to heavy metals is directly linked to the rate of industrialization. The chapter discusses various waste management strategies to overcome the difficulties associated with waste generation and to maintain urban ecology. The chapter falls into following subtopics.

2. Urbanization

Urbanization is defined as moderation of an urban area based on economic and social status towards a more urban area having service sectors and industries (Hensley et al., 2019). In recent years, the developing countries having an economic transition via urbanization and have high demand of energy resources (Shahbaz et al., 2015). The world population keeps on increasing but at a slower pace as compared to 1950 probably due to the reduced fertility level. From an estimated world population of 7.7 billion in 2019, world population is estimated to rise to 8.5 billion in 2013 to 9.7 billion in 2050 and 10.9 billion in 2100 (WPA, 2019). About 55% of world total population lives in urban areas. The urban population will increase 1.5 times reaching six billion by 2045 (World Bank, 2019) (Table 14.1).

2.1 Types of urban wastes

Waste generation will increase by 70% on the basis of current level of generation. The overall global waste generation which was 2.01 billion tons in 2016 is expected to become 3.4 billion tons in the next 30 years (World Bank, 2019). Urban waste includes liquid waste, solid waste, plastic waste, paper waste, tins and metals, ceramics and glass, organic waste, recyclable waste, nondegradable waste, hazardous waste, nonhazardous waste, radioactive waste, sanitary waste, construction and demolition waste.

2.1.1 Solid wastes

Waste generation is rising day by day globally. In 2016 urban population generated 2.01 billion tons solid waste with each person contributing approximately 0.74 kg/day (Khan et al., 2016a,b). With increasing urbanization, the annual generation of wastes is expected to increase by 70% from 2016 to 2050. The waste generation is expected to increase from 2.01 billion tons from 2016 to 3.04 billion tons in 2050 (World Bank, 2019).

2.1.1.1 Garbage/trash

Solid wastes are defined as the wastes which are produced in solid form. Whenever 'waste' term is used most of the people think about municipal solid waste (MSW). MSW

TABLE 14.1   Average annual rate of change of the urban population (%) (World Bank, 2019).

Region	Average annual rate of change of the urban population (%)									
	Time interval									
	2000 −05	2005 −10	2010 −15	2015 −20	2020 −25	2025 −30	2030 −35	2035 −40	2040 −45	2045 −50
World	2.29	2.23	2.04	1.90	1.73	1.58	1.45	1.33	1.22	1.13
Sub-Saharan Africa	4.07	4.14	4.14	3.98	3.82	3.65	3.47	3.27	3.09	2.89
Africa	3.52	3.61	3.70	3.58	3.44	3.32	3.19	3.04	2.89	2.71
Asia	3.06	2.80	2.43	2.16	1.84	1.58	1.35	1.15	0.98	0.84
Europe	0.33	0.46	0.35	0.35	0.30	0.28	0.26	0.25	0.22	0.17
Latin America & Caribbean	1.74	1.61	1.47	1.30	1.15	1.00	0.85	0.72	0.59	0.47
North America	1.13	1.13	0.95	0.95	0.96	0.92	0.84	0.75	0.67	0.62
Oceania	1.35	1.78	1.54	1.42	1.30	1.24	1.18	1.15	1.12	1.07

includes most wastes that is produced by businesses and homes and that is not gas or liquid in nature (EPA, 2017). MSW is generally termed as 'garbage' or 'trash' including wastes from shops, commerce, retailers and offices excluding hazardous, constructional and industrial wastes (Yang et al., 2018). MSW is one of the most produced wastes type of urban lifestyle. The rate of MSW generation exceeds the rate of urbanization (Khan et al., 2016a,b). The overall generation of waste is likely to be 3.40 billion tons by 2050 at a rate greater than the rate of population increase at the same time (World Bank, 2019) (Table 14.2).

2.1.2 Liquid waste

Wastewater s generated from shops, fuel depots, mines, quarries, vessels, factories, offices and homes. There are mainly three sources of wastewater: domestic, industrial and agricultural (DES, 2019). Water scarcity, that is increasing day by day, poses an unavoidable threat to human life. Water pollution on the other hand had become a dangerous problem, especially in developing countries (Zhang et al., 2016a,b).

2.1.2.1 Wastewater

Due to increasing population and urbanization, the generation of wastewater especially in developing countries increases significantly. In developing countries almost 90% of wastewater is left untreated into lakes, rivers and oceans (Ahmad et al., 2019). Six to eight million people die annually from wastewater borne diseases out of which 1.8 million are children below 5 years of age (Khan et al., 2016a,b). Moreover, the gases such as nitrous oxide and methane (CH_4) generated from wastewater also have a negative impact on climate and their levels are estimated to increase 25% and 50%, respectively. Wastewater is the biggest stream of wastewater from petrochemical, paper, agricultural, textile and food industry with highly contaminated inorganic or organic salts and oil (Salgot and Folch, 2018). Demographic growth–associated industrial revolution results in higher demand of fresh water supply. Although there are various methods for treatment such as biological, chemical and physical methods, the reality is freshwater is depleting in nature (Tetteh et al., 2019).

TABLE 14.2 Projected generation of wastes from different regions (million tons/year) (World Bank, 2019).

Region	2016	2030	2050
Middle East and North Africa	129	177	225
Sub-Saharan Africa	174	269	516
Latin America and Caribbean	231	290	369
North America	289	342	396
South Asia	334	466	661
Central Asia and Europe	392	440	490
Pacific and East Asia	468	602	714

2.1.3 Plastic wastes

In consumer market place plastic has become the dominant demand of consumer since the commercial development of plastic in 1930 and 1940. Plastic pollution has a very drastic effect on wildlife and oceans. Higher income countries tend to generate more plastic than the lower income countries (Uekert et al., 2018). According to Our World in Data report (2019) the overall plastic production in 1950 was only two million tons per year. Since then the annual production of plastic became 200-fold greater than the 1950s production, reaching 381 million tons in 2015 (OWID, 2019). The global market of plastic waste has an estimation value of US$ 33,681.2 Mn by 2017, and is expected to reach a value of US$ 45,642.4 Mn by 2026 (Recycling-Magazine, 2019) (Figs 14.1 and 14.2).

2.1.4 Paper waste

Being a consumer product, paper has become the driving force in the sustainability of mankind and society, especially in the areas of communication, sanitation, security and education (EPN, 2018). Paper was first developed in China in 105 AD and since then its demand has increased significantly. This is because of two main reasons: (i) the physiochemical properties of paper such as flexibility and durability that made it possible to attain several kinds of forms (ii) socioeconomic realities of the world such as in modern technology, increasing world population and world's economic prosperity (Ezeudu et al., 2019). Tissue paper is the most used type of paper materials used in newspapers, hardware, electrical hardware, sanitation, cardboard and footwear. The paper waste generation in highly industrialized areas such as United States has gone from 33.5% in 1990 to 65.8% in 2017 and is expected to reach 70% in 2020 (Sharma and Garg, 2018).

2.1.5 Metal waste

In order to regain metal and environmental sustainability and metal recycling from wastes, green processes need to be developed (Clark et al., 2016). The recycling of metals from printed circuit boards is necessary for the development of industries like electronics. The

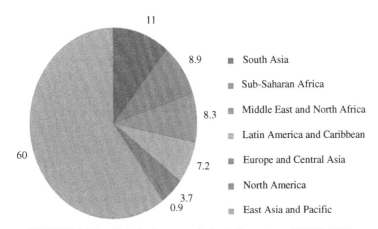

FIGURE 14.1 Global mismanaged plastic by regions (OWID, 2019).

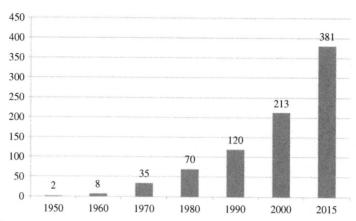

FIGURE 14.2   Annual production of plastic waste in million tons (Recycling-Magazine, 2019).

recycling of metals from printed circuit board (PCB) waste has resulted in many serious environment conditions and critical risk to human health (Chen et al., 2016). PCBs are being widely used; electronic and electric equipment sensing and information industries generate waste (Alippi, 2016; Jin et al., 2016) that contain almost 40 types of metals (e.g., Zn, Cu, Fe, Sn) including hazardous substances such as Cd, Cr, Pb (Zeng et al., 2017). Tin is another valuable metal used in electronic industry. Tin solder is plated over the copper surface during PCB manufacturing which serves as an eth resistant. ECs like expansion slots, capacitors, resistors and chips are also plated over PCB. About 99% of tin can be leached out with HCl and $SnCl_2$ at $60-90^\circ C$ and then tin is recovered by electrodeposition (González et al., 2018). Metal recycling is becoming more important day by day in order to minimize the mineral deficiencies and also as the demand for metals continues to increase along with the increase of urbanization (Yang et al., 2017).

2.1.6 Ceramics and glass wastes

Bioactive glasses are produced by sol-gel technique and melt quenching. In sol-gel processing an inorganic network is synthesized by mixing alkoxides in the form of solution with hydrolysis, gelation associated with firing at low temperature to synthesize glass (Zhitong et al., 2018). Preparation of bioactive glasses via melt quenching is achieved by mixing stoichiometric number of different carbonates and oxides and then melting at higher temperature. This is then cast into moulds to attain the desire shape. On the other hand, ceramics exist in both natural and synthetic form (García-González et al., 2015). The main difference between synthetic and natural ceramics is the presence of phosphate group instead of carbonate group in synthetic ceramics which give them antigenic action, less inflammation and chemical stability (Hajiali et al., 2018).

2.1.7 Organic wastes

The demand for reducing harmful emission generated during disposal of organic waste has enhanced the development of technologies that convert waste into energy. This approach will replace the use of fossil fuels in near future both as a source of materials and energy (Pagliano et al., 2017).

2.1.7.1 Urban organic wastes (UOW)

Biogas is defined as a renewable energy form expected to be used in electricity and heat generation and as vehicle fuel. However, this process still requires advancement in technologies required to produce socioeconomic and environmentally sustainable conversion of urban organic waste (UOW) into biogas (Pereira et al., 2018). UOW such as garden waste, grass and food could help in producing bulk of biogas if codigested with sludge (Fitamo et al., 2017).

2.1.7.2 Residential organic wastes

There is significant amount of food waste (33%–56%) in residual household waste, indicating high energy potential if utilized properly (Edjabou et al., 2014). The urban organic waste constitutes 70% of total waste stream. This waste should be looked as a source of biofuel instead of wastes that could be landfilled or incinerated to get rid of wastes (Muralikrishna and Manickam, 2017).

2.1.8 Recyclable wastes

The wastes generated from residential sources and industrial production and which have potential of recycling after being appropriately used are called recyclable wastes. They include wastes such as waste glass, waste plastic, electronic equipment, waste metal, waste paper and rejected home appliances (Ahsan et al., 2014). Developed countries, especially Japan, are the earliest starters of MSW management and have achieved significant result in the recycling of waste. However, China is still at the early stages of removing hurdles in the way of waste recycling (Xiao et al., 2018) (Fig. 14.3).

2.1.9 Hazardous wastes

Hazardous wastes have unique properties such as persistence, ignitability, reactivity, corrosivity and toxicity. Hazardous wastes pose a continuous threat to our environment and human health. Because of their above mentioned properties, inhalation and ingestion of hazardous wastes pose a danger to human health (Bharadwaj et al., 2015). The hazardous waste rules 2016 state that it is the responsibility of the individual industry to handle, treat and dispose of the hazardous waste. With the growth of industrialization, there is significant increase in the amount of hazardous industrial wastes (Lieberman et al., 2018).

2.1.9.1 Major contributors

The major generators of hazardous wastes are the manufacturer of distilleries, tiles, textiles, engineering, electronics, paints/pigments, pharmaceuticals and chemicals. One of the most dominant producers of hazardous waste is the process of electroplating (Kwikiriza et al., 2019).

FIGURE 14.3 Common recyclable wastes (Xiao et al., 2018).

2.1.9.2 Hazardous hospital waste

Hazardous wastes generated from hospitals consist of medicine packaging, discarded chemicals and radioactive materials. Infectious wastes have a danger of disease outbreak if not properly disposed (Bharadwaj et al., 2015).

2.1.10 Nonhazardous wastes

Nonhazardous waste is defined as the wastes which are not considered as hazardous to human health and our environment. Nonhazardous waste cannot be added to a sewage line or to a dumpster. Sugars, carbonates, bromides and lactic acid are the common examples of nonhazardous waste (Borghi et al., 2018). There is no proper way to collect and dispose these wastes because these depend on individual regulations, for example, the disposal of oil is different from lactic acid disposal. The one possible way for collection and transportation of nonhazardous waste is to classify the wastes (BWS, 2017).

2.1.11 Radioactive wastes

Radioactive wastes are the byproducts of many nuclear reactions such as nuclear fission, nuclear power generation and many other nuclear reactions such as in medicine and research fields. These wastes contain radioactive materials which are injurious to our environment and health (AE, 2019). These kinds of wastes are usually stored to avoid the exposure of radiation to human beings, because the decay time of these wastes is longer than the other kind of wastes around. High level radioactive waste is recommended to be stored for 50 years before being disposed of as compared with low level radioactive waste which requires less storage and which are disposed of immediately (Lee et al., 2019). For the disposal of used fuel wastes, the wastes are stored normally under water for 45 years and then dry-stored. The disposal of radioactive waste in deep geographical areas appears to be the best solution for their disposal (WNA, 2018).

2.1.12 Construction and demolition wastes

Construction and demolition wastes (CDW) consist of debris that comes from construction, renovation and demolition of bridges, roads and buildings (EPA, 2019). According to EPA (2015) 548 MT of CDW is generated which is two times greater than MSW. Construction industries play a significant role in the sustainable development of urban areas. For sustainable construction, the biggest issue is waste handling, since the construction industry requires more natural sources and energy (Huang et al., 2018). The major construction wastes are glass, plastics, wood and steel, surplus mortar, surplus concrete, broken bricks, green wastes (grass, bushes) and excavated soil. With the increase in urbanization, construction and demolition waste (CDW) has many side effects such as environmental degradation (Ulubeylia et al., 2017). The amount of CDW in urban areas is still unknown. An old saying 'you cannot improve what you cannot measure' is the case with CDW. Demolition represents 90% of the CDW while construction represents only 10% of total CDW (EPA, 2019).

2.2 Generation of wastes

There are many kinds of wastes generated in urban areas of the world. Most common among these waste types are residential waste, industrial waste, hospital wastes, MSW, agricultural waste, biomedical wastes and e-waste (Khan et al., 2016a,b).

2.2.1 Municipal solid wastes (MSW)

Municipal solid wastes (MSW) include garbage or trash, consisting of daily use items, which we thrown away after use such as batteries, paints, appliances, newspapers, food scraps, bottles, clothing, furniture, grass clippings and packaging (Khan et al., 2016a,b).

2.2.1.1 Sources of MSW generation

MSW is generated from industries, hospitals, schools and houses (EPA, 2019). MSW is a biomass waste type consisting of glass, food wastes, metals, textiles, wood, plastics and paper. The methods most commonly used for management of MSW are open dumping and landfilling. Both methods have side effects such as environmental contamination, methane gas generation which promote global warming and labor issues (Murtaza et al., 2017). Due to presence of chemical bonding between C−H−O, MSW can generate energy when such bonds are broken (Khan et al., 2016a,b).

In 2016, urban population generated 2.01 billion tons solid waste with each person contributing approximately 0.74 kg/day (Khan et al., 2016a,b). With increasing urbanization, the annual generation of wastes is expected to increase by 70% from 2016 to 2050. The waste generation is expected to increase from 2.01 billion tons from 2016 to 3.04 billion tons in 2050 (World Bank, 2019). Annually, 1.9 billion tons of MSW is generated with each person contributing 218 kg MSW to this grand total (WA, 2019) (Fig. 14.4).

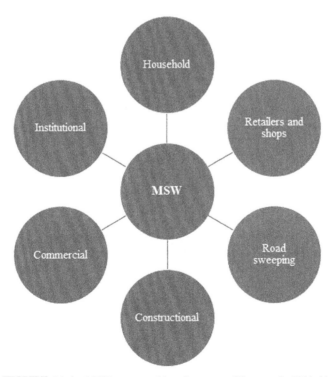

FIGURE 14.4   MSW generated in urban areas (Khan et al., 2016a,b).

2.2.2 Residential/household wastes

Wastes are the byproducts of biota. It is the rule of nature that every living person on earth generate wastes, and when individuals die, they are also considered as wastes (Umeh et al., 2019). The residential wastes are mostly solid in nature. It is estimated that 3.5 billion or 52% of all over the world population does not have access to solid waste management in residential areas (TWC, 2019).

The production of residential wastes depends upon the economy and size of the residential area. Higher the income value and larger the residential area, the more will be production of household wastes (Khan et al., 2016a,b). The problems for the management of household wastes will keep on increasing with increasing urbanization and financial status of any area. The amount of household wastes worldwide is approximately 991,901,800 tons (TWC, 2019).

2.2.3 Agricultural wastes

The residual material produced during the growth and cultivation of agricultural crops, vegetables, fruits, dairy products, meat and poultry are called agricultural wastes. There are other residual material that can be produced from the growth and processing of agricultural products but these byproducts are of less importance and profit from these byproducts are less than the cost of production (Acharya et al., 2018).

Agricultural wastes or agrowastes consist of wastes from food processing, animal waste (animal carcasses, manure), crop waste (pruning, vegetables and fruit droppings, sugarcane bagasse and corn stalks) and hazardous wastes (herbicides, insecticides and pesticides). The estimation of contribution of agricultural waste is rare but in developed areas it constitutes a large proportion (Obi et al., 2016). The expansion of agricultural production is directly proportion to the crop residues, livestock waste and agricultural byproducts. It is estimated that a total of 998 million tons of wastes is generated from agriculture (Acharya et al., 2018). Organic wastes constitute almost 80% of this total solid wastes. On wet weight basis the manure production is 5.27 kg/day/1000 kg live weight (Obi et al., 2016) (Table 14.3).

2.2.4 Biomedical/hospital waste

Biomedical wastes are considered as hazardous and toxic. Waste type which is collected from the vaccination or treatment of living beings which may be in the form of liquid or solid including wastes from clinical laboratories, biological labs, pathological labs and repository such as bowl, syringes and ampulla form the biomedical waste (Kaur et al., 2019). Biomedical waste has become a health hazard in urban areas of developed countries. Indiscriminate and careless disposal of biomedical waste from healthcare laboratories promotes the risk of spreading harmful diseases such as hepatitis or AIDS (Kwikiriza et al., 2019). In urban areas of developed countries, better facilities for healthcare are linked with more production of biomedical waste. In India about 0.33 million tons of hospital waste is generated with 0.5−2 kg/person/day. The estimated volume for hazardous waste is 10%−25% (EPA, 2017).

2.2.5 E-waste

Since the 21st century, the waste generated from electronic devices and equipment has become a very targeting issue for society, technology companies and the government itself. As the rate of urbanization has increased significantly over the past three decades, the growth

TABLE 14.3 Yearwise production (tons) of different wastes.

Year	World	Africa	America	Asia	Europe	Oceania
			Region			
Primary fruit						
2000	576,653,541	68,108,456	139,204,687	279,409,107	83,742,582	6,188,709
2010	755,776,803	93,057,889	150,052,750	429,317,661	76,257,163	7,091,341
2017	865,590,060	109,480,474	161,083,687	510,221,628	76,747,557	8,056,713
Primary vegetables						
2000	684,733,756	44,631,919	72,112,756	480,514,064	84,161,989	3,313,028
2010	923,736,560	66,004,425	75,965,389	690,046,745	88,398,141	3,321,860
2017	1,094,343,707	79,914,814	80,650,030	834,203,458	96,258,802	3,316,603
Cereals crops						
2000	1,859,299,823	105,327,892	521,009,978	814,750,005	383,246,475	34,965,473
2010	2,233,366,176	158,076,791	622,406,863	1,014,377,780	404,076,235	34,428,506
2017	2,723,878,753	189,240,229	753,574,461	1,210,595,808	519,712,908	50,755,347
Citrus fruit						
2000	106,176,458	11,695,207	53,954,193	29,745,626	10,109,284	672,147
2010	129,411,096	15,865,925	46,186,691	55,472,114	11,321,767	564,598
2017	146,599,168	19,672,747	44,968,626	70,766,938	10,643,991	546,867

Data extracted from FAOSTAT. 2019. Food and Agriculture Data. https://greentumble.com/effect-of-pollution-on-plants (Accessed 17 June 2019).

of e-waste has become unprecedented (Balde et al., 2015). The growth rate of e-waste generation will keep on increasing in coming years which is 8% at present. Almost 41.8 million metric tons of e-waste is generated in 2014 which is likely to be 50 million metric tons by 2018. In this era of globalization and advancements in electronic devices such as computers, mobile phones and laptops, the mass generation of e-waste become determined. Today, people are more inclined towards better and latest technology and that's why they reject old devices. For example, the life span of a computer in 1992 was 4.5 years which has shrunk to 2 years in 2005 (Balde et al., 2015; Zhitong et al., 2018).

E-waste is composed of hazardous, precious and valuable elements which can be recovered or recycled through proper recycling system. Dumping of e-waste without any management system would expose hazardous elements to our environment, thus putting biota at danger (Kumar et al., 2017). E-waste consists of both toxic and valuable materials which include 60% nonmetals, 30% metals such as Pb, Au, Cu, Fe, etc. and 3.70% pollutants. The best management strategies for e-waste are recovery and recycling not only for its disposal but also for the recovery of precious metal from electronic scrap (Amankwah-Amoah, 2016).

2.2.6 Industrial waste

With better society and economic status, the deleterious effects of pollution begin to disturb human health and ecological security. The industrialization rate increases day by day, so does the industrial waste which puts human health at risk (Prabakar et al., 2018). The solid waste originating from industrial areas results in soil, water and air pollution, requiring a solid management. During industrial production industrial solid wastes are generated such as waste residues, chemical processing, smelting, fuel residues, ore dressing and mining waste (Liu et al., 2018).

2.2.6.1 Exhaust gases

The exhaust gases originating from industrial wastes are sulfur compounds, CS_2, H_2S and SO_2. The removal of these exhaust gases is necessary for environmental protection and commercial benefits by recovering sulfur (Prabakar et al., 2018). The SO_2 enters through the waste material from power plants, chemical industries, metallurgical industries and petrochemical units. The emission of SO_2 in our environment is 20–30 million tons per year. This large amount of SO_2 is recovered by conversion into H_2SO_4 (Brückner et al., 2015).

2.3 Effects of wastes on life/urban ecology

Toxic waste is defined as the harmful product of different activities such as farming, garages, hospitals, water treatment plants, automatic garages, hospitals, laboratories and manufacturing. This waste can be solid, liquid, sludge, pathogens, radiation, heavy metals and chemicals. On the other hand, households also generate harmful wastes from pesticides, leftover paints, computer equipment and batteries (Javied et al., 2014). Toxic wastes can be very harmful to plants, animals and people no matter where the waste goes, either in streams, air or even in the ground. Some heavy metals like lead and mercury persist and accumulate in our environment for many years. Wildlife and humans can swallow the toxicant via eating fish or other prey (NG, 2019). The waste generated from urban areas have dangerous effect on urban ecology, the effect of urban waste on biota is discussed as:

2.3.1 Effects on animal life

Tons of wastes generated from urban areas have been illegally burned and dumped at agricultural sites. These wastes affect biota badly. Some of the harmful wastes are as follows:

2.3.1.1 Effect of chlorinate family

These wastes release many dangerous chemicals at night time including dioxins and 17 highly lethal molecules of chlorinate family. Among these 17 molecules, there is 2,3,7,8-tetrachlorodibenzo-p-dioxins (TCDD), which by International Agency for Research on Cancer has already been classified as carcinogenic both for humans and animals (NG, 2019).

2.3.1.2 Effect of plastic wastes

Ten percent of household wastes constitute of plastic wastes. Most of these plastic wastes are landfilled while 60%–80% of plastic wastes are found as floating materials on oceanic surface and at beaches. At the Southern California beach approximately 2.4 billion pieces of wastes were collected in 72 h out of which 30,500 kg was plastic wastes (OWID, 2019).

Most of these wastes are polystyrene, miscellaneous fragments, pellets of preproduction and whole materials account for 71%, 14%, 10% and 1% respectively. Eighty-one percent of all these plastic materials were in between 1 and 4.75 mm. Incomplete combustion of polystyrene (PS), polypropylene (PP) and polyethylene (PE) during the process of thermal utilization generates high production of noxious emission and carbon monoxide (CO) (Geyer et al., 2017). On the other hand, the incomplete combustion of PVC generates carbon black, aromatics such as chrysene and pyrene and dioxins. Hazardous emission from such kind of wastes include bromide and pigmented materials that contain heavy metals such as cadmium, lead, selenium, cobalt, copper and chromium (Verma et al., 2016).

2.3.1.3 Effect of industrial residues

The illegal burning of industrial residues, textiles, plastics, tyres represent major concerns for population health and environmental pollution. There is an increased risk of lung, lymphopoietic neoplasms, and liver linked with 2,3,7,8-TCDD, despite considering the controversy that the individuals of such specific diseases are overexposed to the dioxins or 2,3,7,8-TCDD (Mazza et al., 2015). The sheep observed in contaminated areas are at higher risk of developing abnormal fetus, chromosome fragility and higher mortality as compared to those sheep living in noncontaminated areas (Murray et al., 2014). Congenital malformations and cancer mortality have been reported under such environmental pressure. Triassi et al. (2015) reported that there is positive long-term correlation between such environmental stress and diseases like liver and lung cancer in human beings of Caserta and Naples provinces.

2.3.1.4 Effect of metallic elements

Metallic elements in our environment act as intrinsic components. Once they enter our environment it looks quite difficult to remove them from our environment, that's why these components are termed as unique. To a large extent toxic metals are dispersed through organic wastes, power generation, transport, refuse burning and industrial effluents. Toxic metals can be carried miles away from the source of emission via wind depending upon their nature such as particulate or gaseous (Mahurpawar, 2015) (Table 14.4).

TABLE 14.4 Different metals, their sources and affected organs (Mahurpawar, 2015).

Metals	Industries	Target systems
Arsenic	Metal hardening, textile, paints, phosphate and fertilizers	Skin, nervous, pulmonary
Chromium	Tanning, metal plating, photography, rubber	Pulmonary
Cadmium	Electronics, phosphate fertilizer, paints, pigments	Pulmonary, renal, skeletal
Lead	Battery, paints	Renal, hematopoietic, nervous
Nickel	Iron steel, electroplating	Skin, pulmonary
Copper	Electrical, rayon, plating	Renal
Mercury	Scientific instruments, chemicals, chlor-alkali	Renal, pulmonary
Zinc	Steel, plating iron, galvanizing	Immunity

2.3.2 Effects on plant life

Like animals, plants also get affected from different harmful wastes. Some of these wastes are as follows:

2.3.2.1 Effects of pesticide wastes

Soil pollution is caused by improper disposal of wastes from different sources such as pesticides, illegal dumping and oil spills. These sources generate different chemicals which seep into the soil, harm the soil which lacks nutritional fulfillment and fill the soil with metals or chemicals (Devi et al., 2016). These changes result in plant damage and act as retarding factor as the plant fails to obtain proper nutrients. Instead of these the plant cells can be damaged by heavy metals and this process will overall decrease the crop yield (Bruin et al., 2019).

2.3.2.2 Effect of water pollution

Water pollution occurs due to various reasons some of which are biological contamination, farm runoff, industrial spills and sewage leakage. Water pollution has many adverse effects on plant life. Sometimes, due to the excessive nutrient in water, plants can show excessive growth. On the other hand, these excessive nutrients alter the soil pH and kill the plants (GT, 2019).

2.3.2.3 Effect of toxic elements

Nonessential toxic elements are transferred into plant systems same like other essential elements required by the plant for homeostasis. This uptake of mineral nutrition and metal homeostasis of plant has started the process of dissecting accumulated nonessential elements. For example, during early phase of discovery several transporters of nonessential toxic elements were permeable to cadmium. The path of toxic elements into reproductive or vegetative part of the plant depends on various processes like tissue type, developmental state, habitat and concerned specie of plant (Ying-Chu et al., 2018). These elements are taken up by the plant from soil via apoplast pathways or active transport by the help of transporters present on the cell membrane of root cells. The transport of these elements to the shoots depends upon various processes such as mobility of symplast pathway, trapping rate of nonessential elements within different root compartments, movement across the barrier, i.e., endodermis, loading of elements within xylem cells and mobility of elements from xylem cells to the upper parts of shoot (Clemens and Ma, 2018).

2.3.2.4 Effect of alkylphenol ethoxylates (APEO)

Among the most used classes of surfactants are alkylphenol ethoxylates (APEO). APEO are most commonly used in household sectors, food industry and agriculture industry. Nonylphenol ethoxylates (NPEO) are the most commonly used APEOs accounting for 80% of the total used APEOs. These are incomplete biodegradable elements in treated wastewater plants and our environment. The incomplete degradation results in the release of ethoxy groups to form nonylphenol monoethoxylate (NPEO1), nonylphenol diethoxylate (NPEO2) and ultimately the completely deethoxylated product nonylphenol (NP) (Bei-Xing et al., 2018). NP is also the intermediate used in tar and oil refining industries but the major source of environmental NP is the incomplete degradation of APEOs. NP mimics the estrogen (female sex hormone) thus disturbing the sexual life of mammals; apart from these NP persists in our environment and disturbs every kind of life. Flora is exposed to NP from various ways. NP enters into plant systems via sewage treatment plants (Jardak et al., 2016). On the other hand the soil is being polluted by manure or biosolids, pesticides runoff, feeding operations and agricultural fertilizers. Due to its physiochemical properties like high hydrophobicity and low solubility, a greater concentration of NPs are found in sediments, soil, air and water (Clemens and Ma, 2018).

2.4 Approaches to waste management

Waste management, in any municipality, has become one of the key issues. With the increase in world human population, there is greater food consumption linked with greater amount of wastes such as packages, paper and food wastes that have ended up in the constitution of waste streams (Khan et al., 2016a,b). It is the need of time that resources available in these waste streams are utilized properly. The increasing population in modern cities is creating new challenges for environment, the major challenge being waste management that requires new solutions (Filho et al., 2016).

2.4.1 Unregulated disposal

Unregulated disposal has become a major problem in the urban areas of the world. Furniture, mattresses, sofas and other household materials are disposed of in the streets and under the bridges due to the lack of any related regulation. This not only retards the visual appearance, but can also pose a great threat to people's lives. To reduce this kind of open dumping in urban areas, many cities initiate various steps such as monitoring by surveillance camera, executing policies and education programmes. The educated people try to control unregulated disposal themselves. The hotspot areas of illegal dumping in nonresidential areas are controlled on daily basis by city cleaning personnel. On waste detection, trash trucks transport these wastes to dumping sites (IEEE, 2017).

2.4.1.1 Unregulated disposal of MSW

Global production of MSW is nearly 1.3 billion tons per year which is double as compared to the previous decade. The major reasons behind this mass production are increasing urbanization and economic development. With the increase of urbanization and economy, consumption of goods and services will increase corelatedly, and so does the amount of waste. Open dumping of MSW creates problem for our environment and also for citizens lives as well (IEEE, 2017).

2.4.2 Recycling

Recycling is a mechanism for screening out of things after they had been used by the customer (Khan et al., 2016a,b). This mechanism can result in the same thing or any different thing that is valuable to use. Any kind of waste at any dumping site has 8% potential of recycling (Ahsan et al., 2014).

2.4.2.1 Recycling of MSW

A total of 67.8 million tons of MSW have been recycled. The major contributor of this recycled waste was paper and paperboard contributing 67% with metals contributing 12% and wood, glass and plastic contributing between 4% and 5% (EPA, 2018).

2.4.2.2 Recycling of paper waste

Today 72% of raw material for paper industry comes from the recycling of paper and pulp. Paper has become the most recycled product in Europe followed next by North America. The recycling rate of paper industry moved from 50% to 72% between 1998 and 2015 (CEPI, 2018). So, the recycling of paper is beneficial from the economic and environment perspective. However, there is a strong concern related to paper recycling like the migration of harmful manufacturing chemicals from paper to the consumer food (Rivera et al., 2015).

2.4.2.3 Recycling of polymer wastes

Increase in production of polymer waste results in the greater production of solid waste, which has negative impact on the environment. The recycling of polymer waste is comparatively easy due to its high versatility. The recycling of polymer waste consists of four steps, i.e., collection of polymer waste; separation based on size, volume and weight; waste reprocessing and marketing (CEPI, 2018).

2.4.3 Recovery

In the era of production and consumption, recovery of wastes has become an important area of waste management. Thirty percent of waste collected has been subjected to waste recovery worldwide out of which recovery of energy from materials is 19% and recovery of materials is 11%. Recovery takes place in developed urban areas of the world. Recovery of waste means to make use of things that we want to dispose of for other useful purposes (SUEZ, 2018). A great demand has been observed to obtain renewable energy such as secondary staple materials. Transformation of household wastes into secondary resources is an economical as well as environmental issue. Waste recovery is helpful for preserving natural resources, reducing the amount of waste storage and controlling GHGs. There will be +37% expected growth in worldwide energy demand by 2040 (Plastics Europe, 2014).

2.4.3.1 Recovery of plastics wastes

In Europe in 2013 the demand for plastics reached 46.3 million tons, out of which 40% became part of plastic stock. The remaining 54% (25 million tons) have been landfilled (38%), incinerated (36%) and recovered (26%) (Plastics Europe, 2014).

2.4.4 Incineration

Incineration is another waste treatment method that includes thermal treatment by combusting the organic materials present in wastes. Incineration results in heat, gas and ash. The ash might contain particulate or solid lumps dispersed by flue gas. This flue gas generated by incineration must be free from particulate or gaseous pollutants before being dispersed (Ghasemi and Yusuf, 2016). Incineration has a vital positive impact that the heat generated from incineration can be used by electric power stations thus reducing the need of fossil fuels. The process of incineration requires a temperature between 900 and 1200°C. Due to its nontarget destruction such as disease outbreak and requirement of large sample, incineration requires expertise (Khan et al., 2016a,b).

The process of incineration makes use of grate system that combusts the organic material in wastes that are crude. The boilers used in these systems are facilitated with hydraulic rams, which drive wastes into the ignition cubicle (Makarichi et al., 2018). The grate systems act in a monitoring way during the wastes' passage to the burning chamber. The drying of wastes enhanced the overall accuracy of this system, then further burning on abrade produces ash. The gases generating by grate system are transported out in the form of steam at a temperature of 850°C and the heat produced is used for running the grate system itself (Yasin et al., 2017).

2.4.5 Burial/landfilling

Landfilling or burial is among the most widely used method in urban waste management. It is the method of filling the empty land with wastes such as generated from poultry farms, godown, shops, and slaughter houses (Khan et al., 2016a,b). Unlike open dumping it does not result in disease outbreak, but it can happen when wastes are dumped in an underwater environment. Other than this, landfilling can promote global warming by generating methane gas (Matsakas et al., 2017).

2.4.5.1 Landfilling of MSW

A total of 137.7 million tons of MSW had been landfilled in 2015. This included 22% of food wastes, 19% of plastics, 13% of paper and paperboard, 11% of textiles, leather and rubber while others contributed for less than 10% (EPA, 2018).

2.4.6 Anaerobic digestion

Anaerobic digestion is a process of recovering sustainable energy from biodegradable waste products obtained from agricultural residues, sewage sludge, residues from wood industry, municipal and industrial wastewater, livestock manure, brewery and food wastes (Sawatdeenarunat et al., 2015). Anaerobic digestion has many benefits such as reduction in greenhouse gas emission, reduction in organic waste and production of methane gas (Čater et al., 2015). Besides these benefits the main challenge is to process the lignocellulose waste via hydrolysis. This process requires technologies for pretreatment associated with anaerobic digestion (Ozbayram et al., 2017). Specific microorganisms present within the gut systems of certain animals secrete enzymes such as hemi-cellulases, lignin breaking enzymes and cellulases that are responsible for breakdown of lignin (Lazuka et al., 2015). Ruminants including sheep, goat and cattle have intestinal microbial symbionts which enable them to feed on plant fibers. The microbiota within their gut systems is adapted to different types of diet and they meet their metabolic needs by gaining energy (Morgavi et al., 2015).

2.5 Common disposal methods of urban wastes

Different strategies of urban waste disposal in past and present and suggestions for future are described below:

2.5.1 Submergence

During the past, people living near the ocean, chose the ocean for waste disposal such as industrial wastes, chemical disposal, sewage sludge, munitions, trash and radioactive waste. During this kind of disposal, very less attention was given to the aquatic life being targeted. These wastes were disposed near the coastal areas, and ocean water has much ability to dissolve such wastes in it (EPA, 2018).

During the period of 1946 and 1970 approximately 55,000 radioactive waste containers were ocean dumped at Pacific Ocean at three sites (EPA, 2018). Many areas of ocean have been damaged due to continuous deposition of harmful pollutants such as inorganic nutrients, chlorinated petrochemicals and heavy metals. This pollution has result in deficiency of oxygen levels (Khan et al., 2016a,b) (Table 14.5).

2.5.2 Trade

Due to the immense use of plastics, immense plastic waste is produced which creates problem for solid waste management. Single-use items and plastic packages immediately enter the waste stream after being used, and result in retardation of waste stream flow (Brooks et al., 2018).

2.5.2.1 Trade of plastic wastes

Globally 6.3 billion MT plastic wastes are produced each day. Management of such large amount of plastic waste is much difficult in highly populated and better economic development countries. Globally plastic waste has only 8% potential of recycling, 80% of the plastic waste result in contaminating our environment and some are landfilled resulting in 4–12 million MT of plastic waste being ocean dumped (Geyer et al., 2017).

Due to wide variety of use, plastic recycling can be challenging and material properties of plastic can limit the number of times plastic can be recycled. Lack of plastic recycling has made recycling system limited; due to this reason China in 1990 made waste trade system in which plastic waste can be traded to other countries via ships. The shipping prevents plastic waste to be incinerated or landfilled. By 2010, China had implemented more rigid policies on plastic waste import. China invented 'Green Fence' in 2013 by implementing restrictions on the import of plastic waste that contain high amount of pollutants and the relation between China and exporters was disrupted due to this reason. The core goal of 'Green Fence' was the import of good quality plastic waste that contains very less pollutants. In 2017, China permanently banned the import of plastic waste that was generated from different industries (EPA, 2017).

2.5.3 Attenuation

Attenuation in sense of waste production refers to the reduction of threatening signals to their lowest lethality. It is a method of reducing toxic waste into nontoxic wastes (Xiang et al.,

TABLE 14.5   Disposal by submergence of different types of wastes till times (EPA, 2018)

	Type of wastes							
Year	Petroleum products	Chemical wastes	Heavy metals	Organic chemical wastes	Dredged material	Industrial wastes	Sewage sludge	Construction/ demolition wastes
1968	100 MT	2–4 MT	1 MT	0.1 MT	—	—	—	—
1970	—	—	—	—	38 MT	4.5 MT	4.5 MT	0.5 MT

2019). Increase in urban population has resulted in increase in the amount of waste generation. The material properties of commodities affect the life of waste material. For example, the material properties of cemented materials affect the radiation coefficient. The radiation causes genetic or somatic mutations in tissues or organs. The effect of radiation can be neutralized by three methods, i.e., distance, shielding and time. The most important among these is shielding. Shielding is the attenuation of materials used in ceramics to avoid the harmful effects of radiations (Özavci and Çetin, 2016). Vegetables and fruits are playing an important role in human life and diet. Therefore, with changing diet habits and increasing population the demand for such commodities has increased significantly (Vilariño et al., 2017). The attenuation of food wastes is discussed in the following section.

2.5.3.1 Attenuation of food wastes

Food wastes are generated in large quantities by consumers and retailers, cultivators, post-harvesters and farmers. Food wastes constitute the largest share of waste products which are landfilled. Food wasted means loss of money. This landfilled waste also results in global warming by generating methane gas into our environment. Furthermore, the food are exposed to radiation for pest control, preservation, increasing shelf life, to prevent foodborne illness, sprouting and ripening delay (Ying-Chu et al., 2018). It became really very important to know the attenuation characteristic of radiation once they applied. Compositional studies suggest the presence of primary and secondary plant metabolites such as gums, mucilage, oils, glycosides, alkaloids and phenolic. The fruit peels also contain high amount of antioxidants which can be used in nutraceuticals. For determination of effective reference sample for elemental analysis, mass attenuation coefficients of materials used become necessary (Akman et al., 2018).

2.5.4 Isolation

Global demand for alternatives of fossil fuel is increasing day by day, which puts the attentions of scientist towards the biofuel production via biological processes. Butanol production has gained the attention of many industries such as Green Biologics, BP, DuPont and Gevo Inc (Khan et al., 2016a,b). Scientists are trying to produce butanol at an industrial scale, because butanol can replace gasoline. This is due to the fact that butanol is less sensitive to temperature, possesses higher content of energy and requires no modification for usage and being less corrosive. The production of butanol can be achieved by anaerobic fermentation via *Clostridium* known as acetone-butanol-ethanol (ABE) (Al-Shorgani et al., 2016). The fruit wastes generated in abundant amount can be used to isolate several biological molecules.

2.5.4.1 Isolation of fruit wastes

Fruits like banana, rambutan, papaya and jackfruit are liked and consumed worldwide because they are rich in nutrients and provide health benefits. As the fruit production has increased (14%) since the last 5 years, the wastes generated from such fruits also have increased. The disposal of such wastes has become worrisome as it results in environmental pollution. Besides this many vegetables, natural polymers like starch and legumes are gaining agricultural attention for their potential as being natural coagulants. However, starches can be isolated from plant materials which are rich in polysaccharides. During recent times, research has been conducted to isolate starch from cereal crops, fruit seeds, seeds of trees, tuber crops, legumes and weeds. Starch is used as natural coagulants, binders, encapsulation of β-carotene, stabilizers and thickeners in food industry.

2.5.5 3R — reduce, reuse and recycle

The continuous emission of greenhouse gases is responsible for today's environment. These gases are generated by energy power plants consuming fossil fuels such as oil, natural gas and coal, by constructing and manufacturing industries, agriculture, transportation and traffic (Dijkers, 2019). So, technologies are being developed to limit the emissions of CO_2 into the environment by electromobility development generation of energy from renewable sources supported by industries and government policies. Still numerous technologies are responsible for global warming by ejecting CO_2 (Ali and Yusof, 2018).

The energy demand has increased significantly worldwide and the curve for demand and supply shows discomfort with increasing urbanization and limited resources. There is increasing demand day by day with the increasing competition for limited resources (Dijkers, 2019). There are many reasons for such resource pressure. Among them the prime reason is human population which will reach to eight billion by 2030 and nine billion by 2050. Circular economy consists of closed-loop ecosystem for the better usage of resources. This corresponds to the linear use of resources towards the closed-loop ecosystem (Nelson et al., 2016). The model implies 3R system consisting of reduce, reuse and recycle.

- Reduce: reduction in the utilization of nonrenewable resources
- Reuse: reusing the things having potential of utilization after being used
- Recycling: method for screening out of things after they had been used by the customer (Fig. 14.5)

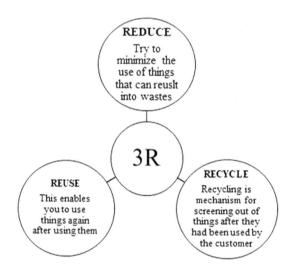

FIGURE 14.5   3R — reduce, reuse and recycle (Ali and Yusof, 2018)

3. Conclusions and recommendations

The challenges which we face today are mainly due to the unsustainable use of resources resulting in exhaustion of raw materials, ecosystem degradation, environmental pollution and climate change. We not only require an immediate action to prevent environmental crises occurring from wastes but also require to change our lifestyle so that waste generation or unsustainable use of resources can be prevented. We must investigate the possible ways to reduce the unsustainable utilization of resources, suggest procedures for limiting the utilization of resources and change our lifestyle to encourage a sustainable enviroment.

References

Acharya, J., Kumar, U., Rafi, P.M., 2018. Removal of heavy metal ions from wastewater by chemically modified. Agricultural Waste Material as Potential Adsorbent-A Review International Journal of Current Engineering and Technology 8 (3), 1–5.

AE, 2019. Radioactive Waste Management & Conditioning. www.ansaldoenergia.com/Pages/Radioactive-Waste-Management-and-Conditioning-.aspx. (Accessed 21 July 2019).

Ahmad, K., Wajid, K., Khan, Z.I., Ugulu, I., Memoona, H., Sana, M., Nawaz, K., Malik, I.S., Bashir, H., Bashir, H., Sher, M., 2019. Evaluation of potential toxic metals accumulation in wheat irrigated with wastewater. Bulletin of Environmental Contamination and Toxicology 102, 822–828.

Ahsan, A., Alamgir, M., El-Sergany, M.M., Shams, S., Rowshon, M.K., Daud, N.N.N., 2014. Assessment of municipal solid waste management system I a developing country. Chinese Journal of Engineering 1, 1–11.

Akman, F., Geçibesler, I.H., Sayyed, M.I., Tijani, S.A., Tufekci, A.R., Demirtas, I., 2018. Determination of some useful radiation interaction parameters for waste foods. Nuclear Engineering and Technology 50, 944–949.

Al-Shorgani, N.K.N., Isa, M.D.M., Yusoff, W.M.W., Kalil, M.S., Hamid, A.A., 2016. Isolation of a clostridium acetobutylicum strain and characterization of its fermentation performance on agricultural wastes. Renewable Energy 86, 459–465.

Ali, M.S.M., Yusof, R.N.R., 2018. Intention to practice reduce, reuse & recycle (3R) among expatriates working in Malaysia. International Journal of Academic Research in Business and Social Sciences 8 (3), 276–295.

Alippi, C.A., 2016. Unique timely moment for embedding intelligence in applications. CAAI Transaction on Intelligence Technology 1 (1), 1–3.

Amankwah-Amoah, J., 2016. Global business and emerging economies: towards a new perspective on the effects of e-waste. Technological Forecasting and Social Change 105, 20–26.

Balde, K., Wang, F., Huisman, J., Kuehr, R., 2015. The Global E-Waste Monitor. United Nations University, IAS – SCYCLE, Bonn, Germany.

Bei-Xing, L., Xiu-Yu, P., Peng, Z., Jin, L., Xiao-Xu, L., Yang, L., Hua, L., Feng, L., Wei, M., 2018. Alcohol ethoxylates significantly synergize pesticides than alkylphenol ethoxylates considering bioactivity against three pests and joint toxicity to *Daphnia magna*. The Science of the Total Environment 644, 1452–1459.

Bharadwaj, A., Yadav, D., Varshney, S., 2015. Non-biodegradable waste – its impact & safe disposal. International Jounal of Advanced Technology in Engineering and Science 3 (1), 184–191.

Borghi, G., Pantini, S., Rigamonti, L., 2018. Life cycle assessment of non-hazardous construction and demolition waste (CDW) management in Lombardy region (Italy). Journal of Cleaner Production 184, 815–825.

Brooks, A.L., Wang, S., Jambeck, J.R., 2018. The Chinese import ban and its impact on global plastic waste trade. Science Advances 4 (6), 1–15.

Brückner, S., Liu, S., Miró, L., Radspieler, M., Cabeza, L.F., Lävemann, E., 2015. Industrial waste heat recovery technologies: an economic analysis of heat transformation technologies. Applied Energy 151, 157–167.

Bruin, W.D., Kritzinger, Q., Bornman, R., Korsten, L., 2019. Occurrence, fate and toxic effects of the industrial endocrine disrupter, nonylphenol, on plants - a review. Ecotoxicology and Environmental Safety 181, 419–427.

BWS, 2017. Non-hazardous Waste Are Still Regulated. http://www.bwaste.com/medical-waste-services.php. (Accessed 21 August 2019).

Čater, M., Fanedl, L., Malovrh, Š., Logar, R.M., 2015. Biogas production from brewery spent grain enhanced by bioaugmentation with hydrolytic anaerobic bacteria. Bioresource Technology 1, 1–35.

CEPI, 2018. CEPI: Paper and Board Production Stable in 2018. https://www.recycling-magazine.com/2019/02/22/cepi-paper-and-board-production-stable-in-2018. (Accessed 19 August 2019).

Chen, M., Ogunseitan, O.A., Wang, J., Chen, H., Wang, B., Chen, S., 2016. Evolution of electronic waste toxicity: trends in innovation and regulation. Environment International 89—90, 147—154.

Clark, J.H., Farmer, T.J., Herrero-Davila, L., Sherwood, J., 2016. Circular economy design considerations for research and process development in the chemical sciences. Green Chemistry 18 (14), 3914—3934.

Clemens, S., Ma, J.F., 2018. Toxic heavy metal and metalloid accumulation in crop plants and foods. Annual Review of Plant Biology 67, 489—512.

DES, 2019. Water Quality and Ecosystem Health. https://environment.des.qld.gov.au/water. (Accessed 8 September 2019).

Devi, K.S., Swammy, D.A.V.V.S., Nilofer, S., 2016. Municipal solid waste management in India — an overview. Asia Pacific Journal Operation Research 1 (39), 118—126.

Dijkers, M., 2019. Reduce, reuse, recycle: good stewardship of research data. Spinal Cord 57, 165—166.

Edjabou, M.E., Jensen, M.B., Götze, R., Pivnenko, K., Petersen, C., Scheutz, C., Astrup, T.F., 2014. Municipal solid waste composition: sampling methodology, statistical analyses, and case study evaluation. Waste Management 1, 1—12.

EPA, 2015. Construction and Demolition Web Resources & Publications. https://www.epa.ie/waste/nwpp/candd. (Accessed 17 August 2019).

EPA, 2017. EPA Year in Review 2017—2018. https://www.epa.gov/newsroom/epa-year-review-2017-2018. (Accessed 11 August 2019).

EPA, 2018. National Overview: Facts and Figures on Materials, Wastes and Recycling. www.epa.gov/facts-and-figures-about-materials-waste-and-recycling/national-overview-facts-and-figures-materials. (Accessed 12 September 2019).

EPA, 2019. Sustainable Management of Construction and Demolition Materials. www.epa.gov/smm/sustainable-management-construction-and-demolition-materials. (Accessed 11 August 2019).

EPN, 2018. Project: Paper Saving & Efficiency. https://environmentalpaper.org/project/paper-saving-efficiency. (Accessed 11 August 2019).

Ezeudu, O.B., Agunwamba, J.C., Ezeasor, I.C., Madu, C.N., 2019. Sustainable production and consumption of paper and paper products in Nigeria: a review. Resources 8 (53), 1—22.

FAOSTAT, 2019. Food and Agriculture Data. (Accessed 17 June 2019). https://greentumble.com/effect-of-pollution-on-plants.

Filho, W.L., Brandli, L., Moora, H., Kruopiene, J., Stenmarck, A., 2016. Benchmarking approaches and methods in the field of urban waste management. Journal of Cleaner Production 112, 4377—4386.

Fitamo, T., Treu, L., Boldrin, A., Sartori, C., Angelidaki, I., Scheutz, C., 2017. Microbial population dynamics in urban organic waste anaerobic co-digestion with mixed sludge during a change in feedstock composition and different hydraulic retention times. Water Research 1, 1—41.

García-González, J., Rodríguez-Robles, D., Juan-Valdés, A., Pozo, J.M., Guerra-Romero, M.L., 2015. Ceramic ware waste as coarse aggregate for structural concrete production. Environmental Technology 36 (23), 3050—3059.

Geyer, R., Jambeck, J.R., Law, K.L., 2017. Production, use, and fate of all plastics ever made. Science Advances 3, 17—27.

Ghasemi, M.K., Yusuf, R.B.M., 2016. Advantages and disadvantages of healthcare waste treatment and disposal alternatives: Malaysian scenario. Polish Journal of Environmental Studies 25 (1), 17—25.

González, A., Norambuena-Contreras, J., Storey, L., Schlangen, E., 2018. Self-healing properties of recycled asphalt mixtures containing metal waste: an approach through microwave radiation heating. Journal of Environmental Management 214, 242—251.

GT, 2019. Effects of Pollution on Plants. https://greentumble.com/effect-of-pollution-on-plants. (Accessed 16 June 2019).

Hajiali, F., Tajbakhsh, S., Shojaei, A., 2018. Fabrication and properties of polycaprolactone composites containing calcium phosphate-based ceramics and bioactive glasses in bone tissue engineering: a review. Polymer Reviews 58 (1), 164—207.

Hensley, C.B., Trisos, C.H., Warren, P.S., MacFarland, J., Blumenshine, S., Reece, J., Katti, M., 2019. Effects of urbanization on native bird species in three southwestern US cities. Frontiers in Ecology and Evolution 1, 1—10.

Huang, B., Wang, X., Kua, H., Geng, Y., Bleischwitz, R., Ren, J., 2018. Construction and demolition waste management in China through the 3R principle. Resources, Conservation and Recycling 129, 36—44.

IEEE, 2017. Unregulated Disposal of Waste. In: https://ieeexplore.ieee.org/xpl/conhome/7960754/proceeding. (Accessed 16 June 2019).

Jardak, K., Drogui, P., Daghrir, R., 2016. Surfactants in aquatic and terrestrial environment: occurrence, behavior, and treatment processes. Environmental Science and Pollution Research International 23, 3195–3216.

Javied, S., Hanan, F., Munawar, S., Qasim, M., Anees, M.M., Ghani, M.U., Azad, A., Khalid, M., Ullah, I., Ansar, A., 2014. Management of municipal solid waste generated in eight cities of Pakistan. Journal of Scientific and Engineering Research 5 (12), 1186–1192.

Jin, H., Chen, Q., Chen, Z., Hu, Y., Zhang, J., 2016. Multi-LeapMotion sensor based demonstration for robotic refine tabletop object manipulation task. CAAI Transaction on Intelligence Technology 1 (1), 104–113.

Kaur, H., Siddique, R., Rajor, A., 2019. Influence of incinerated biomedical waste ash on the properties of concrete. Construction and Building Materials 226, 428–441.

Khan, M.N., Luna, I.Z., Islam, M.M., Sharmeen, S., Salem, K.S., Rashid, T.U., Zaman, A., Haque, P., Rahman, M.M., 2016a. Cellulase in waste management applications. In: Gupta, V.K. (Ed.), New and Future Developments Microbial Biotechnology and Bioengineering, 1st. John Fedor, Netherlands, pp. 238–256.

Khan, M.Z., Nizami, A.S., Rehan, M., Ouda, O.K.M., Sultana, S., Ismail, I.M., Shahzad, K., 2016b. Microbial electrolysis cells for hydrogen production and urban wastewater treatment: a case study of Saudi Arabia. Applied Energy 1, 1–11.

Kumar, A., Holuszko, M., Espinosa, D.C.R., 2017. E-waste: an overview on generation, collection, legislation and recycling practices. Resources, Conservation and Recycling 122, 32–42.

Kwikiriza, S., Stewart, A.G., Mutahunga, B., Dobson, A.E., Wilkinson, E., 2019. A whole systems approach to hospital waste management in rural Uganda. Frontiers in Public Health 7, 1–9.

Lazuka, A., Auer, L., Bozonnet, S., Morgavi, D.P., O'Donohue, M., Hernandez-Raquet, G., 2015. Efficient anaerobic transformation of raw wheat straw by a robust cow rumenderived microbial consortium. Bioresource Technology. 1, 1–27.

Lee, S., Hrma, P., Pokorny, R., Klouzek, J., Eaton, W.C., Kruger, A.A., 2019. Global production rate in electric furnaces for radioactive waste verification. Journal of the American Ceramic Society 1, 1–15.

Lieberman, R.N., Knop, Y., Izquierdo, M., Palmerola, N.M., Rosa, J.D.L., Cohen, H., Muñoz-Quirós, C., Cordoba, P., Querol, X., 2018. Potential of hazardous waste encapsulation in concrete with coal fly ash and bivalve shells. Journal of Cleaner Production 185, 870–881.

Liu, C., Cai, W., Zhang, C., Ma, M., Rao, W., Li, W., He, K., Gao, M., 2018. Developing the ecological compensation criterion of industrial solid waste based on energy for sustainable development. Energy 1, 1–22.

Mahurpawar, M., 2015. Effects of heavy metals on human health. International Journal of Research 1, 1–7.

Makarichi, L., Jutidamrongphan, W., Techato, K., 2018. The evolution of waste-to-energy incineration: a review. Renewable and Sustainable Energy Reviews 91, 812–821.

Matsakas, L., Gao, Q., Jansson, S., Rova, U., Christakopoulos, P., 2017. Green conversion of municipal solid wastes into fuels and chemicals. Electronic Journal of Biotechnology 26, 69–83.

Mazza, A., Piscitelli, P., Neglia, C., Rosa, G.D., Iannuzzi, L., 2015. Illegal dumping of toxic waste and its effect on human health in campania, Italy. International Journal of Environmental Research and Public Health 12, 6818–6831.

Morgavi, D.P., Rathahao-Paris, E., Popova, M., Boccard, J., Nielsen, K.F., Boudra, H., 2015. Rumen microbial communities influence metabolic phenotypes in lambs. Frontiers in Microbiology 6, 1–13.

Muralikrishna, I.V., Manickam, V., 2017. Solid waste management. Environmental Management 1, 431–462.

Murray, I.A., Patterson, A.D., Perdew, G.H., 2014. Aryl hydrocarbon receptor ligands in cancer: friend and foe. Nature Reviews Cancer 14, 801–814.

Murtaza, G., Habib, R., Shan, A., Sardar, K., Rasool, F., Javeed, T., 2017. Municipal solid waste and its relation with groundwater contamination in Multan, Pakistan. International Journal of Applied Research 3 (4), 434–441.

Nelson, P.M., Norman, E.R.V., VanDerHeyden, A., 2016. Reduce, reuse, recycle: the longitudinal value of local cut scores using state test data. Journal of Psychoeducational Assessment 1, 1–12.

NG, 2019. Toxic Waste, Explained. www.nationalgeographic.com/environment/global-warming/toxic-waste.

Obi, F.O., Ugwuishiwu, B.O., Nwakaire, J.N., 2016. Agricultural waste concept, generation, utilization and management. Nigerian Journal of Technology 35 (4), 957–964.

OWID, 2019. Plastic Pollution. https://ourworldindata.org/plastic-pollution. (Accessed 9 July 2019).

Özavci, S., Çetin, B., 2016. Determination of radiation attenuation coefficients in concretes containing different wastes. International Conference on Computational and Experimental Science and Engineering 130 (1), 316–317.

Ozbayram, E.G., Kleinsteuber, S., Nikolausz, M., Ince, B., Ince, O., 2017. Effect of bioaugmentation by cellulolytic bacteria enriched from sheep rumen on methane production from wheat straw. Anaerobe 1, 1–9.

Pagliano, G., Ventorino, V., Panico, A., Pepe, O., 2017. Integrated systems for biopolymers and bioenergy production from organic waste and by-products: a review of microbial processes. Biotechnology for Biofuels 113 (10), 1–22.

Pereira, R.F., Cardoso, E.J.B.N., Oliveira, F.C., Estrada-Bonilla, G.A., Cerri, C.E.P., 2018. A novel way of assessing C dynamics during urban organic waste composting and greenhouse gas emissions in tropical region. Bioresource Technology Reports 3, 35–42.

Plastics Europe, 2014. Plastics – the Facts 2014. www.plasticseurope.org/en/resources/publications/95-plastics-facts-2014. (Accessed 12 June 2019).

Prabakar, D., Manimudi, V.T., Suvetha, S., Sampath, S., Mahapatra, D.M., Rajendran, K., Pugazhendhi, A., 2018. Advanced biohydrogen production using pretreated industrial waste: outlook and prospects. Renewable and Sustainable Energy Reviews 96, 306–324.

Recycling Magazine, 2019. Waste Plastics. www.recycling-magazine.com/waste-plastics. (Accessed 11 June 2019).

Rivera, J.A., López, V.P., Casado, R.R., Hervás, J.S., 2015. Thermal degradation of paper industry wastes from a recovered paper mill using TGA. characterization and gasification test. Waste Management 1, 1–11.

Salgot, M., Folch, M., 2018. Wastewater treatment and water reuse. Current Opinion in Environmental Science and Health 2, 64–74.

Sawatdeenarunat, C., Surendra, K.C., Takara, D., Oechsner, H., Khanal, S.K., 2015. Anaerobic digestion of lignocellulosic biomass: challenges and opportunities. Bioresource Technology 1, 1–27.

Shahbaz, M., Loganathan, N., Sbia, R., Afza, T., 2015. The effect of urbanization, affluence and trade openness on energy consumption: a time series analysis in Malaysia. Renewable and Sustainable Energy Reviews 47, 683–693.

Sharma, K., Garg, V.K., 2018. Comparative analysis of vermicompost quality produced from rice straw and paper waste employing earthworm *Eisenia fetida* (Sav.). Bioresource Technology 250, 708–715.

SUEZ, 2018. Recovery and Waste Management. www.suez.com/en/our-offering/Local-authorities/What-are-you-looking-for/Recovery-and-waste-management. (Accessed 12 June 2019).

Tetteh, E.K., Rathilal, S., Chetty, M., Armah, E.K., Asante-Sackey, D., 2019. Treatment of water and wastewater for reuse and energy generation-emerging technologies. Water and Waste Treatment 1, 1–22.

Triassi, M., Alfano, R., Illario, M., Nardone, A., Caporale, O., Montuori, P., 2015. Environmental pollution from illegal waste disposal and health effects: a review on the 'triangle of death'. International Journal of Environmental Research and Public Health 12, 1216–1236.

TWC, 2019. Household Waste Statistics. www.theworldcounts.com. (Accessed 17 June 2019).

Uekert, T., Kuehnel, M.F., Wakerleya, D.W., Reisner, E., 2018. Plastic waste as a feedstock for solar-driven H_2 generation. Energy and Environmental Science 10, 1–12.

Ulubeylia, S., Kazazb, A., Arslan, V., 2017. Construction and demolition waste recycling plants revisited: management issues. Procedia Engineering 172, 1190–1197.

Umeh, P.P., Nkwocha, K.F., Iheukwumere, S.O., 2019. Geographical analysis of household waste generation and disposal in Taraba state, northeast Nigeria. International Journal of Geography and Geology 8 (2), 58–68.

Verma, R., Vinoda, K.S., Papireddy, M., Gowda, A.N.S., 2016. Toxic pollutants from plastic waste- A review. Procedia Environmental Sciences 35, 701–708.

Vilariño, M.V., Franco, C., Quarrington, V., 2017. Food loss and waste reduction as an integral part of a circular economy. Frontiers Environmental Science 5 (21), 1–5.

WA, 2019. Global Charts. http://www.atlas.d-waste.com. (Accessed 17 June 2019).

WNA, 2018. Storage and Disposal of Radioactive Waste. www.world-nuclear.org/information-library/nuclear-fuel-cycle/nuclear-waste/storage-and-disposal-of-radioactive-waste.aspx. (Accessed 11 September 2019).

World Bank, U.S., 2019. World Development Report, Population Growth. https://data.worldbank.org/indicator/SP.POP.GROW. (Accessed 11 September 2019).

WPA, 2019. United Nation DESA/Population Division. https://population.un.org/wpp/DataQuery. (Accessed 1 August 2019).

Xiang, R., Xu, Y., Liu, Y., Lei, G., Liu, J., Huang, Q., 2019. Isolation distance between municipal solid waste landfills and drinking water wells for bacteria attenuation and safe drinking. Scientific Reports 9, 1–10.

Xiao, S., Dong, H., Geng, Y., Brander, M., 2018. An overview of China's recyclable waste recycling and recommendations for integrated solutions. Resources, Conservation and Recycling 134, 112–120.

Yang, C., Li, J., Tan, Q., Liu, L., Dong, Q., 2017. Green process of metal recycling: coprocessing waste printed circuit boards and spent tin stripping solution. ACS Sustainable Chemistry and Engineering 5, 3524–3534.

Yang, Y., Heaven, S., Venetsaneas, V., Banks, C.J., Bridgwater, A.V., 2018. Slow pyrolysis of organic fraction of municipal solid waste (OFMSW): characterisation of products and screening of the aqueous liquid product for anaerobic digestion. Applied Energy 213 (1), 158–168.

Yasin, H., Usman, M., Rashid, H., Nasir, D.A., Randhawa, D.I.A., 2017. Alternative approaches for solid waste management: a case study in Faisalabad Pakistan. Earth Sciences Pakistan 1 (2), 7–9.

Ying-Chu, C., Yi-Cheng, H., Chung-Ting, W., 2018. Effects of storage environment on the moisture content and microbial growth of food waste. Journal of Environmental Management 214, 192–196.

Zeng, X., Yang, C., Chiang, J.F., Li, J., 2017. Innovating e-waste management: from macroscopic to microscopic scales. The Science of the Total Environment 575, 1–5.

Zhang, M., Xie, L., Yin, Z., Khanal, S.K., Zhou, Q., 2016a. Biorefinery approach for cassava-based industrial wastes: current status and opportunities. Bioresource Technology 1, 1–53.

Zhang, Q.H., Yang, W.N., Ngo, H.H., Guo, W.S., Jin, P.K., Dzakpasu, M., Yang, S.J., Wang, Q., Wang, X.C., Ao, D., 2016b. Current status of urban wastewater treatment plants in China. Environment International 92, 11–22.

Zhitong, Y., Weiping, S., Jie, L., Weihong, W., JunhongT, Ling, T.C., Sarker, P.K., 2018. Recycling difficult-to-treat e-waste cathode-ray-tube glass as construction and building materials: a critical review. Renewable and Sustainable Energy Reviews 81, 595–604.

CHAPTER

15

Nutrient recovery from municipal waste stream: status and prospects

Vaibhav Srivastava[1], Barkha Vaish[1], Anita Singh[2], Rajeev Pratap Singh[1]

[1]Department of Environment and Sustainable Development, Institute of Environment and Sustainable Development, Banaras Hindu University, Varanasi, Uttar Pradesh, India; [2]Department of Botany, Banaras Hindu University, Varanasi, Uttar Pradesh, India

OUTLINE

Urban Ecology
https://doi.org/10.1016/B978-0-12-820730-7.00015-X

1. Introduction

Rapid urbanization and industrialization are the most significant observable global trends with negative implications on environment. The exploding populations of the developing world exacerbate perennial problems such as pollution, land degradation, environmental degeneration and climate change (Fig. 15.1A). Correspondingly, waste and wastewater generation are one of the challenging issues due to anthropogenic activities that need to be managed scientifically without compromising environmental sustainability. According to the World Bank report, in South Asia, around 334 million tonnes (t) of waste was generated in 2016, which will reach 661 million tonnes in 2050. India alone generates 109,598 tonnes of solid waste per day and is projected to reach 376,639 tonnes per day by 2025 (Hoornweg and Bhada-Tada, 2012). Likewise, nearly 38.88 km^3 volume of wastewater is generated in South Asian region, of which only 7% is being treated (Evans et al., 2012). In India, around, 25.41 km^3/year volume of wastewater was generated in 1996, whereas 2.56 km^3/year was treated in 2004 (Evans et al., 2012). In most of the developing countries, urban solid wastes are usually disposed in open dumps, and wastewater is pumped in nearby water bodies without prior treatment and necessary precautions, posing severe threat to environment and human health. Simultaneously, increasing urban population is pushing additional pressure on non-renewable resources for food and energy security (Fig. 15.1A) (Vaish et al., 2016, 2019a). Therefore, it is imperative to manage urban waste and wastewater in such a way that besides managing their quantity, it would help in reducing our dependency over non-renewable resources so as to sustain the environment. During late 18th and early 19th century, rapid population and urbanization led to increased demand of nutrient supply beyond the levels available in natural soil and other organic forms (Cordell et al., 2009). The three essential macronutrients necessary for plant growth are nitrogen (N), phosphorus (P) and potassium (K), which are required in large amount for intensive agriculture. Rising demand for

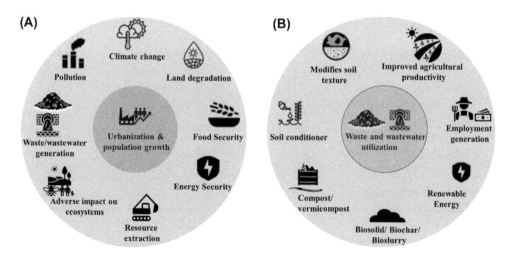

FIGURE 15.1 (A) Contemporary environmental issues due to unprecedented population growth. (B) Resource recovery from waste and wastewater application.

these nutrients resulted in the green revolution that led to increased anthropogenic NPK flows over the past two centuries (Carey et al., 2016). However, there are concerns over long-term availability of these nutrients particularly P and K, which are primarily obtained from mineral deposits. It is believed that P reserves will deplete within the next three centuries at current extraction rate (Gilbert, 2009; Childers et al., 2011; Desmidt et al., 2015).

Many literatures have suggested that solid waste, wastewater and biosolids have high nutritive value and can be utilized in agricultural fields either directly or in other forms such as compost/vermicompost, bioslurry and so on (Srivastava et al., 2015, 2016; Sharma et al., 2017) (Fig. 15.1B). They contain high amount of N, P and K, along with others macro- and micronutrients (Deng et al., 2006; Ali and Schneider, 2008; Liu et al., 2011). Organic fraction of waste could be either applied to fields as compost/vermicompost or anaerobic digestate as they contain considerable amount of soil nutrients (Fig. 15.1B). Similarly, wastewater that comprises high quantity of NPK is also a good source of struvite ($MgNH_4PO_4 \cdot 6H_2O$). Struvite precipitation could be an effective substitution to rock phosphate, which is a finite reserve and rapidly exhausting at a global scale. Recovery of struvite from wastewater closes anthropogenic phosphorus cycle so as to maintain agricultural sustainability (Vaish et al., 2019b). Nutrients recovered from wastewater can be utilized in fertilizer industry and other applications by involving sustainable nutrient recovery technologies. Also, biosolids or domestic wastewater residual produced from different treatment processes contain valuable nutrients and organic matter that may act as a good soil ameliorating agent, especially for soil organic matter (SOM) deficit soils (Acosta-Martinez et al., 2007; Singh and Agrawal, 2008, 2010b, c; Singh et al., 2011). Application of biosolids to agricultural farm alters the physicochemical properties of soil such as soil moisture, organic matter, pH, bulk density, cation exchange capacity (Wang et al., 2008), which consequently enhances plant nutrient balance. Recently, utilization of wastewater and biosolids in agriculture has gained much attention worldwide as it not only provides an economic alternative to waste disposal but also improves physicochemical and biological properties of soil (Sharma et al., 2017). However, waste stream may contain high quantity of toxic pollutants such as heavy metals and several organic micropollutants such as pesticides, pharmaceuticals, personal care products along with various other inorganic and organic pollutants (Luo et al., 2014; Gonzalez-Gil et al., 2016; Srivastava et al., 2017) that critically limit the beneficial usage of waste for irrigation (Blumenthal et al., 2000; Begum and Rasul, 2009; Qadir et al., 2009; Kumar and Chopra2014). Their long-term application could result in toxic accumulation of heavy metals in plants that may enter food chain when consumed by animals (Rutkowski et al., 2007; Kumar and Chopra, 2013, 2014). Hence, much emphasis must be laid on assessing the physicochemical properties of waste prior to agronomic application, which will promote food safety and ensure health and environmental well-being that are the prerequisite for agricultural sustainability (Srivastava et al., 2016). Therefore, waste stream has an immense potential for nutrient recovery in different forms that will help not only in meeting the contemporary nutrient demand but also in reducing the burgeoning amount of waste and wastewater generation. Thus, this chapter sheds light on different strategies of nutrient recovery from wastes and their agronomic responses.

2. Nutrient recovery options through waste biorefineries

Waste can be utilized for production of different value-added by-products through biorefinery, further fueling circular bioeconomy. Biorefineries have great diversity in respect to the feedstock inputs and desired output (Fig. 15.2). Depending on the feedstock material and desired product, array of physical, chemical and biological processes is needed for nutrient recovery. Although complex mix nutrient products such as biosolid, compost, char, etc. play a significant role in sustainable resource management in agrisystems, separate recovery of NH_3–N, PO_4–P and K salts is desirable for their wide range of flexibility and applicability and high market value (Gerardo et al., 2015) (Fig. 15.2). NH_3 and NH_4^+ are the most common recovered form of N species present in solid precipitate forms. Similarly, phosphate is the most recoverable form of P through biorefinery from aqueous streams as solid precipitates and reuse as struvite. Likewise, the major recovered form of K is K-struvite (Fig. 15.2).

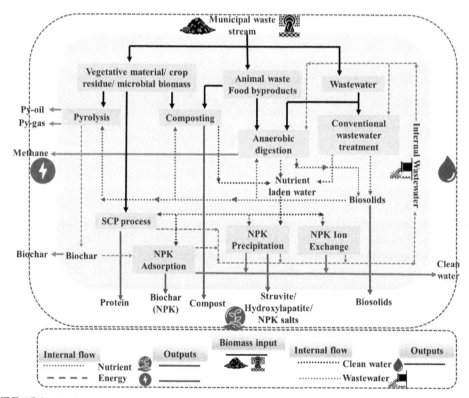

FIGURE 15.2   Nutrient recovery from municipal waste streams: An integrated biorefinery. *Adapted from Carey, D.E., Yang, Y., McNamara, P.J., Mayer, B.K., 2016. Recovery of agricultural nutrients from biorefineries. Bioresource Technology 215, 186–198.bib_Carey_et_al_2016*

2.1 Nutrient selective recovery technologies

2.1.1 Nitrogen selective

2.1.1.1 Biological N recovery

Municipal wastewater and animal wastes contain 34% of the total nitrogen entering in the agricultural system (Matassa et al., 2015), providing an excellent platform for nitrogen (N) recovery. At present, biological wastewater treatment plants convert ammonia to nitrate through nitrification, with potential conversion to nitrogen gas through denitrification, further liberating nitrogen back to the atmosphere. However, by recovering NH_3–N instead of its conversion to nitrate or N_2, a significant amount of energy can be saved by reducing Haber–Bosch energy requirements for fixing N and in nitrification. In addition, single cell proteins (SCPs) have immense potential to be used as feedstock for both animals and human beings (Matassa et al., 2015) (Fig. 15.2). Anaerobic digestion (AD) could provide a good platform for SCP synthesis from biomass or wastewater. During the process, substantial NH_3 and CH_4 are released that can fulfill the N and C needs of methane-oxidizing bacteria used for SCP production (Rittmann and McCarty, 2001). Similarly, algae and fungi can also be used for SCP synthesis using NH_3 from biomass (Anupama and Ravindra, 2000).

2.1.1.2 Physicochemical recovery of N from biomass

N-rich solids and mainstream liquids can be used by N recovery technologies. There are many physicochemical approaches for N recovery, viz., ammonia stripping and distillation, ammonium precipitation as struvite, ion exchange for ammonium and nitrate recovery and ammonia adsorption (Capodaglio et al., 2015; Williams et al., 2015). The concentration of ammonia determines more suitable technique for N recovery. When NH_3 is present in high concentration, then air stripping and distillation are effective, although pH and temperature adjustments are needed (Mehta et al., 2015). Likewise, nitrate can only be extracted using anion exchange resins, followed by precipitation (Samatya et al., 2006). The final N fertilizer product (NH_3/NH_4^+ salts) is obtained by condensation, absorption, or oxidation of the separated NH_3 (Mehta et al., 2015).

2.1.2 Phosphorus selective

The anthropogenic phosphorus cycle is represented by greater loss through diffusion, which results in transfer of P from lithosphere to hydrosphere causing eutrophication. Around 27 million tonne per year of P is applied to the agricultural fields, and only 10%–11% is consumed by human beings and rest is dissipated (van Enk and van der Vee, 2011). Therefore, it is imperative to recover P in an efficient and more sustainable way. One of the promising approaches to P accumulation is by employing phosphate accumulating organisms (PAOs), which uptake and stock high amount of P than typical ones. Thereafter, enhanced biological phosphorus removal (EBPR) select PAOs using alternating anaerobic/aerobic conditions during wastewater treatment. Subsequently, EBPR sludge could have five times greater P in comparison with typical activated sludge treatments (Rittmann and McCarty, 2001), and P can be recovered from EMBR sludge either under anaerobic conditions or through physicochemical processing. Furthermore, different microbes such as cyanobacteria, nonsulfur bacteria and algae have potential to bioaccumulate polyphosphates or proteins (Mehta et al., 2015). Phosphate is the most recoverable form of P when recovered from aqueous stream as solid precipitates. However, phosphorus extraction as struvite ($MgNH_4PO_4 \cdot 6H_2O$) and hydroxyl-apatite is currently favored as compared with mineral fertilizers (Johnston and Richards, 2004) (Fig. 15.2). Although, struvite contains considerable amount of N and Mg, it is used for phosphate recovery and an alternative to rock phosphate as the recovering rate of phosphate is very high through struvite crystallization (Rahman et al., 2014). Wastewater containing high amount of P and N is a good source of struvite. Besides hydroxylapatite used as fertilizer, it can be used as substitution for chemical production of rock phosphate (Cornel and Schaum, 2009). Another innovative approach for phosphorus recovery from wastewater is utilizing algal pond or wetlands. Many researchers state that the preconsideration for P removal from pond, or constructed wetland is effective biomass growth (Korner et al., 2003). Through this process, high uptake of P content up to 2.9% can be achieved (Chaiprapat et al., 2005). Another promising technology for phosphate recovery is hybrid anion exchange (HAIX) resin, although research on this approach is in infancy. The HAIX resins have higher selectivity for phosphate as compared with other competing anions such as sulfate (Blaney et al., 2007; Pan et al., 2009; Sengupta and Pandit, 2011). Thus, these underflow waste stream had huge prospective to recover high phosphorus as fertilizer supplements (Sutton et al., 2011).

2.1.3 Potassium selective

Potassium is another important nutrient for increasing crop yield and enriching soil. Forms of potassium that are accessible to plants from soil are exchangeable potassium and soil solution potassium. SOM present in surface layer is a crucial factor that retains sufficient quantity of potassium, but SOM is rapidly diminishing in soils of tropical region. Adeoye et al. (2001) stated that potential to utilize resource of potash from farm waste is available in ample quantity. They assessed 13 farm wastes for their potassium and other nutrient contents. Likewise, fermentation of molasses or sugar produces K-rich vinasses that makes this waste an excellent feedstock for K recovery (Zhang et al., 2012a). Especially, electrodialysis, membrane filtration and adsorption or ion exchange processes exhibit efficient recovery of potassium. Electrodialysis and strong acid cation exchange resins have been utilized for effective recovery of 99% K products (Zhang et al., 2012b; Mehta et al., 2015). Additionally, regeneration of ion exchange resins can be done using sulfuric acid to produce concentrated potassium solution, which leads to precipitation of K_2SO_4. However, potassium-struvite is the major recovered product that has an advantage of simultaneous recovery of phosphorus.

2.2 Mixed nutrient recovery technologies

In recent days, land application of organic wastes such as biosolids, composted/vermicomposted organic wastes, anaerobic digestate, etc., (Hargreaves et al., 2008; Singh et al., 2011a, Srivastava et al., 2016; Sharma et al., 2017) are primary means of nutrient recovery (Fig. 15.2). However, there is a regional/country-specific restriction based on the presence of heavy metals, pathogens and other organic pollutants such as antibiotics, pharmaceuticals and personal care products for their agricultural application. Likewise, single cell biomass, algae, plant and microbial derived biomass could be directly applied on lands (Mehta et al., 2015). Similarly, char and ash are getting attention for their ability to sequester carbon, immobilizing heavy metals and augmenting soil quality (Mehta et al., 2015).

2.2.1 Anaerobic digestion

During anaerobic digestion (AD) process, digestate is the digested effluent that consists of the feedstock component after abstraction of biogas (Fig. 15.2). In AD, simpler molecules are obtained when complex matter is hydrolyzed, and then organic acids are transformed into methane. During the process, ammonical nitrogen is also released by hydrolysis of organic nitrogen from proteins. From influent to the effluent, concentration of ammonia is likely to increase and is much more readily available than organic nitrogen. Therefore, digestate seems to be more useful fertilizer than manure (Massé et al., 2007; Lansing et al., 2010). Furthermore, phosphorus and potassium content in the digestate are present in considerable amount and readily available to plants. In an experiment conducted by Tani et al. (2006), higher yield was attained when digested cattle slurry was used and compared with manure. However, it must be noticed that few components of the digestate might be unfavorable for the crops. In this respect, Zaldivar et al. (2006) obtained lower yield of lettuce by utilizing digestate from brick masonry as when compared with compost; therefore, Brechelt (2004) strongly recommended dilution of digestate in water before applying so as not to damage crops foliage. The digestate can be most sustainably utilized by recycling of organic matter back to the soil (Alfa et al., 2014). For safe recycling, digestate should be free from undesirable compounds of physical, chemical and biological nature, should be rich in nutrient and must be of high quality. All the while after AD, vital nutrients such as NPK remain in the digestate (Igboro, 2011). Availability of nutrients is more in digestate as when compared with untreated organic waste. For instance, digestate possesses 25% more readily accessible NH_4-N (inorganic nitrogen) and high pH as when compared with untreated liquid manure (Monnet, 2011). Type of feedstocks also determines the quality and components of the dewatered digestate (Mata-Alvarez et al., 2003). Dewatering divides the digestate into fiber and liquid effluent. Furthermore, the fiber can be used as low-quality fertilizer and also as soil conditioner as it is heavy and contains low level of plant nutrients. However, composting process can further convert it into good quality compost. On the other hand, liquid effluent contains large proportion of vital nutrients and can be utilized as excellent fertilizer. Additionally, the high-water content aids its application through conventional irrigation methods. Thus, utilizing fiber and liquid effluent from AD process leads to improved fertilizer application and consequently less consumption of chemical fertilizer during cropping system (Monnet, 2011).

2.2.2 Pyrolysis

Transformation of biomass into bio-oil, biochar and syngas through pyrolysis is quite simple, inexpensive and robust thermochemical process (Fig. 15.2). The robust nature of the pyrolysis technology with a distributed network of small pyrolysis plants would be compatible with existing agriculture and forestry infrastructure, which permits the flexibility in both the variety and quality of biomass feedstock. In the existing industrial boilers, bio-oil can be used as a fuel and biochar could be used instead of pulverized coal. Also, utilizing biochar as soil improver leads to sequestration of large amount of C along with greenhouse gas reduction that consequently causes significant environmental and agronomic benefits. The preliminary study of Laird et al. (2009) indicates that biochar quality is very important in deciding the impact of soil biochar on crop yields.

Application of biochar in soil can enhance soil carbon pool, boost crop productivity and decrease the bioavailability and phytotoxicity of heavy metals (Chan and Xu, 2009; Park et al., 2011). Highly porous structure and large surface area are crucial physical factor of most biochars. This structure can influence beneficial soil microorganisms such as mycorrhizae and bacteria to bind important nutritive cations and anions. The availability of macronutrients such as N and P can be increased due to binding of nutritive cations and anions. The application of biochar also causes alkalization of soil pH and enhancement in electrical conductivity (EC) and cation exchange capacity (CEC) (Atkinson et al., 2010). Similarly, in a study conducted by Liu et al., (2014), use of biochar did not impede the germination of seeds. In his advanced research, he also concluded that biochar has potentiality to serve as fertilizer to satisfy the needs of plant growth in agricultural production. In his analysis the yields of biochar and fixed carbon were 46.3% and 39.9%, respectively. He also concluded that pyrolysis caused enhancement in the total and available P and K concentrations and detraction in total and available N.

2.2.3 Aerobic composting/vermicomposting

Composting is referred as decomposition of biodegradable organic material from various sources into a stabilized product under aerobic conditions with the aid of indigenous microorganisms (thermophile and mesophile) (Srivastava et al., 2015) (Fig. 15.2). In earlier thermophilic phase, temperature rises to 45−65°C results in sanitization followed by mesophilic or maturation phase. During the process, complex organic matter breaks into simpler forms. The nutrient profile and quality of the end product largely depends on the source of waste, the composting design, operational parameters and maturation length during composting (Bernal et al., 2009; Hargreaves et al., 2008). The ready compost has nutrient-rich profile (NPK and carbon) and helps in recycling of plant nutrients. Likewise, vermicomposting is an eco-friendly and biooxidative process, which stabilizes complex biodegradable wastes into valuable vermicompost through the action of earthworms and microorganisms. Earthworms are important drivers that support the growth of microbes by providing accessible surface area, thereby improving enzymatic actions (Fornes et al., 2012). Many epigeic earthworm's species, *Eisenia fetida, Perionyx excavatus, Perionyx sansibaricus* and *Eudrilus eugeniae*, are generally employed for vermicomposting process (Oyedele et al., 2005; Suthar, 2010). Prior to vermicomposting process, precomposting is required to provide a favorable environment to earthworms. The maturation of vermicompost is indicated by significant decrease in C/N ratio and other complex organic forms such as carbohydrate, lignocellulosic and protein whereas increase in humic acid and enzymatic activities.

Both composting and vermicomposting has immense potential that adds value to the waste and convert it in nutrient-rich end product, which has high nutritive value and water holding capacity, low bulk density and good economic value (Soumare et al., 2003). In addition to this, both techniques provide a very promising tool for nutrient recovery from wastes and recycling of organic fraction of municipal solid wastes (MSWs).

3. Fertilizer value of waste

3.1 Municipal solid waste compost/vermicompost

Degradation of agricultural land as a result of poor management is a major environmental and agricultural challenge, which is attributed to low nutrient availability and loss of organic matter leading to decreased productivity (Tejada et al., 2009; Duong et al., 2012). To revert the declining trend of agricultural productivity and to restore the degraded soils, fertilizer application is requisite (Goyal et al., 1999). However, extensive use of inorganic fertilizer without any organic supplements poses risk to soil health (i.e., soil's physicochemical and biological properties) and environment (i.e., water pollution). John et al. (2006) reported MSW as one of the best forms of organic fertilizers due to high organic matter and nutrients such as Ca, S, P, N and K. However, there is a potential menace due to various organic and heavy metal contaminants present in MSW (Srivastava et al., 2016). Nowadays, there is growing interest in composting/vermicomposting of MSW, as it decreases the stabilization time of household waste and sewage sludge (Garcia-Gil et al., 2004; Walter et al., 2006; Hargreaves et al., 2018; Srivastava et al., 2020a,b). Besides this, it helps in reducing the large volume of waste, extending the life span of landfill sites (Singh et al., 2011). Therefore, compost has more significant cumulative effect on availability of nutrients as compared with chemical fertilizers. It also has an advantage over inorganic fertilizers whose uncontrolled use during the past few decades has badly affected the soil's physicochemical and biological properties (Bruun et al., 2006; Mathivanan et al., 2014). Application of MSW compost supplements organic matter content and other nutrients in soil (Evanylo et al., 2008). Due to the presence of macro- and micronutrients such as N, P, K, Cu, Fe and Zn, the compost can be used as soil conditioner. It also increases the soil aeration and water holding capacity through the addition of high organic matter (Ingelmo et al., 2012). Therefore, applications of organic fertilizer such as compost, vermicompost and manure are now becoming more lucrative methods that support sustainability to the system (Paul, 2007; Birkhofer et al., 2008). However, if the composting process is not completed, then there will be problems of nitrate leaching or gaseous losses of nitrous oxide (N_2O), carbon dioxide (CO_2) or methane (CH_4) emissions. These gasses are involved in global warming, so their examination after long-term organic amendment is requisite for a balanced picture of nutrient cycling after compost application. The rate of nutrient release is based upon the microbial diversity of soil (White, 2006). As compared with the chemical fertilizer–treated soil, the organic fertilized soils are more stable (Van Diepeningen et al., 2006). Compost application is directly correlated with the biotic component of the soil. Crecchio et al. (2004) and Walter et al. (2006) have observed that the soluble organic carbon and soil C:N ratio were increased by the frequent use of municipal solid waste compost (MSWC). The MSWC can easily replace the chemical fertilizer as it helps in the formation of steady humus due to the presence of organic mass (Tidsell and Breslin, 1995) and it increases the fertility of soil (Perez et al., 2007). Meena et al. (2016) have evaluated the ability of MSWC to increase the microbiological and chemical properties of soil. They observed that MSWC in combination with chemical fertilizers leads to enhancement in the microbial activity and nutrient availability of the soil under saline condition. Pascual et al. (1999) found increased levels of soil enzymes and soil basal respiration rate in MSW amended soil as compared with control. Garcia-Gil et al. (2000) observed an increase of 10% and 46% in soil microbial biomass when amended with MSW compost at 20 and 80 tonnes/hectare (t/ha), respectively. Likewise, Warman et al. (2011) found increased microbial activity depicted by augmented enzyme activities and microbial biomass carbon. Srivastava et al. (2018) applied MSW vermicompost at 0%, 20%, 40%, 60%, 80% and 100% ratios to the agricultural soil and found increased concentration of total N, P, K and organic carbon along with heavy metals as compared with unamended soil. On beneficial side, MSW compost maintains soil fertility, but on the other hand, it also acts as a potential threat having different kind of pathogens and pollutants (Crecchio et al., 2004; Hargreaves et al., 2008; Singh and Agrawal, 2010a,b; Srivastava et al., 2016). Heavy metal accumulation is a major drawback of MSW application in the agricultural field, and it increases the level of metal concentration at different trophic levels (Singh et al., 2011; Srivastava et al., 2018). Consequently, it is obligatory to check the maturity and physicochemical characteristics of composts/vermicomposts prior to its agricultural application. Permissible limits for heavy metals in compost from different countries are given in Table 15.1.

3.2 Wastewater

Agriculture sector demands more than 70% of potable water (FAO, 2016). To achieve this demand under contemporary water-scarce condition is a tough challenge. Nowadays, the application of wastewater for irrigation purposes has gained attention due to nutrient-rich profile. Wastewater, which contains high amount of NPK, organic matter and other essential macro- and micronutrients, could be used for irrigation of agricultural lands (Deng et al., 2006; Ali and Schneider, 2008; Sarkar et al., 2018). This would further help in reducing demand of freshwater for agricultural purposes and intensive fertilizer inputs and check effluent discharge into water bodies (Pereira et al., 2011; de Oliveira Marinho et al., 2013).

TABLE 15.1 Permissible limits of heavy metals in organic waste compost.

Country	Heavy metals (mg/kg dry compost)						
	Cr	Cu	Cd	Ni	Zn	Pb	Hg
Australia	—	200	3	—	250	200	—
Canada	210	100	3	62	500	150	0.8
Denmark	—	1000	0.8	30	4000	60–120	0.8
European Commission	—	100–150	0.70–1.50	—	200–400	100–150	—
France	—	—	8	200	—	800	8
Germany	150	150	2	50	400	200	1
India	50	300	5	50	1000	100	0.15
United States	100	300	4	50	400	150	0.5
United Kingdom	100	200	1.5	50	400	150	1

Adapted from Sharma B, Sarkar A, Singh P et al., 2017. Agricultural utilization of biosolids: a review onpotential effects on soil and plant grown. Waste Management 64, 117–132.

Numerous researches on application of wastewater to irrigate different crops such as radish (Dantas et al., 2014), lemon (Pedrero et al., 2012), tomato (Cirelli et al., 2012) and lettuce (da Silva Cuba et al., 2015; Varallo et al., 2012) have been done extensively. These researches revealed that in spite of providing essential nutrients to the crops, wastewater could be the source of heavy metal (HM) pollution in the soil and other potent environmental polluting agents. Apart from HM contamination, few other detrimental effects include soil salinization, decrease in soil infiltration capacity or bacterial contamination by *Escherichia coli*. FAO (1985) formulated guidelines for interpretation of water quality index suitable for agricultural irrigation (Table 15.2).

Siebe and Cifuentes (1995) compared the sites under irrigation with wastewater (over 80 years) with that of non-irrigated fields at different periods of time on the growth and nutrient uptake by alfalfa crops. The application rate of wastewater varied between 1500 and 2200 mm/ha/year. A substantial increase of organic carbon, total nitrogen and extractable phosphorus was observed. Extractable P showed remarkable increase from 3 to 34 $\mu g/g$ P in non-irrigated soils to 60–97 $\mu g/g$ P in wastewater-irrigated soils. On the contrary, Ca^{2+} decreased whereas the Na^+ saturation increased in soil. Results for nutrient uptake in alfalfa showed that neither N nor Mg and K uptake increased, whereas Ca uptake decreased significantly.

Angin et al. (2005) found that wastewater-fed irrigation caused increased salinity and reduced pH, may be due to increase in organic matter, nitrogen and available P content. Despite increase in organic matter and NPK, lowering of pH might cause loss of nutrients. Therefore, application rate of wastewater must be monitored thoroughly so as to minimize the potential risk of soil degradation. In a similar experiment on lettuce crop, Urbano et al. (2017) showed significant increase of soil nutrients such as K, Ca, Al and S when irrigated with treated wastewater. Also, contamination of *E. coli* bacteria was not found on lettuce leaves or in soil. Lettuce production (fresh weight) and exchangeable sodium percentage ($6.44 \pm 0.52\%$) were higher in wastewater-fed soil as compared with soil irrigated with drinking water with conventional fertilizers. Urbano et al. (2015) found that red latosol soil did not suffer any physiochemical damage after irrigation with wastewater. However, soil acidification was reported after one and subsequent year of cultivation (Singh et al., 2012; Varallo et al., 2012). This may be attributed to increased amount of ammonia applied from wastewater.

Bedbabis et al. (2015) conducted an experiment on olive tree development irrigated with treated wastewater and found sudden increase of soil pH. No negative effects were reported during the entire experiment. Significant increase in EC values, available P and K contents suggested that wastewater has high organic matter adsorption and high soluble P and K contents. Application of wastewater after 5 and 10 years revealed considerable increase in Cl and Na. The high Na concentration could be due to antagonistic activity of either K^+ or NH_4^+ or may be due to high Ca supply that increased the selectivity for the uptake and transport of K^+ over Na^+. Also, Zn, Mn and Fe content were increased significantly after wastewater irrigation. However, no severe environmental problem or phytotoxicity effects could be noticed during experimental period. In a similar experiment on wheat (AKW-1071), gram (Jacky-9218), palak (pusa jyoti), methi (kasuri) and berseem (multicut) were conducted by Singh et al. (2012) and found improvement in physico-chemical properties of soil, better yield

TABLE 15.2 Guidelines for interpretation of water quality for irrigation (FAO, 1985).

Potential irrigation problem		Units	Degree of restriction on use		
			None	Slight to moderate	Severe
Salinity (affects crop water availability)					
EC		dS/m	<0.7	0.7–3.0	>3.0
TDS		mg/L	<450	450–2000	>2000
Infiltration (affects infiltration rate of water into the soil. Evaluate using EC and SAR together)					
SAR = 0–3			EC > 0.7	EC = 0.7–0.2	EC < 0.2
SAR = 3–6			EC > 1.2	EC = 1.2–0.3	EC < 0.3
SAR = 6–12			EC > 1.9	EC = 1.9–0.5	EC < 0.5
SAR = 12–20			EC > 2.9	EC = 2.9–1.3	EC < 1.3
SAR = 20–40			EC > 5.0	EC = 5.0–2.9	EC < 2.9
Specific ion toxicity (affects sensitive crops)					
Sodium (Na)					
	Surface irrigation	SAR	<3	3–9	>9
Sprinkler irrigation	me/L		<3	>3	
Chloride (Cl)					
	Surface irrigation	me/L	<4	4–10	>10
Sprinkler irrigation	me/L		<3	>3	
Boron (B)		mg/L	<0.7	0.7–3.0	>3.0
Miscellaneous effects (affect susceptible crops)					
Nitrogen (NO$_3$–N)		mg/L	<5	5–30	>30
Bicarbonate (HCO$_3$)		me/L	<1.5	1.5–8.5	>8.5
pH			Normal range 6.5–8.4		

Item id Caption title	Creditline changes
Table 15.2 Guidelines for interpretation of water quality for irrigation	Food and Agriculture Organization of the United Nations, [1985], Reproduced with permission.

EC, electrical conductivity; *SAR*, sodium adsorption ratio.

and enhanced nutritional status on soil and plants irrigated with domestic wastewater as compared with soil irrigated with groundwater and conventional fertilization.

Undeniably, if applied at appropriate rates, wastewater could provide ecological and economic benefits by reducing the usage of synthetic fertilizers (Varallo et al., 2012; da Silva Cuba et al., 2015).

3.3 Biosolids

Biosolids, which is also a good source of organic matter and other essential micro- and macronutrients, could provide good alternative to synthetic fertilizers. Application of biosolids as supplement to inorganic fertilizers helps in enriching soil nutrient profile (Singh et al., 2011b). It also enhances agromorphological attributes and yields of different crops. Biosolids contain substantial quantity of essential micronutrients such as Cu, Zn and Ni. However, long-term application of biosolids must be monitored cautiously as it might contain toxic metal residues along with other organic and inorganic pollutants (Jia-yin et al., 2006) that may end up in contaminating soil ecosystem and eventually could be taken up by plants (Jia-yin et al., 2006).

Soil microbial fauna along with associated enzymatic activity also gets disturbed due to disruption of nutrient cycling processes (Fernández et al., 2009; Srivastava et al., 2016). Therefore, long-term application of biosolids could degrade soil quality, affects nutrient uptake by crops and may further contaminate food chain. Hence, appropriate dosage and continuous monitoring for application of biosolids in different climatic regimes and for different crops must be studied extensively. The permissible limits of organic pollutants in sewage sludge for agricultural application are given in Table 15.3.

García-Gil et al. (2004) studied the residual effect of sewage sludge (at 40 Mg/ha) after 9 and 36 months from application to barley cultivated land. The results showed significant increase in microbial biomass carbon (MBC), basal soil respiration (BSR), metabolic quotient (qCO$_2$) and enzymatic activities after 9 months due to increased microbial activity and improved mineralization rate. However, the values of these parameters decreased after 36 months of investigation and were closed to the control due to depletion of organic substrates.

Kizilkaya and Bayrakli (2005) conducted an experiment to study the effect of different biosolids amendment rates (0, 100, 200 and 300 t/ha dry wt.) with C/N ratios (3:1, 6:1 and 9:1) on different enzymatic activities (alkaline phosphatase, arylsulphatase, b-glycosidase and urease) in a clay loam soil. Addition of biosolids increased enzymatic activity significantly as compared with unamended soil. β-Glucosidase activity was highest in soil with highest amended dose and highest C/N ratio. However, for enzymatic activities of alkaline

TABLE 15.3 Permissible limits of organic pollutants in sludge for agricultural application.

Compound	Permissible value[a] (mg/kg)
PCB (polychlorinated biphenols)	0.8
PAH (polycyclic aromatic hydrocarbons)	6
NPE (nonylphenolethoxylates)	50
DEHP (di-2-ethylhexyl phthalates)	100
AOX (absorbable organic halogens)	500
LAS (linear alkybenzene sulphonates)	2600

[a]EEC-Sludge Rule (2000), European Commission.

phosphatase, urease and arylsulphatase showed an increase during first 30 days of incubation; however, a pronounced decrease was noticed thereafter. Similarly, Roig et al. (2012) examined the effect of aerobically digested biosolids as fertilizer supplement for 16 years. They studied the effect of biosolids on soil physicochemical, functional and ecotoxicological properties. The results revealed enhanced C and N mineralization, increase in microbial activity and considerable increase in dehydrogenase activity. This could be attributed to enhanced SOC content due to biosolid application (Roig et al., 2012). Likewise, Xue and Huang (2014) examined the effect of biosolid compost on *Paeonia suffruticosa*. The impact was seen on soil physicochemical, microbial and biochemical properties of soil and plant at 0%—75% amendment dose. Soil enzymatic activities (invertase, polyphenoloxidase, phosphatase, urease and proteinase), soil basal respiration (Rmic) and soil microbial biomass (Cmic) increased significantly at doses ranged between 15% and 45%. Thereafter, decrease from 45% to 75% was noticed. A similar trend was found in ratios of (Cmic/Corg) and (Cmic/Rmic) that increased significantly and then decreased at dosage of 45%. Decrease in SOC, total Kjeldahl N and total P was noticed above 45% amendment rate. Therefore, the overall results revealed that a dosage around 45% could be beneficial for soil ecosystem and growth of tree peony.

Angin et al. (2017) studied the effect of varying treatments of sewage sludge (SS) (0.0, 2.5, 5.0, 7.5 and 10.0 kg of dry matter per plant) on soil physicochemical characteristics in a 3-year field experiment. Soil pH decreased significantly due to the release of organic acids during mineralization process. SOM, N and P contents increased with increasing doses of SS. CEC was found to be highest at 10.0 kg of dry matter per plant. This can be attributed to elevated macroelements (except Mg) and increased adsorption surface, presumably due to organic matter content of SS. Likewise, Na, K and Ca content increased in rates of 20.0%, 1.6% and 0.2%, respectively, as compared with control under highest application rate. No detrimental effects were seen on the cultivated plants throughout the experiment.

Kayikcioglu et al. (2019) used four treatments of treated municipal sewage sludge (TSS) i.e., chemical fertilizer control (CF), 10 Mg/ha/a (TSS_1), 20 Mg/ha/a (TSS_2) and 30 Mg/ha/a (TSS_3) on a dry weight basis to assess its impact on soil physicochemical and biological properties in a 3-year experiment. Soil pH increased slightly following TSS application, which would not only reduce metal bioavailability but will also improve overall pH buffering capacity of soil. There was no significant difference in EC and carbonate content before and after 3 years. SOC increased significantly with increasing doses of TSS due to the presence of high OM in sludge, which plays an important role for soil microbial activity. The microbial activity was comparatively lower in unamended soil. MBC increased following TSS amendments with maximum increase in TSS_3 (211.75 µg Cmic/g). The BSR was not significantly different among different TSS amendments, whereas soil enzyme activities increased with increasing levels of TSS.

In conclusion, MSW compost/vermicompost, wastewater irrigation and biosolid application in agriculture could be a promising approach for substitution of inorganic fertilizer. However, the presence of heavy metals is one of the major disadvantages for its safer application. There are many legislations that govern permissible limits of heavy metals in water, soil and plants given in Table 15.4.

TABLE 15.4 Guidelines for safe limits of heavy metals.

Sample	Standards	Cr	Cu	Cd	Ni	Mn	Pb	Zn
Water (mg/L)	FAO (1985)	0.10	0.20	0.01	0.20	0.20	5	2
	European Union Standards (EU, 2002)	–	–	–	–	–	–	–
	USEPA (2000)	–	1	0.005	–	–	0.015	2
	Indian Standard (Awashthi, 2000)	0.05	0.05	0.01	–	0.10	0.10	5
Soil (mg/kg)	WHO/FAO (2007)	–	–	–	–	–	–	–
	European Union Standards (EU, 2002)	150	140	3	75	–	300	300
	USEPA (2000)	–	50	3	–	–	300	200
	Indian Standard (Awashthi, 2000)	–	135–270	3–6	75–150	–	250–500	300–600
Plant (mg/kg)	WHO/FAO (2007)	–	40	0.20	–	–	5	60
	Commission regulation (EU, 2006)	–	–	0.20	–	–	0.30	–
	USEPA (2000)	–	–	–	–	–	–	–
	Indian Standard (Awashthi, 2000)	20	30	1.5	1.5	–	2.5	50

4. Agronomic response of waste stream

4.1 Municipal solid waste compost/vermicompost

Application of MSWC/vermicompost enriches soil nutrient level, soil texture, buffering capacity, water holding capacity, bulk density, porosity, microbial activity (Srivastava et al., 2018). Many studies showed positive impact of MSW compost/vermicompost on yield attributes of various crops such as potato and sweet corn (Mkhabela and Warman, 2005), lettuce (Fagnano et al., 2011), wheat (Lakhdar et al., 2011), maize (Carbonell et al., 2011; Onwudiwe et al., 2014), winter squash (Warman et al., 2009), spring triticale (Weber et al., 2014).

Ozores-Hampton et al. (1994) have observed the role of MSW application on qualitative and quantitative characteristics of tomato (*Lycopersicon esculentum* Mill.) and squash (*Cucurbita maxima* Duch. Ex Lam.) plants. On both the plants, beneficial impact of MSW varied considerably between years. They had a higher canopy volume than the control treatment. The fruit yield and level of Mg and Zn increased with the application of compost. Bhattacharyya et al. (2003) have reported that different doses (2.5, 10, 20 and 40 Mg/ha) of MSW compost increased the microbial biomass C and soil respiration that gives an idea about the index of general metabolic properties of soil microbial level. Shabani et al. (2011) have studied the impact of MSWC (50, 100, 150 and 200 t/ha) on quality and quantity of eggplant. Maximum yield was achieved at 150 Mg/ha treatments compost. The yield of plant was increased by improving physicochemical properties such as water holding capacity, cation exchange capacity, bulk density and percentage organic carbon content and microbial biomass. The level of Ca, P and Mg in root and leaf of the plants was also found to be increased by the application of MSWC (Shabani et al., 2011). Haghighi et al. (2016) assessed the response of MSWC on tomato growth under hydroponic system. As compared with the control, the numbers of fruits increased by applying 25% of MSWC in hydroponic culture. Cherif et al. (2009) studied the impact of MSWC on wheat growth at the application rate of 40 (C1) and 80 Mg/ha (C2). C1 and C2 displayed a significant increase in grain yield of wheat (234.1% and 241.1%, respectively) as compared with the control (17.65 Mg/ha). Likewise, Hargreaves et al. (2008) assessed the effect of MSWC on fruit quality and yield of strawberry grown in sandy loam soil at Nova Scotia, Canada, during 2005–06. The soil was amended by MSWC with the application rate of 150 and 75 kg/ha in first and subsequent year, whereas IF was applied at the rate of 150,70 and 70 kg/N/ha during 2004, 2005 and 2006. No significant difference was found in sugar content during the experimental period, whereas mean fruit yield was found to be 1639 g/m^2 in MSWC-treated soil as compared with 1182 g/m in IF during the study. Naderia and Ghadirib (2010) examined combined effect of urban waste compost (UWC) on growth of corn. Results showed increased plant dry matter, height, diameter, area and number of leaves, and level of N as compared with control. There are several reports to improve the soil properties and nutrient quality of vegetable crops and fruits by the addition of MSWC (Table 15.5).

4.2 Wastewater

Several studies have been done by national and international research communities emphasizing the beneficial usage of wastewater for irrigation and also to be cautious of their impact if overused (da Silva Cuba et al., 2015; Fonseca et al., 2007; Pedrero et al., 2012; Pereira et al., 2011; Singh et al., 2012; Urbano et al., 2015; Varallo et al., 2012) (Table 15.6). The studies concluded that if irrigated with proper amended rate for specific period of time, the harmful effect of wastewater utilization could be avoided. In this respect, Kumar et al. (2017) conducted an experiment to assess the effect of secondary treated municipal wastewater at different amendment rates (0%, 10%, 20%, 40%, 60%, 80% and 100%) on hybrid cultivar of okra (*Hibiscus esculentus* L. var. JK 7315). The maximum agronomic production was found in soil treated with 60% wastewater in both the seasons. The heavy metal content of Cd, Cr, Cu, Mn and Zn increased, although they were below the maximum standards permitted for soils in India (BIS, 2012). At 60% concentration, maximum plant height (160.5 and 168.7 cm), root length (23.8 and 26.7 cm), dry weight (105.5 and 110.6 g), chlorophyll content (4.7 and 4.8 mg/g.f.wt) and LAI/plant (4.6 and 4.7) were observed at both the seasons. Therefore, wastewater diluted with 60% concentration was most suitable for vegetative growth of *H. esculentus*. Likewise, many researches have reported considerable growth of plant, increase in chlorophyll content, enhanced biomass production and increased leaf area index if irrigated with recycled wastewater (Çalisir et al., 2005; Sunilson et al., 2008). Similar results were found by Çalisir et al. (2005) on *H. esculentus* L. when applied with 25% and 50% concentration of municipal wastewater. In a similar study conducted by Castro et al. (2013) on lettuce crop (*Lactuca sativa*, var. libano), the seedlings were treated with wastewater from the wastewater treatment plant of Alcáazar de San Juan for extracting essential nutrients such as NPK and OM. The long-term application of wastewater could cause accumulation of HM in soil and eventually in plants. In this experiment, soils treated with wastewater

TABLE 15.5 Agronomic Responses of MSW compost/vermicompost.

Source of MSW	Soil type and properties	Crop	Doses	Effect/Response	References
Mornag, Tunisia	Clayey loamy/pH 7.64/Total K, Mg and organic C were about 5650 and 3380 mg/kg, and 0.93%, respectively	Wheat (Triticum aestivum)	80 mg/ha	Wheat grain yield increased significantly	Cherif et al. (2009)
Madrid, Spain	Sandy loam soil/pH 6.4/EC 0.1 dS/m. OC and total N were 8.0 and 0.7 g/kg Similarly, P, K, Ca, Mg and Na were 0.03, 0.2, 1.5, 0.2 and 0.01 g/kg, respectively	Barley (Hordeum vulgare)	20 and 80 t/ha	Increased microbial activity, improved soil buffering capacity	Garcia-Gil et al. (2000)
Naples, Italy	Sandy loam/total carbonates and assimilable P_2O_5 were 520 g/kg and 46 mg/kg, respectively	Lettuce (Lactuca sativa)	10, 30 and 60 Mg/ha	Quality and yield of lettuce plant increased	Fagnano et al. (2011)
Bazyafte zobaleh co. Rasht (Capital of Guilan province), Iran	Sandy loam/pH 6.8; containing total N (3%), total C (1.5%), C/N ratio 0.5; Ca, P and K were 12, 68, 100 mg/kg, respectively	Eggplant (Solanum melongena)	50, 100, 150 and 200 t/ha	Yield and number of leaves per plant increased. Reduction in soil-borne disease. The best of yield achieved at 50 t/ha fertilizer level	Shabani et al. (2011)
Beja, Tunish	pH 7.95, EC 263 µS/cm, N 0.11%, C 1.2%. Zn, Cu and Ni were 70.0, 32.0 and 50.0 µg/g Pot/glasshouse	Wheat (Triticum durum)	40, 100, 200 and 300 t/ha	Increased photosynthetic rate, stomatal conductance, RubisCO activity and biomass gain were noticed. Optimal result was noticed in 100 t/ha	Lakhdar et al. (2012)
Madrid province, Spain	Fluvisol, pH 8.3, EC 0.19 dS/m. Total organic carbon (TOC), total N and carbonates were 13.08, 1.4 and 88 g/kg, respectively; and available P was 25.6 mg/kg	–	160 Mg/ha	The application significantly increased microbial biomass carbon (MBC). Enzyme activity remains stable throughout experimental period	Jorge-Mardomingo et al. (2013)
Municipal solid waste (MSW) and cattle manure	Different proportion of composting of MSW mixed with different proportion of cattle manure	–	1:1.5, 1.5:1 and 1:1	MSW mixed with different proportion of cattle manure significantly influenced the compost quality and process dynamics	Varma et al. (2013)
MSW Chania, Greece	Pot experiment with clay loam/pH 7.7/EC 0.1 dSm	Lettuce (Lactuca satira) and tomato (Lycopersicon esculentum)	0, 50 and 100 t/ha	Lower doses of MSW compost (MSWC) was more beneficial	Giannakis et al. (2014)

(Continued)

TABLE 15.5 Agronomic Responses of MSW compost/vermicompost.—cont'd

Source of MSW	Soil type and properties	Crop	Doses	Effect/Response	References
Nsukka, Nigeria	Sandy loam Ultisol	Maize (*Zea mays*)	MSWC 0,1000,1500 and 2000 kg/ha	Leaf area, harvest index and yield increased significantly	Onwudiwe et al. (2014)
Composted tannery sludge	Fluvisol, pH 6.5, EC 0.63 dS/m long term/ 5-year field experiment	Cowpea (*Vigna unguiculata* L.)	0, 2.5, 5, 10 and 20 Mg/ha	5 Mg/ha composted tannery sludge showed highest values for soil MBC, microbial biomass nitrogen and soil respiration	Araujo et al. (2015)
MSW, Kerala, India	A pot experiment was conducted having laterite soil with pH 5.5, water holding capacity was 42.3%	Cassava (*Manihot esculenta* Crantz)	0, 2.5, 5, 10 and 20 t/ha	With the increasing dose of MSWC, available N, residual C and decomposition rate significantly increased	Byju et al. (2015)
MSW from Chania, Greece	Sandy soil: pH 8.35, EC 436 μS/cm; TOC and total N were 4.24 and 0.41 g/kg, respectively; Clayey soil: pH 7.82, EC 744 μS/cm, TOC and TN were 21.28 and 2.12 g/kg, respectively	Spiny chicory (*Cichorium spinosum*)	0, 60 and 150 t/ha	Yield was higher in the sandy than in clayey soil even in absence of compost application. Bioavailability of Cu, Zn, Fe, Mn, Cr, Ni, Pb and Cd in both soils was increased, but content was below toxic level in edible part; sandy soil with 60 t/ha is recommended dose	Papafilippaki et al. (2015)
Composted vegetal wastes	The pH of the soil is 8.5, its electric conductivity is 250 mScm^{-1}, its water holding capacity is 40.2 g 100 g^{-1} and the bulk density is 2.57 gcm^{-3}	–	150 and 300 t/ha	Increase in microbial biomass and more notable activity than the addition of composted waste	Torres et al. (2015)
MSW vermicompost (MSWVC), India	pH 8.05, EC 0.26 mS/cm, TOC and TKN were 0.63% and 0.24% respectively. TP and K were 4.14 and 5.85 g/kg	Ladyfinger (*Abelmoschus esculentus* L.)	0%, 20%, 40%, 60%, 80% and 100%	Positive effect on soil nutrient profile represented by increased organic carbon and NPK. MSWVC showed positive effect on biochemical, physiological and yield responses of *A. esculentus* (lady's finger) up to 60% MSWVC-amended soils. Recommended dose up to 60%	Srivastava et al. (2018)

TABLE 15.6 Agronomic Responses of wastewater and sewage sludge.

Source of waste water/ sewage sludge	Wastewater (WW)/sewage sludge (SS) characteristics	Experimental object (plant/soil)	Doses	Effect	Reference
Ramtha, Jordan,	WW: pH 7.5, TDS 1225 mg/L, BOD5 290 mg/L, PO$_4$ 49 mg/L, NH$_4$ 118 mg/L, NO$_3$ 2.9 mg/L	Soil	540 and 675 mm of treated wastewater for corn, and 74 mm and 93 mm for vetch	Increased levels of N, P, K, Fe and Mn in the soil, while changes in Cu and Zn concentration were insignificant. Negatively affected phosphatase activity and soil microorganisms	Mohammad and Mazahreh (2003)
Central Mexico	WW: pH 8.4, EC 1483 µS/cm, Ca^{2+} 2.66 meq/L, Mg^{2+} 2.1 meq/L, Na$^+$ 1081 meq/L, Cl$^-$ 6.43 meq/L. Total solids 1488, N–NO$_3$ 0.015 meq/L	Soil and plant; Alfalfa,	1500 and 2200 mm/ha per year, according to crop requirements, soil textures and depths	Soil organic carbon, total nitrogen, P and Na content increased significantly. P and Na content were high in alfalfa, while no changes were observed for N, Mg, and K uptakes.	Siebe and Cifuentes (1995)
Alcazar de San Juan, Central Spain	WW: pH 7.1, EC 2085 µS/cm, COD 60.2 mg/L, BOD 8 mg/L, TP 3.3 mg/L, TKN 20.8 mg/L, TOC 33.9 mg/L, Na 318 mg/L, K 106 mg/L, Ca 138 mg/L, Zn 0.4 mg/L Fe, Pb, Cd, Cr and Ni were 2314, 334, 11, 284 and 221 µg/L, respectively	Soil and plant; Lettuce (Lactuca sativa L.)	Mixed wastewater: 70% of domestic origin and 30% of industrial	Significant increase in macronutrients (P and Mg) in soil; heavy metal content was relatively low (<0.2 mg/kg) and constant throughout the study except Cr and Pb content. Dry and fresh weight, average height and diameter were significantly higher, while N and P content in plant tissue were low	Castro et al. (2013)
Erzurum plain, Turkey	WW: pH 8.26, EC 1.73 dS/m, TDS 1402 mg/L. Total N, PO$_4$, Cu, Mn, Zn, Fe, Ca, Mg, Na and K were 1502, 3.0, 0.11, 0.12, 0.17, 0.20, 55, 45, 221 and 51.5 mg/L, respectively. Sodium adsorption ratio (SAR) 6.47	Soil and plant; Cabbage (Brassica oleracea var. Capitate cv. Yalova-1) and potato (Solanum tuberosum)	Irrigation is done by flooding	pH decreased, whereas salinity and nutrients increased in the soil. Likewise, N, P, Fe, Mn, Zn, Cu, B, Mo and Cd contents increased in the crops	Angin et al. (2005)
Colorado, US	WW: pH 8.1, EC 0.84 dS/m SAR 3.1, TDS 614 mg/L	Soil	—	Elevated pH and higher concentrations of extractable Na, B and P were reported in recycled wastewater treated soil. Similarly, around 187 and 481% hike were observed in EC and SAR	Qian and Mecham (2005)

(Continued)

TABLE 15.6 Agronomic Responses of wastewater and sewage sludge.—cont'd

Source of waste water/ sewage sludge	Wastewater (WW)/sewage sludge (SS) characteristics	Experimental object (plant/soil)	Doses	Effect	Reference
Haridwar, India	WW: pH 8.8, BOD 896.4, COD 1968.7, Na^+ 582.5, K^+ 515.8, TKN 845.3, PO_4^{3-} 264.2, SO_4^{2-} 985.3 mg/L Fe 24.5, Cd 9.5, Cr 3.9, Cu 22.8 mg/L	Soil and plant; Okra (*Hibiscus esculentus* L. var. JK 7315)	10%, 20%, 40%, 60%, 80% and 100% along with the control (groundwater) were used for the irrigation	EC, OC, Na^+, K^+, Ca^{2+}, Mg^{2+}, Fe^{2+}, TKN, PO_4^{3-}, SO_4^{2-} Cd, Cr, Cu, Mn and Zn content of the soil was significantly and positively correlated with different concentrations of the municipal wastewater. The contamination factor of Mn was the highest, whereas that of Cr was the lowest. The maximum agronomic performance of the *H. esculentus* was recorded with 60% concentration. The seed germination of the *H. esculentus* was noted to be negatively correlated with different concentrations of the municipal wastewater	Kumar et al. (2017)
Riyadh, Saudi Arabia	WW: pH 7.48–7.84, TDS 992 –3982 mg/L, SAR from 2.35 to 6.43. Heavy trace metals such as Ni, Pb, Co, Cd and Mo were not traceable	Plant; Corn (*Zea mays* L.) and forage sorghum (*Sorghum vulgare* Pers)	1.32 L of water was applied per pot per irrigation	Corn—Mg, Na, N, K, P, Cu, Mn, Zn and Mo increased with increase in concentration of the respective element in the irrigation water Sorghum—Ca, Mg, Na and N increased, whereas K and P decreased with increase in irrigation water salinity	Al-Jaloud et al. (1995)
West Africa	WW: pH 9.2, EC 2.3 dS/m, TCOD 885 mg/L, total N 4.4 mg/L, Cl^{-1} 0.17 mEq/L, P 19 mg/L, K 53.9 mg/L, Na 21.9 mEq/L	Soil and plant; Spinach (*Spinacia oleracea* L.)	Irrigation with an average amount of 12 L/m thrice a week	Reduced soil pH and increased electrical conductivity (EC). Fresh matter of marketable spinach leaves was not affected by wastewater irrigation. A higher soil EC of spinach fields was related to a decreased abundance of Auchenorrhyncha, Diptera and Hymenoptera but to a stronger occurrence of Formicidae	Stenchly et al. (2017)

Location	Soil and plant	SS properties	Treatment	Results	Reference
Dinapur STP, Varanasi, India	Soil and plant; Spinach (*Beta vulgaris* var. Allgreen H-1)	SS: pH 7, EC 2.28 mS/cm, OC 5.52%, N 1.73%, available P 716.7 mg/kg Ni, Zn, Mn, Pb, Cr, Cd and Cu were 47.17, 785.3, 186.2, 60, 35.5, 154.5 and 317.7 mg/kg, respectively	20% (w/w) and 40% (w/w)	Increased concentrations of Zn, Cd, Cr, Cu, Pb and Ni in sewage sludge (SS) amended soil and palak plants. The concentrations of Cd, Ni and Zn were above the permissible limit in edible part Negative impact on physiology of palak Yield was decreased by 7% and 22%, respectively	Singh and Agrawal (2007)
Dinapur STP, Varanasi, India	Soil and plant; Lady's finger (*Abelmoschus esculentus* L. var Varsha)	SS: pH 7, EC 2.28 mS/cm, OC 5.52%, N 1.73%, available P 716.7 mg/kg Ni, Zn, Mn, Pb, Cr, Cd, and Cu were 47.17, 785.3, 186.2, 60, 35.5, 154.5, and 317.7 mg/kg, respectively.	20% (w/w) and 40% (w/w)	Soil pH decreased whereas EC, OC, total N, available P and K and exchangeable Na$^+$ and K$^+$ contents increased due to SS amendment (SSA) Concentrations of Cd, Cu, Mn and Zn in soil were highest at 40% SSA, whereas Ni, Pb and Cr concentrations were highest at 20% SSA. Cd crossed the permissible limit. The concentrations of Ni, Cd and Pb in fruits were found to be above the Indian safe limits SSA below 20% is recommended Yield increased by 75% and 135% in 20% and 40% SSA, respectively	Singh and Agrawal (2009)
Dinapur STP, Varanasi, India	Soil and plant; Rice (*Oryza sativa* L. cv. Pusa sugandha 3)	-SS: pH 7, EC 2.28 mS/cm, OC 5.52%, N 1.73%, available P 716.7 mg/kg Ni, Zn, Mn, Pb, Cr, Cd, and Cu were 47.17, 785.3, 186.2, 60, 35.5, 154.5, and 317.7 mg/kg, respectively.	0, 3, 4.5, 6, 9 and 12 kg/m^{-2}	Soil pH decreased, whereas EC, OC, total N, available P and total Fe contents of soil increased due to SSA Concentration of Zn was highest in SSA soil followed by Mn, Cu, Pb, Ni, Cd and Cr Plants showed tolerance under elevated heavy metal concentrations by increased rate of photosynthesis and chlorophyll content and various antioxidant levels	Singh and Agrawal (2010c)

(Continued)

TABLE 15.6 Agronomic Responses of wastewater and sewage sludge.—cont'd

Source of waste water/sewage sludge	Wastewater (WW)/sewage sludge (SS) characteristics	Experimental object (plant/soil)	Doses	Effect	Reference
Dinapur STP, Varanasi, India	SS: pH 7, EC 2.28 mS/cm, OC 5.52%, N 1.73%, available P 716.7 mg/kg Mn, Cu, Zn, Cr, Cd, Ni and Pb were 104.5, 17.66, 57.7, 9.76, 2.23, 14.85 and 11.67 mg/kg, respectively	Soil and plant; Mung bean (*Vigna radiata* L.)	6, 9 and 12 kg/m^{-2}	Increased organic C, N, available P and heavy metals in soil. SSA increased N, P, K and Fe and Ca contents of seeds, but protein content declined Heavy metal concentrations were also higher in seeds of plants grown at different SSA rates Yield was increased by 39%, 76%, 60% SSA Recommended dose was below 6 kg/m	Singh and Agrawal (2010b)
Bhagwanpur STP, Varanasi, India	SS: pH 6.16, EC 2.7 dS/m, organic C 12.6% total N, P, K and S content were 1.6%, 1.3%, 0.8% and 2.1%, respectively Cd, Cr, Ni and Pb contents were 32.3, 44.3, 54.7 and 28.5 mg/kg, respectively	Soil and plant; Wheat (*Triticum aestivum* L. cv. Malviya 234)	10, 20, 30 and 40 t/ha	Residual effect of higher doses of sludge (30 and 40 t/ha) in terms of fertility buildup is evident as higher N, K, S and Zn contents in soil Yield was increased by 37%, 48%, 63% and 68%, respectively	Latare et al. (2014)
Greece	SS: pH 6.5, EC 2.55 mS/cm; Kjeldahl N, NH$_4$–N, NO$_3$–N, total P and total K were 17.65, 1.59, 1.06, 2.45 and 15.37 g/kg, respectively	Plant; Wheat (*Triticum aestivum* L. cv. Centauro)	20, 40 and 60 mg dry weight/ha	Increase of 7%, 23% and 73% in yield, respectively	Koutroubas et al. (2014)
Tunisia	SS: pH 6.8, mS/cm; OM, TOC, total N and total P were 68.5%, 3.68%, 2.44% and 1.2% Cd, Cu, Pb, Ni, Zn, Cr and Hg were 0.66, 120, 38.6, 4.7, 470, 70.5 and 218 mg/kg, respectively	Soil and plant; Triticale (X *Triticosecale* Wittmack).	6, 12 and 18 t/ha	Application of sewage sludge improved soil nutrient profile and crop yield SS increased straw yield more than 123% compared with control and nearly 57% compared with mineral fertilizer at highest dosing of SS, i.e., 18 t/ha. No toxic effects were noticed	Kchaou et al. (2018)
Turkey	SS: pH 6.23, EC 1.23 mS/cm; OM 43.5%, total N 34,300 mg/kg, total P 9400 mg/kg. Fe, Cu, Mn, Zn, Ni, Pb, Cd and Hg were 11, 2.3, 3.9, 2.98, 0.26, 1.5, 0.07 and 0.002 g/kg	Soil and plant; Raspberry (*Rubus ideaus* L.)	0.0, 2.5, 5.0, 7.5 and 10.0 kg of dry matter per plant	Improvement of vegetative growth, yield, soil and plant chemical properties of raspberry in light textured soils The most effective application rate was found as 7.5 kg per plant	Angin et al. (2017)

were found to have increased salt and sodium contents, which could consequently lead to the problem of salinity or sodicity in the long run. Also, lettuce irrigated with wastewater does not meet the nutrient demand of the crop because of low uptake of essential nutrients due to Na^+ accumulation, which indicates fertilizer supplement is needed by the crop. However, plants treated with wastewater showed higher growth parameters (dry and fresh weight, average height and diameter). Similarly, Angin et al. (2005) noticed significant increase in N, P, Fe, Mn, Zn, Cu, B, Mo and Cd contents of cabbage (*Brassica oleracea* var. Capitate cv. Yalova-1) and potato (*Solanum tuberosum*) plants when irrigated with wastewater by flooding.

Similar experiment on forest ecosystem was conducted by Mavrogianopoulos and Kyritsis (1995) in Greece using three treatments (no irrigated plants, ordinary irrigation water, irrigation with treated municipal wastewater). The forest species used as test plants were *Eucalyptus* sp., *Bombus variabilis*, *Forsythia* sp., *Moringa arborea* and *Nerium oleander*. The results revealed that plants irrigated with wastewater attained greater heights and diameter in *Eucalyptus* sp. and *B. variabilis*, whereas *Forsythia* sp. showed no significant difference in the treatments applied. It was overall concluded that due to high nutrient content, growth of plants was favored significantly. Likewise, Segura et al. (2001) conducted an experiment on melon (*Cucumis melo* L. cv. Galia) irrigated with wastewater and compared with groundwater that is normally used for irrigation. No significant differences were found for different treatments, the obtained yield for both the treatments were 8.97 and 7.98 kg/m^2, respectively. Obtained concentration of N, P, K, Ca, Mg and Na in leaves showed no significant differences. Also, no contamination of heavy metal was found in both the treatments for Fe (11–14 mg/kg), Cu (7.23–8.68 mg/kg), Zn (3.09–3.38 mg/kg), Mn (1.91–2.96 mg/kg), Pb (1.83–2.23 mg/kg) and Ni (0.10–0.18 mg/kg). Foliar concentrations of heavy metal for Mn (30 ppm), Zn (45–47 ppm) and Cu (20 ppm) were below the phytotoxic levels. No contamination of *E. coli*, an indicator microorganism, was found during microbiological analysis of fruit. Similar experiments on lettuce (*Lactuca sativa* L.) (Manas et al., 2009) and alfalfa (*Medicago sativa*) (Palacios-Díaz et al., 2009) were done to evaluate the effect of wastewater using subsurface drip irrigation. Although the concentration of wastewater was saline (EC: 2.24 dS/m) and sodic (sodium adsorption ratio: 6.9), it did not negatively affect the crop yields. Similar positive effects on growth and fruit quality of orange trees were found when treated with wastewater in Navelina for 3 years. It was concluded that foliar concentration of chlorides, sodium and boron did not reach toxicity levels (Reboll et al., 2000).

Similarly, Urbano et al. (2017) conducted a greenhouse experiment on lettuce plant in São Paulo, Brazil. Drip method was used to irrigate the crop with two treatments that are wastewater with partial conventional fertilization (T1) and groundwater with conventional fertilization (T2). The results obtained showed significant increase of fresh weight when irrigated with T1 as when compared with those of T2. Meanwhile, to reduce the dependency on potable water resource and decrease the negative impact of over exploitation of fossil fuels, Zema et al. (2012) conducted an experiment on three energy crops, viz., *Typha latifolia*, *Arundo donax* and *Phragmites australis*, irrigated it for 2 years with the effluents of an urban wastewater. Evaluation of energy yield and biomass of the crops were done periodically and found that irrigation with wastewater increased the biomass yield. This may be attributed to high fertilizer value of wastewater. No significant changes in soil's physicochemical properties were noticed, whereas only an insignificant reduction in hydraulic conductivity was noticed. Heating values were found to be higher for *T. latifolia*; nevertheless, *A. donax* achieved highest energy yield per unit of cultivated area that was much higher than *T. latifolia* and *P. australis*. Likewise, Bedbabis et al. (2015) conducted an experiment on olive orchard grown on sandy—silty soil under two irrigation methods, i.e., well water (WW) and treated wastewater (TWW) for 10 years. Results obtained revealed that irrigation with wastewater caused increase in soil pH, EC, OM, essential nutrients, fruit water content and the concentrations of β-carotene and tocopherols (α, β and γ) and yield. It was also noticed that free acidity, K232, K270, VOO and other oil content were not affected significantly, whereas chlorophyll content, total phenols, induction time and δ-tocopherol decreased significantly after 10 years of irrigation with TWW.

4.3 Biosolids

With the ever-rising population and urbanization, there is tremendous increase in generation of biosolids. Managing such huge amount of biosolids still remains a major challenge all over the world. Recently, land application of biosolids has become a widely accepted practice due to many benefits such as improvement in plant nutrient uptake and increased soil physicochemical properties that help to enhance the growth of plant and yield. To study the effect of biosolids amendments, Singh and Agrawal (2007) conducted an experiment to check the response on *Beta vulgaris* var. Allgreen H-1. Results obtained showed significant decrease in morphological parameters such as root length decreased by 45% along with shoot length and number of leaves at 40% amendment. Root biomass showed significant decrease at 20% and 40% amendment ratios, whereas shoot, total biomass and yield decreased significantly only at 40% amendment ratio. In this respect, Singh and Agrawal (2010b) further conducted the experiment on mung bean (*Vigna radiata* L. cv. Malviya janpriya (HUM 6)) at 6, 9 and 12 kg m^2 sewage sludge amendment (SSA) rates. Increment in shoot length, leaf area, plant biomass, pod weight and yield was observed maximum in 9 kg m^2 SSA, followed by 12 kg m^2 and 6 kg m^2 amendment rates. The crop was tolerant to heavy metal as higher yield was observed at high heavy metal accumulation. Silva et al. (2010) conducted an experiment on two types of composted tannery sludge, viz., tannery sludge + sugarcane straw and cattle manure mixed in the ratio 1:3:1 (v:v:v) and tannery sludge + carnauba straw and cattle manure in the ratio 1:3:1 (v:v:v) on capsicum. Insignificant variation on morphological parameters was observed for both the treatments. However, 100% compost treatment showed highest chlorophyll content, number of leaves and yield. Antolín et al. (2005) in his experiment on barley (*Hordeum vulgare* L.) var. Sunrise found significant increase in yield of crop when treated with cumulative biosolids application, i.e., repeated applications of 15 t/ha every year. Belhaj et al. (2016) reported that different amendments of sewage sludge (2.5%, 5% and 7.5%) had positive impact on biomass of sunflower (*Helianthus annus*).

Angin et al. (2017) studied the effect of varying treatments of sewage sludge (0.0, 2.5, 5.0, 7.5 and 10.0 kg of dry matter per plant) on vegetative growth and yield attributes of raspberry (*Rubus ideaus* L.). Results showed increased macro- and microelements in all the treatments with maximum increase at 7.5 and 10.0 kg dry matter per plant amendment rates. Also, leaf area, shoot number, shoot length and shoot thickness were greater in 7.5 kg of dry matter per plant. Application rate above this showed negative affect on vegetative growth of plants. Similar trends were noticed for yield response with maximum yield in 7.5 kg of dry matter per plant. Kchaou et al. (2018) investigated the effects of SS application on growth and yield attributes of triticale (X *Triticosecale* Wittmack). Five different treatments, viz., control, mineral fertilizer treated, 6, 12 and 18 t/ha were used in the study. SS application positively affected leaf area index, tillering capacity, plant height and aboveground biomass. The highest amendment rate, i.e., 18 t/ha, was suggested as recommended dose due to two-fold increase in dry matter as compared with control. Cu content increased in straw tissues but was below the permissible limit and had no toxic effects on plant growth. The agronomic responses of biosolid/sludge are given in Table 15.6.

5. Conclusion

With the increase in urban growth, waste generation is also increasing; hence, innovative ways to manage urban waste need to be devised. Using nutrient recovery methods for MSW is one such approach in the direction of urban sustainability. Organic fraction of MSW has immense potential for nutrient recovery, and therefore, it is imperative to capitalize the untapped potential of MSW for nutrient recovery. Land application of biosolids, biodigestate, biochar and compost/vermicompost of waste origin not only helps in alleviating the escalating pressure on nonrenewable resources for nutrient recovery (P and K) but would also help in maintaining the soil fertility in a more sustainable way. Also, this will improve the productivity of soil by augmenting soil physicochemical and biological properties. There are many biorefinery platforms that can be utilized for nutrient recovery in an efficient way. Depending upon the feedstock and desired nutrient recovery, a number of physical, chemical and biological techniques can be used to recover either mixed nutrient or N-, P- and K-specific fertilizers. Therefore, it can be concluded that organic fraction of MSW has great potential for macronutrients recovery that should be capitalized in an efficient way; therefore, much emphasis should be paid in this direction by the scientists, government and policy makers.

Acknowledgement

The authors are thankful to the Dean & Head, Department of Environment and Sustainable Development and Director, Institute of Environment and Sustainable Development, Banaras Hindu University, for providing necessary facilities. RPS is thankful to the Department of Science & Technology for providing financial support (DST-SERB P07-678). VS is thankful to the Indian Council of Medical Research for providing Junior and Senior Research Fellowships. BV is also thankful to the Council of Scientific & Industrial Research for awarding the Senior Research Fellowship.

References

Acosta-Martinez, V., Cruz, L., Sotomayor-Ramirez, D., Pérez-Alegría, L., 2007. Enzyme activities as affected by soil properties and land use in a tropical watershed. Applied Soil Ecology 35 (1), 35–45 (s).

Adeoye, G.O., Sridhar, M.K.C., Ipinmoroti, R.R., 2001. Potassium recovery from farm wastes for crop growth. Communications in Soil Science and Plant Analysis 32 (15–16), 2347–2358.

Alfa, M.I., Adie, D.B., Igboro, S.B., Oranusi, U.S., Dahunsi, S.O., Akali, D.M., 2014. Assessment of biofertilizer quality and health implications of anaerobic digestion effluent of cow dung and chicken droppings. Renewable Energy 63, 681–686.

Ali, M.I., Schneider, P.A., 2008. An approach of estimating struvite growth kinetic incorporating thermodynamic and solution chemistry, kinetic and process description. Chemical Engineering Science 63 (13), 3514–3525.

Al-Jaloud, A.A., Hussain, G., Al-Saati, A.J., Karimulla, S., 1995. Effect of wastewater irrigation on mineral com-position of corn and sorghum plants in a pot experiment. Journal of Plant Nutrition 18 (8), 1677–1692.

Angin, I., Yaganoglu, A.V., Turan, M., 2005. Effects of long-term wastewater irrigation on soil properties. Journal of Sustainable Agriculture 26 (3), 31–42.

Angin, I., Aslantas, R., Gunes, A., Kose, M., Ozkan, G., 2017. Effects of Sewage sludge amendment on some soil properties, growth, yield and nutrient content of raspberry (Rubus idaeus L.). Erwerbs-Obstbau 59 (2), 93–99.

Antolín, M.C., Pascual, I., García, C., Polo, A., Sánchez-Díaz, M., 2005. Growth, yield and solute content of barley in soils treated with sewage sludge under semiarid Mediterranean conditions. Field Crops Research 94 (2–3), 224–237.

Anupama, P., Ravindra, P., 2000. Value-added food: single cell protein. Biotechnology Advances 18, 459–479.

Araujo, A.S.F., Miranda, A.R.L., Oliveira, M.L.J., Santos, V.M., Nunes, L.H.L.P., Melo, W.J., 2015. Soil microbial properties after 5 years of consecutive amendment with composted tannery sludge. Environmental Monitoring and Assessment 187, 4153.

Atkinson, C.J., Fitzgerald, J.D., Hipps, N.A., 2010. Potential mechanisms for achieving agricultural benefits from biochar application to temperate soils: a review. Plant and Soil 337, 1–18.

Awashthi, S.K., 2000. Prevention of Food Adulteration Act No 37 of 1954. Central and State Rules as Amended for 1999, Ashoka Law House, New Delhi..

Bedbabis, S., Trigui, D., Ahmed, C.B., Clodoveo, M.L., Camposeo, S., Vivaldi, G.A., Rouina, B.B., 2015. Long-terms effects of irrigation with treated municipal wastewater on soil, yield and olive oil quality. Agricultural Water Management 160, 14–21.

Begum, S., Rasul, M.G., 2009. Reuse of stormwater for watering gardens and plants using green gully: a new storm-water quality improvement device (SQID). Water, Air, and Soil Pollution: Focus 9 (5–6), 371.

Belhaj, D., Elloumi, N., Jerbi, B., et al., 2016. Effects of sewage sludge fertilizer on heavy metal accumulation and consequent responses of sunflower (Helianthus annuus). Environ Sci Pollut Res 23, 20168–20177. https://doi.org/10.1007/s11356-016-7193-0.

Bernal, M.P., Alburquerque, J.A., Moral, R., 2009. Composting of animal manures and chemical criteria for compost maturity assessment. A review. Bioresource Technology 100, 5444–5453.

Bhattacharyya, P., Chakrabarti, K., Chakraborty, A., 2003. Effect of MSW compost on microbiological and biochem-ical soil quality indicators. Compost Science and Utilization 11, 220–227.

Birkhofer, K., Bezemer, T.M., Bloem, J., Bonkowski, M., Christensen, S., Dubois, D., et al., 2008. Long-term organic farming fosters below and aboveground biota: implications for soil quality, biological control and productivity. Soil Biology and Biochemistry 40, 2297–2308.

BIS, 2012. Indian Standards for Drinking Water – Specification (BIS 10500:2012 Revised in 2012). Available on-line at. http://cgwb.gov.in/Documents/WQ-standards.pdf.

Blaney, L.M., Cinar, S., SenGupta, A.K., 2007. Hybrid anion exchanger for trace phosphate removal from water and wastewater. Water Research 41 (7), 1603–1613.

Blumenthal, U.J., Mara, D.D., Peasey, A., Ruiz-Palacios, G., Stott, R., 2000. Guidelines for the microbiological quality of treated wastewater used in agriculture: recommendations for revising WHO guidelines. Bulletin of the World Health Organization 78, 1104–1116.

Brechelt, A., 2004. Manejo Ecológico del Suelo. Red de Acción en Plaguicidas y sus Alternativas para América Latina (RAP-AL). Santiago del Chile, Chile (in Spanish).

Bruun, J.M., Hansen, T.L., Christensen, T.H., Magid, J., Jensena, L.S., 2006. Application of processed organic munic-ipal solid waste on agricultural land-scenario analysis. Environmental Modeling and Assessment 11, 251–265.

Byju, G., Haripriya Anand, M., Moorthy, S.N., 2015. Carbon and nitrogen mineralization and humus composition following municipal solid waste compost addition to laterite soils under continuous cassava cultivation. Commu-nications in Soil Science and Plant Analysis 46, 148–168.

Çalısır, S., Özcan, M., Hacıseferoğulları, H., Yıldız, M.U., 2005. A study on some physico-chemical properties of Turkey okra (Hibiscus esculenta L.) seeds. Journal of Food Engineering 68 (1), 73–78.

Capodaglio, A.G., Hlavinek, P., Raboni, M., 2015. Physico-chemical technologies fornitrogen removal from wastewa-ters: a review. Revista Ambiente and Água 10, 481–498.

Carbonell, G., de Imperial, R.M., Torrijos, M., Delgado, M., Rodriguez, J.A., 2011. Effects of municipal solid waste compost and mineral fertilizer amendments on soil properties and heavy metals distribution in maize plants (Zea mays L.). Chemosphere 85 (10), 1614–1623.

Carey, D.E., Yang, Y., McNamara, P.J., Mayer, B.K., 2016. Recovery of agricultural nutrients from biorefineries. Bioresource Technology 215, 186–198.

Castro, E., Mañas, P., De Las Heras, J., 2013. Effects of wastewater irrigation in soil properties and horticultural crop (*Lactuca sativa* L.). Journal of Plant Nutrition 36 (11), 1659–1677.

Chaiprapat, S., Cheng, J.J., Classen, J.J., Liehr, S.K., 2005. Role of internal nutrient storage in duckweed growth for swine wastewater treatment. Transactions of the ASAE 48 (6), 2247–2258.

Chan, K.Y., Xu, Z., 2009. Biochar: nutrient properties and their enhancement. Biochar for Environmental Management: Science and Technology 1, 67–84.

Cherif, H., Ayari, F., Ouzari, H., Marzorati, M., Brusetti, L., Jedidi, N., Hassen, A., Daffonchio, D., 2009. Effects of municipal solid waste compost, farmyard manure and chemical fertilizers on wheat growth, soil composition and soil bacterial characteristics under Tunisian arid climate. European Journal of Soil Biology 45, 138–145.

Childers, D.L., Corman, J., Edwards, M., Elser, J.J., 2011. Sustainability challenges of phosphorus and food: solutions from closing the human phosphorus cycle. BioScience 61, 117–124.

Cirelli, G.L., Consoli, S., Licciardello, F., Aiello, R., Giuffrida, F., Leonardi, C., 2012. Treated municipal wastewater reuse in vegetable production. Agricultural Water Management 104, 163–170.

Cordell, D., Drangert, J.O., White, S., 2009. The story of phosphorus: global food security and food for thought. Global Environmental Change 19, 292–305.

Cornel, P., Schaum, C., 2009. Phosphorus recovery from wastewater: needs, technologies and costs. Water Science and Technology 59 (6), 1069–1076.

Crecchio, C., Curci, M., Pizzigallo, M.D., Ricciuti, P., Ruggiero, P., 2004. Effects of municipal solid waste compost amendments on soil enzyme activities and bacterial genetic diversity. Soil Biology and Biochemistry 36, 1595–1605.

da Silva Cuba, R., do Carmo, J.R., Souza, C.F., Bastos, R.G., 2015. Potential of domestic sewage effluent treated as a source of water and nutrients in hydroponic lettuce. Ambiente e Agua-An Interdisciplinary Journal of Applied Science 10 (3), 574–586.

Dantas, I.L.D.A., Faccioli, G.G., Mendonça, L.C., Nunes, T.P., Viegas, P.R.A., Santana, L.O.G.D., 2014. Viability of using treated wastewater for the irrigation of radish (*Raphanus sativus* L.). Revista Ambiente and Água 9 (1), 109–117.

de Oliveira Marinho, L.E., Tonetti, A.L., Stefanutti, R., Coraucci Filho, B., 2013. Application of reclaimed wastewater in the irrigation of rosebushes. Water, Air, and Soil Pollution 224 (9), 1669.

Deng, L.W., Zheng, P., Chen, Z.A., 2006. Anaerobic digestion and post-treatment of swine wastewater using IC–SBR process with bypass of raw wastewater. Process Biochemistry 41 (4), 965–969.

Desmidt, E., Ghyselbrecht, K., Zhang, Y., Pinoy, L., Van der Bruggen, B., Verstraete, W., Rabaey, K., Meesschaert, B., 2015. Global phosphorus scarcity and full-scale P-recovery techniques: a review. Critical Reviews in Environmental Science and Technology 45, 336–384.

Duong, T.T., Penfold, C., Marschner, P., 2012. Amending soils of different texture with six compost types: impact on soil nutrient availability, plant growth and nutrient uptake. Plant and Soil 354 (1–2), 197–209.

EEC-Sludge Rule, 2000. Working Document of Sludge. ENV.E.3/LM. Brussels April, 2000.

EU, 2006. Commission regulation (EC) No. 1881/2006 of 19 December 2006 setting maximum levels for certain contaminants in foodstuffs. Official Journal of European Union L364/5.

EU, 2002. Heavy Metals in Wastes. European Commission on Environment. http://ec.europa.eu/environment/waste/studies/pdf/heavy metals report.pdf.

Evans, A.E., Hanjra, M.A., Jiang, Y., Qadir, M., Drechsel, P., 2012. Water quality: assessment of the current situation in Asia. International Journal of Water Resources Development 28 (2), 195–216.

Evanylo, G., Sherony, C., Spargo, J., Starner, D., Brosius, M., Haering, K., 2008. Soil and water environmental effects of fertilizer-, manure-, and compost-based fertility practices in an organic vegetable cropping system. Agriculture, Ecosystems and Environment 127, 50–58.

Fagnano, M., Adamo, P., Zampella, M., Fiorentino, N., 2011. Environmental and agronomic impact of fertilization with composted organic fraction from municipal solid waste: a case study in the region of Naples, Italy. Agriculture, Ecosystems and Environment 141, 100–107.

FAO, 1985. Water Quality for Agriculture. UNESCO, Publication, Rome. Paper No. 29 (Rev. 1.

FAO, F., 2016. Agriculture Organization, 2014. Food and Agriculture Organization of the United Nations. Livestock Primary.

292

Fernández, J.M., Plaza, C., García-Gil, J.C., Polo, A., 2009. Biochemical properties and barley yield in a semiarid Mediterranean soil amended with two kinds of sewage sludge. Applied Soil Ecology 42 (1), 18–24.

Fonseca, A.F.D., Herpin, U., Paula, A.M.D., Victória, R.L., Melfi, A.J., 2007. Agricultural use of treated sewage effluents: agronomic and environmental implications and perspectives for Brazil. Scientia Agricola 64 (2), 194–209.

Fornes, F., Mendoza-Hernández, D., Garcia-de-la-Fuente, R., Abad, M., Belda, R.M., 2012. Composting versus vermicomposting: a comparative study of organic matter evolution through straight and combined processes. Bioresource Technology 118, 296–305.

Garcia-Gil, J.C., Plaza, C., Soler-Rovira, P., Polo, A., 2000. Long-term effects of municipal solid waste compost application on soil enzyme activities and microbial biomass. Soil Biology and Biochemistry 32, 1907–1913.

Garcia-Gil, J.C., Ceppi, S., Velasca, M., Polo, A., Senesi, N., 2004. Long-term effects of amendment with municipal solid waste compost on the elemental and acid functional group composition and pH-buffer capacity of soil humic acid. Geoderma 121, 135–142.

Gerardo, M.L., Aljohani, N.H.M., Oatley-Radcliffe, D.L., Lovitt, R.W., 2015. Moving towards sustainable resources: recovery and fractionation of nutrients from dairy manure digestate using membranes. Water Research 80, 80–89.

Giannakis, G.V., Kourgialas, N.N., Paranychianakis, N.V., Nikolaidis, N.P., Kalogerakis, N., 2014. Effects of municipal solid waste compost on soil properties and vegetables growth. Compost Science and Utilization 22, 116–131.

Gilbert, 2009. The disappearing nutrient. Nature 461, 716–718.

Goyal, S., Chander, K., Mundra, M.C., Kapoor, K.K., 1999. Influence of inorganic fertilizers and organic amendments on soil organic matter and soil microbial properties under tropical conditions. Biology and Fertility of Soils 29 (2), 196–200.

Haghighi, M., Barzegar, M.R., da Silva, J.A.T., 2016. The effect of municipal solid waste compost, peat, perlite and vermicompost on tomato (Lycopersicum esculentum L.) growth and yield in a hydroponic system. International Journal of Recycling of Organic Waste in Agriculture 5 (3), 231–242.

Hargreaves, J.C., Adl, M.S., Warman, P.R., 2008. A review of the use of composted municipal solid waste in agriculture. Agriculture, Ecosystems and Environment 123, 1–4.

Hargreaves, A.J., Vale, P., Whelan, J., Alibardi, L., Constantino, C., Dotro, G., et al., 2018. Impacts of coagulation-flocculation treatment on the size distribution and bioavailability of trace metals (Cu, Pb, Ni, Zn) in municipal wastewater. Water Research 128, 120–128.

Hoornweg, D., Bhada-Tata, P., 2012. What a Waste: A Global Review of Solid Waste Management.

Igboro, S.B., 2011. Production of biogas and compost from cow dung in Zaria, Nigeria. An Unpublished PhD Dissertation in the Department of Water Resources and Environmental Engineering. Ahmadu Bello University, Zaria Nigeria.

Ingelmo, F., José Molina, M., Desamparados Soriano, M., Gallardo, A., Lapeña, L., 2012. Influence of organic matter transformations on the bioavailability of heavy metals in a sludge-based compost. Journal of Environmental Management 95, 104–109.

Jia-yin, D., Ling, C., Jian-fu, Z., Na, M., 2006. Characteristics of sewage sludge and distribution of heavy metal in plants with amendment of sewage sludge. Journal of Environmental Sciences 18, 1094–1100.

John, N.M., Edem, S.O., Ndaeyo, N.U., Ndon, B.A., 2006. Physical composition of municipal solid waste and nutrient contents of its organic component in Uyo Municipality, Nigeria. Journal of Plant Nutrition 29, 189–194.

Johnston, A.E., Richards, I.R., 2004. Effectiveness of different precipitated phosphates as phosphorus sources for plants. Phosphorus Research Bulletin 15, 52–59.

Jorge-Mardomingo, I., Soler-Rovira, P., Casermeiro, M.Á., de la Cruz, M.T., Polo, A., 2013. Seasonal changes in microbial activity in a semiarid soil after application of a high dose of different organic amendments. Geoderma 206, 40–48.

Kayikcioglu, H.H., Yener, H., Ongun, A.R., Okur, B., 2019. Evaluation of soil and plant health associated with successive three-year sewage sludge field applications under semi-arid biodegradation condition. Archives of Agronomy and Soil Science 65 (12), 1659–1676.

Kchaou, R., Baccar, R., Bouzid, J., Rejeb, S., 2018. Agricultural use of sewage sludge under sub-humid Mediterranean conditions: effect on growth, yield, and metal content of a forage plant. Arabian Journal of Geosciences 11 (23), 746.

Kizilkaya, R., Bayrakli, B., 2005. Effects of N-enriched sewage sludge on soil enzyme activities. Applied Soil Ecology 30 (3), 192–202.

Körner, S., Vermaat, J.E., Veenstra, S., 2003. The capacity of duckweed to treat wastewater. Journal of Environmental Quality 32 (5), 1583–1590.

Koutroubas, S.D., Antoniadis, V., Fotiadis, S., Damalas, C.A., 2014. Growth, grain yield and nitrogen use efficiency of Mediterranean wheat in soils amended with municipal sewage sludge. Nutrient Cycling in Agroecosystems 100 (2), 227−243.

Kumar, V., Chopra, A.K., 2013. Response of sweet sorghum after fertigation with sugar mill effluent in two seasons. Sugar Tech 15 (3), 285−299.

Kumar, V., Chopra, A.K., 2014. Accumulation and translocation of metals in soil and different parts of French bean (*Phaseolus vulgaris* L.) amended with sewage sludge. Bulletin of Environmental Contamination and Toxicology 92 (1), 103−108.

Kumar, V., Chopra, A.K., Srivastava, S., Singh, J., Thakur, R.K., 2017. Irrigating okra with secondary treated municipal wastewater: observations regarding plant growth and soil characteristics. International Journal of Phytoremediation 19 (5), 490−499.

Laird, D.A., Brown, R.C., Amonette, J.E., Lehmann, J., 2009. Review of the pyrolysis platform for coproducing bio-oil and biochar. Biofuels, Bioproducts, and Biorefining 3, 547−562.

Lakhdar, A., Falleh, H., Ouni, Y., Oueslati, S., Debez, A., Ksouri, R., Abdelly, C., 2011. Municipal solid waste compost application improves productivity, polyphenol content, and antioxidant capacity of Mesembryanthemum edule. Journal of Hazardous Materials 191 (1−3), 373−379.

Lakhdar, A., Slatni, T., Iannelli, M.A., Debez, A., Pietrini, F., Jedidi, N., Massacci, A., Abdelly, C., 2012. Risk of municipal solid waste compost and sewage sludge use on photosynthetic performance in common crop (Triticum durum). Acta Physiologiae Plantarum 34 (3), 1017−1026.

Lansing, S., Martin, J., Botero, R., Nogueira da Silva, T., Dias da Silva, E., 2010. Wastewater transformations and fertilizer value when codigesting differing ratios of swine manure and used cooking grease in low-cost digesters. Biomass and Bioenergy 34, 1711−1720.

Latare, A.M., Kumar, O., Singh, S.K., Gupta, A., 2014. Direct and residual effect of sewage sludge on yield, heavy metals content and soil fertility under rice−wheat system. Ecological Engineering 69, 17−24.

Liu, Y., Rahman, M.M., Kwag, J.H., Kim, J.H., Ra, C., 2011. Eco-friendly production of maize using struvite recovered from swine wastewater as a sustainable fertilizer source. Asian-Australasian Journal of Animal Sciences 24 (12), 1699−1705.

Liu, Z., Chen, X., Jing, Y., Li, Q., Zhang, J., Huang, Q., 2014. Effects of biochar amendment on rapeseed and sweet potato yields and water stable aggregate in upland red soil. Catena 123, 45−51.

Luo, Y., Guo, W., Ngo, H.H., Nghiem, L.D., Hai, F.I., Zhang, J., et al., 2014. A review on the occurrence of micropollutants in the aquatic environment and their fate and removal during wastewater treatment. The Science of the Total Environment 473, 619−641.

Manas, P., Castro, E., de las Heras, J., 2009. Irrigation with treated wastewater: effects on soil, lettuce (*Lactuca sativa* L.) crop and dynamics of microorganisms. Journal of Environmental Science and Health Part A 44 (12), 1261−1273.

Massé, D.I., Croteau, F., Masse, L., 2007. The fate of crop nutrients during digestion of swine manure in psychrophilic anaerobic sequencing batch reactors. Bioresource Technology 98, 2819−2823.

Mata-Alvarez, J., Mace, S., Llabres, P., 2003. Anaerobic digestion of organic solid wastes. An overview of research achievements and perspectives. Bioresource Technology 74, 3−16.

Matassa, S., Batstone, D.J., Hülsen, T., Schnoor, J., Verstraete, W., 2015. Can direct conversion of used nitrogen to new feed and protein help feed the world? Environmental Science and Technology 49, 5247−5254.

Mathivanan, K., Rajaram, R.D., 2014. Anthropogenic influences on toxic metals in water and sediment samples collected from industrially polluted Cuddalore coast, Southeast coast of India. Environmental Earth Sciences 72 (4), 997−1010.

Mavrogianopoulos, G., Kyritsis, S., 1995. Use of Municipal Wastewater for Biomass Production. Project report of Agricultural University of Athens.

Meena, M.D., Joshi, P.K., Jat, H.S., Chinchmalatpure, A.R., Narjary, B., Sheoran, P., Sharma, D.K., 2016. Changes in biological and chemical properties of saline soil amended with municipal solid waste compost and chemical fertilizers in a mustard−pearl millet cropping system. Catena 140, 1−8.

Mehta, C.M., Khunjar, W.O., Nguyen, V., Tait, S., Batstone, D.J., 2015. Technologies to recover nutrients from waste streams: a critical review. Critical Reviews in Environmental Science and Technology 45, 385−427.

Mkhabela, M.S., Warman, P.R., 2005. The influence of municipal solid waste compost on yield, soil phosphorus availability and uptake by two vegetable crops grown in a Pugwash sandy loam soil in Nova Scotia. Agriculture, Ecosystems & Environment 106 (1), 57−67.

Mohammad, M.J., Mazahreh, N., 2003. Changes in soil fertility parameters in response to irrigation of forage crops with secondary treated wastewater. Communications in Soil Science and Plant Analysis 34 (9–10), 1281–1294.

Monnet, F., 2003. An Introduction to Anaerobic Digestion of Organic Waste, Being a Final Report Submitted to Remade. Scotland. Available from: http://www.remade.org.uk/media/9102/anintroductiontoanaerobicdigestionnov2003. pdf. (Accessed May 2011).

Naderia, R., Ghadirib, H., 2010. Urban waste compost, manure and nitrogen fertilizer effects on the initial growth of corn (Zea mays L.). Desert 15, 159–165.

Onwudiwe, N., Benedict, O.U., Ogbonna, P.E., Ejiofor, E.E., 2014. Municipal solid waste and NPK fertilizer effects on soil physical properties and maize performance in Nsukka, Southeast Nigeria. African Journal of Biotechnology 13 (1), 68–75.

Oyedele, D.J., Schjonning, P., Amussan, A.A., 2005. Physicochemical properties of earthworm casts and uningested parental soil from selected sites in southwestern Nigeria. Ecological Engineering 20, 103–106.

Ozores-Hampton, M., Bryan, H.H., Schaffer, B., Hanlon, E.A., 1994. Nutrient concentrations, growth, and yield of tomato and squash in municipal solid-waste-amended soil. HortScience 29 (7), 785–788.

Palacios-Díaz, M.P., Mendoza-Grimón, V., Fernández-Vera, J.R., Rodríguez-Rodríguez, F., Tejedor-Junco, M.T., Hernández-Moreno, J.M., 2009. Subsurface drip irrigation and reclaimed water quality effects on phosphorus and salinity distribution and forage production. Agricultural Water Management 96 (11), 1659–1666.

Pan, B., Wu, J., Pan, B., Lv, L., Zhang, W., Xiao, L., et al., 2009. Development of polymer-based nanosized hydrated ferric oxides (HFOs) for enhanced phosphate removal from waste effluents. Water Research 43 (17), 4421–4429.

Papafilippaki, A., Paranychianakis, N., Nikolaidis, N.P., 2015. Effects of soil type and municipal solid waste compost as soil amendment on Cichorium spinosum (spiny chicory) growth. Scientia Horticulturae 195, 195–205.

Park, J.H., Choppala, G.K., Bolan, N.S., Chung, J.W., Chuasavathi, T., 2011. Biochar reduces the bioavailability and phytotoxicity of heavy metals. Plant and Soil 348, 439.

Pascual, J.A., García, C., Hernandez, T., 1999. Lasting microbiological and biochemical effects of the addition of municipal solid waste to an arid soil. Biology and Fertility of Soils 30 (1–2), 1–6.

Paul, E.A., 2007. Soil microbiology, ecology and biochemistry in perspective. In: Paul, E.A. (Ed.), Soil Microbiology, Ecology and Biochemistry, third ed. Academic Press, San Diego, CA, pp. 3–24.

Pedrero, F., Allende, A., Gil, M.I., Alarcón, J.J., 2012. Soil chemical properties, leaf mineral status and crop production in a lemon tree orchard irrigated with two types of wastewater. Agricultural Water Management 109, 54–60.

Pereira, B.F.F., He, Z.L., Stoffella, P.J., Melfi, A.J., 2011. Reclaimed wastewater: effects on citrus nutrition. Agricultural Water Management 98 (12), 1828–1833.

Perez, D.V., Alcantra, S., Ribeiro, C.C., Pereira, R.E., Fontes, G.C., Wasserman, M.A., Venezuela, T.C., Meneguelli, N.A., Parradas, C.A.A., 2007. Composted municipal waste effects on chemical properties of Brazilian soil. Bioresource Technology 98, 525–533.

Qadir, M., Scott, C.A., 2009. Non-pathogenic trade-offs of wastewater irrigation. Wastewater Irrigation and Health. Routledge, pp. 127–152.

Qian, Y.L., Mecham, B., 2005. Long-term effects of recycled wastewater irrigation on soil chemical properties on golf course fairways. Agronomy Journal 97 (3), 717–721.

Rahman, M.M., Salleh, M.A.M., Rashid, U., Ahsan, A., Hossain, M.M., Ra, C.S., 2014. Production of slow release crystal fertilizer from wastewaters through struvite crystallization—A review. Arabian Journal of Chemistry 7 (1), 139–155.

Reboll, V., Cerezo, M., Roig, A., Flors, V., Lapeña, L., García-Agustín, P., 2000. Influence of wastewater vs groundwater on young Citrus trees. Journal of the Science of Food and Agriculture 80 (10), 1441–1446.

Rittmann, B.E., McCarty, P.L., 2001. Environmental Biotechnology: Principles and Applications. McGraw Hill, New York.

Roig, N., Sierra, J., Martí, E., Nadal, M., Schuhmacher, M., Domingo, J.L., 2012. Long-term amendment of Spanish soils with sewage sludge: effects on soil functioning. Agriculture, Ecosystems and Environment 158, 41–48.

Rutkowski, T., Raschid-Sally, L., Buechler, S., 2007. Wastewater irrigation in the developing world—two case studies from the Kathmandu Valley in Nepal. Agricultural Water Management 88 (1–3), 83–91.

Samatya, S., Kabay, N., Yuksel, U., Arda, M., Yuksel, M., 2006. Removal of nitrate from aqueous solution by nitrate selective ion exchange resins. Reactive and Functional Polymers 66, 1206–1214.

Sarkar, A., Das, S., Srivastava, V., Singh, P., Singh, R.P., 2018. Effect of wastewater irrigation on crop health in the Indian agricultural scenario. In: Emerging Trends of Plant Physiology for Sustainable Crop Production. Apple Academic Press, pp. 357–371.

Segura, M.L., Moreno, R., Martinez, S., Perez, Z., Moreno, J., 2001. Effect of wastewater irrigation on melon growth under greenhouse conditions. Acta Horticulturae 559, 345–352.

Sengupta, S., Pandit, A., 2011. Selective removal of phosphorus from wastewater combined with its recovery as a solid-phase fertilizer. Water Research 45 (11), 3318–3330.

Shabani, H., Peyvast, G.A., Olfati, J.A., Kharrazi, P.R., 2011. Effect of municipal solid waste compost on yield and quality of eggplant. Comunicata Scientiae 2, 85–90.

Sharma, B., Sarkar, A., Singh, P., et al., 2017. Agricultural utilization of biosolids: a review onpotential effects on soil and plant grown. Waste Management 64, 117–132.

Siebe, C., Cifuentes, E., 1995. Environmental impact of wastewater irrigation in central Mexico: an overview. International Journal of Environmental Health Research 5 (2), 161–173.

Silva, J.D., Leal, T.T., Araújo, A.S., Araujo, R.M., Gomes, R.L., Melo, W.J., Singh, R.P., 2010. Effect of different tannery sludge compost amendment rates on growth, biomass accumulation and yield responses of Capsicum plants. Waste Management 30, 1976–1980.

Singh, R.P., Agrawal, M., 2007. Effects of sewage sludge amendment on heavy metal accumulation and consequent responses of Beta vulgaris plants. Chemosphere 67 (11), 2229–2240.

Singh, R.P., Agrawal, M., 2008. Potential benefits and risks of land application of sewage sludge. Waste Management 28, 347–358.

Singh, R.P., Agrawal, M., 2009a. Potential benefits and risks of land application of sewage sludge. Waste Management 28, 347–358.

Singh, R.P., Agrawal, M., 2009b. Use of sewage sludge as fertilizer supplement for Abelmoschus esculentus plants: physiological, biochemical and growth responses. International Journal of Environment and Waste Management 3, 91–106.

Singh, R.P., Agrawal, M., 2010a. Biochemical and physiological responses of Rice Oryza sativa L grown on different sewage sludge amendments rates. Bulletin of Environmental Contamination and Toxicology 23, 606–612.

Singh, R.P., Agrawal, M., 2010b. Effect of different sewage sludge applications on growth and yield of Vigna radiata L. field crop: metal uptake by plant. Ecological Engineering 36, 969–972.

Singh, R.P., Agrawal, M., 2010c. Variations in heavy metal accumulation, growth and yield of rice plants grown at different sewage sludge amendment rates. Ecotoxicology and Environmental Safety 73, 632–641.

Singh, R.P., Singh, P., Ibrahim, M.H., Hashim, R., 2011. Land application of sewage sludge: physico-chemical and microbial response. Reviews of Environmental Contamination and Toxicology 214, 41–61.

Singh, P.K., Deshbhratar, P.B., Ramteke, D.S., 2012. Effects of sewage wastewater irrigation on soil properties, crop yield and environment. Agricultural Water Management 103, 100–104.

Soumare, M., Tack, F., Verloo, M., 2003. Characterisation of Malian and Belgian solid waste composts with respect to fertility and suitability for land application. Waste Management 23, 517–522.

Srivastava, V., Ismail, S.A., Singh, P., et al., 2015. Urban solid waste management in the developingworld with emphasis on India: challenges and opportunities. Reviews in Environmental Science and Biotechnology 14, 317–337.

Srivastava, V., de Araujo, A.S.F., Vaish, B., et al., 2016. Biological response of using municipal solid waste compost in agriculture as fertilizer supplement. Reviews in Environmental Science and Biotechnology 15, 677–696.

Srivastava, V., Sarkar, A., Singh, S., Singh, P., de Araujo, A.S., Singh, R.P., 2017. Agroecological responses of heavy metal pollution with special emphasis on soil health and plant performances. Frontiers in Environmental Science 5, 64.

Srivastava, V., Gupta, S.K., Singh, P., Sharma, B., Singh, R.P., 2018. Biochemical, physiological, and yield responses of lady's finger (Abelmoschus esculentus L.) grown on varying ratios of municipal solid waste vermicompost. International Journal of Recycling of Organic Waste in Agriculture 7 (3), 241–250.

Srivastava, V., Goel, G., Thakur, V.K., Singh, R.P., de Araujo, A.S.F., Singh, P., 2020a. Analysis and advanced characterization of municipal solid waste vermicompost maturity for a green environment. Journal of Environmental Management 255, 109914.

Srivastava, V., Vaish, B., Singh, R.P., Singh, P., 2020b. An insight to municipal solid waste management of Varanasi city, India, and appraisal of vermicomposting as its efficient management approach. Environmental Monitoring and Assessment 192, 191. https://doi.org/10.1007/s10661-020-8135-3.

Stenchly, K., Dao, J., Lompo, D.J.P., Buerkert, A., 2017. Effects of waste water irrigation on soil properties and soil fauna of spinach fields in a West African urban vegetable production system. Environmental Pollution 222, 58–63.

Sunilson, J.A.J., Jayaraj, P., Mohan, M.S., Kumari, A.A.G., Varatharajan, R., 2008. Antioxidant and hepatoprotective effect of the roots of *Hibiscus esculentus* Linn. International Journal of Green Pharmacy 2 (4), 200–203.

Suthar, S., 2010. Recycling of agro-industrial sludge through vermitechnology. Ecological Engineering 36, 703–712.

Sutton, M.A., Howard, C.M., Erisman, J.W., Billen, G., Bleeker, A., Grennfelt, P., et al. (Eds.), 2011. The European Nitrogen Assessment: Sources, Effects and Policy Perspectives. Cambridge University Press.

Tani, M., Sakamoto, N., Kishimoto, T., Umetsu, K., 2006. Utilization of anaerobically digested dairy slurry combined with other wastes following application to agricultural land. International Congress Series 1293, 331–334.

Tejada, M., Hernandez, M.T., Garcia, C., 2009. Soil restoration using composted plant residues: effects on soil properties. Soil and Tillage Research 102 (1), 109–117.

Tidsell, S.E., Breslin, V.T., 1995. Characterization and leaching of elements from municipal solid waste compost. Journal of Environmental Quality 24, 827–833.

Torres, I.F., Bastida, F., Hernández, T., García, C., 2015. The effects of fresh and stabilized pruning wastes on the biomass, structure and activity of the soil microbial community in a semiarid climate. Applied Soil Ecology 89, 1–9.

Urbano, V.R., Mendonça, T.G., Bastos, R.G., Souza, C.F., 2015. Physical-chemical effects of irrigation with treated wastewater on Dusky Red Latosol soil. Revista Ambiente and Água 10 (4), 737–747.

Urbano, V.R., Mendonça, T.G., Bastos, R.G., Souza, C.F., 2017. Effects of treated wastewater irrigation on soil properties and lettuce yield. Agricultural Water Management 181, 108–115.

USEPA, 2000. Risk-based Concentration Table. United State Environmental Protection Agency, Washington, DC.

Vaish, B., Srivastava, V., Singh, P., Singh, A., Singh, P.K., Singh, R.P., 2016. Exploring untapped energy potential of urban solid waste. Energy, Ecology and Environment 1 (5), 323–342.

Vaish, B., Sharma, B., Srivastava, V., Singh, P., Ibrahim, M.H., Singh, R.P., 2019a. Energy recovery potential and environmental impact of gasification for municipal solid waste. Biofuels 10 (1), 87–100.

Vaish, B., Srivastava, V., Kumar Singh, P., Singh, P., Pratap Singh, R., 2019b. Energy and nutrient recovery from agrowastes: rethinking their potential possibilities. Environmental Engineering Research.

van Diepeningen, A.D., de Vos, O.J., Korthals, G.W., van Bruggen, A.H.C., 2006. Effects of organic versus conventional management on chemical and biological parameters in agricultural soils. Applied Soil Ecology 31, 120–135.

van Enk, R.J., van der Vee, G., 2011. The Phosphate Balance. Current Developments and Outlook. Ministry for Economic Affairs, Agriculture and Innovation of Netherlands.

Varallo, A.C., Souza, C.F., Santoro, B.D.L., 2012. Physical and chemical characteristics changes of a red-yellow latosol after implementation of water reuse in the culture of curly lettuce (*Lactuca sativa* L.). Engenharia Agrícola 32 (2), 271–279.

Varma, V., Sudharsan Ajay S, K., 2013. Composting of municipal solid waste (MSW) mixed with cattle manure. International Journal of Environmental Sciences 3, 2068–2079.

Walter, I., Martinez, F., Cuevas, G., 2006. Plant and soil responses to the application of composted MSW in a degraded, semiarid shrubland in central Spain. Compost Science and Utilization 14, 147–154.

Wang, H., Brown, S.L., Magesan, G.N., Slade, A.H., Quintern, M., Clinton, P.W., Payn, T.W., 2008. Technological options for the management of biosolids. Environmental Science and Pollution Research International 15 (4), 308–317.

Warman, P.R., Rodd, A.V., Hicklenton, P., 2009. The effect of MSW compost and fertilizer on extractable soil elements and the growth of winter squash in Nova Scotia. Agriculture, Ecosystems and Environment 133 (1–2), 98–102.

Warman, P.R., Rodd, A.V., Hicklenton, P., 2011. The effect of MSW compost and fertilizer on extractable soil elements, tuber yield, and elemental concentrations in the plant tissue of potato. Potato Research 54 (1), 1–11.

Weber, J., Kocowicz, A., Bekier, J., Jamroz, E., Tyszka, R., Debicka, M., et al., 2014. The effect of a sandy soil amendment with municipal solid waste (MSW) compost on nitrogen uptake efficiency by plants. European Journal of Agronomy 54, 54–60.

White, R.E., 2006. Principles and Practice of Soil Science: The Soil as a Natural Resource. Blackwell Publishing, Malden, MA.

WHO/FAO (World Health Organization/Food and Agriculture Organization), 2007. Joint FAO/WHO Food Standard Programme Codex Alimentarius Commission 13th Session. Report of the Thirty Eight Session of the Codex Committee on Food Hygiene.

Williams, A.T., Zitomer, D.H., Mayer, B.K., 2015. Ion exchange-precipitation for nutrient recovery from dilute wastewater. Environmental Science: Water Research and Technology 1, 832–838.

Xue, D., Huang, X., 2014. Changes in soil microbial community structure with planting years and cultivars of tree peony (*Paeonia suffruticosa*). World Journal of Microbiology and Biotechnology 30 (2), 389–397.

Zaldivar, A., Siura, S., Delgado, J., 2006. Efecto de diferentes fuentes de abonos orgánicos y úrea sobre el rendimiento de lechuga (*Lactuca sativa* L.) en la Molina. Congreso Peruano de Horticultura, Arequipa, Peru (in Spanish).

Zema, D.A., Bombino, G., Andiloro, S., Zimbone, S.M., 2012. Irrigation of energy crops with urban wastewater: effects on biomass yields, soils and heating values. Agricultural Water Management 115, 55–65.

Zhang, P.J., Zhao, Z.G., Yu, S.J., Guan, Y.G., Li, D., He, X., 2012a. Using strong acid-cation exchange resin to reduce potassium level in molasses vinasses. Desalination 286, 210–216.

Zhang, X., Lei, H., Chen, K., Liu, Z., Wu, H., Liang, H., 2012b. Effect of potassium ferrate (K_2FeO_4) on sludge dewaterability under different pH conditions. Chemical Engineering Journal 210, 467–474.

Gonzalez-Gil, L., Papa, M., Feretti, D., Ceretti, E., Mazzoleni, G., Steimberg, N., et al., 2016. Is anaerobic digestion effective for the removal of organic micropollutants and biological activities from sewage sludge? Water Research 102, 211–220.

Determinants of soil carbon dynamics in urban ecosystems

Shweta Upadhyay, A.S. Raghubanshi

Integrative Ecology Laboratory (IEL), Institute of Environment & Sustainable Development
(IESD), Banaras Hindu University (BHU), Varanasi, Uttar Pradesh, India

1. Introduction

The urbanization has caused urban population explosion, whereby modified all different ecological processes in urban ecosystems. As per Sustainable Development Goal 11, about 6.5 billion peoples will be occupying $\sim 3\%$ of the land surface area, and it is a huge counterpart of about 70% carbon emissions. Moreover, about 90% of urbanization will be in the developing world such as countries like India. Therefore, nature-based climate solutions can contribute to a third of CO_2 reduction (Foley et al., 2005). However, its impacts are affecting various ecosystems in the world (Vasenev and Kuzyakov, 2018). Urban ecosystems are most

Urban Ecology
https://doi.org/10.1016/B978-0-12-820730-7.00016-1

heterogeneous land-use systems that are evolving as a potential landscape actively consuming large area continuously, whereas the increasing urban population and urban expansion have led to the loss of forest, agricultural areas and open spaces (Foley et al., 2005; Vasenev et al., 2016). Urban ecosystems are characterized by a large number of introduced plant species, including lawn grasses. An understanding of these land-use changes due to rapid urbanization is important to understand the process of greenhouse gas (GHG) emissions and the design of subsequent measures for climate change mitigation. The land-use change in the urban ecosystem is promoting the conversion of the plantation into agriculture or grassland and then into urban lawns for recreational and aesthetic values (Vasenev and Kuzyakov, 2018).

Urban soils vary drastically in composition and cover (Fig. 16.1). Various land-use practices in urban areas such as surface soil removal, clearing, grading and construction activities lead to degradation of urban soil quality in terms of soil fertility, soil erosion, scarce vegetation cover, higher soil bulk densities, lower infiltration rates and interpreted regional and local carbon cycle (Pouyat et al., 2010). These soils are under constant disturbance due to land-use change in the urban ecosystem, which contributed substantially to the soil CO_2 efflux from the conversion of one land-use to other (Pouyat et al., 2006; Wiesmeier et al.,

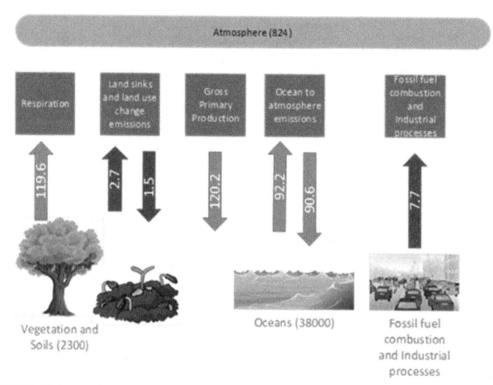

FIGURE 16.1 Modified from IPCC 2007, the carbon cycle with natural and anthropogenic carbon flux with sources and sinks (*green arrow*: gray arrow in printed version = natural flux, *brown arrow*: dark gray arrow in printed version = anthropogenic flux sources).

2015). The understanding of soil carbon stock is essential to analyze the global carbon flux and its role in GHG-mediated climate change. In the recent past, soil CO_2 efflux (also known as soil respiration) study has drawn huge attention, the reason being escalating atmospheric CO_2 concentration and its connection with global warming, biogeochemical cycles and climate change (IPCC, 2015; Scharlemann et al., 2014). Vasenev and Kuzyzkov (2018) have observed two- to threefolds higher soil C content under urban soil than its natural soil system. This may be a reason why urban soils have 50% higher CO_2 efflux than the surrounding nonurban areas (Koerner and Klopatek, 2002). Sarzhanov et al. (2017) investigated the soil CO_2 efflux in urban soil and concluded that urbanization increases the soil CO_2 emission rate significantly. Ng et al. (2015) investigated urban tropical grasslands where soil CO_2 efflux was higher relative to natural grasslands but comparable with magnitude to the few other previous urban lawn studies.

Soil CO_2 efflux is the metabolic ecological process, which acts as one of the sensitive indicators for soil carbon dynamics (López-López et al., 2012). It comprises two major pathways, viz., autotrophic (via plants roots) and heterotrophic (through the soil microbial activities). Both of these fluxes are influenced by multitudes of environmental factors (Zhou et al., 2013). Climate, vegetation and soil biophysical parameters are considered as the key determinants of spatiotemporal variability in the soil CO_2 efflux. Soil pH, soil organic carbon (SOC), soil porosity, particle size distribution and soil bulk density have been reported to influence CO_2 production and diffusion in urban soils (Pouyat et al., 2010; Riveros-Iregui et al., 2012; Srivastava et al., 2015, 2016a,b). Particularly, soil parameters such as temperature and moisture are demonstrated as the most important regulatory factors of the variation in soil CO_2 efflux (Raich et al., 2002; Curiel Yuste et al., 2007). Weissert et al. (2015) studied the soil CO_2 efflux in urban forest, parklands and sports fields to examine the role of soil biophysical properties on soil CO_2 efflux from various urban land uses. They observed that the soil properties and vegetation characteristics are significantly affecting the soil CO_2 efflux. Similarly, leaf area index, litterfall and fine root biomass are the major plant-related parameters regulating the variability in soil CO_2 efflux. The relationship between increased soil CO_2 efflux and increasing temperature is well established, which provides some feedback for global warming (Kim et al., 2015).

The current extent of land-use change in an urban ecosystem has contributed substantially for the conversion of SOC to soil CO_2 efflux (Pouyat et al., 2006; Wiesmeier et al., 2015). Land-use change is considered as the second most important factor (after fossil fuel burning) in increasing relative soil CO_2 emission leading to climate change (Raich et al., 2002). Several independent studies have been carried out for individual land uses in urban tree plantation (Chen, 2015; Taylor et al., 2015), agriculture systems (González-Ubierna et al., 2014) and grassland ecosystem (Darrouzet-Nardi et al., 2015). The combination of huge and rapid change in carbon flux makes cities one of the hot spot for soil carbon studies, which showed early signs of the contemporary climate change (Fragkias et al., 2013). The urban ecosystem sustainable carbon management has been highlighted for the reduction of soil CO_2 efflux and soil carbon sequestration (Decina et al., 2016; Weissert et al., 2016; Schwendenmann and Macinnis-Ng, 2016).

As urban soils are a significant source of CO_2 to the atmosphere, it requires proper ecological understanding. In this chapter, a brief insight has been given on various aspects of urban ecosystems in terms of their soil carbon dynamics (pools and fluxes). Urban soils are anthropogenically modified to various degrees due to physical disturbances and various soil management practices such as fertilization and irrigation (Pouyat et al., 2002). Various activities such as soil compaction that impact the soil bulk density, which on the contrary affects the SOC stock. Such urban soils may, therefore, show huge variability in SOC stocks within and between different land uses (such as parklands, residential gardens, forests and sports fields). Furthermore, indigenous topography and its variance in different land use and management may also contribute to variability in SOC stocks (Vasenev et al., 2016).

Critical components of spatiotemporal variability in soil CO_2 efflux in urban areas are still poorly understood. The spatial and temporal variations in the magnitude of SOC storage and soil CO_2 efflux are determined by various biotic and abiotic factors. For example, soil temperature and soil moisture constitute the major abiotic determinants of variation in soil CO_2 efflux at a temporal scale (Davidson et al., 1998). Above- and below-ground litter (Bowden et al., 1993), plant photosynthesis, vegetation type (Metcalfe et al., 2011) and soil nutrient availability (Singh et al., 2019) are the major regulators of the variation in soil CO_2 efflux at spatial scale. In particular, there is evidence of greater SOC and soil CO_2 efflux in continuously fertigated lawns (Pouyat et al., 2006). Therefore, recent studies have highlighted the significance of urban SOC stock and soil CO_2 efflux in different land uses (Weissert et al. 2016). This chapter will provide an insight for various key regulators of soil carbon dynamics and soil CO_2 efflux under different land uses in urban ecosystems. Moreover, it will also suggest possible management practices to obtain a carbon balance in urban landscapes.

1.1 Urban soil as CO_2 source or sink

Carbon dioxide (CO_2) is one of the most important GHGs emitted by the anthropogenic activities such as fossil fuel and biomass burning during transport and energy production, which have particular significance in terms of radiative forcing (IPCC, 2013). Soil CO_2 efflux is an imperative component of carbon cycling in an urban ecosystem. Thus, for understanding the soil carbon dynamics in the urban ecosystem, detailed investigation of the regulatory factors of the soil CO_2 efflux is very crucial. The climate, vegetation and soil factors control the spatiotemporal variability in the soil CO_2 efflux in an urban ecosystem. Soil texture and SOC are the two most important soil-related variables for soil CO_2 efflux. The terrestrial soil CO_2 efflux gets affected by climate, vegetation and soil factors contributing to the spatial and temporal variabilities in the urban ecosystem. The plant root and microbial respiration rate get affected by temperature and precipitation rate. The rising concentration of CO_2 in the atmosphere has led to an increase in soil CO_2 efflux from the soil. It is a well-established fact that urbanization alters the local climatic conditions by creating a heat islands effect (Oke, 1982), which in turn decrease human comfort and result in the increased energy costs for cooling, for example. On the contrary, vegetated areas have been reported to mitigate urban heat through evapotranspiration (Sailor, 1998), even under urban areas. Vegetation (such as urban parks, private gardens, lawns, scattered trees, etc.) act as scrubbers for removing CO_2 from the atmosphere through photosynthesis during the daytime. However, the relationship between urban vegetation cover and urban CO_2 emissions has been less explored. In general, urban areas are the major sources of CO_2 emissions due to enhanced anthropogenic origin, which have a considerable impact on the global C cycle and public health (Tong and Soskolne, 2007). Thus, studies revealed that there is a net CO_2 exchange from urban ecosystems (Pouyat et al., 2006).

Net CO_2 exchange depends on the climate conditions and anthropogenically mediated built environment, which regulates the temporal variability of biological as well as fossil fuel emissions of CO_2 (Zegras, 2010). However, measurements of soil CO_2 fluxes have been carried out for a limited number of cities around the world (Table 16.1) (Vogt et al., 2006). For example, emissions of 85, 100, 125 and 128 t CO_2 ha^{-1}/year have been reported from urban residential areas of Melbourne (Coutts et al., 2007), Tokyo (Moriwaki and Kanda, 2004), Copenhagen metropolitan area (Soegaard and Moller-Jensen, 2003) and Mexico City (Velasco et al., 2005), respectively. In the next few sections, major regulatory parameters determining the soil carbon dynamics in the urban ecosystems are described in brief.

1.2 Urban soil carbon dynamics: a brief insight

Urbanization has transformed the cities into a concrete jungle, which comprises sealed soils surface and construction due to infrastructure development (Lehmann and Stahr, 2007). Despite this fact, we still need to restore the urbanized area underlying green infrastructures (Edmondson et al., 2012). Urban areas also face various challenges in the form of food sufficiency, urban heat island and flood mitigation (Jenerette et al., 2011). To handle such issues, every land use shall be used efficiently for a long run (Gómez- Baggethun and Barton, 2013). Several studies have been done on specific aspects such as SOC storage in cities and global climate regulation, management of the urban heat island and green infrastructures of urban soil to demarcate the urban soil carbon dynamics (Davies et al., 2011; Rhea et al., 2014).

There are two schools of thought according to soil physiological hypothesis, which encompasses responses in both heterotrophic and autotrophic communities; soil CO_2 efflux is primarily regulated by soil temperature, soil moisture and substrate availability. One attempt to describe the interactions among these individual hypotheses has been the Dual Arrhenius Michaelis Menten (DAMM) model (Davidson et al., 2012). Temperature is a fundamental ecosystem property driving different biophysical processes (Davidson and Janssans, 2006). However, the temperature sensitivity of soil CO_2 efflux depends on enzymatic processes that are also regulated by soil moisture and substrate availability. Soil moisture regulates soil CO_2 efflux at low soil moisture directly by limiting microbial and root activities or indirectly by limiting diffusion of substrates, whereas high soil moisture limits soil oxygen concentration, thereby constraining soil CO_2 efflux. Peak rates of soil CO_2 efflux levels typically occur at intermediate soil moisture (Davidson et al., 2012). The relationships of soil CO_2 efflux with substrate availability have been described with models using Michaelis–Menten enzyme kinetics that express soil CO_2 efflux as a saturating function of substrate concentration (Eberwein et al., 2015). The biogeochemical explanations of these physiological drivers predict that sites with higher soil temperature and substrate availability and intermediate soil moisture levels will have higher rates of soil CO_2 efflux (Davidson et al., 2012). However, substrate diffusion covaries with temperature and soil moisture to create a dynamic environment where each driver can cancel or work synergistically with the effects of other drivers (Davidson et al., 1998). Furthermore, while the activation energy of enzymatic reactions is always positive, at low soil moisture and soil temperature, it becomes virtually irrelevant in predicting soil CO_2 efflux (Davidson et al., 2012). These effects have consequences on biogeochemical hot spots. At the local ecosystem scale, increasing availability of soil moisture and substrate may lead to the reduced importance of hot spots, whereas increasing temperature positively influences hot spot distributions (Jenerette and Larsen, 2006). These relationships are further complicated by a higher level of control such as climate and land use.

In addition, landscape position, which includes climate and land use distributions, also influences variation in soil CO_2 efflux (Singh et al., 2017, 2019). Landscape regulation of soil CO_2 efflux occurs through a direct influence to the physiological drivers of soil CO_2 efflux and may also have indirect effects by influencing the sensitivity of soil CO_2 efflux to the physiological drivers (Jenerette and Chatterjee, 2012; Singh et al., 2019). Regional-scale climate patterns have a direct influence on soil temperature and moisture conditions and are further associated with variation in soil organic matter (SOM) distributions (Zhou et al., 2009). Furthermore, land cover and land use can influence physiological drivers through processes such as irrigation, resource amendments and modification of local temperatures (Kaye et al., 2005). Additionally, long-term variation in soil environmental conditions associated with landscape patterns can alter sensitivities to environmental drivers, through temperature acclimation (Luo et al., 2001), altered wetting sensitivity (Jenerette and Chatterjee, 2012) or altered carbon use efficiency (Eberwein et al., 2015). The interactions between land use and physiological regulation may further vary in response to changes in climate. Aridity may increase differences in soil temperature, soil moisture and SOM between managed and wildland land uses. As a consequence, in arid climates, extensive irrigation may increase rates of soil CO_2 efflux, whereas in more mesic climates, irrigation may reduce rates of soil CO_2 efflux in urban areas compared with associated nonurban areas.

Seasonal differences may further influence how variation in land use and climate affects soil CO_2 efflux. Soil CO_2 efflux in urban and agricultural ecosystems is often decoupled with seasonal precipitation patterns and instead responds more to a warming of the soil environment (Kaye et al., 2005). In contrast, arid and semiarid wildland ecosystems experience the potentially higher magnitude of soil CO_2 efflux than baseline following precipitation (Sponseller, 2007). Counteracting the reduced precipitation sensitivities in urban and agricultural systems and resource amendments, including topsoil and fertilizers, may increase both recalcitrant and labile carbon availability (Jenerette and Larsen, 2006), which generally increases soil CO_2 efflux in response to water amendments (Kaye et al., 2005). However, agricultural and urban land uses may decrease summer soil temperatures in arid systems (Kaye et al., 2005), which can then decrease soil CO_2 efflux under optimal moisture and substrate conditions (Davidson et al., 2012).

1.3 Factors affecting the urban soil carbon concentration and stock

Urban SOC paves its direct impact on soil properties, i.e., bulk density, total soil nitrogen, soil pH and soil texture along with land-use change (Fang et al., 2005). Pedogenic processes are modified by anthropogenic activities, human settlement and landscape management (Lehmann and Stahr, 2007), which resulted in the spatiotemporal variation in SOC stocks along the urban horizons (Wei et al., 2014). The soil microbial biomass gets affected by urban landscape patterns, the distribution of heat and pollution (Zhao et al., 2012). The SOC concentration in urban soils is regulated by a number of factors such as climatic (precipitation, temperature, evapotranspiration) and anthropogenic (tillage, built area, land-use changes) factors (Fig. 16.2). For example, for any given rate of annual precipitation, SOC storage generally increases with the decrease in potential evaporation ratio, which is the ratio between annual precipitation and evapotranspiration (Post et al., 1982).

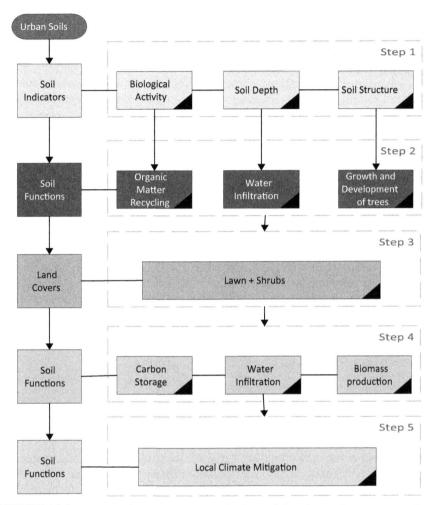

FIGURE 16.2 Diagram showing integration of urban soil functions and ecosystem services.

The soil temperature and precipitation are widely accepted critical climatic factors regulating the soil CO_2 efflux at the global scale (Chen et al., 2014; Hursh et al., 2017). In addition, SOC and soil N concentrations, and soil pH, can also influence soil CO_2 efflux directly or indirectly via regulating substrate availability and microbial activity (Chen et al., 2014). Previous studies reported that soil CO_2 efflux increased with total, above- and below-ground net/gross primary production and leaf area index (Chen et al., 2014). A detailed account on different factors is given in the following subsections.

1.3.1 Soil CO_2 efflux

As described previously, soil CO_2 efflux is a process of microbial and root respiration being controlled by soil temperature and soil moisture. Moreover, SOM quantity and quality also play an important role. It usually considers soil temperature being a crucial central control and soil moisture as the secondary control on soil CO_2 efflux (Risk et al., 2002). The soil CO_2 efflux generally increases with increasing soil temperature (Raich et al., 2002) and soil moisture (Davidson et al., 2000) conditions. However, there is an optimal value of soil temperature for soil CO_2 efflux, and if the temperature increased above this value, a decrease in the soil microbial metabolism led to a decrease in soil CO_2 efflux. Furthermore, under water-saturated areas, soil moisture may become the limiting factor for soil CO_2 efflux (Happell and Chanton, 1993). On the contrary, during extreme warm summer months (high soil temperature), soil moisture is seldom low, and therefore, the interactions of soil moisture and temperature regulate soil CO_2 efflux (Welsch and Hornberger, 2004). Doran et al. (1984) observed optimal soil biological activity at 50%–80% of soil saturation levels. While Melling et al. (2005) reported optimal soil CO_2 efflux at soil moisture level at 45%–57%. The soil CO_2 efflux rates decline sharply with too wet soil, due to reduction in the gas diffusivity rates as soil pores become water-filled (Davidson et al., 2000; Andrews and Schlesinger, 2001). Previous research has also related soil C:N ratio to soil respiration. Soils having lower C:N ratios will decompose relatively faster as compared with higher C:N ratios (Brady and Weil, 2002). It can distinguish soil CO_2 efflux into three different components of a source or sink in various ecosystems:

1. Soil CO_2 efflux comprises aerobic and anaerobic microbial respiration. Autotrophic (via root) respiration on an average contributes to ~50% of total soil respiration; depending on the vegetation type and season, it might reach up to 95% as well (Hanson et al., 2000).
2. Above-ground plant respiration, in addition, contributes to the ecosystem respiration.
3. Net ecosystem exchange (NEE) is the difference between photosynthesis and ecosystem respiration. Soil may act as a source or sink of soil CO_2 efflux depending on the positive or negative net ecosystem exchange (NEE) rate.

1.3.2 Soil texture

Soil texture is a summation of proportions of sand, silt and clay content. Soil texture is a very stable characteristic that influences soil biophysical properties. Soil texture is interrelated with the soil fertility and quality in the long term. The soil texture is associated with soil porosity, which in turn regulates the water holding capacity, gaseous diffusion and water movement that determines the soil health. Thus, the gaseous diffusion and water infiltration triggers the survival of microbial propagules and supply of moisture and air for microbial growth, shows diversity with the soil texture and, therefore, affects the soil CO_2 production in clay loam soil ~50% higher than sandy soil (Kowalenko and Ivarson, 1978). Soil texture also determines the rooting system and, thus, regulates the soil CO_2 efflux. Therefore, soil with lower water storage capacity, unsaturated hydraulic conductivity and lower fertility have slower root growth in coarser texture than in finer texture (Högberg and Högberg, 2002). In addition, soil texture also determines the extent of root litter decomposition and its rhizospheric microbial respiration, i.e., higher for clay soil than for sandy loam soil (Silver et al., 2005).

1.3.3 Soil bulk density and porosity

Soil bulk density is the mass of soil per unit volume in its natural field state and includes air space and mineral plus organic material and is represented as g cm^{-3}. It gives useful information in assessing the potential for lacking nutrient, erosion and crop productivity. The bulk density is lower than particle density due to the presence of pore space (the portion of the soil body is not occupied by particles) and humus. The bulk density of the soil is a variable property due to the effects of weathering, cultivation and effects of the animal activity.

Soil bulk density is considered as a key indicator for assessing the soil health with respect to soil functioning such as aeration and infiltration (Reynolds et al., 2009). Since bulk density is negatively correlated with SOM, loss of SOC via increased decomposition mediated by elevated temperatures (Weil and Magdoff, 2004; Davidson and Janssens, 2006) may lead to increase in bulk density. Thus, increase in bulk density via land management activities and climate change stresses may lead the soil more prone to compaction, which further regulates the soil CO_2 efflux, in general, and soil carbon dynamics in particular.

1.3.4 Soil pH

Soil pH generally varies from 4.0 to 8.0 under dry tropical soils, but in exceptional cases, values can be higher or lower. Soil pH is considered as one of the key chemical indicators of soil health (Schindelbeck et al., 2008) as different microbial activities (such as nutrient availability and cycling and biological activity), crop performance and physical conditions (such as acidification and salinization) can be characterized by a range in soil pH conditions (Dalal and Moloney, 2000). The effect of soil pH on plant growth is generally indirect but significant. However, the toxic effect of metals, such as aluminium and manganese, which become quite soluble at low pH (less than 6.5), is direct. The most significant direct effect of pH is on the availability of nutrients. Major nutrients such as nitrogen and phosphorous are most available at and around neutral pH (6.0–8.0). Iron and boron are most available at strongly acidic to moderately acidic pH (4.0–6.0), whereas the availability of molybdenum increases with increasing pH. As described earlier, such changes in soil nutrient availability regulate microbial-mediated soil CO_2 efflux.

1.3.5 Soil moisture

Soil moisture is one of the most important abiotic factors that regulate the production and transportation of soil CO_2 efflux in the soil matrix (Srivastava et al., 2015). Soil moisture regulates access to oxygen, carbon mass transport and SOC bioavailability (Davidson et al., 2012). It is directly simulated by global climate models, and the outputs from these models may help in finding the possible directions of change. The climate change—mediated changes in precipitation may lead to the changes in soil moisture at local levels, which further influence the soil carbon dynamics. Soil moisture is a very critical component that determines the rate of soil CO_2 efflux. The fundamental relationship defines that soil CO_2 efflux is minimum under dry and high moisture conditions due to anaerobic condition that prevails to decline the microbial aerobic activities, and it reaches its peak in the intermediate soil moisture condition. Thus, the optimum water content in a soil matrix is near its field capacity. Moreover, soil CO_2 efflux gets triggered by soil moisture through various physiological processes of microorganisms and roots along with the movement of oxygen and its substrate availability.

1.3.6 Soil organic matter

SOM comprised a range of living and nonliving components. It is one of the most complex and heterogeneous components of soils, which vary in their properties, functions and turnover rates (Weil and Magdoff, 2004). SOM consists of materials in various stages of decomposition from recent inputs of organic residues (plant, animal and microbial origin) to highly refractory materials. It is a highly complex mixture of carbon compounds, which contains nitrogen, sulphur and phosphorous. It is further considered as one of the key indicators of soil health and regulates several ecosystem services of soil such as nutrient cycling, provides microbial and faunal habitat and substrates, as well as affects aggregate stability, water retention and hydraulic properties (Weil and Magdoff, 2004). As SOM regulates majority of soil functions, a decrease in SOM can lead to a decrease in soil biological activities, which further resulted in the loss of soil structure and related soil physical properties (such as water holding capacity, erosion, bulk density, porosity and compaction) (Weil and Magdoff, 2004). Thus, the land use and management practices that lead to building up of SOM may further help in global climate change mitigation via a reduction in atmospheric CO_2 concentration.

1.3.7 Interactions of multiple factors

Soil various factors such as SOM, soil texture, soil pH and substrate supply have important effects in soil CO_2 efflux. Moreover, soil temperature and moisture are the major controlling factors for the variation of soil CO_2 efflux. Soil CO_2 efflux is a multifunctional parameter affected by the interaction of various assisting soil physicochemical properties, and it is challenging to segregate these interactions from one another. Soil CO_2 efflux gets affected by all the various parameters but at most by the limiting factor (like that of plants and microbes). Soil CO_2 efflux is very sensitive for soil temperature and its soil moisture content, but its interactions may vary depending on the existing environmental conditions. Soil CO_2 efflux is least sensitive towards moisture under low temperatures (less the 5°C) and for soil temperature under low moisture (less than 7.5%). On the contrary, it is more responsive at high temperature (more than 15 C) and high moisture content (15%), respectively. Moreover, when both temperature and moisture are not towards the extremes, other factors influence the soil CO_2 efflux via interactive effects of soil temperature and moisture conditions. This condition can be explained by giving an example of coniferous forest soil seasonal variability in soil CO_2 efflux with temperature, moisture and soil pH. At constant soil temperature (14°C) conditions, soil CO_2 efflux is regulated by soil moisture and pH conditions. However, when soil moisture is constant, $\sim$60% of the soil CO_2 efflux is controlled by soil pH and organic matter. The substrate availability, particularly soil nitrogen concentration, further interacts with other factors to regulate the soil CO_2 efflux (Srivastava et al., 2015, 2016a,b). It is generally assumed that the labile carbon (substrate supply) is more sensitive to temperature variability as compared with the recalcitrant carbon. Therefore, it can be concluded that the variation in soil CO_2 efflux is higher during the active growing seasons (more microbial activity) than its dormant seasons, similar is the case with elevated than that of ambient CO_2.

Overall, the changes in urban land-use pattern and the climate change—mediated changes in the urban ecosystems are having considerable influence on soil carbon dynamics. These changes are affecting different soil biophysical properties, which further regulate the soil CO_2 efflux and overall soil carbon dynamics.

2. Conclusion

Urban soil CO_2 efflux is a very important contributor to global CO_2 emissions, which further influence the soil carbon dynamics. Human-induced land-use changes are modifying the existing urban ecosystem at a very rapid rate; such urban plantation is getting transformed into agricultural field and lawns causing urban heat island effect. The changes in soil moisture and temperature conditions in urban ecosystems are influencing the decomposition of soil organic matter, which is one of the key indicators of soil carbon dynamics. Moreover, variability in its soil carbon fluxes would enhance the carbon emission by many folds from the urban ecosystems. Hence, further quantification and incorporation of the relationship between soil CO_2 efflux and its key regulatory variables from the diverse ecosystem is needed to achieve soil carbon balance in urban ecosystems.

TABLE 16.1 Adapted from Björkegren and Grimmond (2018). Selection of urban CO_2 net emission and concentration studies since 2002. An overview of urban CO_2 concentration studies after year 2005.

References	City	Study duration	Soil CO_2 efflux (μmol/ m^2 s^{-1})
Velasco et al. (2005)	Mexico City, Mexico	2003/097−119	−5 to 36.4
Coutts et al. (2007)	Melbourne, Australia	2004/February−2005/June	2−11.5
Schmidt et al. (2008)	Münster, Germany	2006/August−2006/September	−9 to 29
Vesala et al. (2008)	Helsinki, Finland	2005/December−2006/August	−10 to 17
Burri et al. (2009)	Cairo, Egypt	2007/November−2008/February	6.18 (1.19−9.86)
Velasco et al. (2009)	Mexico City, Mexico	2006/March−2009/February	2−25
Sparks and Toumi (2010)	London, UK	2008/156−195, 2008/354−2009/288	18.6 (8−35)
Christen et al. (2011)	Vancouver, Canada	2008/May−2010/April	17.7
Crawford et al. (2011)	Baltimore, USA	2002−06	−6 to 11
Pawlak et al. (2011)	Łódź, Poland	2006/July	0−30
Helfter et al. (2011)	London, UK	2006/October−2008/May	7−47
Gioli et al. (2012)	Florence/Firenze, Italy	2005/September	26.2 (9.7−39.4)
Liu et al. (2012)	Beijing, China	2006−09	15.8 (8.86−31.82)
Lietzke and Vogt (2013)	Basel, Switzerland	2009/October−2011/March	8.2
Velasco et al. (2013)	Telok Kurao, Singapore	2010/October−2012/June	(0.3−7.4)
Velasco et al. (2014)	Mexico City, Mexico	2011/June−2012/September	16.8 (5−32)
Hirano et al. (2015)	Tokyo, Japan	2012/November−2013/October	2.3−21.6
Ward et al. (2015)	London, UK	2011/January−2013/April	0−125
Ward et al. (2015)	Swindon, UK	2011/January−2013/April	−7 to 19
Ao et al. (2016)	Shanghai, China	2012/December−2013/November	2−60
Helfter et al. (2016)	London, UK	2011/September−2014/December	28.2 (11.8−41.9)

References

Andrews, J.A., Schlesinger, W.H., 2001. Soil CO_2 dynamics, acidification, and chemical weathering in a temperate forest with experimental CO_2 enrichment. Global Biogeochemical Cycles 15 (1), 149–162.

Ao, X., Grimmond, C.S.B., Chang, Y., Liu, D., Tang, Y., Hu, P., Wang, Y., Zou, J., Tan, J., 2016. Heat, water and carbon exchanges in the tall megacity of Shanghai: challenges and results. International Journal of Climatology 36 (14), 4608–4624.

Björkegren, A., Grimmond, C.S.B., 2018. Net carbon dioxide emissions from central London. Urban Climate 23, 131–158.

Bowden, R.D., Nadelhoffer, K.J., Boone, R.D., Melillo, J.M., Garrison, J.B., 1993. Contributions of aboveground litter, belowground litter, and root respiration to total soil respiration in a temperate mixed hardwood forest. Canadian Journal of Forest Research 23 (7), 1402–1407.

Brady, N.C., Weil, R.R., 2002. The Nature and Properties of Soils, thirteenth ed. Prentice Hall, Upper Saddle River.

Burri, S., Frey, C., Parlow, E., Vogt, R., 2009. CO_2 fluxes and concentrations over an urban surface in Cairo, Egypt (Doctoral dissertation, Institut für Meteorologie, Klimatologie und Fernerkundung).

Chen, S., Zou, J., Hu, Z., Chen, H., Lu, Y., 2014. Global annual soil respiration in relation to climate, soil properties and vegetation characteristics: summary of available data. Agricultural and Forest Meteorology 198, 335–346.

Chen, W.Y., 2015. The role of urban green infrastructure in offsetting carbon emissions in 35 major Chinese cities: a nationwide estimate. Cities 44, 112–120.

Christen, A., Coops, N.C., Crawford, B.R., Kellett, R., Liss, K.N., Olchovski, I., Tooke, T.R., Van Der Laan, M., Voogt, J.A., 2011. Validation of modeled carbon-dioxide emissions from an urban neighborhood with direct eddy-covariance measurements. Atmospheric Environment 45 (33), 6057–6069.

Crawford, B., Grimmond, C.S.B., Christen, A., 2011. Five years of carbon dioxide fluxes measurements in a highly vegetated suburban area. Atmospheric Environment 45 (4), 896–905.

Coutts, A.M., Beringer, J., Tapper, N.J., 2007. Characteristics influencing the variability of urban CO_2 fluxes in Melbourne, Australia. Atmospheric Environment 41 (1), 51–62.

Curiel Yuste, J., Baldocchi, D.D., Gershenson, A., Goldstein, A., Misson, L., Wong, S., 2007. Microbial soil respiration and its dependency on carbon inputs, soil temperature and moisture. Global Change Biology 13 (9), 2018–2035.

Dalal, R.C., Moloney, D., 2000. Sustainability Indicators of Soil Health and Biodiversity. Management for Sustainable Ecosystems. Centre for Conservation Biology, Brisbane, pp. 101–108.

Darrouzet-Nardi, A., Reed, S.C., Grote, E.E., Belnap, J., 2015. Observations of net soil exchange of CO_2 in a dryland show experimental warming increases carbon losses in biocrust soils. Biogeochemistry 126 (3), 363–378.

Davidson, E.A., Verchot, L.V., Cattanio, J.H., Ackerman, I.L., Carvalho, J.E.M., 2000. Effects of soil water content on soil respiration in forests and cattle pastures of Eastern Amazonia. Biogeochemistry 48 (1), 53–69.

Davidson, E.A., Janssens, I.A., 2006. Temperature sensitivity of soil carbon decomposition and feedbacks to climate change. Nature 440 (7081), 165.

Davidson, E.A., Samanta, S., Caramori, S.S., Savage, K., 2012. The Dual Arrhenius and Michaelis–Menten kinetics model for decomposition of soil organic matter at hourly to seasonal time scales. Global Change Biology 18, 371–384. https://doi.org/10.1111/j.1365-2486.2011.02546.x.

Davidson, E., Belk, E., Boone, R.D., 1998. Soil water content and temperature as independent or confounded factors controlling soil respiration in a temperate mixed hardwood forest. Global Change Biology 4 (2), 217–227.

Davies, Z.G., Edmondson, J.L., Heinemeyer, A., Leake, J.R., Gaston, K.J., 2011. Mapping an urban ecosystem service: quantifying above-ground carbon storage at a city-wide scale. Journal of Applied Ecology 48 (5), 1125–1134.

Decina, S.M., Hutyra, L.R., Gately, C.K., Getson, J.M., Reinmann, A.B., Gianotti, A.G.S., Templer, P.H., 2016. Soil respiration contributes substantially to urban carbon fluxes in the greater Boston area. Environmental Pollution 212, 433–439.

Doran, J.W., Wilhelm, W.W., Power, J.F., 1984. Crop residue removal and soil productivity with No-till corn, sorghum, and soybean 1. Soil Science Society of America Journal 48 (3), 640–645.

Eberwein, J.R., Oikawa, P.Y., Allsman, L.A., Jenerette, G.D., 2015. Carbon availability regulates soil respiration response to nitrogen and temperature. Soil Biology and Biochemistry 88, 158–164.

Edmondson, J., Davies, Z., McHugh, N., Gaston, K., Leake, J., 2012. Organic carbon hidden in urban ecosystems. Scientific Reports 2, 963.

Escolar, C., Maestre, F.T., Rey, A., 2015. Biocrusts modulate warming and rainfall exclusion effects on soil respiration in a semi-arid grassland. Soil Biology and Biochemistry 80, 9–17.

Fang, S., Gertner, G.Z., Sun, Z., Anderson, A.A., 2005. The impact of interactions in spatial simulation of the dynamics of urban sprawl. Landscape and Urban Planning 73 (4), 294–306.

Foley, J.A., DeFries, R., Asner, G.P., Barford, C., Bonan, G., Carpenter, S.R., Chapin, F.S., Coe, M.T., Daily, G.C., Gibbs, H.K., Helkowski, J.H., 2005. Global consequences of land use. Science 309 (5734), 570–574.

Fragkias, M., Lobo, J., Strumsky, D., Seto, K.C., 2013. Does size matter? Scaling of CO2 emissions and US urban areas. PLoS One 8 (6), e64727.

Gioli, B., Toscano, P., Lugato, E., Matese, A., Miglietta, F., Zaldei, A., Vaccari, F.P., 2012. Methane and carbon dioxide fluxes and source partitioning in urban areas: The case study of Florence, Italy. Environmental Pollution 164, 125–131.

Gómez-Baggethun, E., Barton, D.N., 2013. Classifying and valuing ecosystem services for urban planning. Ecological Economics 86, 235–245.

González-Ubierna, S., de la Cruz, M.T., Casermeiro, M.Á., 2014. Climate factors mediate soil respiration dynamics in mediterranean agricultural environments: an empirical approach. Soil Research 52 (6), 543–553.

Hanson, P.J., Edwards, N.T., Garten, C.T., Andrews, J.A., 2000. Separating root and soil microbial contributions to soil respiration: a review of methods and observations. Biogeochemistry 48 (1), 115–146.

Happell, J.D., Chanton, J.P., 1993. Carbon remineralization in a north Florida swamp forest: effects of water level on the pathways and rates of soil organic matter decomposition. Global Biogeochemical Cycles 7 (3), 475–490.

Helfter, C., Famulari, D., Phillips, G.J., Barlow, J.F., Wood, C., Grimmond, C.S.B., Nemitz, E., 2011. Controls of carbon dioxide concentrations and fluxes above central London. Atmospheric Chemistry and Physics 11 (5), 1913–1928.

Helfter, C., Tremper, A.H., Halios, C.H., Kotthaus, S., Bjorkegren, A., Grimmond, C.S.B., Barlow, J.F., Nemitz, E., 2016. Spatial and temporal variability of urban fluxes of methane, carbon monoxide and carbon dioxide above London, UK. Atmospheric Chemistry and Physics 16 (16), 10543–10557.

Hirano, T., Sugawara, H., Murayama, S., Kondo, H., 2015. Diurnal variation of CO_2 flux in an urban area of Tokyo. Sola 11, 100–103.

Högberg, M.N., Högberg, P., 2002. Extramatrical ectomycorrhizal mycelium contributes one-third of microbial biomass and produces, together with associated roots, half the dissolved organic carbon in a forest soil. New Phytologist 154 (3), 791–795.

Hursh, A., Ballantyne, A., Cooper, L., Maneta, M., Kimball, J., Watts, J., 2017. The sensitivity of soil respiration to soil temperature, moisture, and carbon supply at the global scale. Global Change Biology 23 (5), 2090–2103.

Intergovernmental Panel on Climate Change (IPCC), 2015. Climate Change 2014: Mitigation of Climate Change, vol. 3. Cambridge University Press.

Intergovernmental Panel on Climate Change (IPCC), 2013. Climate Change 2013: The Physical Science Basis, Contribution of Working Group I to the Fifth Assessment Report of the Intergovernmental Panel on Climate Change. Cambridge University Press, Cambridge, United Kingdom/New York, NY, USA.

Jenerette, G.D., Chatterjee, A., 2012. Soil metabolic pulses: water, substrate, and biological regulation. Ecology 93 (5), 959–966.

Jenerette, G.D., Larsen, L., 2006. A global perspective on changing sustainable urban water supplies. Global and Planetary Change 50 (3–4), 202–211.

Jenerette, G.D., Harlan, S.L., Stefanov, W.L., Martin, C.A., 2011. Ecosystem services and urban heat riskscape moderation: water, green spaces, and social inequality in Phoenix, USA. Ecological Applications 21 (7), 2637–2651.

Kaye, J.P., McCulley, R.L., Burke, I.C., 2005. Carbon fluxes, nitrogen cycling, and soil microbial communities in adjacent urban, native and agricultural ecosystems. Global Change Biology 11 (4), 575–587.

Koerner, B., Klopatek, J., 2002. Anthropogenic and natural CO_2 emission sources in an arid urban environment. Environmental Pollution 116, S45–S51.

Kim, S., Liu, Y.Y., Johnson, F.M., Parinussa, R.M., Sharma, A., 2015. A global comparison of alternate AMSR2 soil moisture products: why do they differ? Remote Sensing of Environment 161, 43–62.

Kowalenko, C.G., Ivarson, K.C., 1978. Effect of moisture content, temperature and nitrogen fertilization on carbon dioxide evolution from field soils. Soil Biology and Biochemistry 10 (5), 417–423.

Lehmann, A., Stahr, K., 2007. Nature and significance of anthropogenic urban soils. Journal of Soils and Sediments 7 (4), 247–260.

Lietzke, B., Vogt, R., 2013. Variability of CO2 concentrations and fluxes in and above an urban street canyon. Atmospheric environment 74, 60–72.

Liu, Y., Kang, C., Gao, S., Xiao, Y., Tian, Y., 2012. Understanding intra-urban trip patterns from taxi trajectory data. Journal of geographical systems 14 (4), 463–483.

López-López, G., Lobo, M.C., Negre, A., Colombàs, M., Rovira, J.M., Martorell, A., Reolid, C., Sastre-Conde, I., 2012. Impact of fertilisation practices on soil respiration, as measured by the metabolic index of short-term nitrogen input behaviour. Journal of Environmental Management 113, 517–526.

Luo, Y., Wan, S., Hui, D., Wallace, L.L., 2001. Acclimatization of soil respiration to warming in a tall grass prairie. Nature 413 (6856), 622.

Melling, L., Hatano, R., Goh, K.J., 2005. Soil CO_2 flux from three ecosystems in tropical peatland of Sarawak, Malaysia. Tellus B: Chemical and Physical Meteorology 57 (1), 1–11.

Metcalfe, D.B., Fisher, R.A., Wardle, D.A., 2011. Plant communities as drivers of soil respiration: pathways, mechanisms, and significance for global change. Biogeosciences 8 (8), 2047–2061.

Moriwaki, R., Kanda, M., 2004. Seasonal and diurnal fluxes of radiation, heat, water vapor, and carbon dioxide over a suburban area. Journal of Applied Meteorology 43 (11), 1700–1710.

Nelson, D.E., Sommers, L.E., 1982. Total carbon, organic carbon and organic matter. In: Page, A.L., Miller, R.H., Keeney, R.D. (Eds.), Methods of Soil Analysis. Part 2: Chemical and Microbiological Properties. American Society of America. Soil Science Society of American Journal, Madison, Wisonsin, pp. 539–758.

Ng, B.J.L., Hutyra, L.R., Nguyen, H., Cobb, A.R., Kai, F.M., Harvey, C., Gandois, L., 2015. Carbon fluxes from an urban tropical grassland. Environmental Pollution 203, 227–234.

Oke, T.R., 1982. The energetic basis of the urban heat island. Quarterly Journal of the Royal Meteorological Society 108 (455), 1–24.

Pawlak, W., Fortuniak, K., Siedlecki, M., 2011. Carbon dioxide flux in the centre of Łódź, Poland—analysis of a 2-year eddy covariance measurement data set. International Journal of Climatology 31 (2), 232–243.

Post, W.M., Emanuel, W.R., Zinke, P.J., Stangenberger, A.G., 1982. Soil carbon pools and world life zones. Nature 298 (5870), 156.

Pouyat, R., Groffman, P., Yesilonis, I., Hernandez, L., 2002. Soil carbon pools and fluxes in urban ecosystems. Environmental Pollution 116, S107–S118.

Pouyat, R.V., Szlavecz, K., Yesilonis, I.D., Groffman, P.M., Schwarz, K., 2010. Chemical, physical, and biological characteristics of urban soils. Urban ecosystem ecology 55, 119–152.

Pouyat, R.V., Yesilonis, I.D., Nowak, D.J., 2006. Carbon storage by urban soils in the United States. Journal of Environmental Quality 35 (4), 1566–1575.

Raich, J., Potter, C., Bhagawati, D., 2002. Interannual variability in global soil respiration, 1980–94. Global Change Biology 8, 800–812.

Reynolds, W.D., Drury, C.F., Tan, C.S., Fox, C.A., Yang, X.M., 2009. Use of indicators and pore volume-function characteristics to quantify soil physical quality. Geoderma 152 (3–4), 252–263.

Rhea, L., Shuster, W., Shaffer, J., Losco, R., 2014. Data proxies for assessment of urban soil suitability to support green infrastructure. Journal of Soil and Water Conservation 69 (3), 254–265.

Risk, D., Kellman, L., Beltrami, H., 2002. Carbon dioxide in soil profiles: production and temperature dependence. Geophysical Research Letters 29 (6), 11-1.

Riveros-Iregui, D.A., McGlynn, B.L., Emanuel, R.E., Epstein, H.E., 2012. Complex terrain leads to bidirectional responses of soil respiration to inter-annual water availability. Global Change Biology 18 (2), 749–756.

Sailor, D.J., 1998. Simulations of annual degree day impacts of urban vegetative augmentation. Atmospheric Environment 32 (1), 43–52.

Sarzhanov, D.A., Vasenev, V.I., Vasenev, I.I., Sotnikova, Y.L., Ryzhkov, O.V., Morin, T., 2017. Carbon stocks and CO_2 emissions of urban and natural soils in Central Chernozemic region of Russia. Catena 158, 131–140.

Scharlemann, J.P., Tanner, E.V., Hiederer, R., Kapos, V., 2014. Global soil carbon: understanding and managing the largest terrestrial carbon pool. Carbon Management 5 (1), 81–91.

Schindelbeck, R.R., van Es, H.M., Abawi, G.S., Wolfe, D.W., Whitlow, T.L., Gugino, B.K., Idowu, O.J., Moebius-Clune, B.N., 2008. Comprehensive assessment of soil quality for landscape and urban management. Landscape and Urban Planning 88 (2–4), 73–80.

Schmidt, A., Wrzesinsky, T., Klemm, O., 2008. Gap filling and quality assessment of CO_2 and water vapour fluxes above an urban area with radial basis function neural networks. Boundary-Layer Meteorology 126 (3), 389–413.

Schwendenmann, L., Macinnis-Ng, C., 2016. Soil CO_2 efflux in an old-growth southern conifer forest (Agathis australis)—magnitude, components and controls. Soils 2 (3), 403–419.

Silver, W.L., Thompson, A.W., McGroddy, M.E., Varner, R.K., Dias, J.D., Silva, H., Crill, P.M., Keller, M., 2005. Fine root dynamics and trace gas fluxes in two lowland tropical forest soils. Global Change Biology 11 (2), 290–306.

Singh, R., Singh, H., Singh, S., Afreen, T., Upadhyay, S., Singh, A.K., Srivastava, P., Bhadouria, R., Raghubanshi, A.S., 2017. Riparian land uses affect the dry season soil CO_2 efflux under dry tropical ecosystems. Ecological Engineering 100, 291–300.

Singh, R., Singh, A.K., Singh, S., Srivastava, P., Singh, H., Raghubanshi, A.S., 2019. Geomorphologic heterogeneity influences dry-season soil CO_2 efflux by mediating soil biophysical variables in a tropical river valley. Aquatic Sciences 81, 43.

Soegaard, H., Møller-Jensen, L., 2003. Towards a spatial CO_2 budget of a metropolitan region based on textural image classification and flux measurements. Remote Sensing of Environment 87 (2–3), 283–294.

Sparks, N., Toumi, R., 2010. Remote sampling of a CO_2 point source in an urban setting. Atmospheric Environment 44 (39), 5287–5294.

Sponseller, R.A., 2007. Precipitation pulses and soil CO_2 flux in a Sonoran desert ecosystem. Global Change Biology 13 (2), 426–436.

Srivastava, P., Raghubanshi, A.S., Singh, R., Tripathi, S.N., 2015. Soil carbon efflux and sequestration as a function of relative availability of inorganic N pools in dry tropical agroecosystem. Applied Soil Ecology 96, 1–6.

Srivastava, P., Singh, P.K., Singh, R., Bhadouria, R., Singh, D.K., Singh, S., Afreen, T., Tripathi, S., Singh, P., Singh, H., Raghubanshi, A.S., 2016a. Relative availability of inorganic N-pools shifts under land use change: an unexplored variable in soil carbon dynamics. Ecological Indicators 64, 228–236.

Srivastava, P., Singh, R., Bhadouria, R., Tripathi, S., Singh, P., Singh, H., Raghubanshi, A.S., 2016b. Organic amendment impact on SOC dynamics in dry tropics: a possible role of relative availability of inorganic-N pools. Agriculture, Ecosystems and Environment 235, 38–50.

Taylor, M.S., Wheeler, B.W., White, M.P., Economou, T., Osborne, N.J., 2015. Research note: urban street tree density and antidepressant prescription rates—a cross-sectional study in London, UK. Landscape and Urban Planning 136, 174–179.

Tong, S., Soskolne, C.L., 2007. Global environmental change and population health: progress and challenges. EcoHealth 4 (3), 352–362.

Vasenev, V., Kuzyakov, Y., 2018. Urban Soils as Hot Spots of Anthropogenic Carbon Accumulation: Review of Stocks, Mechanisms and Driving Factors. Land Degradation & Development.

Vasenev, V.I., Castaldi, S., Vizirskaya, M.M., Ananyeva, N.D., Shchepeleva, A.S., Mazirov, I.M., Ivashchenko, K.V., Valentini, R., Vasenev, I.I., September 2016. Urban soil respiration and its autotrophic and heterotrophic components compared to adjacent forest and cropland within the Moscow megapolis. In: International Conference on Landscape Architecture to Support City Sustainable Development. Springer, Cham, pp. 18–35.

Velasco, E., Pressley, S., Allwine, E., Westberg, H., Lamb, B., 2005. Measurements of CO_2 fluxes from the Mexico City urban landscape. Atmospheric Environment 39 (38), 7433–7446.

Velasco, E., Pressley, S., Grivicke, R., Allwine, E., Coons, T., Foster, W., Jobson, T., Westberg, H., Ramos, R., Hernández, F., Molina, L.T., 2009. Eddy covariance flux measurements of pollutant gases in urban Mexico City.

Velasco, E., Perrusquia, R., Jiménez, E., Hernández, F., Camacho, P., Rodríguez, S., Retama, A., Molina, L.T., 2014. Sources and sinks of carbon dioxide in a neighborhood of Mexico City. Atmospheric environment 97, 226–238.

Velasco, E., Roth, M., Tan, S.H., Quak, M., Nabarro, S.D.A., Norford, L., 2013. The role of vegetation in the CO_2 flux from a tropical urban neighbourhood.

Vesala, T., Järvi, L., Launiainen, S., Sogachev, A., Rannik, Ü., Mammarella, I., Ivola, E.S., Keronen, P., Rinne, J., Riikonen, A.N.U., Nikinmaa, E., 2008. Surface—atmosphere interactions over complex urban terrain in Helsinki, Finland. Tellus B: Chemical and Physical Meteorology 60 (2), 188–199.

Vogt, R.D., Seip, H.M., Larssen, T., Zhao, D., Xiang, R., Xiao, J., Luo, J., Zhao, Y., 2006. Potential acidifying capacity of deposition: experiences from regions with high NH4+ and dry deposition in China. The Science of the Total Environment 367 (1), 394–404.

Ward, H.C., Kotthaus, S., Grimmond, C.S.B., Bjorkegren, A., Wilkinson, M., Morrison, W.T.J., Evans, J.G., Morison, J.I.L., Iamarino, M., 2015. Effects of urban density on carbon dioxide exchanges: Observations of dense urban, suburban and woodland areas of southern England. Environmental Pollution 198, 186–200.

Wei, Z., Wu, S., Yan, X., Zhou, S., 2014. Density and stability of soil organic carbon beneath impervious surfaces in urban areas. PLoS One 9 (10), e109380.

Weil, R.R., Magdoff, F., 2004. Significance of soil organic. In: Soil Organic Matter in Sustainable Agriculture, pp. 1–2.

Weissert, L.F., Salmond, J.A., Schwendenmann, L., 2015. Variability in soil CO_2 efflux across distinct urban land cover types. Geophysical Research Abstracts 17. EGU2015-3032. http://meetingorganizer.copernicus.org/EGU2015/EGU2015-3032.pdf.

Weissert, L.F., Salmond, J.A., Schwendenmann, L., 2016. Variability of soil organic carbon stocks and soil CO2 efflux across urban land use and soil cover types. Geoderma 271, 80–90.

Weissert, L.F., Salmond, J.A., Turnbull, J.C., Schwendenmann, L., 2016. Temporal variability in the sources and fluxes of CO_2 in a residential area in an evergreen subtropical city. Atmospheric Environment 143, 164–176.

Welsch, D.L., Hornberger, G.M., 2004. Spatial and temporal simulation of soil CO_2 concentrations in a small forested catchment in Virginia. Biogeochemistry 71 (3), 413–434.

Wiesmeier, M., Munro, S., Barthold, F., Steffens, M., Schad, P., Kögel-Knabner, I., 2015. Carbon storage capacity of semi-arid grassland soils and sequestration potentials in northern China. Global Change Biology 21 (10), 3836–3845.

Zegras, P.C., 2010. Transport and land use in China: introduction to the special issue. Journal of Transport and Land Use 3 (3), 1–3.

Zhao, D., Li, F., Wang, R., Yang, Q., Ni, H., 2012. Effect of soil sealing on the microbial biomass, N transformation and related enzyme activities at various depths of soils in urban area of Beijing, China. Journal of Soils and Sediments 12 (4), 519–530.

Zhou, X., Wan, S., Luo, Y., 2007. Source components and interannual variability of soil CO_2 efflux under experimental warming and clipping in a grassland ecosystem. Global Change Biology 13 (4), 761–775.

Zhou, L.M., Li, F.M., Jin, S.L., Song, Y., 2009. How two ridges and the furrow mulched with plastic film affect soil water, soil temperature and yield of maize on the semiarid Loess Plateau of China. Field Crops Research 113 (1), 41–47.

Zhou, J., Zhang, Z., Sun, G., Fang, X., Zha, T., McNulty, S., Chen, J., Jin, Y., Noormets, A., 2013. Response of ecosystem carbon fluxes to drought events in a poplar plantation in Northern China. Forest Ecology and Management 300, 33–42.

Cities: healthy, smart and sustainable

Urban ecology and human health: implications of urban heat island, air pollution and climate change nexus

Nidhi Singh, Saumya Singh, R.K. Mall

DST-Mahamana Centre of Excellence in Climate Change Research, Institute of Environment and Sustainable Development, Banaras Hindu University, Varanasi, Uttar Pradesh, India

O U T L I N E

1. Introduction

The world experienced an upsurge in human population during the Industrial Revolution in the 19th century (Grimm et al., 2008) which accelerated the process of urbanization since then. The unregulated urbanization has caused intense alterations in the physical land surface properties (roughness, thermal inertia and albedo) (Fan et al., 2017). These changes have correspondingly affected the surface-atmosphere coupling including the exchange of water, momentum and energy in regions undergoing urbanization (Li et al., 2017). This further influences regional meteorological variables such as temperature, wind speed, and planetary boundary layer (PBL) height and air quality (Civerolo et al., 2007). In an urban area, the roads and buildings are manufactured from impervious materials, e.g., asphalt, concrete. Due to low albedo and high solar absorption, these surfaces absorb high solar radiation and reradiate it later (Wang et al., 2017). In addition, the materials used in construction have elevated thermal inertia and heat storage that is released during the night-time causing decrease in the difference between day and night temperature and thus consequent decrease in diurnal temperature range (Hardin and Vanos, 2018). The increase in overall temperature due to the above process is responsible for *urban heat island* (*UHI*) *effect*, extensively observed over urban dwellings. Street canyons referred as a U-shaped space between the adjoining building structures can also trap the longwave radiation attributed to reduction in sky view factors (SVFs) (Qiao et al., 2013) and increase the temperature around. On the other hand, pervious surfaces such as urban parks and lawns help in creating cooling effect due to evapotranspiration referred to as *urban oasis effect*. The water content in the soil governs the ground heat fluxes that in turn regulate surface temperatures. But the continuous decrease in the open and green spaces in urban areas cannot counter the accelerated increase in surface temperature. The intensification of urbanization has played the biggest role in the transformation of social, economic and environmental processes. Importantly, among all the others, urbanization has a profound and multifaceted effect on environment manifested at several spatial scales (Parrish and Stockwell, 2015).

Apart from UHI, unregulated urbanization has created piles of environmental challenges including air pollution; changes in biogeochemical cycles and water pollution; changes in land-use and ecosystem functions; solid waste management and climate. Among them, air quality and UHI are likely to pose the greatest threat to the environment (Rosenzweig et al., 2010; Parrish and Stockwell, 2015). Urbanization brings changes to urban meteorology that in turn influences ambient air pollutants. The changes brought by meteorology in emission rates, chemical reaction rates, gas—particle-phase partitioning of semi-volatile species, pollutant dispersion, and deposition play an important role in determining air pollutant concentrations (e.g., biogenic volatile organic compounds (BVOCs) and evaporative emissions of gasoline). Among air pollutants, particulate matter ($PM_{1.0}$, $PM_{2.5}$ and PM_{10}), nitrogen oxides (NOx) and ozone (O_3) pollution are major public health concerns. Like UHI, anthropogenic pollutant emissions peak in urban region, and thus the pollutant load is much higher in urban environment than rural. It is interesting to notice that the two phenomena are driven by urbanization, and the UHI and air pollution influence each other too. UHI has been found to exacerbate air pollution. Increased near-surface temperature and presence of primary pollutants due to vehicular emission pave the way for formation of secondary pollutants such as O_3

in urban area. The turbulent fluxes upstream have the potential to transport the pollutant load to the downstream region where they are trapped in night time due to inversion and disperse during day deteriorating the ambient air quality as well as weather in case of surface aerosol transfer. Sarrat et al. (2006) applied mesoscale model with urban parametrization to state that UHI modifies the dynamics of atmospheric boundary layer causing turbulent fluxes to drive the spatial distribution of primary (NOx) and secondary pollutants (O_3) in urban areas. It was observed that UHI and air pollution are responsible for large health impacts. According to a report by World Health Organization, indoor air pollution causes an estimated 3.8 million deaths every year and around 4.2 million deaths each year are attributed to ambient (outdoor) air pollution (WHO, 2016). Moreover, the population living in places where air quality index exceeds WHO guideline limits is estimated to be around 91% (WHO, 2016). Therefore, regulating urbanization can have two-way benefits.

In a similar manner, the number of people exposed to heatwaves is increasing year by year. A total of around 125 million people have been affected by heatwave exposure between 2000 and 2016, particularly in the year 2015. These heatwave events can last for many consecutive days to weeks and can cause excess mortality. For example, the heat waves of 2003 killed around 70,000 people between June and August in Europe (Robine et al., 2008). Further, in the Russian Federation a 44-day heatwave in 2010 caused 55,000 excess deaths during the event (Barriopedro et al., 2011).

It is projected that climate change may worsen air quality through increasing temperatures, seasonal shifts, extreme events and increased episodes of sporadic rainfall may increase public health issues further (Fann et al., 2016). UHI also acts as a driver for extreme weather events. The present chapter discusses the phenomena of urbanization, how it drives the UHI effect and its impact on human health. Further, the chapter discusses the effect of urbanization on air pollution, the mechanism and process behind it, the sources of UHI and air pollution and their effect on public health. In the latter section we have discussed the synergistic effect of UHI and air pollution in the era of climate change and concluded with the mitigation and adaptation strategies that can be planned to increase the population's resilience.

2. Urbanization, Urban Heat Island (UHI) and its effect

2.1 Urbanization and UHI

More than half of the world's population is residing in urban areas with Asia alone accommodating 54% of the urban population (United Nations, 2018). Urban sprawl owing to population growth and increased migration has a profound impact on the changing dynamics of urban landscapes (Feng et al., 2014). Kalnay and Cai (2003) attributed the decrease in diurnal temperature and increase in surface warming to land-use changes and urbanization. Declining natural vegetation and increase in concrete and asphalt surface which release the heat absorbed during day back to the atmosphere at night have a significant impact on warming of the environment (Wang et al., 2007). Bek et al. (2018) studied the impact of unplanned urbanism on the thermal behaviour of two areas in Egypt and found a difference of 1−4°C between a planned area with green spaces and a highly populated unplanned settlement. Urbanization has a profound impact on the characteristics of rainfall as well. A meta-analysis on the role of urbanization in modification of rainfall pattern by Liu and Niyogi (2019) found that urbanization enhances mean precipitation by 16% over the urban areas while over 18% downwind of the city. Thus, urbanization plays a key role in modification of local climate.

The urban environment differs significantly from the rural settings on the account of its infrastructure, geometry, energy balance, air and water quality, etc. (Taha, 1997). These structures and processes modify the local climate of the urban areas and can be perceived as a manifestation of anthropogenic influences on the climate (Kalnay and Cai, 2003). Urban Heat Island formation is primarily attributed to an increase in sensible heat and reduction in latent heat fluxes due to reduced evapotranspiration in the urban areas. Rising heat transport due to mechanical turbulence caused by highrise buildings and dense infrastructural setup and surge in anthropogenic heat emissions contribute to UHI. The urban canyons having low SVF with limited radiative capacity interfere with the heat exchanges rising the temperature many folds inside the buildings, which may shade the streets during the day. SVF is a ratio between the visible sky and the dominant hemisphere over the location (Oke, 1982). On the other hand, heat released during the night from these built-up patches raise the temperature of the surrounding significantly (Arnfield, 2003) and thus leads to diverse and adverse environmental consequences.

2.1.1 Global scenario

UHI and its impact have been studied globally using UHI intensity (UHII), which refers to the temperature difference between urban and surrounding suburban/rural areas (Zhou et al., 2015). UHI intensity is measured as atmospheric urban heat island (AUHI) and surface urban heat island (SUHI) intensities. While lack of consistent weather station data impedes the AUHI measurement, satellite-based radiometers measure the land surface temperature also known as the skin temperature of the earth, which is the widely used metric to estimate SUHI intensity (Li et al., 2017). Yang et al. (2017) noted the lack of a comprehensive understanding in the effect of SUHI intensity and so studied the relationship of SUHI effect with landscape pattern in 332 Chinese urban areas located in different climatic zones. It was found that SUHI intensity exhibited seasonal, diurnal and climatic variation and changes with landscape characteristics. Li et al. (2017) assessed variation in SUHI intensity with expansion in 5000 urban regions of United States and found an increase of 0.7°C with a twofold increase in the size of urban area. Similarly, Imhoff et al. (2010) studied 38 cities of USA and found seasonal and diurnal variation in SUHI intensity to be higher in summer by 3°C compared to winter. Zhao et al. (2017) found that periodic changes in solar radiation and difference in rate of evapotranspiration contributed to the temporal variation of SUHI in rural and urban region. Zhou et al. (2015) on the other hand emphasized the need to understand both the magnitude and extent of UHI, i.e., UHI footprint. They quantified UHI footprints in terms of UHI intensity and urban size in 32 major cities of China. The UHI intensity varied spatiotemporally reaching 3.9 times during the night as compared to 2.3 during the day depending on the size of the urban area. However, this approach still needs to be explored to open new avenues in understanding and quantifying the impact of UHI effect.

2.1.2 Indian scenario

As urbanization in India is on the rise, the country is very much prone to UHI effect. As a result, several studies have been done to investigate UHI effects in highly urbanized cities of

India (Mohan et al., 2011; Borbora and Das, 2014; Singh et al., 2017). Nesarikar-Patki and Ray-kar-Alange (2012) investigated the role of land use land cover change in influencing the temperature variability in Pune during 1999−2006 and observed a tremendous rise of 1−4°C with UHI effect being higher over urbanized areas as compared to vegetated areas. Mohan et al. (2011) studied the annual and seasonal temperature trends over the rapidly urbanizing National Capital Region of Delhi and observed an increasing mean minimum temperature attributed to UHI. Sharma and Joshi (2014) used a land surface temperature anomaly−based approach to study the seasonal and temporal variations of UHI effect in Delhi. The study found annual UHI intensity was highest in the industrial and commercial zones due to high-density built-up area, concrete surface and less green cover. Similarly, Joshi et al. (2015) also found UHI effect to be more pronounced in industrial and urbanized areas and seasonally higher in summers in Ahmedabad city. Singh et al. (2017) observed that UHI effect was positively correlated with normalized difference vegetation index (NDVI) and urban thermal field variance index (UTFVI) with highest UHI effect observed over the urbanized central area. This gives us an overview that UHI effect has similar spatial and seasonal variations globally.

2.2 UHI and health effects

UHI has a profound impact on human health which manifests as a function of extreme temperature, the influence of heatwave and air pollution (Singh et al., 2019; Mall et al., 2017). High ambient temperature, absence of nighttime heat discharge, lack of surface moisture and vegetations with reduced diurnal temperature variations increase the risk of heat-related morbidity and mortality of urban inhabitants (Fig. 17.1). The thermoregulatory mechanism of human body responds to increased temperature by dissipating heat in order to resist any changes in the internal state. However, in the event of unusually high ambient

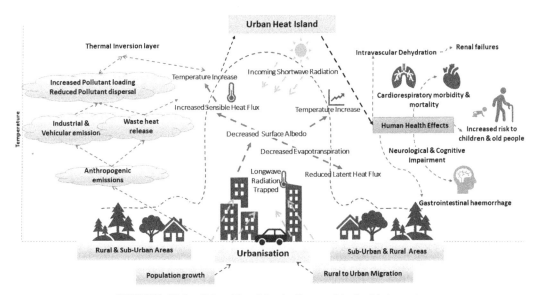

FIGURE 17.1   Urban Heat Island effects and its health impacts.

temperature body fails to maintain the homeostasis and suffers from severe dysfunctionalities. Reduced cardiac output and intravascular volume inhibit effective heat transfer in older people and puts them at a heightened risk. Vasodilation lowers the blood pressure and it is found that with 1°C rise in ambient temperature there is a reduction of 0.659 and 0.368 mmHg in systolic and diastolic blood pressures, respectively. Increase in indoor temperatures which is a characteristic of UHI phenomenon has significantly higher impact on change in human blood pressure than the outdoor temperature putting the residents at higher risk (Kim et al., 2012). Heat strokes alone attribute to 40%–64% mortality and increase the risk every time body temperature reaches above 40°C (Shahmohamadi et al., 2011). Apart from cardiorespiratory system other organs also suffer from severe impact of extreme temperature. Intravascular dehydration causing dizziness, nausea and unconsciousness can trigger renal failure and imbalance in electrolytes may cause hypoglycaemia, cerebral ischaemia and sepsis (Ev and Sharma, 2019). Hyperthermia induces neurological and cognitive dysfunction adversely affecting memory, attention and information processing. In a recent study it was found that at an elevated temperature of 38.8°C memory impairment occurred (Walter and Carraretto, 2016).

Heatwaves show a synergistic impact with UHI increasing the risk (Heaviside et al., 2016). Founda and Santamouris (2017) studied the synergistic behaviour of UHI and heat waves during the 2012 heat wave in Athens and found that heatwaves tend to intensify the UHI intensity. During the heatwave period, excess radiative heat released during night time in the urban areas raises the UHI intensity by 1.5°C on an average thereby prohibiting the nighttime cooling (Founda and Santamouris, 2017). A WRF model simulation-based study revealed that 52% of the mortality was attributed to the enhanced ambient temperature due to UHI effect during West Midlands heatwave episode in August 2003 (Heaviside et al., 2016). Kouis et al. (2019) reported mortality risk to be a function of extreme temperature rising steeply at a threshold of 33°C with an increase of 4.4% and 5.9% in cardiovascular and respiratory mortality risk in Greece. This threshold varies depending on the latitude where generally higher thresholds are observed in regions at lower latitude as the population is adapted to extreme temperatures.

3. Urbanization, air pollution and its effects

3.1 Air pollution: a general introduction

Air pollution is defined as the mixture of solid, liquid and gaseous particles suspended in the atmosphere. Small air particles such as black carbon (BC), nitrogen oxides (NO_x), ozone (O_3), sulphur dioxide (SO_2) and particulate matter (PM_{10}, $PM_{2.5}$, $PM_{1.0}$) constitute major air pollutants (Rao et al., 2018). The problem of air pollution is most prevalent in cities that witness emissions from a variety of sources. The pollutants tend to concentrate in the cities as the high-rise buildings prevent the air pollutants' dispersion. The problem tends to be basically consistent with large cities in low- and middle-income group countries. The World Health Organization (WHO) has listed Karachi, New Delhi, Beijing, Lima and Cairo as some of the world's most polluted cities. But the problem has now extended its domain to some of the most populated cities in developed nations like Los Angeles, California (NGS, 2011).

Most often the air pollutants from the local or regional sources affect human health, but some atmospheric conditions make the pollutants travel across boundaries up to long distances and affect people elsewhere. Therefore, local, regional and global actions are needed to mitigate the cross-boundary transportation and solve it at local sites only.

3.2 Sources of air pollution

Based on its source, air pollution can be divided into indoor and outdoor air pollution.

Indoor air pollution: It consists of pollutants released from substances such as kerosene, wood and coal used for cooking and heating the house; radon released naturally from earth's surface; wall paint; construction materials; insulation materials. Further, air conditioners create a dampness and the place becomes favourable for the growth of moulds that could spread with air and can act as allergens that can trigger asthma (Leung, 2015).

Outdoor air pollution: It consists of air pollutants contributed by burning of fossil fuels by power plants, automobiles, nonroad equipment, burning and incineration of municipal and agricultural wastes and industrial facilities (Leung, 2015).

The outdoor and indoor air pollutants altogether kill about 7 million people each year worldwide, where South-East Asia and the Western Pacific Regions account for largest health burden due to air pollution (WHO, 2016). Table 17.1 gives an account of the major air pollution episodes attributed to anthropogenic activities. In the southwestern United States, wildfire, heat events and dry periods can place an extra burden on air quality standards even as efforts are undertaken to reduce emissions.

3.3 Air pollutants and health effects

Among developing economy, India and China contributes majority of the air pollution related deaths. In India, air pollution accounts for 0.65 million deaths per year and in China, it is more than the combined mortality from AIDS, malaria and tuberculosis (Lelieveld et al., 2015). People suffer from different health effects due to air pollution depending upon their existing health condition, exposure duration (short-term and long-term effects), exposure concentration and type of pollutants (Table 17.2). Those with underlying health conditions (lung disease, diabetes, cardiovascular diseases (CVD), asthma and cancer), children and elderly, and low socioeconomic group are at greater risk on account of their weak immune system (Singh et al., 2019). Air pollution may also induce diabetes, autism and could cause lower IQ (Guo et al., 2018). The coarse particulate matter (PM_{10}, particles < 10 microns in diameter) causes irritation in nasal passages and upper respiratory tract. Fine particulate matter ($PM_{2.5}$, particles < 2.5 microns in diameter; ultrafine particles) can penetrate deep into the lungs and blood that leads to heart attacks and strokes, asthma and bronchitis, and is responsible for several premature deaths and can interfere with brain development in children (Lelieveld et al., 2015). Based on the duration of exposure, health effects can be categorized into:

3.3.1 Short-term effects

In short-term health effects, people may suffer from pneumonia or bronchitis. Other harmful health effects include irritation in eyes, nose, throat and skin; headaches; dizziness and nausea (NGS, 2011).

TABLE 17.1 Episodes of major urban air pollution in the world.

Episode	Composition	Causes
Acid Rain	Precipitation of pH below 5.2 due to presence of strong acids, sulphuric acid and nitric acid formed by atmospheric reaction of sulphur dioxide (SO_2) and nitrogen oxides (NO_x; the combination of NO and NO_2) with water. (Casiday and Frey, 1998)	After World War II, Europe and eastern North America showed large increase in consumption of fossil fuels emitting huge amounts of sulphur dioxide and nitrogen oxide
Great Smog of London (1952) (Sulphurous smog)	A combination of cold weather with anticyclone and windless conditions mixed with airborne pollutants (mostly SO_2) coming from coal formed a thick layer of smog over London (Laskin, 2006)	To keep themselves warm during cold weather Londoners burnt more coal of relatively low-grade sulphurous variety that increased the amount of sulphur dioxide in the smoke and also large number of coal-fired power stations in greater London area contributed to deadly smog (ApSimon, 2005)
Los Angeles Smog 1943 (Photochemical smog)	The major component was ozone (O_3). Emission of nitrogen oxides from vehicular exhausts and gaseous hydrocarbons from cars and oil refineries in presence of sunlight formed ozone and photochemical smog (Gardner, 2018)	City built for cars during wartime led Los Angeles to become the largest car market and the unprecedented emission from vehicles and factories led to the deadly Los Angeles Smog (Jacobs and Kelly, 2008)
Malaysian Haze 2005	The haze was dominated by fine particles in which secondary inorganic aerosols (SIA, such as SO_4^{2-} and NH_4^+) and organic substances (such as levoglucosan, LG) were dominant (Latif et al., 2018)	The practice of slash and burn for farming and peat fires from Indonesia were major causes (Latif et al., 2018)

3.3.2 Long-term effects

Long-term health effects may cause cardiorespiratory disorder such as lung cancer, emphysema and chronic obstructive pulmonary disorder (COPD). It may also affect vital body organs such as brain, kidney, liver and may also cause birth defects (NGS, 2011).

3.4 Pathway of air pollution effect

Air pollutants are composed of chemical particles that may take several other forms such as solids, liquids or gases. These oxidants at low doses may mediate physiological effects, while in high doses mediate toxicity (Xia et al., 2009). The other hypothesis suggests the role of redox-active mediators, for example, reactive oxygen species (ROS) and reactive nitrogen species are site-specific mediators of cell signalling and prime regulators of the inflammatory response, where both interact in a multifaceted manner (Daiber et al., 2017). There is a possibility that the pollutants may damage endogenous redox signalling and/or activation of endogenous ROS sources that may result in enhanced responses. Factors such as size, shape, structure, surface reactivity, solubility, biopersistence and 'leachable' components may affect particle toxicity (Nel et al., 2006). For example, the smaller sized particles tend to have larger surface-to-mass ratio and possess large reactive components and therefore can easily penetrate lower respiratory airways and induce toxicity (Oberdörster 2012; Oberdorster et al., 1994). Ultrafine particles like metal ions, polycyclic aromatic hydrocarbons (PAHs) produced from the burning of fossil fuel (coal, oil, gasoline), trash, wood, charcoal-broiled meat and tobacco possess reactive components on their surface that possibly mediate systemic effects (Oberdorster and Utell, 2002). The studies show that organic carbon and sulphates are known to have strong associations with the CVD (Maynard et al., 2007). The strength of impact lies in the duration, concentration and toxicity of the exposure pollutant, and susceptibility of the exposed subjects. On entering the human body, pollution particles can be sequestrated, distributed intracellularly and systemically transmitted to others (Simkhovich et al., 2008). They can also activate the autonomic nervous system and produce adverse effects (Calderon-Garciduenas et al., 2008). Pollutants can enter from the olfactory nerve to the olfactory bulb and to central nervous system and may induce inflammation (Rao et al., 2018).

3.5 Urbanization and anthropogenic air pollution

The reform policies of the 1990s opened the path for rapid expansion and urbanization. Increasing urban population, expanding urban land and private vehicles, and increasing economic growth have undoubtedly posed a great threat to the environment. However, since the urbanization pattern and magnitude are not consistent, therefore it can have a variable effect on environment over changing time and variable emission sources at different stages (Kelly and Zhu, 2016). Densely populated cities experience a very high pollution load, which is responsible for a high morbidity and mortality burden. Serious efforts to improve air quality has led to better air quality in cities of North America, Europe and Latin America but situation is worse in developing nations like India and China. The primary and secondary gaseous and particulate pollutants take part in heterogeneous reactions in the atmosphere thus forming complex air pollution mixtures. Air pollution has become a life-threatening challenge disturbing overall earth systems' stability, ecosystems and human health, and is a potential driver for global climate change (Rockström et al., 2009). Furthermore, as the long-term air pollution data are limited, its spatial and temporal effects remain largely unknown.

The air temperature plays an important role in accelerating chemical reactions. It was seen that variable air temperature and vegetation regulate the production of biogenic volatile organic compounds (BVOCs), a precursor of ground-level O_3 (Guenther et al., 2006). The photolysis reaction rates are accelerated by high air temperatures enhancing the production of tropospheric O_3, secondary inorganic aerosols (e.g., nitrate, sulphate and ammonium aerosols) and secondary organic aerosols (SOAs) (Aw and Kleeman, 2003). The land surface roughness helps in pollutant deposition which increases pollutant concentrations over the land surface that may affect regional meteorology in urban regions (Kalnay and Cai, 2003). However, the evidence of impact of land surface changes on regional air quality is sparsely available and limited to ozone (Zhang et al., 2019). In one such study, Chen et al. (2018) showed that alternation in urban land surfaces was responsible for increases in lower air surface temperature and PBL (Planetary Boundary Layer) height, that has caused increases in surface O_3 concentrations but decreases (16.6 μgm^3) in concentrations of $PM_{2.5}$.

In another study, the increasing temperature, PBL height, reduced wind speed and increasing daytime ozone together contributed by urbanization is equivalent to increase in emissions of around 20% (Yu et al., 2012). One major study gap found in past studies was that previous studies could not find any robust driving process that could describe the interaction between land surface change, air pollution and meteorology. Further, they assumed that urban properties remain homogenous throughout the city. Therefore, the urbanization brings a multifaceted problem associated with unplanned growth, deteriorating air quality and changing meteorology (Bai et al., 2017). Scientific research and controlling policies implemented so far suggest that an integrated and comprehensive solution is required, which integrates urban planning, clean energy, energy efficiency and innovation in transportation.

4. UHI, air pollution and human health nexus in the era of climate change

Urbanization is known to drive the problems of UHI and air pollution, which are most prominently visible in cities. Further, UHI and air pollution are known to drive one another. UHI effect increases the energy demand and consumption particularly during summer, when temperatures rise. The use of different electric appliances to keep the rooms cool and methods of conventional energy generation to meet the increased energy demand cause increase in greenhouse gas emissions (EPA, 2008). The spatial temperature difference between cities and suburban areas creates intensive heat islands causing local Hadley type of circulation (Rao et al., 2014). This causes the rising air from the city centre to move towards suburban areas following temperature gradient areas and again move to city interiors. This recirculation process causes elevated inversion and inhibits dispersion of air pollutants upward in urban areas (Oke, 1974, 1977). The greenhouse gases and particulate pollutants absorbs some parts of the long-wavelength infrared radiation emitted from the earth's surface and reradiate it back to the earth's surface that warms the ambient air and enhances low level stability and higher pollutant concentrations. The ambient pollutants now will generate heat island effect and will alter the structure of vertical temperature profile that will hinder the pollutant dispersion (Patterson, 1969). The potential greenhouse gases include carbon dioxide (CO_2), methane(CH_4), nitrous oxide (NOx) and fluorinated gases that have their source as natural and artificial both. All of them are found in high concentration in urban areas (Li et al., 2017). Apart from air pollution and the greenhouse effect, UHI significantly contributed to warming in the 20th century (Estrada et al., 2017). It was found responsible for the warming of about 0.6°C between 1950 and 2015. The researchers anticipated that UHI can further contribute to 2°C to warming in some highly populated cities by 2050 (Estrada et al., 2017). What makes the situation worse is that climate change has brought new challenges to mankind that is threatening to its own existence. Climate change is said to be responsible for high atmospheric temperature; long, severe and frequent heat waves and deadly cold spells; frequent floods, droughts and cyclones. Urban population is augmenting the effect of climate change by contributing to air pollution and UHI effect. This complex nexus of climate change-air pollution-UHI driven by urbanization would prove devastating to human health as discussed in the above sections (Heaviside et al., 2016)

5. Policy recommendations

With 68% of the population about to reside in urban regions in future, urbanization and its impacts are inevitable (UNITED NATIONS, 2018). In this scenario taking the route of mitigation and adaptation is the only way out (Mall et al., 2019). The rapid urban population growth has led to more energy consumption, and increased household utility and vehicles emission. Despite climate change presenting increasing challenges to meet air quality standards, policy and action to mitigate these impacts have been surprisingly absent. Certain measures, including urban design and infrastructure improvements, adequate provision for green spaces, adaptation and mitigation action at multiple policy levels, can strengthen urban reliance and reduce the impact from UHI effect and air quality.

5.1 Urban design and infrastructure

Transitioning towards sustainable urban infrastructure in urban planning and design are at the core of mitigation and adaptation strategies to address UHI impact (Table 17.3). The urban configuration is instrumental in increasing UHI Intensity and has a key role in mitigating the impact of UHI. Yue et al. (2019) studied the urban configuration of 36 cities across China and found that the small built-up patches had lower UHI intensity as compared to large built-up areas. Buildings account for 40% of all energy consumption, globally accounting for substantial carbon dioxide emission in the environment leaving a significant footprint on urban ecology (Ürge-Vorsatz et al., 2007, Ching et al., 2019). Optimizing building structure and orientation using materials having low thermal conductivity can enable reduction in solar radiation, improved ventilation, efficient energy consumption, waste heat emissions and reduction in thermal discomfort. Green buildings offer a solution to UHI mitigation. Shin et al. (2017) reported that Leadership in Energy and Environmental Design (LEED) certified green buildings have the potential to bring down the UHI intensity by 0.26−0.48°C. A major share of mitigation and adaptation can help to increase the surface albedo and water retention that in turn increase thermal emissivity and reduce surface temperature. Adopting these can significantly lower the impact of UHI (Table 17.3).

5.2 Green cover and water body rejuvenation

Urban constructions require land which is acquired by encroaching the water bodies, forest cover and agricultural lands and disrupting the ecological balance (Zhao et al., 2017). Increasing impervious surface and shrinking traditional water reservoirs alter the natural hydrological cycle making urban flood a common disaster along with impairing the groundwater recharge in these areas (McGrane, 2016). Increasing the green cover and rejuvenating the water bodies improve the heat dissipation and water recharge. Green roofs and rooftop and vertical gardens increase the surface albedo and latent heat flux in the urban area and add to the greenery in the urban region (Sanchez and Reames, 2019). Rainwater harvesting, vegetated rooftops, tree plantation, infiltration and bioretention systems are some of the ways to tap the excess runoff, hold the peak flows, remove the contamination and

TABLE 17.2 The list of most common air pollutants, their sources and related health effects.

Pollutant	Sources and health effects of air pollutants
Particulate matter (PM_{10}, $PM_{2.5}$, $PM_{1.0}$)	**Sources:** Combustion of diesel and petrol in automobile engines, combustion of coal, heavy oil and biomass for energy production in households and factories, combustion process in industrial and manufacturing processes (construction and building materials, mining, ceramic and bricks, and smelting).
	Effect: $PM_{2.5}$ and PM_{10} can easily enter lungs and irritate the respiratory tract. Fine ($PM_{2.5}$) and ultrafine particles ($PM_{1.0}$) can penetrate bloodstream and may cause inflammation of heart and lung.
Nitrogen oxides (NO, NO_2)	**Sources:** Theses gases are primarily contributed by automobiles, power generation and industry.
	Effects: The gas causes irritation of airways and exacerbates chronic symptoms like asthma and bronchitis and increases the risk of cardiac disorder, increased respiratory infections and reduced lung function.
Carbon monoxide (CO)	**Sources:** Dominantly contributed by automobile exhaust and burning of fossil fuels.
	Effects: This gas is known to have more binding affinity with haemoglobin and can interfere with the oxygen uptake in blood. A reduction in oxygen supply can severely affect brain, heart and other vital organs.
Black carbon (BC)	**Sources:** Black carbon are short-lived climate pollutants emitted from burning of fossil fuel (especially diesel, wood, coal and others).
	Effects: Black carbon is associated with heart attack and stroke on exposure to long periods. It may cause hypertension, chronic obstructive pulmonary disease (COPD) or cancer and may trigger bronchitis and asthma
Ozone (O_3)	**Source:** Tropospheric ozone is a short-lived secondary pollutant, contributed indirectly by combustion of fossil fuels. NOx; volatile organic compounds (VOCs) emitted by automobiles, industries and chemical factories; and methane released by waste, fossil fuel and agricultural industry reacts in sunlight to form O_3.
	Effect: It acts as a respiratory irritant. Short-term exposure to ozone can cause chest pain, coughing and throat irritation, while long-term exposure can cause reduced lung function, asthma and COPD, causing damaging impact on immune activation.
Sulphur dioxide (SO_2)	**Source:** SO_2 is released by burning of fossil fuels containing sulphur from coal, metallurgical process and mineral ores smelting, ship engines, and heavy diesel equipment.
	Effect: Sulphur dioxide is known to cause irritation in eye, may aggravate asthma and bronchitis, lung inflammation and increase the risk of respiratory infections and cardiovascular disorder.

Source: Modified from Landrigan et al., 2018; WHO, 2019. Ambient Air Pollution: Pollutants. Air Pollution. World Health Organization. Accessed at: https://www.who.int/airpollution/ambient/pollutants/en/.

sediment load from the stormwater and augment infiltration in urban areas (McGrane, 2016). These measures not only mitigate the impact of UHI but also make up for the loss in ecosystem services such as stormwater management, restoring the biodiversity and carbon sequestration and contributing to the aesthetic beauty of the region.

TABLE 17.3 UHI mitigation and adaptation techniques and their local and global effects.

UHI Mitigation and adaptation measures	Building-based measures	Community-based measures	Local effects	Global effects
Green infrastructure	1. Green building 2. Green roofs 3. Vertical gardens	1. Reflective surfaces 2. Cool pavements 3. Green pathways	1. Regulation of ambient temperature profile 2. Increase in evapotranspiration 3. Improved energy exchanges	1. Increase in carbon sequestration 2. Reduced greenhouse gas emission 3. Improved microclimate
Sustainable urban planning	1. Green building 2. Green roofs 3. Vertical gardens	1. Less dense urban configuration 2. Permeable pavements 3. Efficient public transport 4. Green spaces/ Urban parks 5. Land-use monitoring and management	1. Reduced UHI intensity 2. Storm water management 3. Improved convection for pollutant dispersion 4. Improved public health	1. Increased resilience towards extreme weather events such as heat waves and urban floods. 2. Improved air quality 3. Health risk reduction
Heat mitigation system	1. Building orientation change for solar exposure 2. Use of low thermal conductivity materials 3. Use of reflective surfaces 4. Installing energy efficient systems	1. Urban greening 2. Osmotic pool 3. Regulating ponds	1. Reduction in anthropogenic heat emissions 2. Reduction in heat-related morbidity and mortality 3. Reduced energy demands	1. Reduced greenhouse gas emission 2. Health risk reduction

5.3 Air quality mitigation and adaptation strategies

The synergy of UHI and anthropogenic emissions in urban areas pose a great risk to the health of its inhabitants. An extensive network of monitoring stations, emission inventories serve as the first step to assess the air quality of the region and formulate appropriate measures to be implemented. Air quality modelling, wind profile study and application of plume dispersion modelling techniques help in identifying regions which are the most suitable for industrial setup and have minimum impact on the health of residents. Stringent monitoring on the use of vehicle-based emission control devices, avoiding the use of transport for short distances and promoting the use of public transport can reduce ground-level emission and also congestion on the roads (Harlan and Ruddell, 2011). Using clean and renewable energy-based fuels curtails the emission, lowers the pollutant load, reduces the temperature and helps in creating a cleaner environment. Making the urban landscape greener by including green roofs, green walls, parks, hedges, etc., leads to dilution of pollution by promoting dispersion and higher deposition of pollutants especially particulate matter on the vegetation. Also, trees act as barrier between source and receptor, causing turbulence by disrupting the streamline flow of plumes. Modelling the distribution of trees in high emitting pockets of urban areas is a mitigation measure to be adopted during urban landscape designing stage (Hewitt et al., 2019).

6. Conclusion

As the episodes of extreme weather events, air pollution and related health effects are becoming more intense, frequent, it is indeed mandatory to strengthen the adaptive capacity of the cities. Given the current climate scenario and complexity of the urban environment, adaptation strategies need to cover the entire cycle from capacity building of citizens against the possible hazard in the first place to immediate response to extreme events. Effective communication between the stakeholders and active participation of masses in the decision-making are a must to track the tipping point for development trajectories of the urban region. However, economic disparity, lack of good governance, lack of resources and lack of political will and preparedness are some of the challenges faced by middle and low-income countries which ironically are also witnessing rampant urbanization. The problem of urban heat island, anthropogenic air pollution and urban climate change have attracted global attention as global initiatives to promote climate actions at the city level have been taken up rapidly in recent decades. The Urban Climate Change Research Network is one such international collaborative decision support system to provide insight into climate-based actions and policies for urban administration. Such approaches are needed to put a strong foot forward in combating the climate-air-ecological threat posed by urbanization across the globe.

7. Funding information

The authors thank the Climate Change Programme, Department of Science and Technology, New Delhi, for financial support (DST/CCP/CoE/80/2017(G)).

References

ApSimon, H., 2005. The air over London. In: Hunt, J. (Ed.), London's environment: prospects for a sustainable world city. Imperical College Press, pp. 83–98.
Arnfield, A.J., 2003. Two decades of urban climate research: a review of turbulence, exchanges of energy and water, and the urban heat island. International Journal of Climatology 23 (1), 1–26.
Aw, J., Kleeman, M.J., 2003. Evaluating the first-order effect of intraannual temperature variability on urban air pollution. Journal of Geophysical Research: Atmosphere 108 (D12).

Bai, X., McPhearson, T., Cleugh, H., Nagendra, H., Tong, X., Zhu, T., Zhu, Y.G., 2017. Linking urbanization and the environment: conceptual and empirical advances. Annual Review of Environment and Resources 42, 215–240.

Barriopedro, D., Fischer, E.M., Luterbacher, J., Trigo, R.M., García-Herrera, R., 2011. The hot summer of 2010: redrawing the temperature record map of Europe. Science 332 (6026), 220–224.

Bek, M.A., Azmy, N., Elkafrawy, S., 2018. The effect of unplanned growth of urban areas on heat island phenomena. Ain Shams Engineering Journal 9 (4), 3169–3177.

Borbora, J., Das, A.K., 2014. Summertime urban heat island study for Guwahati city, India. Sustainable Cities and Society 11, 61–66.

Calderón-Garcidueñas, L., Solt, A.C., Henríquez-Roldán, C., Torres-Jardón, R., Nuse, B., Herritt, L., Villarreal-Calderón, R., Osnaya, N., Stone, I., Garcia, R., Brooks, D.M., 2008. Long-term air pollution exposure is associated with neuroinflammation, an altered innate immune response, disruption of the blood-brain barrier, ultrafine particulate deposition, and accumulation of amyloid β-42 and α-synuclein in children and young adults. Toxicologic Pathology 36 (2), 289–310.

Casiday, R., Frey, R., 1998. Acid Rain. Inorganic Reactions Experiment, Washington University. Retrieved from: http://www.chemistry.wustl.edu/∼edudev/LabTutorials/Water/FreshWater/acidrain.html.

Chen, L., Zhang, M., Zhu, J., Wang, Y., Skorokhod, A., 2018. Modeling impacts of urbanization and urban heat island mitigation on boundary layer meteorology and air quality in Beijing under different weather conditions. Journal of Geophysical Research: Atmosphere 123 (8), 4323–4344.

Ching, J., Aliaga, D., Mills, G., Masson, V., See, L., Neophytou, M., Middel, A., Baklanov, A., Ren, C., Ng, E., Fung, J., 2019. Pathway using WUDAPT's digital synthetic city tool towards generating urban canopy parameters for multi-scale urban atmospheric modeling. Urban Climate 28, 100459.

Civerolo, K., Hogrefe, C., Lynn, B., Rosenthal, J., Ku, J.Y., Solecki, W., Cox, J., Small, C., Rosenzweig, C., Goldberg, R., Knowlton, K., 2007. Estimating the effects of increased urbanization on surface meteorology and ozone concentrations in the New York City metropolitan region. Atmospheric Environment 41 (9), 1803–1818.

Daiber, A., Oelze, M., Steven, S., Kröller-Schön, S., Münzel, T., 2017. Taking up the cudgels for the traditional reactive oxygen and nitrogen species detection assays and their use in the cardiovascular system. Redox Biology 12, 35–49.

Estrada, F., Botzen, W.W., Tol, R.S., 2017. A global economic assessment of city policies to reduce climate change impacts. Nature Climate Change 7 (6), 403.

Ev, O., Sharma, S., 2019. Physiology, Temperature Regulations. 10–14. StatPearls Publishing, Treasure Island (FL).

Fan, C., Myint, S., Kaplan, S., Middel, A., Zheng, B., Rahman, A., Huang, H.P., Brazel, A., Blumberg, D., 2017. Understanding the impact of urbanization on surface urban heat islands—a longitudinal analysis of the oasis effect in subtropical desert cities. Remote Sensing 9 (7), 672.

Fann, N., Brennan, T., Dolwick, P., Gamble, J.L., Ilacqua, V., Kolb, L., Nolte, C.G., Spero, T.L., Ziska, L., 2016. Ch. 3: Air Quality Impacts. The Impacts of Climate Change on Human Health in the United States: A Scientific Assessment. U.S. Global Change Research Program, Washington, DC, pp. 69–98.

Feng, H., Zhao, X., Chen, F., Wu, L., 2014. Using land use change trajectories to quantify the effects of urbanization on urban heat island. Advances in Space Research 53 (3), 463–473.

Founda, D., Santamouris, M., 2017. Synergies between urban heat island and heat waves in Athens (Greece), during an extremely hot summer (2012). Scientific Reports 7 (1), 10973.

Gardner, S., July 14, 2018. Smog: The Battle against Air Pollution. Marketplace.org. American Public Media.

Grimm, N.B., Faeth, S.H., Golubiewski, N.E., Redman, C.L., Wu, J., Bai, X., Briggs, J.M., 2008. Global change and the ecology of cities. Science 319 (5864), 756–760. https://doi.org/10.1126/science.1150195.

Guenther, A., Karl, T., Harley, P., Wiedinmyer, C., Palmer, P.I., Geron, C., 2006. Estimates of global terrestrial isoprene emissions using MEGAN (model of emissions of gases and aerosols from nature). Atmospheric Chemistry and Physics 6 (11), 3181–3210.

Guo, Z., Xie, H.Q., Zhang, P., Luo, Y., Xu, T., Liu, Y., Fu, H., Xu, L., Valsami-Jones, E., Boksa, P., Zhao, B., 2018. Dioxins as potential risk factors for autism spectrum disorder. Environment International 121, 906–915.

Hardin, A.W., Vanos, J.K., 2018. The influence of surface type on the absorbed radiation by a human under hot, dry conditions. International Journal of Biometeorology 62 (1), 43–56.

Harlan, S.L., Ruddell, D.M., 2011. Climate change and health in cities: impacts of heat and air pollution and potential co-benefits from mitigation and adaptation. Current Opinion in Environmental Sustainability 3 (3), 126–134.

Heaviside, C., Vardoulakis, S., Cai, X.M., 2016. Attribution of mortality to the urban heat island during heatwaves in the West Midlands, UK. Environmental Health 15 (1), S27.

Hewitt, C.N., Ashworth, K., MacKenzie, A.R., 2019. Using green infrastructure to improve urban air quality (GI4AQ). Ambio 1–12.

Imhoff, M.L., Zhang, P., Wolfe, R.E., Bounoua, L., 2010. Remote sensing of the urban heat island effect across biomes in the continental USA. Remote Sensing of Environment 114 (3), 504–513.

Jacobs, C., Kelly, W., 2008. Smogtown: the lung-burning history of pollution in Los Angeles. Abrams.

Joshi, R., Raval, H., Pathak, M., Prajapati, S., Patel, A., Singh, V., Kalubarme, M.H., 2015. Urban heat island characterization and isotherm mapping using geo-informatics technology in Ahmedabad city, Gujarat state, India. International Journal of Geosciences 6 (03), 274.

Kalnay, E., Cai, M., 2003. Impact of urbanization and land-use change on climate. Nature 423 (6939), 528.

Kelly, F.J., Zhu, T., 2016. Transport solutions for cleaner air. Science 352 (6288), 934–936.

Kim, Y.M., Kim, S., Cheong, H.K., Ahn, B., Choi, K., 2012. Effects of heat wave on body temperature and blood pressure in the poor and elderly. Environmental Health and Toxicology 27.

Kouis, P., et al., 2019. The effect of ambient air temperature on cardiovascular and respiratory mortality in Thessaloniki, Greece. Science of the Total Environment 647, 1351–1358. https://doi.org/10.1016/j.scitotenv.2018.08.106.

Landrigan, P.J., Fuller, R., Acosta, N.J., Adeyi, O., Arnold, R., Baldé, A.B., Bertollini, R., Bose-O'Reilly, S., Boufford, J.I., Breysse, P.N., Chiles, T., 2018. The lancet commission on pollution and health. The Lancet 391 (10119), 462–512.

Laskin, D., 2006. The great London smog. Weatherwise 59 (6), 42–45. https://doi.org/10.3200/WEWI.59.6.42-45.

Latif, M.T., Othman, M., Idris, N., Juneng, L., Abdullah, A.M., Hamzah, W.P., Khan, M.F., Sulaiman, N.M.N., Jewaratnam, J., Aghamohammadi, N., Sahani, M., 2018. Impact of regional haze towards air quality in Malaysia: a review. Atmospheric Environment 177, 28–44.

Lelieveld, J., Evans, J.S., Fnais, M., Giannadaki, D., Pozzer, A., 2015. The contribution of outdoor air pollution sources to premature mortality on a global scale. Nature 525 (7569), 367.

Leung, D.Y., 2015. Outdoor-indoor air pollution in urban environment: challenges and opportunity. Frontiers in Environmental Science 2, 69.

Li, X., Zhou, Y., Asrar, G.R., Imhoff, M., Li, X., 2017. The surface urban heat island response to urban expansion: a panel analysis for the conterminous United States. The Science of the Total Environment 605, 426–435.

Liu, J., Niyogi, D., 2019. Meta-analysis of urbanization impact on rainfall modification. Scientific Reports 9 (1), 7301.

Mall, R.K., Singh, N., Prasad, R., Tompkins, A., Gupta, A., 2017. Impact of climate variability on human health: a pilot study in tertiary care hospital of Eastern Uttar Pradesh, India. Mausam 68 (3), 429–438.

Mall, R.K., Srivastava, R.K., Banerjee, T., Mishra, O.P., Bhatt, D., Sonkar, G., 2019. Disaster risk reduction including climate change adaptation over south Asia: challenges and ways forward. International Journal of Disaster Risk Science 10 (1), 14–27.

Maynard, D., Coull, B.A., Gryparis, A., Schwartz, J., 2007. Mortality risk associated with short-term exposure to traffic particles and sulfates. Environmental Health Perspectives 115 (5), 751–755.

McGrane, S.J., 2016. Impacts of urbanisation on hydrological and water quality dynamics, and urban water management: a review. Hydrological Sciences Journal 61 (13), 2295–2311.

Mohan, M., Kandya, A., Battiprolu, A., 2011. Urban heat island effect over national capital region of India: a study using the temperature trends. Journal of Environmental Protection 2 (04), 465.

Nel, A., Xia, T., Mädler, L., Li, N., 2006. Toxic potential of materials at the nanolevel. Science 311 (5761), 622–627.

Nesarikar-Patki, P., Raykar-Alange, P., 2012. Study of influence of land cover on urban heat islands in pune using remote sensing. Journal of Mechanical and Civil Engineering 3, 39–43.

NGS, 2011. Air Pollution. Resource Library, Encyclopedic Entry. National Geographic Society. Retrieved from: https://www.nationalgeographic.org/encyclopedia/air-pollution/.

Oberdörster, G., 2012. Nanotoxicology: in vitro–in vivo dosimetry. Environmental Health Perspectives 120 (1) a13–a13.

Oberdörster, G., Utell, M.J., 2002. Ultrafine particles in the urban air: to the respiratory tract–and beyond? Environmental Health Perspectives 110 (8), A440–A441.

Oberdörster, G., Ferin, J., Lehnert, B.E., 1994. Correlation between particle size, in vivo particle persistence, and lung injury. Environmental Health Perspectives 102 (Suppl. 5), 173–179.

Oke, T.R. Review of urban climatology 1968 -1973, Tech. Note No. 134, WMO No. 383, *World Meteorological Organization*, Geneva, 1974, 132.

Oke, T.R. The urban atmosphere as an environment for air pollution dispersion, WMO Special Environmental Report, 1977.

Oke, T.R., 1982. The energetic basis of the urban heat island. Quarterly Journal of the Royal Meteorological Society 108 (455), 1–24.

Parrish, D.D., Stockwell, W.R., 2015. Urbanization and air pollution: then and now. Eos 96.

Patterson, J.T. The climate of cities: a survey of recent literature. NAPCA Pub. No. AP59. U.S. Govt. Printing Office, Washington. D.C., 1969, 48.

Qiao, Z., Tian, G., Xiao, L., 2013. Diurnal and seasonal impacts of urbanization on the urban thermal environment: a case study of Beijing using MODIS data. ISPRS Journal of Photogrammetry and Remote Sensing 85, 93–101.

Rao, V.L., 2014. Effects of urban heat island on air pollution concentrations. International Journal of Current Microbiology and Applied Sciences 3 (10), 388–400, 0.

Rao, X., Zhong, J., Brook, R.D., Rajagopalan, S., 2018. Effect of particulate matter air pollution on cardiovascular oxidative stress pathways. Antioxidants and Redox Signaling 28 (9), 797–818.

Robine, J.M., Cheung, S.L.K., Le Roy, S., Van Oyen, H., Griffiths, C., Michel, J.P., Herrmann, F.R., 2008. Death toll exceeded 70,000 in Europe during the summer of 2003. Comptes Rendus Biologies 331 (2), 171–178.

Rockström, J., Steffen, W., Noone, K., Persson, Å., Chapin, F.S., Lambin, E.F., et al., 2009. A safe operating space for humanity. Nature 461, 472–475.

Rosenzweig, C., Solecki, W., Hammer, S.A., Mehrotra, S., 2010. Cities lead the way in climate-change action. Nature 467 (7318), 909–911.

Sanchez, L., Reames, T.G., 2019. Cooling Detroit: a socio-spatial analysis of equity in green roofs as an urban heat island mitigation strategy. Urban Forestry and Urban Greening 44, 126331.

Sarrat, C., Lemonsu, A., Masson, V., Guedalia, D., 2006. Impact of urban heat island on regional atmospheric pollution. Atmospheric Environment 40 (10), 1743–1758.

Shahmohamadi, P., Che-Ani, A.I., Etessam, I., Maulud, K.N.A., Tawil, N.M., 2011. Healthy environment: the need to mitigate urban heat island effects on human health. Procedia Engineering 20, 61–70.

Sharma, R., Joshi, P.K., 2014. Identifying seasonal heat islands in urban settings of Delhi (India) using remotely sensed data−An anomaly based approach. Urban Climate 9, 19–34.

Shin, M., Kim, H., Gu, D., Kim, H., 2017. LEED, its efficacy and fallacy in a regional context—an urban heat island case in California. Sustainability 9 (9), 1674.

Simkhovich, B.Z., Kleinman, M.T., Kloner, R.A., 2008. Air pollution and cardiovascular injury: epidemiology, toxicology, and mechanisms. Journal of the American College of Cardiology 52 (9), 719–726.

Singh, P., Kikon, N., Verma, P., 2017. Impact of land use change and urbanization on urban heat island in Lucknow city, Central India. A remote sensing based estimate. Sustainable Cities and Society 32, 100–114.

Singh, N., Mhawish, A., Ghosh, S., Banerjee, T., Mall, R.K., 2019. Attributing mortality from temperature extremes: a time series analysis in Varanasi, India. Science of The Total Environment 665, 453–464.

Taha, H., 1997. Urban climates and heat islands: albedo, evapotranspiration, and anthropogenic heat. Energy and Buildings 25 (2), 99–103.

UNITED NATIONS, 2018. World Urbanization Prospects 2018. Retrieved from: https://population.un.org/wup/.

Urban, E.P.A., 2008. Heat Island Basics (Online) October. Retrieved from: www.epa.gov https://www.epa.gov/heat-islands/heat-island-compendium.

Ürge-Vorsatz, D., Danny Harvey, L.D., Mirasgedis, S., Levine, M.D., 2007. Mitigating CO_2 emissions from energy use in the world's buildings. Building Research and Information 35 (4), 379–398.

Walter, E.J., Carraretto, M., 2016. The neurological and cognitive consequences of hyperthermia. Critical Care 20 (1), 199.

Wang, K., Wang, J., Wang, P., Sparrow, M., Yang, J., Chen, H., 2007. Influences of urbanization on surface characteristics as derived from the Moderate-Resolution Imaging Spectroradiometer: a case study for the Beijing metropolitan area. Journal of Geophysical Research: Atmosphere 112 (D22).

Wang, K., Jiang, S., Wang, J., Zhou, C., Wang, X., Lee, X., 2017. Comparing the diurnal and seasonal variabilities of atmospheric and surface urban heat islands based on the Beijing urban meteorological network. Journal of Geophysical Research: Atmosphere 122 (4), 2131–2154.

WHO, 2016. Air Pollution. World Health Organization. Retrieved from: https://www.who.int/airpollution/en/.

WHO, 2016. World Health Statistics data visualizations dashboard. Retrieved from: https://apps.who.int/gho/data/node.sdg.3-9-viz-1?lang=en (on 7th May 2020).

WHO, 2019. Ambient Air Pollution: Pollutants. Air Pollution. World Health Organization. Retrieved from: https://www.who.int/airpollution/ambient/pollutants/en/.

Xia, T., Li, N., Nel, A.E., 2009. Potential health impact of nanoparticles. Annual Review of Public Health 30, 137–150.

Yang, Q., Huang, X., Li, J., 2017. Assessing the relationship between surface urban heat islands and landscape patterns across climatic zones in China. Scientific Reports 7 (1), 9337.

Yu, M., Carmichael, G.R., Zhu, T., Cheng, Y., 2012. Sensitivity of predicted pollutant levels to urbanization in China. Atmospheric Environment 60, 544–554.

Yue, W., Liu, X., Zhou, Y., Liu, Y., 2019. Impacts of urban configuration on urban heat island: an empirical study in China mega-cities. Science of The Total Environment 671, 1036–1046.

Zhang, J., Li, Y., Tao, W., Liu, J., Levinson, R., Mohegh, A., Ban-Weiss, G., 2019. Investigating the urban air quality effects of cool walls and cool roofs in Southern California. Environmental Science and Technology 53 (13), 7532–7542.

Zhao, Z.Q., He, B.J., Li, L.G., Wang, H.B., Darko, A., 2017. Profile and concentric zonal analysis of relationships between land use/land cover and land surface temperature: case study of Shenyang, China. Energy and Buildings 155, 282–295.

Zhou, D., Zhao, S., Zhang, L., Sun, G., Liu, Y., 2015. The footprint of urban heat island effect in China. Scientific Reports 5, 11160.

Cities management and sustainable development: monitoring and assessment approach

André C.S. Batalhão[1], Denilson Teixeira[2]

[1]Environmental Sciences, Center for Environmental and Sustainability Research – CENSE/
Nova Lisbon University, Caparica, Portugal; [2]Environmental Engineering, Federal University of
Goiás, Goiânia, Brazil

1. Introduction and objectives

Research initiatives and new development paradigms for cities have adopted sustainability as their background, conducting a critical analysis about the lack of the participation of different stakeholders in this territory. In this context, the term 'sustainable' has appeared with remarkable regularity.

While some sectors of society, including local public managers, have questioned the reasons behind this widespread popularity (Verma and Raghubanshi, 2018; Tanguay et al., 2010; Mori and Christodoulou, 2012), we cannot doubt that sustainable development (SD) is now a fixed and often mandatory issue on city governments' agendas. Although the sustainability of cities has been widely debated in the past three decades, it is still a poorly agreed-upon concept, empty and devoid of a complete and integral definition.

The SD has a broad and nonconsensual definition, giving rise to different interpretations (Tanguay et al., 2010). Despite the lack of a consensus definition of SD, as well as interpretations and methods (Pope et al., 2017), the term sustainability assessment has been used to refer to ex-prospective and prospective ex-ante approaches (Pope et al., 2004), characterizing the level of sustainability, covering all SD dimensions. SD provides a general direction for assessing and monitoring urban structures, not forgetting policies (Pupphachai and Zuidema, 2017). Urban sustainability includes different topics such as biodiversity, energy, material balance, air pollution, heat island, noise pollution and others (Verma and Raghubanshi, 2018). Measuring sustainability for politically defined urban areas may be irrelevant, considering that urban areas extend beyond their political or administrative boundaries (Fiala, 2008). However, political and administrative boundaries provide a way for the delivery of sustainable policies to the general society (Verma and Raghubanshi, 2018).

There are many sustainability assessment initiatives and systems for cities, establishing sets of indicators that help put new ways of planning cities into practice. Indeed, the use of indicators as a means of measuring local sustainability performance has become popular, with governments at different levels and funding agencies directing substantial resources to the development of new indicators. Therefore, the monitoring of this development is essential through sustainability indicators (SIs) that report the efficiency of the implemented public policies and actions that support the incorporation of sustainability in local public management.

Around the world the idea of a sustainable city has massively enhanced, even if rhetorically. These initiatives have generated rankings where those cities considered most sustainable appear top-ranked, and indicators play an important role in this assessment process. As current example, we can cite the Arcadis Sustainable Cities Index (SCI) (Arcadis, 2018) that uses the people, profit and planet (PPP) approach to build a ranking according to the sustainability level of cities around the world. The SCI has been exploring the dimensions of city sustainability, developing an understanding of the underlying issues that enable some to outperform their peers. An earlier effort was the study conducted by Becker et al. (1987), where different cities were ranked depending on a subjective weighting scheme. Notably, there are different priorities for different issues. It is important to note that the conceptual meaning of being 'sustainable' varies depending on where and who is using it, a divergent point between nations and international organizations.

In this chapter, we have given an overview of how sustainability can be managed and measured in a city context. We have focused on the main pillars of sustainability, discussing the main themes of the indicators within each pillar. Then, we have provided some guidance on how to build an assessment system and choose appropriate indicators.

2. Sustainability management in cities

The essence of SD is driven by the concept of sustainability to meet fundamental human needs while preserving the critical life support systems of our entire planet (Mörtberg et al., 2013; Kates et al., 2001; Mori and Christodoulou, 2012). In general, it is defined as a type of development that must be balanced in all its dimensions for a continuous and long period (Sartori et al., 2014; Olawumi and Chan, 2018). In debates about SD, the role of cities and the urbanization process is critically analyzed to manage political interests, natural resources, social actions and economic capital, considering the transversality of the institutional role (Phillis et al., 2017; Verma and Raghubanshi, 2018). Thus, cities have a key role to play in accelerating sustainability, where a total of two-thirds of the world's population will be living in cities by 2050 (UN, 2019).

Urbanization is one of the most significant social processes, playing a key role at the local and global scales (Hiremath et al., 2013). In the past two decades, there have been significant changes influencing contemporary society, including social reconfiguration, territorial planning, innovations in economic mechanisms and political action (Dijst et al., 2018; Zhang et al., 2018). Different factors contribute to such changes, in which specific dimensions of a broader and more general phenomenon are identified, starting from a smaller territorial range, such as cities, to the impacts felt at the national level.

Disoriented urban growth has caused different socioeconomic and environmental crises in cities across the world, with temporary and permanent consequences (Yigitcanlar et al., 2019; Kamruzzaman et al., 2018). Some examples, as cited by Huang et al. (2018), include loss of biodiversity, widening socioeconomic disparities, lack of waste management, decreased levels of knowledge and culture of the population and slow response of public administration to citizen needs. In addition, a large number of cities around the world are experiencing these consequences directly. These can create even greater persistent difficulties and complex challenges for city management (Mapar et al., 2017).

Megacities are an urban or metropolitan area with a total population of over 10 million (Westfall and Villa, 2001). Cities currently house more than half of the world's population (Dempsey et al., 2012; ISO 37120, 2018), whereas 34 megacities (Demographia World Urban Areas, 2015) have a population of approximately 580 million. Increasing levels of urbanization and population agglomeration in megacities (Huang et al., 2016; Mapar et al., 2017) have detrimental consequences for citizens associated with sustainability issues (Singh et al., 2012). This requires the adoption of sustainability assessment processes, continuous monitoring and generation of urban sustainability data and information (Zhou et al., 2015).

To harness the potential of SD at the city level, we must be aimed at understanding how factors of social, economic, environmental and institutional dimensions are managed at the local scale (Spangenberg et al., 2002). From this perspective, cities should be understood as human communities in which these dynamic factors interact and generate conclusive elements about the mutual interdependence between the dimensions of the SD. Thinking of cities as urban spaces, with well-defined physical spaces, and a collection of inhabitants, sustainability issues and concerns are deep and increasingly in need of investigation to identify the specifics of the territory (Shen and Guo, 2014). The subjectivities of each city should be included in the public government management process, providing good medium- and long-term planning and transparent management linked to citizens' needs.

The sustainability of the city has been explored by several approaches. Examples of global networks include the International Council for Local Environmental Initiatives (ICLEI) and the C40 Cities (Climate Leadership Group), which has focused on measuring sustainability at the local level and provides guidance on effective ways to achieve local goals, focusing on regional, national and global sustainability goals. ICLEI, also called Local Governments for Sustainability, is a global network of more than 1750 governments in cities, towns and regions committed to progressing towards sustainable urban development in over 124 countries, covering more than 25% of the global urban population (ICLEI, 2019). This network aims to support local governments in the operationalization of sustainability principles and goals (Bhagavatula et al., 2013) and to influence local actions and public policies, based on nature, concern with resilient bases and circular development (ICLEI, 2019; Jamil et al., 2015). The C40, on the other hand, consists of a global network of 94 megacities, representing more than 700 million citizens, where the economies of these cities represent 25% of global gross domestic product. City mayors are committed to complying with the Paris Agreement at the local level and collaborating on initiatives that share knowledge and support meaningful and measurable action on climate risk and change, enhancing the well-being of urban citizens (C40 Cities, 2019). Initiatives such as the UN Habitats Urban indicator program, applied in different versions by more than 200 cities (UN, 2019), and the European Common Indicators, tested in more than 42 cities (European Union, 2019), are also examples of efforts to building a more mature and complete management and assessment structure.

Sustainable Development Goal (SDG) 11, one of the 17 SDGs and 169 targets of the post-2015 2030 Agenda (UN, 2015), maintains its focus on sustainable cities and communities, as well as the importance of integrating the elements that influence sustainability into a city. Therefore, the specific context of the sustainable city should emphasize the interdependence between city structure and the level of citizenship found in these spaces (UN, 2015; Michael et al., 2014). Also in this context, ISO 37120 indicators, known as Sustainable Development of Communities, has been adopted to assess the sustainability of cities and communities. This standard defines and establishes methods for using a set of indicators to guide and measure urban service performance and quality of life. This standard follows the principles set out in ISO 37101 and can be used in conjunction with other standards from the same perspective. This document may be applied to any city, county or local government that is committed to measuring its performance in a comparable and verifiable manner, regardless of size and location (International Standards Organization, 2019). The different approaches mentioned show that city administration was part of a sustainability initiative movement collaborating with other public and private actors and public authorities at different territorial levels. Therefore, it is a fact that local public administrations are becoming widely recognized as pioneers and sustainability leaders compared with other levels of government (Domingues et al., 2015).

Due to the goal of achieving and sustaining SD, including adding performance from socioeconomic, institutional and environmental levels, policymakers, researchers and governments of different territorial levels should seek solutions to improve the environment of urban areas, with the prospect of progress of welfare of the human population, not forgetting to maintain the natural resources for the maintenance of life in this space.

The development of sustainable cities is considered a contemporary problem of local public management, revealing an issue that is not being clearly measured, and presents several challenges for improving and revising existing initiatives. The process of measuring, monitoring and quantifying sustainability has limitations and has clearly shown that it is difficult to measure all aspects and elements of SD. There is a trade-off between the complexities of cities and the simplifications needed to make SIs more reliable and meaningful. Many local governments have endeavoured to develop complete evaluation systems taking into account the dimensions of SD (Holman, 2009; Mascarenhas et al., 2010); however, many submit incomplete evaluation proposals in this regard (Singh et al., 2009). Current research and initiatives show that the sustainability assessment process helps identify weaknesses and threats that need to be improved or optimized by facilitating sectoral corrective actions (Ramos, 2019).

SIs, based on well-defined conceptual bases and robust methodological procedures, can be an important instrument to guide human actions and decisions towards the direction of the SD. For sustainability management in cities, we have found in the literature indicators that emphasize environmental elements over other dimensions of SD (Little et al., 2016). This may cause an imbalance in the actual SD proposal. However, choosing which dimension to evaluate is the task of municipal or other territorial decision-makers that influence local management and planning. Besides, it is necessary to invest in strategies to improve the quality of decision-making and public communication of results, focusing on improving the quality of life for citizens.

Along these lines, for good local management, the development of an urban sustainability assessment system is critical for assessing past and current decisions, public policies, state of existing local infrastructure, local economic aspects, monitoring of resource use and any other elements that contribute to the development of the city. Also, it is noteworthy that by analyzing the sustainability of a city, it is possible to identify the main driving forces and specificities that should promote the participation of local government organizations to improve the habitability and democracy in this space.

The application of methods that identify sustainable or nonsustainable cities should be aware that often local specificities and particularities are not included in the evaluation process, generating a short-sighted indicator that does not capture the real needs and bottlenecks for public administration. Thus, the methods should contain quantitative and qualitative approaches, acting in a complementary manner, deepening the sustainability assessment of cities.

Selecting the indicators for a city sustainability management tool is a crucial step, as the conclusions of the outcome analysis depend on this step. The selected indicators have the role of demonstrating how human activities have influenced, influence and can influence urban dynamics, notably by pointing out alternative scenarios and routes for local planning, reducing uncertain but not eliminating them. First and foremost, it is necessary to identify the political interests of the city government to assist in choosing the types of data and information needed to build endogenous databases as a starting point.

The prosperity of cities, regardless of their geographical location, must be in line with the integration of their economic goals, environmental objectives, social concerns and institutional positioning. Based on this vision, many cities around the world have developed new initiatives and guidelines to incorporate sustainability into their local planning and urban development processes.

The literature shows that most research focuses on aspects of sustainable city development and planning, while governance issues seem to be overlooked (Zhu and Simarmata, 2015). Planning for the development of these cities should include decentralizing decision-making in their management, involving more local government organizations and different stakeholders so that urban sustainability is achieved by cities around the world. For this reason, a robust and continuous connection must be created between urban planning of cities and the elements of SD, where the two are born together and inseparable to form an integrated mechanism to achieve sustainable results. It is also necessary to promote efforts to overcome the different challenges in the elaboration, implementation and routine establishment of a system of indicators capable of accompanying and indicating the pathways to urban sustainability.

3. Indicators to monitoring the sustainable development

Sustainability, currently an aspiration of global scope, takes concrete features at all territorial levels from its peculiarities and responds to the problems and opportunities of each space (Mascarenhas et al., 2015). The choice of SD indicators reflects the situations and specificities of each country while pointing to the need for regular production of statistics on the topics addressed in subnational spaces (Herrera-Ulloa et al., 2003).

Indicators are symbolic representations developed to communicate the ownership or tendency of a complex entity or system (Moldan and Dahl, 2007). According to Ramos and Caeiro (2010), the development of SI, when analyzed as a system where different processes occur, includes a series of actions and decisions with various data and information flows. SI approaches should define several key components to ensure a coherent development process. The same authors categorize these components as (1) planning and conceptualization, including all design components and processes; (2) implementation, including the entire process of data collection, processing and analysis; (3) operation and action, where results are presented through communication tools, leading to different types of reactions, such as policy measures; closer relationship between institutions at local, regional, national and international levels and with strategic tools/instruments (policies, plans or programs) and (4) follow-up of results, oriented to update and revision, mainly based on a metaperformance evaluation process.

In the SI usability sphere, Bell and Morse (2003, 2018a,b) established certain required criteria for these metrics:

- be relevant to an issue according to the conceptual definition employed;
- be available from the public, scientific or institutional sources;
- cover the whole field of sustainability according to the definition used.

To support the use of indicators, several international authors and organizations have provided many arguments for finding ways to standardize indicators and frameworks for comparing sustainability levels (Ramos and Caeiro, 2010; Tanguay et al., 2010; Mascarenhas et al., 2010; Yigitcanlar and Lönnqvist, 2013; Mapar et al., 2017). Importantly, the vast majority of SI efforts are still deductive, top-down approaches. This is because consensus-based definitions of sustainability have been established, leaving these definitions to guide the selection of indicators. Few initiatives use an inductive or bottom-up approach, where significant indicators emerge from the analysis as powerful statistical predictors (Parris and Kates, 2003).

The main advantage of SI is the apprehension of domains involving analysis of economic, social, environmental and institutional processes, expanding the development horizon, providing information support and guiding legal arrangements, government and private institutions and public policies at all territorial levels. Bell and Morse (2012) argued that they are not convinced that narrow approaches to sustainability can work without reducing complexity and without excluding valid and legitimate worldviews. It may reduce the area of concern for an issue that may not represent yet another latent sustainability concerning.

The construction, interpretation and monitoring of indicators are processes that compose daily practices and theoretical activities, from an extensive set of different scientific fields, which relate to socioeconomic and environmental development (Pissourios, 2013). An indicator project is characterized by a space of conscious exploration and should be seen as a rationalizing project, integrating divergent agendas needs with common sense (Bell and Morse, 2018a,b). An indicator project should consider its potential use in public policymaking, where specific communication methods for resource allocation needed for well-being and sustainability need to be developed as indexes and indicator sets are constantly challenged by changes in social, natural conditions and scientific discoveries, which open new questions and policy changes (Rinne et al., 2013).

Planning and management for cities are essential and strategic instruments for the effectiveness of public policies and actions aimed at the search for SD. Government planning, at all levels, precedes the management process, and in the space of construction in the present reality, of the demands and needs of society, for the delimitation of what can be implemented in a future reality. Ingestam (1987) defines planning as a way for society to exercise its power over the future, rejecting resignation and turning to initiatives that can better define its future.

Management aims to combine actions and resources to achieve the desired objectives, operationalizing what was developed in the planning process. Generally speaking, management acts on the means (ways to achieve results) and objectives. Control from performance appraisal systems is required as strategic support (Alecian and Foucher, 2001; Vasu et al., 1998) to monitor what has been planned and what is being executed. City managers and societal organizations increasingly require consistent and objective information about the themes that cover the DS regarding their territory. In this context, the use of sustainability assessment tools, which are supplied by indicators, strengthens the public policy archetype by providing mechanisms to understand the identity of the system under consideration (Hoko and Hertle, 2006). It is worth emphasizing that each spatial cut has its subjectivities and requires a set of indicators consistent with the needs and objectives to be evaluated. Indicators are not only desirable for policy purposes, but they also help to achieve and adjust those policies. The development of indicators cannot be purely a technical or scientific process (Valentin and Spangerberg, 2000), and the actual activities carried out and the limitations on the balance of human and ecological well-being must be transparent (Malovics et al., 2009).

Fig. 18.1 shows some assessment methods and the dimensions of the SD where each approach relies upon, with the aim of contributing to sustainability. Each assessment method has unique characteristics and may be appropriate for a particular type of situation. On the other hand, it is important to understand that each one must be in conformity with its

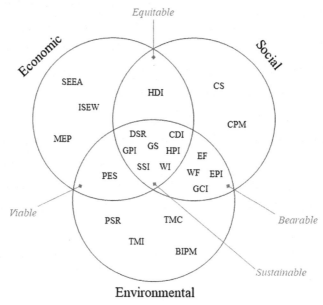

FIGURE 18.1 Diagram of sustainability indicators of the three dimensions of sustainability based on Venn approach (Brusseau et al., 2019).

precepts, theoretical and empirical orientations about the formulation of an indicator system. Demonstrating some initiatives, it is clear that indicator systems abound, which were developed for different purposes and from different conceptual frameworks.

4. Acronyms/publication year/developer or source

BIPM (*Bureau International des Poids et Mesures*)/2005/International System of Measures
CDI (City Development Index)/1997/United Nations-Habitat
CPM (Capability Poverty Measure)/1997/McKinley
CS (Compass Index of Sustainability)/2001/Atkisson and Hatcher
DSR (Driving-Force, State, Response)/1995/United Nations—Center of Sustainable Development
EF (Ecological Footprint)/1992/Wackernagel and Rees
EPI (Environmental Performance Index)/2006/Yale University and Columbia University
GCI (Green City Index)/2009/Economic Intelligence Unit and Siemens
GPI (Genuine Progress Indicator)/1994/Redefining Progress
GS (Genuine Savings)/1999/World Bank
HDI (Human Development Index)/1990/United Nations Development Programme
HPI (Happy Planet Index)/2006/New Economics Foundation
ISEW (Index of Sustainable Economic Welfare)/1989/Daly and Coob
MEP (Monitoring Environmental Progress)/1995/World Bank
PES (Payment for Environmental Services)/2005/Wunder

PSR (Pressure, State, Response)/1993/Organization for Economic Cooperation and Development
SEEA (System of Integrating Environmental and Economic Accounting)/1995/United Nations Statistical Division
SSI (Sustainable Society Index)/2006/Sustainable Society Foundation
TMC (Total Material Consumption)/1995/Weizsäcker et al.
TMI (Total Material Input)/1995/Weizsäcker et al.
WF (Water Footprint)/2002/Hoekstra and UNESCO-Institute for Water Education
WI (Wellbeing Index)/2001/IUCN and International Development Research Centre

The above initiatives reveal that there are different indicator systems acting on specific dimensions of sustainability contributing to the measurement of development. We can approach issues relating to society and the environment by discussing the concepts of strong and weak sustainability. Both are based on the fact that humanity must preserve capital for future generations. Natural capital is constituted by the base of renewable and nonrenewable natural resources, biodiversity and the absorption capacity of waste from ecosystems. Within the concept of strong sustainability, all levels of resources must be maintained and not reduced, and under the concept of weak sustainability, the exchange between different types of capital is allowed, as long as its stock is kept constant (Turner et al., 1993).

The management model and institutional cooperation is a fundamental component of SI, identifying the institutions and their roles and the leadership structures. It is essential to an understanding of the feasibility and societal influence of the indicators system (Ramos and Caeiro, 2010). The most complete definition of SD using forward-looking ex ante approaches, aiming to assess the sustainability performance (e.g., SDGs by UN, 2015) has four dimensions: environmental, economic, social and institutional, reflecting in four categories of indicators. In this approach, institutions not only are understood as organizations but also represent the system of rules guiding the interaction of members of society. The segregation of one of these four dimensions is highly detrimental and tends to hide connections with reality, making a global assessment difficult. While all four dimensions represent established interests, the constitutive integration of sustainability policy should take place as a direct connecting activity, to facilitate monitoring of the system in question (Spangenberg et al., 2002) (Fig. 18.2).

In this context, processes initiated by different organizations characterize institutions as an executing and dynamic element for SD. Institutional mechanisms work across the SD dimensions across the board, making city sustainability feasible. Therefore, city institutions should act as rule systems to cover the actions of individuals or different stakeholders within society. In this sense, the institutional dimension plays a key role in the management and assessment of sustainability in cities, where it also has the guiding role in helping to structure policy choices to comply with rules implicit in the SD process.

It is important to highlight that the institutional dimension must bring together all groups of society (stakeholders), as well as provide information on the administrative functioning and legal order of cities. From the standpoint of SI, a high degree of participation is largely desirable, without superparticipation by some groups in relation to others, without any social, gender or ethnic discrimination (Domingues et al., 2018; Kely and Moles, 2002; Valentin and Spangenberg, 2000; Dahl, 2012).

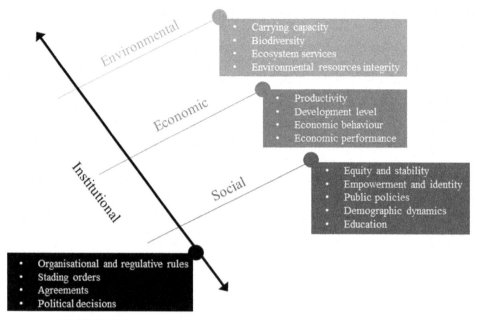

FIGURE 18.2   Dimensions to assess and enhance the sustainability of cities.

For the future research agenda with SI, Bell and Morse (2012, 2018a,b) argue that an essential element for future work on this topic should be reflective practice. These same authors highlight the apparent absence of explicit learning about past problems and errors with SI. In their opinion, it seems that many authors cite those who wish to agree and ignore the rest and that a frightening element is an extreme speed with which this literature is evolving and developing. However, this evolution occurs due to minimal self-reflection by the authors. There is often a mismatch between the expectation of the scientist's society and the reality of what the scientist feels and knows. In addition, Bell and Morse (2012) recognized three vulnerabilities found in SI work: learning is continuous; new contexts need to be tried and the object of our study is part of us (involvement).

According to Ramos (2019), in the context of the current critical point of view, a set of key questions about the future of SI is needed to represent an integrated view of the relevant challenges and opportunities.

5. Findings and discussion

5.1 Formulating indicators assessment based on sustainable development dimensions

The biggest difficulty in assessing the sustainability of cities is the challenge of exploring and analyzing a holistic system, as noted by Gallopin (2018). A holistic system requires the integration and evolution between complex and interdependent dimensions, as well as understanding the interaction and dynamics between them. This interaction can increase the complexity of the roles of each dimension for SD. It is noteworthy that all these dimensions are included in the evaluation process, while central to many but not consensual approaches to sustainability, introducing more complexity in the selection and interpretation of indicators, as well as their conclusions.

Efforts should be directed at capturing the complexity of each dimension, identifying key themes and metrics to reach the full extent of the assessment. A key point in city sustainability management is the degree of reliance on sustainability assessment on the necessary themes and indicators in the underlying context: Cities are local decision-making units, and there is great heterogeneity around cities, even within different individual contexts and current political models of cities. Fig. 18.3 provides a visual reading of a multidimensional assessment analytical map for city management.

Practical approaches and SI research need to be flexible and transdisciplinary to include emerging issues and address aspects that have been overlooked in previous initiatives. SI- and SD-related tools have long been used to assess sustainability, and during this period, much progress has been made. However, it has become necessary to begin to rethink its functions and applicability, as stressed by Ramos (2019), including in the local context.

The literature on sustainability assessment cities shows that sustainability dimensions have a common framework for selecting SIs; however, the SD dimension set has not been uniformly applied to all dimensions, generating an incomplete conceptualization of the SD (see, for example, the work of Cohen, 2017). It has become clear that adopting more integrative and principled assessment frameworks is crucial not to neglect key elements. In the

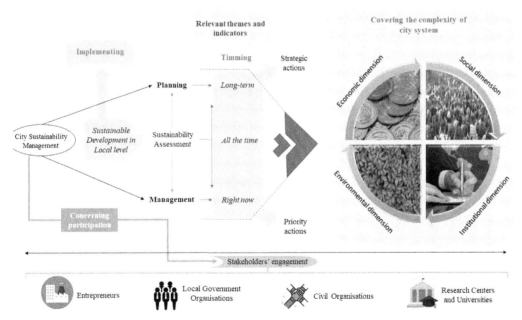

FIGURE 18.3   Tiered structure with basic orients to the SD implementing in the city level. *SD*, sustainable development.

following, we will present the dimensions of the DS, as well as suggested themes and common indicators that have been used recurrently in the international literature for the elaboration of a city-level SD assessment.

5.2 Environmental dimension

The environmental perspective for sustainability is the concern with the impacts of human activities on the environment, degrading with natural capital. From an ecological perspective, the concern is to expand the capacity to maintain ecosystem services and their potential for use. In this line, the environmental dimension is concerned with the issues of conservation and preservation of the environment, as well as environmental services.

More generally, three groups of environmental themes can be used: (1) goal-oriented themes to be achieved in the evaluation system (Bockstaller et al., 2007); (2) themes related to local impacts, with consequences expressed through indicators (Halberg et al., 2005) and (3) themes that cover the ultimate objective of evaluation (Alkan Olsson et al., 2009).

These issues come up in topics such as atmosphere, water, soil, coastal areas, biodiversity, sanitation, waste, energy, recycling, mining, agriculture, fishing and degradation.

Suggested indicators for environmental management in cities:

- Anthropic emissions of greenhouse gases
- Industrial and agricultural consumption of ozone-depleting substances
- Concentration of air pollutants in urban areas
- Irregular burns in the urban perimeter
- Deforestation of environmental protection areas within urban space
- Inland and surface water quality
- Coastal area bathing (oceans and seas)
- Energy consumption by alternative energy matrix (wind, solar, etc.)
- Energy consumption
- Extinct and endangered animal and plant species
- Environmental areas protected by law (land and sea)
- Invasive plant species in urban space
- Access to water supply system
- Collection and treatment of sewage
- Household waste collection service
- Garbage recycling fee
- Radioactive waste

5.3 Economic dimension

As related by Arrow et al. (2012), the economic dimension must address the distribution and allocation of economic resources, considering natural resource stocks and capital flows on an appropriate scale. This dimension is not restricted to finance capital alone but is open to environmental and social aspects, as found by Bossel (2007). Micro- and macroeconomic policies should guide the development process, with measurement models that warn the economic direction towards the SD. Commonly required themes for this dimension are degree of public debt, public investment, gross domestic product, poverty, employment, inflation, access to products and services and trade balance.

Suggested indicators for managing the economic dimension in cities:

- Gross domestic product per capita
- Investment rate
- Trade balance
- Degree of indebtedness
- Mineral consumption per capita
- Inflation
- Unemployed population
- Poverty level population

Although economic dimension measurement commonly uses most of these economic indicators, a broader range of indicators can be inserted to capture other economic properties of cities and improve the analysis and conclusion of the local sustainability assessment.

5.4 Social dimension

The social dimension corresponds to social justice, the satisfaction of human needs, improvement of quality of life and the presence of the citizen in the ecosphere. The social condition of the individual and the means used to increase their quality of life must be built with equity and responsibility. In the social dimension, there is great importance at the level of social relationships. This is related to society's demands, depending on its values and concerns, as emphasized by Lebacq et al. (2013). Clearly, the objective should be to identify current differences between the various levels of society in the territory of cities and to develop sectoral actions to improve the living conditions of local citizens. The different aspects and elements of this dimension can be organized by the themes such as physical and mental health, mortality, fertility, population size, education and knowledge, communication, beliefs and values, rights and freedoms, civil planning, gender and ethnic group equity.

Suggested indicators for social dimension management in cities:

- Population growth rate
- Total fertility rate
- Gini index of income distribution
- Per capita household income
- Average monthly income
- Women in formal work
- Life expectancy at birth
- Child mortality rate
- Prevalence of child malnutrition
- Immunization against childhood infectious diseases
- Supply of basic health services
- Diseases related to inadequate environmental sanitation
- AIDS incidence rate
- School attendance rate

- Literacy rate
- Adult education rate
- Adequate housing
- Homicide mortality coefficient
- Transport and vehicles' accident mortality coefficient

5.5 Institutional dimension

The institutional dimension, as shown by Ramos and Caeiro (2010) and Ramos (2009), is a fundamental and transversal element that operates within which sustainability can be developed. This dimension ensures that the other dimensions are implemented, making SD at the city level effects, such as those proposed by Spangenberg et al. (2002). The ability of a city to moving towards SD is determined by the ability of institutions to develop institutional arrangements with participatory mechanisms, ensuring the inclusion of multistakeholders and facilitating policy articulations and agreements.

Institutional themes do not yet have an extensive statistical production timeline, resulting in fewer data available for the construction of indicators for temporal and comparative analysis. For this reason, there are many gaps, including the engagement of local society stakeholders in the formulation and implementation of public policies based on social and environmental responsibility and eco-efficiency mechanisms. Some topics may be suggested for this dimension, such as governance, research and development; signing of international agreements; program participation, and sustainability projects, citizen participation mechanism, public hearings, local councils, participation in river basin committees.

Suggested indicators for managing the institutional dimension in cities:

- Ratification of global agreements
- Consolidated environmental legislation
- Municipal councils of environment
- Participation in river basin committees
- Civil society organizations
- Research and development expenses
- Municipal environment fund
- Interinstitutional articulations of the municipalities

For sustainability in cities to exist, there must be a reference to some institution to legitimize and facilitate the process of developing and implementing indicators, because institutions can stimulate the sensitivity of local politicians and decision-makers. Sustainable city programs depend on the engagement of well-defined organizations that encourage stakeholders to participate by pointing out clear motivations. The greater the number of institutions involved, the greater the usability of the evaluation system and indicators. By focusing on attitude and institutional planning, it is possible to develop and select clearer and more insightful indicators.

6. Approach to assessing SD at the local level

With the increasing adherence of SD targets in cities, there is a growing need to use metrics to measure issues that identify human and natural contexts and therefore the interface of different dimensions. To measure the sustainability of cities, it is necessary to follow a conceptual approach, containing development and selection criteria to identify the most appropriate indicators. Therefore, it is important to establish indicators that inform common demands and problems in cities, strengthening the diversity of issues and difficulties within the territory, as explored by Verma and Raghubanshi (2018).

The process of building indicator systems may involve different methodological resources, consisting of several steps. Fig. 18.4 demonstrates an approach with seven preestablished steps for building and implementing an applicable sustainability assessment and monitoring proposal for cities.

Each step must meet the assessment aim and be aligned with the aspirations of the proposal. The following is a brief description of each step, highlighting the key targets to move on to the next steps, to be used as a checklist method:

Step1: Strategy formulation and assessment purpose
- The objectives of the assessment should be clear and politically relevant
- To identify the need for city-level assessment
- Take clear the priority themes for city management and planning
- To direct to who will use the outcomes
Step 2: Scope definition and assessment guide
- Based on the intended objectives, identifying the concerns should be considered to get a proper sense on the subject
- To highlight the objectives to support the dimensions of sustainability
Step 3: Choice of analysis dimensions and indicators
- To select the dimensions and criteria that will be included
- To investigate if any dimension has been neglected
- To strengthen any components to ensure a good assessment

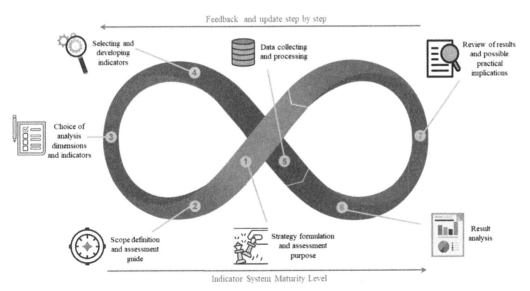

FIGURE 18.4   Approach to sustainability assessment at the local level.

Step 4: Selecting and developing indicators
- To stress the use of indicators representative, reliable, viable and measurable
- To motivate the use of the standardized indicators with validated methodological procedures
- To include quantitative and qualitative indicators

Step 5: Data collecting and processing
- To collect data properly
- To collect data enough to proceed
- To identify temporary interruption in local databases
- To treat the recorded data

Step 6: Result analysis
- To identify the strengths and weaknesses of the assessment
- To identify the dimensions that have the greatest influence on the overall result
- To realize the indicators that have the greatest influence on the overall indicator network

Step 7: Review of results and possible practical implications
- To realize the indicators that are priorities
- To point out the indicators that can be replaced
- To highlight the opportunities that you realized
- To indicate the strengths, weaknesses and challenges to be overcome
- To react face on the main practical and theoretical implications

The combination of sensitivity analysis methods and bivariate and multivariate statistical analyses can help to evaluate the robustness level of the assessment system, as well as the built composite indicator, increasing the transparency of the application models and facilitating propositions and discussions around the aim of the assessment proposal.

The growing critical need for a coupled systems approach to understanding causal relationships encompassing the environmental, economic, social and institutional dimensions can be vividly illustrated with countless examples of unintended conclusions from well-meaning sectoral policies. Approaches to assessing sustainability in cities, such as SI, management tools and rating systems, are simple and easy to use but may not capture the causal relationships necessary to understand the overall behaviour of a complex system.

While today's set of sustainability assessment initiatives approach cities to provide the data and information needed to manage and plan a city, including temporal and spatial data, it is difficult to envision possible long-term trends so that corrective action can be taken immediately. City management will need to identify and define local priorities, as well as maintain and develop new institutional structures to improve multidimensional integration at the system level. Thus, we recognize that this represents a practical challenge that can take decades and that much can be gained by working towards a common-based sustainability assessment framework while preserving local subjectivities.

7. Final remarks

The field of sustainability management is fragmented, with a wide range of city managers and researcher groups pursuing similar goals, but without effective interdisciplinary coordination. New governmental structures are needed to improve local-level integration in the sustainability dimensions, including the guidance of public organizations to provide practical action.

This chapter has undertaken to identify the issue of the sustainability management of cities. It furthermore provides inductive exploratory research with an analytical foundation for an improved system by suggesting an appropriate definition of city management and a development procedure for SI. Using this foundation unveils a need for local linked to targets and sustainable objectives, which can serve as the main basis for the development performance indicators for the dimensions of SD. On a reflexive basis for city management as suggested by this chapter, many competing indicator systems could be developed, tested and validated for their practical usefulness.

The assessment approach needs a consistent conceptual definition of SD as well as a new organizational structure. It represents a daunting challenge that may take many years, and the goal of achieving sustainability may be completed over several decades. Much could be gained by working toward a common assessment approach, respecting the subjectivities. At the same time, different approaches can influence the creation and selection of indicators for city management. Concerning sustainability, the indicators may be derived from objective data, but the selection of the data used is subjective. This can directly influence the conclusions and can generate divergent discourses.

References

Alecian, S., Foucher, D., 2001. Guia de gerenciamento no setor público. Tradução de Márcia Cavalcanti. ENAP, Rio de Janeiro: Revan, Brasília.

Alkan Olsson, J., Bockstaller, C., Turpin, N., Therond, O., Bezlepkina, I., Knapen, R., 2009. Indicator Framework, Indicators, and Up-Scaling Methods Implemented in the Final Version of SEAMLESS-IF. Deliverable 2.13 of the EU FP6 Project SEAMLESS.

Arcadis, 2018. Citizen Centric Cities: The Sustainable Cities Index 2018. Available at: https://www.arcadis.com/media/1/D/5/%7B1D5AE7E2-A348-4B6E-B1D7-6D94FA7D7567%7DSustainable_Cities_Index_2018_Arcadis.pdf.

Arrow, K.J., Dasgupta, P., Goulder, L.H., Mumford, K.J., Oleson, K., 2012. Sustainability and the measurement of wealth. Environment and Development Economics 17 (03), 317–353. https://doi.org/10.1017/S1355770X12000137.

Becker, R.A., Denby, L., McGill, R., Wilks, A.R., 1987. Analysis of data from the Places Rated Almanac. The American Statistician 41 (3), 169–186. https://doi.org/10.1080/00031305.1987.10475474.

Bell, S., Morse, S., 2003. Measuring the Sustainability: Learning by Doing. EarthScan, London.

Bell, S., Morse, S., 2012. Sustainability Indicators: Measuring the Immeasurable?, second ed. EarthScan, London.

Bell, S., Morse, S., 2018a. What next? In: Bell, S., Morse, S. (Eds.), S. Routledge Handbook of Sustainability Indicators. Routledge, Taylor & Francis Group, London, UK, and New York, USA, pp. 522–543.

Bell, S., Morse, S., 2018b. Sustainability indicators past and present: what next? Sustainability 10, 1688. https://doi.org/10.3390/su10051688.

Bhagavatula, L., Garzillo, C., Simpson, R., 2013. Bridging the gap between science and practice: an ICLEI perspective. Journal of Cleaner Production 50, 205–211. https://doi.org/10.1016/j.jclepro.2012.11.024.

Boockstaller, C., Bellon, S., Brouwer, F., Géniaux, G., Girardin, P., Pinto Correia, T., Stapleton, L.M., Alka-Olsson, J., 2007. Developing an indicator framework to assess sustainability of farming systems. In: Donatelli, M., Hatfield, J., Rizzoli, A. (Eds.), Farming System Design 2007: An International Symposium on Methodologies for Integrated Analysis of Farm Production Systems, Catania, Italy, 10–12 September 2007, pp. 137–138 book 2.

Bossel, H., 2007. Systems and Models: Complexity, Dynamics, Evolution, Sustainability. Books on Demand, Norderstedt, Germany.

Brusseau, M.L., Gerba, C.P., Pepper, I.L., 2019. Environmental and Pollution Science. Elsevier, Academic Press, USA. https://doi.org/10.1016/B978-0-12-814719-1.09988-2.

C40 Cities, 2019. About C40 Cities. Available at: https://www.c40.org/about (Accessed 13 August 2019).

Cohen, M., 2017. A systematic review of urban sustainability assessment literature. Sustainability 9, 2048. https://doi.org/10.3390/su9112048.

Dahl, A.L., 2012. Achievements and gaps in indicators for sustainability. Ecological Indicators 17, 14–19. https://doi.org/10.1016/j.ecolind.2011.04.032.

Demographia world Urban Areas, 2015. Word Megacities. Available at: http://www.demographia.com/db-megacity.pdf.

Dempsey, N., Brown, C., Bramley, G., 2012. The key to sustainable urban development in UK cities? The influence of density on social sustainability. Progress in Planning 77 (3), 89–141. https://doi.org/10.1016/j.progress.2012.01.001.

Dijst, M., Schenkel, W., Thomas, I., 2018. Governing Cities on the Move. Routledge, New York.

Domingues, A.R., Pires, S.M., Caeiro, S., Ramos, T.B., 2015. Defining criteria and indicators for a sustainability label of local public services. Ecological Indicators 57, 452–464. https://doi.org/10.1016/j.ecolind.2015.05.016.

Domingues, A.R., Lozano, R., Ramos, T.B., 2018. Stakeholders-driven initiatives using sustainability indicators. In: Bell, S., Morse, S. (Eds.), Routledge Handbook of Sustainability Indicators. Routledge, London, UK, pp. 379–391.

European Union, 2003. European Common Indicators. Available online. https://www.gdrc.org/uem/footprints/eci_final_report.pdf. (Accessed 13 August 2019).

Fiala, N., 2008. Measuring sustainability: why the ecological footprint is bad economics and bad environmental science. Ecological Economics 67 (4), 519–525. https://doi.org/10.1016/j.ecolecon.2008.07.023.

Gallopin, G., 2018. The socio-ecological system (SES) approach to sustainable development Indicators. In: Bell, S., Morse, S. (Eds.), Routledge Handbook of Sustainability Indicators. Routledge, London, pp. 329–346.

Halberg, N., van der Werf, H., Basset-Mens, C., Dalgaard, R., de Boer, I.J.M., 2005. Environmental assessment tools for the evaluation and improvement of European livestock production systems. Livestock Production Science 96 (1), 33–50. https://doi.org/10.1016/j.livprodsci.2005.05.013.

Herrera-Ulloa, A.F., Charles, A.T., Lluch-Cota, S.E., Hermán, H., Hernández-Váques, N., Ortega-Rubio, A., 2003. A regional-scale sustainable development index: the case of Baja California Sur, Mexico. The International Journal of Sustainable Development and World Ecology 10 (4), 353–360. https://doi.org/10.1080/13504500309470111.

Hiremath, R.B., Balachandra, P., Kumar, B., Bansode, S.S., Murali, J., 2013. Indicator-based urban sustainability—a review. Energy for Sustainable Development 17 (6), 555–563. https://doi.org/10.1016/j.esd.2013.08.004.

Hoko, Z., Hertle, J., 2006. An evaluation of the sustainability of a rural water rehabilitation project in Zimbabwe. Physics and Chemistry of the Earth 31 (15–16), 699–709. https://doi.org/10.1016/j.pce.2006.08.038.

Holman, N., 2009. Incorporating local sustainability indicators into structures of local governance: a review of the literature. Local Environment 14 (4), 365–375. https://doi.org/10.1080/13549830902783043.

Huang, C.W., McDonald, R.I., Seto, K.C., 2018. The importance of land governance for biodiversity conservation in an era of global urban expansion. Landscape and Urban Planning 173, 44–50. https://doi.org/10.1016/j.landurbplan.2018.01.011.

Huang, L., Yan, L., Wu, J., 2016. Assessing urban sustainability of Chinese megacities: 35 years after the economic reform and open-door policy. Landscape and Urban Planning 145, 57–70. https://doi.org/10.1016/j.landurbplan.2015.09.005.

ICLEI, 2019. Local Government for Sustainability. Available at: https://iclei.org/en/Home.html (Accessed 30 August 2019).

Ingestam, L., 1987. La planificación Del desarrollo a largo prazo: notas sobre su esencia y metodología. Revista de la CEPAL 31, 69–76. https://repositorio.cepal.org/bitstream/handle/11362/11648/031069075_es.pdf?sequence=1&isAllowed=y.

International Standards Organization. ISO 37120 Smart City Data. Available online: https://www.dataforcities.org/wccd (Accessed 13 August 2019).

Jamil, N.I., Baharuddin, F.N., Maknu, T.S.R., 2015. Factors mining in engaging students learning styles using exploratory factor analysis. Procedia Economics and Finance 31 (15), 722–729. https://doi.org/10.1016/S2212-5671(15)01161-2.

Kamruzzaman, M., Deilami, K., Yigitcanlar, T., 2018. Investigating the urban heat island effect of transit-oriented development in Brisbane. Journal of Transport Geography 66, 116–124. https://doi.org/10.1016/j.jtrangeo.2017.11.016.

Kates, R.W., Clark, W.C., Corell, R., Hall, J.M., Al, E., 2001. Sustainability science. Science 292 (5517), 641−642. https://doi.org/10.1126/science.1059386.

Kelly, R., Moles, R., 2002. The Development of local Agenda 21 in the mid-west region of Ireland: a case study in interactive research and indicator development. Journal of Environmental Planning and Management 45, 889−912. https://doi.org/10.1080/096405602200002439.

Lebacq, T., Baret, P.V., Stilmant, D., 2013. Sustainability indicators for livestock farming. A review. Agronomy for Sustainable Development 33, 311−327. https://doi.org/10.1007/s13593-012-0121-x.

Little, J.C., Hester, E.T., Carey, C.C., 2016. Assessing and enhancing environmental sustainability: a conceptual review. Environmental Science and Technology 50, 6830−6845.

Malovics, G., Toth, M., Gebert, J., 2009. A critical analysis of sustainability indicators and their applicability on the regional level. In: Bucek, M., et al. (Eds.), Cers 2009 − 3rd Central European Conference in Regional Science, International Conference Proceedings − Young Scientists Articles, pp. 1186−1192.

Mapar, M., Jafari, M.J., Mansouri, N., Arjmandi, R., Azizinejad, R., Ramos, T.B., 2017. Sustainability indicators for municipalities of megacities: integrating health, safety and environmental performance. Ecological Indicators 83, 271−297. https://doi.org/10.1016/j.ecolind.2017.08.012.

Mascarenhas, A., Coelho, P., Subtil, E., Ramos, T.B., 2010. The role of common local indicators in regional sustainability assessment. Ecological Indicators 10 (3), 646−656. https://doi.org/10.1016/j.ecolind.2009.11.003.

Mascarenhas, A., Nunes, L.M., Ramos, T.B., 2015. Selection of sustainability for planning: combining stakeholders participation and data reduction techniques. Journal of Cleaner Production 92, 295−307. https://doi.org/10.1016/j.jclepro.2015.01.005.

Michael, F.L., Noor, Z.Z., Figueroa, M.J., 2014. Review of urban sustainability indicators assessment −case study between Asian countries. Habitat International 44, 491−500. https://doi.org/10.1016/j.habitatint.2014.09.006.

Moldan, B., Dahl, A.L., 2007. Challenges to sustainability indicators. In: Hák, T., Moldan, B., Dahl, A.L. (Eds.), Sustainability Indicators: A Scientific Assessment, pp. 1−24. Washington: SCOPE 67.

Mori, K., Christodoulou, A., 2012. Review of sustainability indices and indicators: towards a new City Sustainability Index (CSI). Environmental Impact Assessment Review 32, 94−106. https://doi.org/10.1016/j.eiar.2011.06.001.

Mörtberg, U., Haas, J., Zetterberg, A., Franklin, J.P., Jonsson, D., Deal, B., 2013. Urban ecosystems and sustainable urban development: analysing and assessing interacting systems in the Stockholm region. Urban Ecosystems 16 (4), 763−782. https://doi.org/10.1007/s11252-012-0270-3.

Olawumi, T.O., Chan, D.W., 2018. A scientometric review of global research on sustainability and sustainable development. Journal of Cleaner Production 183, 231−250. https://doi.org/10.1016/j.jclepro.2018.02.162.

Parris, T.M., Kates, R.W., 2003. Characterizing and measuring sustainable development. Annual Review of Environment and Resources 28, 559−586. https://doi.org/10.1146/annurev.energy.28.050302.105551.

Phillis, Y.A., Kouikoglou, V.S., Verdugo, C., 2017. Urban sustainability assessment and ranking of cities. Computers, Environment and Urban Systems 64, 254−265. https://doi.org/10.1016/j.compenvurbsys.2017.03.002.

Pissourios, I.A., 2013. An interdisciplinary study on indicators: a comparative review of quality-of-life, macroeconomic, environmental, welfare and sustainability indicators. Ecological Indicators 34, 420−427. https://doi.org/10.1016/j.ecolind.2013.06.008.

Pope, J., Bond, A., Hugé, J., Morrison-Saunders, A., 2017. Reconceptualising sustainability assessment. Environmental Impact Assessment Review 62, 205−215. https://doi.org/10.1016/j.eiar.2016.11.002.

Pope, J., Annandale, D., Morrison-Saunders, A., 2004. Conceptualising sustainability assessment. Environmental Impact Assessment Review 24, 595−616. https://doi.org/10.1016/J.EIAR.2004.03.001.

Pupphachai, U., Zuidema, C., 2017. Sustainability indicators: a tool to generate learning and adaptation in sustainable urban development. Ecological Indicators 72, 784−793. https://doi.org/10.1016/j.ecolind.2016.09.016.

Ramos, T.B., 2019. Sustainability assessment: exploring the frontiers and paradigms of indicator approaches. Sustainability 11 (3), 824. https://doi.org/10.3390/su11030824.

Ramos, T.B., Caeiro, S., 2010. Meta-performance evaluation of sustainability indicators. Ecological Indicators 10 (2), 157−166. https://doi.org/10.1016/j.ecolind.2009.04.008.

Ramos, T.B., 2009. Development of regional sustainability indicators and the role of academia in this process: the Portuguese practice. Journal of Cleaner Production 17, 1101−1115. https://doi.org/10.1016/j.jclepro.2009.02.024.

Rinne, J., Lyytimäki, J., Kautto, P., 2013. From sustainability to well-being: lessons learned from the use of sustainable development indicators at national and EU level. Ecological Indicators 35, 35−42. https://doi.org/10.1016/j.ecolind.2012.09.023.

Sartori, S., Latronico, F., Campos, L.M.S., 2014. Sustainability and sustainable development: a taxonomy in the field of literature. Ambiente and Sociedade 17, 1–22. https://doi.org/10.1590/1809-44220003491.

Shen, L., Guo, X., 2014. Spatial quantification and pattern analysis of urban sustainability based on a subjectively weighted indicator model: a case study in the city of Saskatoon, SK, Canada. Applied Geography 53, 117–127. https://doi.org/10.1016/j.apgeog.2014.06.001.

Singh, R.K., Murty, H.R., Gupta, S.K., Dikshit, A.K., 2012. An overview of sustainability assessment methodologies. Ecological Indicators 15 (1), 281–299. https://doi.org/10.1016/j.ecolind.2011.01.007.

Singh, R.K., Murty, H.R., Gupta, S.K., Dikshit, A.K., 2009. An overview of sustainability assessment methodologies. Ecological Indicators 9 (2), 189–212. https://doi.org/10.1016/j.ecolind.2008.05.011.

Spangenberg, J., Pfahl, S., Deller, K., 2002. Towards indicators for institutional sustainability: lessons from an analysis of Agenda 21. Ecological Indicators 2, 61–77. https://doi.org/10.1016/j.jclepro.2018.02.162.

Tanguay, G.A., Rajaonson, J., Lefebvre, J.F., Lanoie, P., 2010. Measuring the sustainability of cities: an analysis of the use of local indicators. Ecological Indicators 10 (2), 407–418. https://doi.org/10.1016/j.ecolind.2009.07.013.

Turner, R.K., Pearce, D., Bateman, I., 1993. Environmental Economics: An Elementary Introduction. Johns Hopkins University Press, Baltimore.

UN – United Nations, 2015. The Sustainable Development Agenda – 17 Goals to Transform Our World. Available at: https://www.un.org/sustainabledevelopment/development-agenda/ (Accessed 30 August 2019).

UN – United Nations. Habitat UN Habitats Urban Indicator Programme. Available online: https://unhabitat.org/urbanindicators-guidelines-monitoring-the-habitat-agenda-and-the-millennium-development-goals/#. (Accessed 13 August 2019).

UN – United Nations, 2019. Department of Economic and Social Affairs, Population Division (2019). World Urbanization Prospects: The 2018 Revision (ST/ESA/SER.A/420). United Nations, New York.

Valentin, A., Spangenberg, J.H., 2000. A guide to community sustainability indicators. Environmental Impact Assessment Review 20, 381–392.

Vasu, M.L., Stewart, D.W., Garson, G.D., 1998. Organizational Behaviour and Public Management. Marcel Dekker, New York.

Verma, P., Raghubanshi, A.S., 2018. Urban sustainability indicators: challenges and opportunities. Ecological Indicators 93, 282–291. https://doi.org/10.1016/j.ecolind.2018.05.007.

Westfall, M.S., Villa, V.A., 2001. Urban Indicators for Managing Cities. Asian Development Bank. Available at: https://www.adb.org/sites/default/files/publication/30020/urban-indicators-managing-cities.pdf.

Yigitcanlar, T., Kamruzzaman, M., Foth, M., Sabatini-Marques, J., Costa, E., Ioppolo, G., 2019. Can cities become smart without being sustainable? A systematic review of the literature. Sustainable Cities and Society 45, 348–365. https://doi.org/10.1016/j.scs.2018.11.033.

Yigitcanlar, T., Lönnqvist, A., 2013. Benchmarking knowledge-based urban development performance: results from the international comparison of Helsinki. Cities 31, 357–369. https://doi.org/10.1016/j.cities.2012.11.005.

Zhang, X., Bayulken, B., Skitmore, M., Lu, W., Huisingh, D., 2018. Sustainable urban transformations towards smarter, healthier cities: theories, agendas and pathways. Journal of Cleaner Production 173, 1–10. https://doi.org/10.1016/j.jclepro.2017.10.345.

Zhou, J., Shen, L., Song, X., Zhang, X., 2015. Selection and modeling sustainable urbanization indicators: a responsibility-based method. Ecological Indicators 56, 87–95.

Zhu, J., Simarmata, H.A., 2015. Formal land rights versus informal land rights: governance for sustainable urbanization in the Jakarta metropolitan region, Indonesia. Land Use Policy 43, 63–73. https://doi.org/10.1016/j.landusepol.2014.10.016.

19

Challenges in assessing urban sustainability

Álvaro Corredor-Ochoa[1], Carmen Antuña-Rozado[2], José Fariña-Tojo[3], Juho Rajaniemi[1]

[1]Tampere University, Tampere, Finland; [2]VTT Research Centre of Finland Ltd., Espoo, Finland; [3]Universidad Politécnica de Madrid, Madrid, Spain

1. Introduction

Urbanization, the movement of people from rural to urban areas, has constantly increased since the concept was measured for the first time (United Nations, 2018). Ever since the days of the Industrial Revolution (Rees and Wackernagel, 1996), people from rural areas have moved into the cities in search of a better life and economic opportunities (Grimm et al., 2008; Shen et al., 2011). Globalization of agricultural systems has had a negative effect on the rural areas; hence, urbanization is partly a consequence of unequal economic progress around the world

(Meyerson et al., 2007). Cities are now responsible for most of the environmental impacts on a global scale, including carbon emissions, air pollution, potable water consumption, solid waste produced, biodiversity reduction and material depletion (United Nations, 1987; Meyerson et al., 2007; Grimm et al., 2008). In light of the estimate that by 2050%, 68% of the total world population will live in urban areas (United Nations, 2018), the urge for sustainability solutions has focused on urban areas, which have greatly increased in recent decades (Grimm et al., 2008; Shen et al., 2011) and in which consumption tends to be greatest (Rees and Wackernagel, 1996; Newman, 2006; Grimm et al., 2008). The relationship between population growth, resources and environment was already investigated in the early 1970s, and this human expansion has directed an environment deterioration and resource depletion (Ehrlich and Holdren, 1971). Ruralization, defined as *'the changing lifestyle towards functional rurality; and effective and efficient rural conditions (physical, social, economic, and environmental changes) resulting from human socio-spatial behaviours, migration, and population dynamics'* (Chigbu, 2015) or population reduction does not reduce problems of any kind related to urban and global sustainability (Newman, 2006). Other measures need to be considered.

Nowadays, the overpopulation of cities has created a necessity to identify challenges to urban sustainability (Verma and Raghubanshi, 2018) and the methods to evaluate it have been developed over decades in order to solve all the problems associated with the rapid growth of cities. The issue has also been a subject of controversy ever since it was identified (Davison, 1996).

Until the last decades of the twentieth century ecological knowledge, awareness and performance were concentrated in nonurban areas, when the importance of focusing on the ecology of urban areas arose (Naredo, 1996; Grimm et al., 2008) and evaluations for purposes of urban environmental sustainability or urban ecology began. In parallel, awareness of the impacts caused by humankind and consumerism increased and became a matter of debate in the 1960s and 1970s (Carson, 1962; Ehrlich and Holdren, 1971; Meadows et al., 1972). Before the concept of urban sustainability became a global issue, other concepts like urban ecology — the study of urban ecosystems in natural and social sciences (Grimm et al., 2008) — were mainly utilized.

Rees (1992) elaborated on the ecological footprint concept, calculating the impact of human activities on the surrounding environment, scrutinizing the production of goods and services for society. He therefore concentrated on the relationship between consumption and environmental resources. Rees and Wackernagel (1996) noted the importance of the cities and their populations for global sustainability and proposed a formula (trade-corrected consumption = production + imports − exports) for calculating the ecological footprint per capita. Since then, comparisons between cities and countries have been feasible. The notion of urban sustainability and its evaluation was initially applied at the beginning of the 1990s in several communities in North America. These were just local reports on various environmental, economic and social aspects in different points in North America and Europe (Maclaren, 1996). These evaluations did not include the perspective of global sustainability and were limited to several facets to establish series of objectives. Some years later, the topic became subject matter for debate, research and several policies in the European Union.

Sustainable development was not a new concept in the 1990s, but very similar to the concept of 'ecodevelopment' that emerged in the 1970s (Naredo, 1996). Some elements of sustainability were discussed at the UN Conference on Human Environment in Stockholm in 1972 (Newman and Kenworthy, 1999), and sustainable development was first introduced at the World Commission of Environment and Development (WCED) in 1987: *'Humanity has the ability to make development sustainable to ensure that it meets the needs of the present without compromising the ability of future generations to meet their own needs'*. The WCED or the Brundtland Report (United Nations, 1987, 15) is a critical document that stipulates the relationship between three aspects — economic, social and environmental — needed to achieve sustainable development. The report made an urgent appeal to build 'a new era of economic growth', especially in developing countries, in order to sustain and expand environmental resources. It also bespeaks the close connection between economy and environment, later demonstrated in the Stern (2006). The concept 'sustainable development' was an immediate success due to the inclusion of the word 'development' and the elimination of the word 'eco', thereby satisfying both *developers* and *environmentalists* (Naredo, 1996). Therefore, 20 years after the report for the Club of Rome (Meadows et al., 1972) questioned the notions of growth and development used in economics, a renewed desire for making them 'sustainable' emerged and the physical environment linked to the original concept was abandoned (Naredo, 1996). The Brundtland Report focuses on six aspects to work on in order to achieve the goal: population and human resources, food security, species and ecosystems, energy, industry and cities. It stresses on the cities located in developing countries, with considerably less economic resources, fewer qualified professionals and high social inequality. Cities are the main pillars of national developments; urban development nevertheless consumes a great deal of the world's resources and energy and produces an enormous amount of waste and pollution. The report also mentions various issues cities must consider, such as drinking water provision, sanitation, education facilities, public transportation and proper housing among others. All these issues are a part of urban sustainability and necessary in achieving sustainable development. As Newman mentions (1999, 220): *'It is possible to define the goal of sustainability in a city as the reduction of the city's use of natural resources and production of wastes while simultaneously improving its livability, so that it can better fit within the capacities of the local, regional and global ecosystems'*. The application of the principles of sustainability to cities was later debated at the Earth Summit in 1992 in Rio de Janeiro (United Nations, 1992), where Local Agenda 21 Plans were drawn up to achieve sustainability. This new focus on the local scale that started to emerge as sustainable development can only be achieved by a shift from a global to a circular economy (Rees and Wackernagel, 1996). In this context, urban sustainability can be considered as the capacity to maintain and strengthen sustainable development in urban environments. This capacity has a set of measurable properties occurring at both local and city regional level. In other words, sustainable development must be achieved in sustainable urban environments.

Through this literature review has found that urban sustainability is a complex and a controversial issue. In principle, the existence of sustainability indicators is both useful and necessary. However, the problems arise when seeking consensus on the concept of sustainability or sustainable development. This conceptual barrier could be removed by assimilating the planetary limits related to the individual and social promotion possibilities mentioned in the Club of Rome Report (Meadows, 1972). However, this investigation refers to many other relevant aspects. Thus, the need to understand that each city is different from all others clashes with the possibility of comparing some figures, as each city should have its own system of indicators. The necessity is not only from the absolute point of view, but also from the relative point of view — the percentage of improvement on a particular indicator — because at baseline and hence the percentages of variation are very different.

Furthermore, there are problems with the unit of analysis. This is a crucial issue as seen throughout the investigation. The importance of some indicators for the different urban scales, from the building to the region, is also under discussion. It is essential to establish a basic unit of analysis, leaving the rest as subsidiaries. As seen in the literature consulted, the city-region — sometimes also called ecoregion or bioregion — could be the core of the system, generating a net of relationships throughout the built space: including the district, the neighbourhood, the building or even specific indicators. Probably the utility of a system of indicators is basically strategic, political and social rather than executive, especially regarding indices and general indicators. Conversely, the specific indicators related to a concrete field (trade, environment or others) would increase the chances of being executive. Hence, the difficulties faced by commercial certification bodies and the vast amount of criticism expressed when they have tried to certify urban sustainability. In spite of this, this is a promising and essential field, which requires further investigation to achieve concrete results that can be adopted by the scientific community.

2. Methodology

There is abundant literature on urban sustainability, on the methods for achieving it, discussions and debates. The literature for this article was mostly selected from Scopus data sources (1970–2018), searching for studies using 'urban sustainability' as keywords and limited to the subject areas of environmental science, engineering and energy. The search yielded 5149 documents (Fig. 19.1) and later it is limited to books, book chapters and journals in English and Spanish with a result of 74 potential documents (Fig. 19.2). Frequent articles cited in the preselected material were also included in the final reference list.

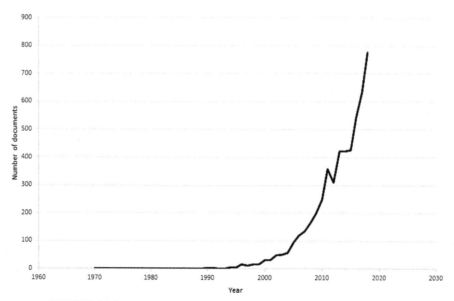

FIGURE 19.1   Documents per year with 'urbansustainability' as keywords.

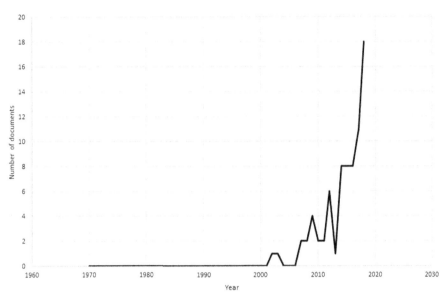

FIGURE 19.2 Documents per year with 'urbansustainability' as keywords limited to books, book chapters and journals in English and Spanish.

3. Evolution of urban sustainability

Different bodies like the Organization for Economic and Cultural Development (OCDE), the World Bank and the European Community decided to create programmes for sustainable cities as the implementation of the principles of the Earth Summit Local Agenda 21 in Rio de Janeiro, but how to achieve the sustainability was unclear (Newman, 1999). Furthermore, some theories claiming the impossibility for cities to achieve sustainability on their own were later released (Rees and Wackernagel, 1996).

3.1 Urban sustainability in Europe

In 1991 the European Union established an expert committee to include sustainable urban policies for achieving urban sustainability (Simón-Rojo and Hernández-Aja, 2011), and the adoption of urban sustainable policies was required of urban planners, stakeholders and other actors needing to make decisions on planning and renewing the municipalities. In 1994, during the European Conference on Sustainable Cities and Towns organized by The International Council of Local Environmental Initiatives (ICLEI), the Charter of European Cities and towns towards sustainability — the Aalborg Charter — was created and signed by the representatives of many cities, small towns and regions in Europe (García-Sánchez and Prado-Lorenzo, 2008). Inspired by the Rio Summit, the Aalborg Charter made a commitment to implement a Local Agenda 21 to achieve sustainable development (ICLEI, 1994). This is considered to be the first executive plan to achieve sustainability in cities at the European level (Devuyst, 2000; García-Sánchez and Prado-Lorenzo, 2008) and it is still ongoing despite the obstacles encountered (García-Sánchez and Prado-Lorenzo, 2008).

According to Blowers (1993), achieving urban sustainability requires a combination of factors:

1. cooperation.
2. understanding and participatory mechanisms between the various stakeholders and actors involved.
3. effort in searching for common interest between actors and stakeholders that support the future generations.
4. efficiency in the use of resources.
5. government's will to accomplish implementation.
6. intake from highly skilled professionals, and
7. patience to examine the long-term effects.

Davison (1996) concludes that performance-oriented plans combining flexibility, focus, efficiency effectiveness and commitment are required to achieve urban sustainability, although the difficult choices that need to be made in the pursuit of some sustainable issues may have negative effects thus necessitating a strategic choice. Grimm et al. (2008) suggest that urban sustainability is necessary to change the relation between climate and urbanization. Devuyst (2000) states that urban sustainability is necessary to ensure that human activities lead to a more sustainable society and that it needs to be implemented in local programmes and evaluated with assessment tools.

Environmental Impact Assessments (EIA) and Strategic Environmental Assessments (SEA) are the origin of the first evaluations of sustainability (Pope et al., 2004). Furthermore, EIAs were considered to be tools for promoting sustainable development (Devuyst, 2000) and were already available at the 1992 Earth Summit in Rio de Janeiro (United Nations, 1992) in principle 17: *'Environmental impact assessment, as a national instrument, shall be undertaken for proposed activities that are likely to have a significant adverse impact on the environment and are subject to a decision of a competent national authority'*. In the Aalborg Charter (1994), towns and cities agree to base *'our policy-making and controlling efforts, in particular our environmental monitoring, auditing, impact assessment, accounting, balancing and reporting systems, on different types of indicators, including those of urban environmental quality, urban flows, urban patterns, and, most importantly, indicators of urban systems sustainability'* (ICLEI, 1994).

In the European context most of the indicators of urban sustainability have been developed by local authorities or national institutions within the European research framework programmes: FP5, FP6, FP7 and Horizon 2020. The Fifth Framework Programme FP5 (1998—2002) funded 77 projects related to urban sustainability and two principal projects investigating indicators of urban sustainability: PETUS — Practical Evaluation Tools for Urban Sustainability (2002—05) — and CRISP — City Related Indicators of Sustainability Project — (2000—02) (Simón Rojo and Hernández Aja, 2011). Simón-Rojo and Hernández-Aja (2011) have classified the projects used for PETUS according to different criteria: the scale (building, block, neighbourhood, city, region and country), the life-cycle phase (design or operation), the object of assessment application (strategy, building, urban planning, policy-making), the type of tool (guideline, checklist, indicators, GIS), the agent who accomplished the project (private consultancies, local authorities, universities and research centres) and the nature of the sustainability investigated (environmental, social or economic). Therefore, according to Simón-Rojo and Hernández-Aja (2011), projects related to indicators of urban sustainability on a city and neighbourhood scale are EUROGISE (1998–2001), valid for the design phase, and Ã–KOSTADT 2000 (*Eco-Team of Graz*, 1999), Moland (European Comission Project, 2004) and ISTAT Environmental Indicators Set (ISTAT, 2010), valid for the operation phase.

During the Sixth FS6 (2002–06) and Seventh FS7 (2007–13) Framework Programmes, the projects related to urban sustainability decreased drastically due to a growing interest in particular aspects of sustainability rather than in the concept as a whole. Such aspects include energy efficiency, sustainable transportation or GHG mitigation. Moreover, only a few projects related to urban sustainability were launched. These research projects were STATUS (Sustainability Tools and Targets for the Urban Thematic Strategy. 2005–06), SENSOR (Sustainability Impact Assessment Tools for Environmental, Social and Economic Effects of Multifunctional Land Use in European Regions, 2004–09), TISSUE (Trends and indicators for monitoring the EU thematic strategy on sustainable development of urban environment. 2004–05)and RAISE (Raising citizens and stakeholders awareness, acceptance and use of new regional and urban sustainability approaches in Europe. 2004–06). Furthermore, the indicators from all the projects from FS6 and FS7 are either highly specific in the field of sustainability or are geared to monitoring or achieving new policies. The ongoing Horizon 2020 programme is clearly aimed at GHG mitigation, smart cities and sustainable transportation. The idea of assessing the city globally seems to have been forsaken. A research framework for assessing the city as a whole has been abandoned by the EU and there is no interest on the part of public bodies to embark on new research assessing urban sustainability as a holistic concept.

4. Recent development

The evaluation of urban sustainability uses methods and tools similar to those used for assessing sustainable development. How to evaluate sustainability, different methods and tools, objectives and applications are topics discussed since the end of the twentieth century. However, at the beginning of 2000s this was still unclear, especially at the local level (Devuyst, 2000). There been debates on what should be assessed: whether general sustainable progress or policies for sustainability achievement (Devuyst, 2000), whether human and environmental issues together (Newman, 2006), or whether if there is a need to define and clarify the concept of sustainability in the field in which the assessment is to be performed (Pope et al., 2004).

Over the years various arguments have been put forward as to why and how to assess urban sustainability and why these assessments should be performed in the first place. Newman (2006) considers that urban sustainability assessments should be utilized to achieve solutions to the negative environmental impact, but at the same time they should afford positive human development possibilities due to the vast potential for policymaking embodied in the evaluation of urban sustainability. Some authors agree that the assessment of urban sustainability should be included in a report that is used later to set sustainability objectives (Devuyst, 2000) while others argue that sustainable objectives should be listed in advance so that the assessment can be utilized to monitor the goals and to facilitate communication among the various actors (Kuik and Verbruggen, 1991; Davison, 1996; Verma and

Raghubanshi, 2018). Pope et al. (2004, 20) conclude that evaluations of sustainability should both assess whether an action is sustainable as well as the target set. Goals for sustainability assessments include simultaneous evaluations for environmental, social and economic facts 'to achieve a "net benefit" outcome in each area, with minimal trade-offs' (Newman, 2006). Moreover, urban sustainability assessments need to be focused on a tool that ensures that the ongoing development moves in the right sustainable direction rather than answering whether an urban area is sustainable or not, as Pope et al. (2004, 13) suggest for all sustainable assessments. This perspective means that urban sustainability assessments should be performed in areas where an urban action is taking place.

Urban sustainability evaluations should be performed using a tool that tests urban sustainability, reports a future vision and establishes urban sustainability goals in a defined urban area or a community. These goals include guidelines, checklists, frameworks, sustainability indicators and distinctive assessment tools (Davison, 1996; Shen et al., 2011).

Verma and Raghubanshi (2018, 286–289) proposed a top-down approach to assessing urban sustainability. It includes 10 steps: preliminary assessment, determining outcomes and goals, selecting a performance indicator, setting the performance baseline, selecting targets, application, evaluation, reporting findings, applying findings and sustaining indicator framework. Guidelines contain general indications normally included in policymaking. They can be very useful and are normally easy to understand although they are usually of a rather general nature. Checklists summarize guidelines with attributed weighting of dubious origin and are normally conducted in an unscientific manner, including some questions related to a scoring system (Devuyst, 2000). Many commercial sustainable certifications use this method because it is easy to implement.

5. Indicators of urban sustainability

Sustainable development, as well as urban sustainability, is normally assessed with a wide range of indicators in use across different cities and regions but varying according to their respective needs and goals (Brandon and Lombardi, 2005). Many sustainability indicators and indices are utilized most commonly as a method, but there are great difficulties in reconciling globally valid indicators and also their calculations. There is moreover the challenge of determining thresholds or demonstrating the sustainability of the value for each indicator or index.

Elemental notions: indicators, indices.

Basic concepts of indicators and indices are brilliantly explained by Tanguay et al. (2010, 408). An indicator can be a figure or a threshold of certain aspects of sustainability, allowing the evaluation of some aspect of urban sustainability during the assessment.

- Indicator

An indicator is a variable or revised datum that assesses a predetermined phenomenon, while an index is composed of a combination of indicators.

- Indices

Indices help to simplify the complexity of an indicator (Brandon and Lombardi, 2005) but are usually more difficult to calculate. Therefore, indices need an aggregation of different indicators to be assessed. For instance, in order to assess carbon dioxide emissions in a defined urban area, it is necessary to evaluate the emissions coming from different sources, such as transportation, buildings, industry and many others. All those different foci form different subindices. A subindex may also have different subindices until a simple form of an indicator is obtained. In this manner, private car carbon dioxide emissions can be considered an indicator of urban sustainability for the whole carbon dioxide emission, but many other indicators may be necessary to evaluate all the emissions, which makes it decidedly difficult to obtain a valid figure.

What is said above is a simplification of a complex phenomenon and therefore not capable of generating a general vision for the problem. However, it makes it possible to set a correct sustainable direction and a policy for achieving it (Maclaren, 1996; Brandon and Lombardi, 2005; Newman, 2006).

Weighting gives attributes to the correct quantity or value for each indicator or index. However, the arbitrariness and irrationalism of the weight values have been criticized in the literature (e.g., Martínez-Alier et al., 1998).

5.1 Threshold, critical value, target value and relative performance

Besides obtaining a general value, there are other possibilities for weighting: the threshold, the critical value, the target value and the relative performance (Tanguay et al., 2010).

- Threshold

The threshold is a scientifically demonstrated minimum value of a quantity causing a certain effect (Maclaren, 1996; Tanguay et al., 2010).

- Target value

The target value is similar to a threshold, but the levels are to be achieved in the future. The value is not scientifically demonstrated and comes from several individual or group interests (Tanguay et al., 2010).

- Relative performance

Finally, a relative performance is based on values taken from other areas (Tanguay et al., 2010).

According to Maclaren (1996, 5), the most important classes are targets — levels to be achieved in the future — and thresholds. It is important to consider different methods for obtaining data, as they may affect the result of an indicator and complicate comparisons of the results (Brandon and Lombardi, 2005). The use of software linked to this practice has recently improved the calculation of sustainability indicators by making the process simpler and easier to understand as well as more reliable (Brandon and Lombardi, 2005).

The number of indicators and indices needed and the areas of sustainability affected have been studied in different European research projects with varying results. Unfortunately, as most of the results from these projects are not available, they cannot be consulted. Therefore, the resources obtained and the possible knowledge are in danger of being lost (Simón-Rojo and Hernández-Aja, 2011).

Various indicators of urban sustainability have been proposed for various urban scales, life-cycle phases and tools. The ecological footprint is one of the indicators that has had a substantial impact, not least because of the clear calculation it includes. A selection of core indicators needs to be chosen even if a wider selection of them necessitates an enormous amount of resources (Brandon and Lombardi, 2005).

To assess the sustainability of any physical product it is necessary to perform the evaluation for the whole life cycle. Sustainability assessments for buildings are performed similar to those for any other product, like a yogurt, a chair or a jacket. The major challenge is the large number of products contained in a building and their different life spans, sometimes rendering life-cycle assessment really complicated. Therefore, a building life-cycle assessment (LCA) contains the LCAs of every single product contained on it, from sewage pipes to elevators, ventilation machines, toilets or doors or windows. The building has a life cycle—based evaluation of sustainability and there are enough examples on sustainable assessments done on buildings based on LCAs. Nevertheless, on the next city scale, the neighbourhood, there are no longer any sustainability assessments based on LCAs, perhaps due to the complexity of the evaluation process.

6. Discussion

Most of the literature reviewed agreed on the critical need to achieve sustainable development and that cities and urban areas have to reach it. There has been a continuous debate on the ambiguity of the concept, a fact that contributed to the success of the term (Naredo, 1996). To approach urban sustainability, there should be an assessment of the actual urban sustainability state and a list of new objectives to improve some of the aspects assessed (Maclaren, 1996). Furthermore, there is a need to question the necessity for arranging comparisons between different communities both on the assessment of the present situation and on the plan of similar objectives to be able to monitor communities and compare their evolution. There is also consensus on the indicators of urban sustainability concerning the most adequate method of evaluating sustainable development and urban sustainability. On the other hand, there is a considerable debate about which indicators should be considered core indicators, as well as the risks of oversimplification of indicators and the selection of indicators depending on the interests of the actors developing them. Additionally, the authors have located eight major challenges in assessing urban sustainability that will be explained in detail later on. United Nations indicators of sustainable development (UN, 2001; UN, 2007) are the most widespread, known and implemented and, therefore, should be used as starting point model in all national agendas (Brandon and Lombardi, 2005). Verma and Raghubanshi (2018, 286—289) discuss categorizing the challenges inherent in indicators of urban sustainability into internal and external; defining internal challenges as those related to development methodology and external challenges as those not conducive to sustainable indicator frameworks.

Once built, urban areas are hard to demolish or even to rearrange. While stakeholders set their own goals, usually related to economic profit and provoking external pressures (Davidson, 1996; Isaksson and Storbjörk, 2012), urban planners and decisionmakers are forced to prioritize targets which will improve the economy of the municipality, particularly in smaller urban areas. Furthermore, urban sustainability is seldom carefully considered by urban planners, due either to lack of knowledge (Devuyst, 2000) or to external pressures. Davidson (1996, 446) also explains the challenge of achieving urban sustainability by factors external to planning, such as the local economy, government desires or demographic trends.

These are the eight major challenges we consider in assessing urban sustainability confusion between sustainable development and urban sustainability, complexity within urban planning, difficulties in finding indicators of urban sustainability valid for all urban scales, lack of indicators of urban sustainability valid for the entire life-cycle phase, actors to whom the indicators are addressed, challenges in using indicators and indices, commercial urban certifications and insufficient and incomparable data (Fig. 19.3).

1. Confusion between sustainable development and urban sustainability

For many, there is still considerable confusion between the concepts of sustainable development and sustainable urban environment. Some authors call for wider agreement on what urban sustainability means before evaluating it (Alberti, 1996; Ludin and Morrison, 2002). It is clearly impossible to achieve sustainable development in an unsustainable urban environment. Therefore, urban sustainability also needs to be applied to urban environments. We have found many examples of indicators of urban sustainability that cannot be applied to the urban environment itself but instead should be applied to society. For example, the reduction of unemployment or the gross national product is a common indicator for sustainable development, applied to society and the governance of the city, but not to the city as a physical structure. Thus, a clear distinction should be made between strategies, goals and frameworks for urban development and urban planning. Objectives, strategies, frameworks and indicators for achieving urban sustainability should be applied by urban planners, stakeholders and actors involved in the whole urban planning process while plans and goals for sustainable development should be achieved through governance, decisionmaking and politics.

Given the situation, a clear distinction between sustainable development and urban sustainability should be made. In the author's view, sustainable development contributes to human welfare and prosperity, avoiding severe social inequality and respecting environmental resources. Urban sustainability is just a part of this process. Therefore, sustainable development is a global and a somewhat abstract goal, whereas urban sustainability refers to phenomena on a more concrete and more local scale.

2. Complexity within urban planning

Urbanism and the city are extremely complex concepts, very difficult to define, assess, investigate and predict as such. Despite the attempt of Western countries to globalize, extend and renew cities in an analogous manner, cities continue to possess their own identity, and are difficult to classify globally. Cities have different topographical, climatic, environmental, social, demographic, economic, political, governmental,

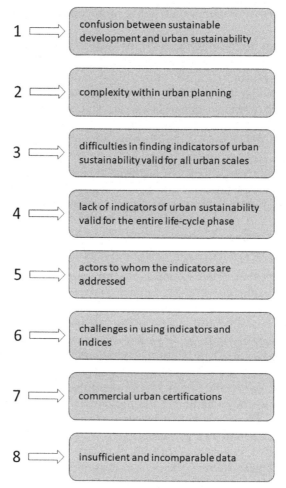

FIGURE 19.3   Eight major challenges in assessing urban sustainability.

cultural and historical conditions. Furthermore, different countries have different urban planning systems and legislations, which complicates the assessment of urban sustainability. It is precisely all these varying conditions that may justify the selection of different sets of core indicators for different cities, even though this is a debatable issue because, since by definition core indicators are fewer, it is often assumed that they must be suitable for all cities, which presents difficulties for their 'local adaptation'. However, since this is such a broad topic, it will not be dealt in detail in this article.

3. Difficulties in finding indicators of urban sustainability valid for all urban scales

After decades of investigation, city sustainability assessment started to be successful on its smallest scale: the building. Funded by the Seventh Framework Programme, the

SuPerBuildings project — Sustainability and performance assessment and benchmarking of buildings — (Häkkinen et al., 2012) conducted extensive research on the existing indicators of sustainability for the building scale and discussed assessment methods, barriers and drivers. More specifically, SuperBuildings:

developed and selected sustainability indicators for buildings; improved the understanding about performance levels considering new and existing buildings, different building types and different national and local requirements; developed methods for the assessment and benchmarking of sustainable buildings; and made recommendations for the effective use of benchmarking systems as instruments of steering and in different stages of building projects.

The city scale is fundamental to assessing urban sustainability. Hence a sustainable building does not necessitate having a sustainable neighbourhood; ensuring a sustainable district does not imply a sustainable metropolis and a sustainable city does not guarantee the existence of a sustainable region. Therefore, it may be that methods of urban sustainability assessment are applicable to all city scales and sustainability guidelines for neighbourhoods may be useless throughout the whole urbanity. Furthermore, values or thresholds of sustainability indicators for buildings may be ineffective throughout the city region. Actually, the city region is in many respects, such as commuting, shopping behaviour, city economics or migration, a sensible scale on which to survey urban sustainability. Among other reasons, indicators for smaller scales are more effective than for bigger ones, in which data may not be obtained for the same time period and the reports may quickly lose relevance (Verma and Raghubanshi, 2018).

4. **Lack of indicators of urban sustainability valid for the entire life-cycle phase**

The life cycle is applied in the building scale, but it is completely ignored in bigger scales. Assessments for sustainable development do not use a life span, because normally their goals are set for the near future. In the building scale the procedure is different. As previously mentioned, it is possible currently to perform the life-cycle assessment of a construction element within a building, but there are difficulties in evaluating the life cycle of an entire building. New buildings are often given an almost standardized life span of 50 years, even though there are uncertainties in relation to this figure. A number of buildings in a city may have a life span of hundreds of years while others are demolished after a few decades. A wide variety of circumstances affecting the life span of the urban development may arise in a neighbourhood, in a district or in a city: natural disasters, political will or others. The variation of occurrences is endless. In addition, there will be completely unpredictable demolition techniques within several decades, and the end of a building's serviceable life may differ from what we now plan. Nevertheless, the life-cycle assessment of a building, as the smallest scale of the city, is already problematic. Due to all these facts, moving to bigger scales (neighbourhood, city, region) is increasingly challenging but nevertheless necessary to a considerable extent. Despite the absence of relevant literature on urban life cycle, investigations of life-cycle assessments on neighbourhood scale (Trigaux et al., 2014; Peuportier, 2016; Ochoa-Sosa et al., 2017) are too few to have a common position. Furthermore, the authors of this article advocate only three valid stages during the urban life cycle: design, renewal or construction and maintenance and operation. Moreover, an open debate on life cycle and urban development seems very necessary.

5. Actors to whom the indicators are addressed

Indicators of urban sustainability should be the result of the contribution of different stakeholders (Maclaren, 1996; Brandon and Lombardi, 2005) and addressed to different actors, who are needed throughout the entire process. Stakeholders can sometimes provide the needed funding, especially when governmental finance is not forthcoming (Davidson, 1996) and, therefore, a common understanding for achieving common goals is indispensable. In all cases a clear distinction should be made between the actors to whom each indicator is addressed. Indicators developed by a wider range of actors are the most reliable, weighty and coherent (Innes, 1990).

Actors and stakeholders are also different depending on the life cycle stage, and it might vary the whole evaluation of the sustainability assessment, especially in the building scale. AlWaer and Clements-Croome (2010, 800) state that during the operation and maintenance stage there are different issues influencing the behaviour of the building. These issues are the products (goods, materials, fabrics, facilities, equipment) and the users (customers, workers and owners) (Fig. 19.4). Therefore, it is necessary to consider all actors and issues, and not just some of them, when performing any kind of sustainability assessment at the operation and maintenance stage. For example, if we want to use an index for carbon emissions in a selected neighbourhood, we can

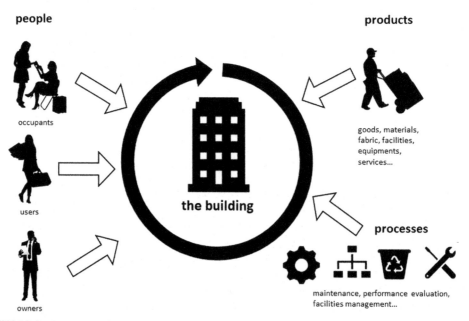

FIGURE 19.4   Key sustainability actors during the building life span according to AlWaer and Clements-Croome (2010).

calculate the emissions produced by every single individual, but then we should calculate the emissions from the products on sale or the buildings located in the area as well. The complexity of including all actors involved in urban environments is enormous. There is an actor classification in the building scale and in different life-cycle phases (Lebègue et al., 2013), but there is no actor classification for bigger urban scales. Reaching consensus among different actors and stakeholders on what urban sustainability indicators, values, thresholds and even sectors of sustainability are valid or important has proven to be truly laborious. The lack of consensus due to the conflict of interests between the different actors has often led to the failure of agreements on urban sustainability indicators. If this type of agreement is challenging within a city, consensus among several municipalities is even more so.

6. Challenges in using indicators and indices

There is agreement on the important role of indicators for achieving urban sustainability and there are several cases where those indicators are effectively in use. However, the know-how gained has not been sufficiently shared and utilized in new sustainable urban developments or in sustainable policymaking (Shen et al., 2011). Discussions about indicators and indices are still ongoing. The utilization of single sets of indicators facilitates monitoring and comparisons with other urban areas. Yet the use of too many indicators may make the evaluation impractical and cities are not interested in long sets of indicators (Jakutyte-Walangitang et al., 2016). Simplification by combining different indicators into indices seems to be the optimal procedure, even if it must be investigated in order to avoid oversimplifying a complex system (Brandon and Lombardi, 2005). Furthermore, a monitored index that progresses adequately can include indicators that regress, and therefore there should be an awareness of possible simplifications. The selection of which indicators and indices need to be evaluated has proven to be more effective with stakeholder participation (Tanguay et al., 2010; Brandon and Lombardi, 2005). Unfortunately, stakeholder participation can vary a lot in different urban areas, thereby rendering comparisons complicated.

Targets and thresholds are usually present in EU policy goals, and the levels for both indices and indicators are decided by scientists or policymakers according to the values which they consider making the system sustainable (Verma and Raghubanshi, 2018). Some authors recommend comparing the values with examples considered to be good practices (Devuyst, 2000). At the same time, there is enormous ambiguity regarding the value and significance of each indicator and index of urban sustainability (Tanguay et al., 2010; Shen et al., 2011). Some of the indicators, such as water or air quality, can be scientifically proven (Maclaren, 1996), some thresholds can be set as percentages and some of them just cannot be proved or quantified. Therefore, selected indicators should be measurable, simple and applied to updated data (Brandon and Lombardi, 2005). The simplification of a complex problem like urban sustainability into simple specific ratios has also been criticized (Du-Plessis, 2009) as has the selection of indicators according to an actor's own interests in order to find a correspondence with the expected decisions (Davidson, 1996; Isaksson and Storbjörk, 2012).

7. Commercial urban certifications

Sustainability schemes, such as the British BREEAM for Communities, the Japanese CASBEE for Urban Development and the American LEED for Neighborhood

Development, are already being used in the certification and benchmarking of urban areas. Sustainable communities are promoted as a desirable policy goal and local authorities are encouraged to contribute, in particular to climate change mitigation through urban planning (Bulkeley and Betsill, 2005; Gunnarson-Östling and Höjer, 2011). Nevertheless, there are contradictory discussions on environmental sustainability and a lack of certainty about what it might mean in practice (Svane and Weingaertner, 2006). Urban planners often appear to struggle with the issue of how to promote area-specific urban environmental sustainability through municipal land-use planning (Runhaar et al., 2009; Kyttä et al., 2013) as well as how to assess different interventions in urban areas, as the above-mentioned ecolabels are only capable of assessing master plans and do not cover different life-cycle stages (Simón Rojo and Hernández Aja, 2011).

Urban planners and policymakers must be aware of the use of these commercial sustainable certifications, as it might lead in an incorrect direction.

8. Insufficient and incomparable data

Urban phenomena are extremely complex, not just in terms of multiplicity in a particular moment, but also in a temporal perspective linked to societal changes. This means that it is very difficult to have enough high-quality data (Wang and Zou, 2010; Severo et al., 2016; Sorokine et al., 2016) for assessing and comparing different urban locations. This concerns especially data on social behaviour. In addition, the data provided by different municipalities and different cities vary widely. Comparisons between different cities, not to mention different countries, may likewise be arbitrary for this reason. Consequently, a researcher is often obliged to develop new methods and indicators for each project.

To achieve a breakthrough in the study of the evaluation of urban sustainability and to take into account the eight challenges described, we understand that the values for each indicator or index of urban sustainability can be different depending on the scale, the life cycle, the type of predominant buildings, as well as the related actors and products. We understand that it is necessary to calculate these values in different case studies to obtain sustainability thresholds.

7. Conclusion

Indicators of urban sustainability are the most appropriate method for assessing urban sustainability due to the great acceptance and many options in combination with software. However, despite the large number of different assessment methods, they have hardly ever been applied to city development processes (Newman, 2006). The scientific community should moreover investigate the reasons for this. There is also a need to focus research on urban sustainability indicators that only evaluate certain aspects of sustainability, following the latest research guidelines issued by the European Union and considering the complexity of globalized indicators of urban sustainability. Moreover, it is essential to concentrate on researching urban sustainability indicators that are quantifiable.

There is a considerable difference between the methods and indicators used to assess the sustainability of city regions and those used for buildings. Therefore, in order to be able to monitor and compare urban areas of similar characteristics, the authors propose that each

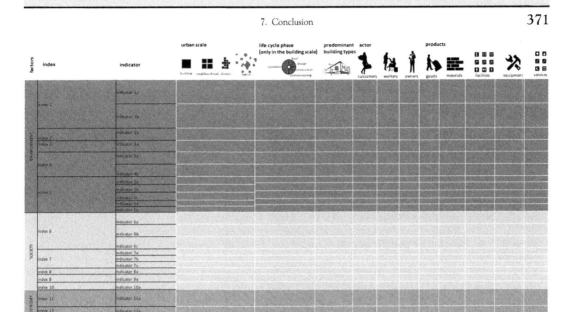

FIGURE 19.5   Sustainability assessment framework.

indicator is applied to a specific urban scale, a defined actor and a certain life-cycle stage as shown in the sustainability assessment framework (Fig. 19.5). The lack of thresholds (or values of sustainability) could be resolved by comparing values for different urban areas in different locations with similar urban environments. In short, sharing information seems really important to learn from each other's experiences and set valid thresholds if these are lacking.

A balance needs to be struck between allowing a certain margin for cities to develop their own definitions of sustainability in response to their specific conditions (geographic, climatic, etc.), and using a common framework that will make comparison possible. More on the varying conditions requiring local adaptation can be found in Antuña-Rozado et al. (2018). However, even if a common framework for the assessment of urban sustainability worldwide were to be developed and agreed upon, the fact that some cities are not only much more advanced than others along the path towards sustainability, but they also have more financing available for sustainable urban development and regeneration, remains a challenge to be resolved.

The cities, urban structures and environments are physical parts of urban development. To achieve sustainable development it is necessary also to arrive at sustainable urban forms. Consequently, future investigations should be addressed to urban planners and urban planning departments, the key actors in achieving a sustainable city. However, urban planners and local governance have serious difficulties in selecting and planning urban areas in a sustainable manner. Actually, they need knowledge and simple and flexible tools to enable them to make more sustainable choices and to plan more sustainable urban areas. These flexible tools should be integrated with the software used by planners and designers (e.g., BIM or GIS based) as they are not able to do everything by themselves (Davison, 1996).

References

Antuña-Rozado, C., García-Navarro, J., Mariño-Drews, J., 2018. Facilitation processes and skills supporting EcoCity development. Energies 11, 777.

Alberti, M., 1996. Measuring urban sustainability. Environmental Impact Assessment Review 16 (4–6), 381–424.

ALwaer, H., Clements-Croome, D.J., 2010. Key performance indicators (KPIs) and priority setting in using the multi-attribute approach for assessing sustainable intelligent buildings. Building and Environment 45, 799–807.

Blowers, A., 1993. Planning for a Sustainable Environment – A Report by the Town and the Country Planning Association. Earthscan, Great Britain.

Brandon, P.S., Lombardi, P., 2005. Evaluating Sustainable Development in the Built Environment. Wiley-Blackwell, United Kingdom.

Bulkeley, H., Betsill, M., 2005. Rethinking sustainable cities: multilevel governance and the urban politics of climate change. Environmental Politics 14, 42–63.

Carson, R., 1962. Silent Spring. Houghton Mifflin, Cambridge, Massachusetts, USA.

Chigbu, U.E., 2015. Ruralisation: a tool for rural transformation. Development in Practice 25 (7), 1067–1073.

Davison, F., 1996. Planning for performance: requirements for sustainable development. Habitat International 20 (3), 445–462.

Devuyst, D., 2000. Linking impact assessment and sustainable development at the local level - the introduction of sustainability assessment systems. Sustainable Development 8, 67–78.

Du Plessis, C., 2009. An Approach to Studying Urban Sustainability from within an Ecological Worldview. PhD dissertation. School of the Built Environment, University of Salford, Salford.

Eco-Team of Graz (1999) Eco-City 2000 – Evaluation. Expertise Report. City of Graz. Available at: https://www.umweltservice.graz.at/infos/la21/eval1_97.pdf.

Ehrlich, P.R., Holdren, J.P., 1971. Impact of population growth. Science 171, 1212–1217.

García-Sánchez, I., Prado-Lorenzo, J.M., 2008. Determinant factors in the degree of implementation of local agenda 21 in the European union. Sustainable Development 16, 17–34.

Grimm, N.B., Faeth, S.H., Golubiewski, N.E., Redman, C.L., Wu, J., Bai, X., Briggs, J.M., 2008. Global change and the ecology of cities. Science 319, 756–760.

Gunnarsson-Östling, U., Höjer, M., 2011. Scenario planning for sustainability in Stockholm, Sweden: environmental justice considerations. International Journal of Urban and Regional Research 35, 1048–1067.

Häkkinen, T., Antuña-Rozado, C., Mäkeläinen, T., Lützkendorf, T., Balouktsi, M., Immendörfer, M., Nibel, S., Bosdevigie, B., Lebert, A., Fies, B., Hernandez-Iñarra, P., Lupíšek, A., Hajek, P., Supper, S., Alsema, E., Delem, L., Van-Dessel, J., 2012. Sustainability and Performance Assessment and Benchmarking of Building – Final Report (SuPerBuildings), vol. 72. VTT Technology (Espoo).

International Council for Local Environmental Initiatives (ICLEI), 1994. Charter of European Cities & Towns towards Sustainability.

Innes, J.E., 1990. Knowledge and Public Policy: The Search for Meaningful Indicators, Second Expanded Edition. Transaction Publishers, New Brunswick USA and London UK.

Isaksson, K., Storbjörk, S., 2012. Strategy making and power in environmental assessments. Lessons from the establishment of an-out-town shopping centre in Västerås, Sweden. Environmental Impact Assessment Review 34, 65–73.

Istituto Nazionale di Statistica (ISTAT) (2010) Urban Environmental Indicators. Available at: https://www.istat.it/it/files//2011/01/Indicatori.pdf

Jakutyte-Walangitang, D., Neumann, H.M., Airaksinen, M., Bosch, P., Homeier, I., Huovila, A., Jiménez, A., Kontinakis, N., Kotakorpi, E., Malnar-Neralic, S., Pangerl, E., Pinto-Seppä, I., Recourt, E., Sarasa-Funes, D., Van der Heijden, R., 2016. CITYkeys Experience: recommendations from cities to cities. Deliverable 3.1 CITYkeys Project. [WWW document]. URL. http://www.citykeys-project.eu/citykeys/resources/general/download/CITYkeys-D3-1-Recommendations-from-cities-to-cities-user-handbook-WSWE-AJENS4. accessed 12 November 2018.

Kuik, O., Verbruggen, H., 1991. Search of Indicators of Sustainable Development. Springer-Science+Business Media, Dordrecht, Germany.

Kyttä, M., Broberg, A., Tzoulas, T., Snabb, K., 2013. Towards contextually sensitive urban densification: location-based softGIS knowledge revealing perceived residential environmental quality. Landscape and Urban Planning 113, 30–46.

Lebègue, E., Soubra, S., Huovila, P., Antuña-Rozado, C., Hyvärinen, J., Lommi, J., 2013. Analysis of present construction business models in relation to eco-innovation. In: Ecobim: Value Driven Life Cycle Based Sustainable Business Models, Deliverable D.1.1. VTT Research Centre of Finland, Espoo.

Ludin, M., Morrison, G.M., 2002. A life cycle assessment based procedure for development of environmental sustainability indicators for urban water systems. Urban Water 4, 145–152.

Maclaren, V.W., 1996. Urban sustainability reporting. Journal of the American Planning Association 62 (2), 184–202.

Martinez-Alier, J., Munda, G., O'Neill, J., 1998. Weak comparability of values as a foundation for ecological economics. Ecological Economics 26, 277–286.

Meadows, D.H., Meadows, D.L., Randers, J., Behrems III, W.W., 1972. The Limits to Growth: A Report for the Club of Rome's Project on the Predicament of Mankind. Potomac Associates, Washington D.C.

Meyerson, F.A.B., Merino, L., Durand, J., 2007. Migration and environment in the context of globalization. Frontiers in Ecology and the Environment 5 (4), 182–190.

Naredo, J.M., 1996. About the unsustainability of current conurbanizations and how to stop them. | [Sobre la insostenibilidad de las actuales conurbanizaciones y el modo de pararlas]. La construcción de la ciudad sostenible. Primer Catálogo Español de Buenas Prácticas, Ministerio de Obras Públicas, Transporte y Medio Ambiente, Madrid, Spain.

Newman, P., 1999. Sustainability and cities: extending the metabolism model. Landscape and Urban Planning 44 (4), 219–226.

Newman, P., Kenworthy, J., 1999. Sustainability and Cities: Overcoming Automobile Dependence. Island Press, Washington DC.

Newman, P., 2006. The environmental impact of the cities. Environment and Urbanization 18.2, 275–295.

Ochoa-Sosa, R., Hernández-Espinoza, A., Garfias-Royo, M., Morillón-Gálvez, D., 2017. Life cycle energy and costs of sprawling and compact neighborhoods. International Journal of Life Cycle Assessment 22 (4), 618–627.

Peuportier, B., 2016. Life Cycle Assessment Applied to Neighbourhoods. Eco-Design of Buildings and Infrastructure. CRC Press, London, pp. 15–27.

Pope, J., Annandale, D., Morrison-Saunders, 2004. Conceptualising sustainability assessment. Environmental Impact Assessment Review 24, 595–616.

Rees, W.E., 1992. Ecological footprints and appropriated carrying capacity: what urban economics leaves out. Environment and Urbanization 4.2, 121–130.

Rees, W., Wackernagel, M., 1996. Urban Ecological footprints: why cities cannot be sustainable – and why they are a key to sustainability. Environmental Impact Assessment Review 16, 223–248.

Runhaar, H., Driessen, P.P.J., Soer, L., 2009. Sustainable urban development and the challenge of policy integration: an assessment of planning tools for integrating spatial and environmental planning in The Netherlands. Environment and Planning 36, 417–431.

Severo, M., Feredej, A., Romele, A., 2016. Soft data and public policy: can social media offer alternatives to official statistics in urban policymaking? Policy & Internet 8 (3), 354–372.

Shen, L.-Y., Ochoa, J.J., Shah, M.N., Zhang, X., 2011. The application of urban sustainability indicators – a comparison between various practices. Habitat International 35, 17–29.

Simón Rojo, M., Hernández Aja, A., 2011. Sustainability assessment tools for urban design at neighbourhood scale | [Herramientas para evaluar la sostenibilidad de las intervenciones urbanas en barrios]. Informes de la Construcción 63 (521), 5–15.

Sorokine, A., Karthik, R., King, A., Budhendra, B., 2016. BigSpatial '16 Proceedings of the 5th ACM SIGSPATIAL International Workshop on Analytics for Big Geospatial Data, pp. 34–41.

Stern, N., 2006. Stern Review on the Economics of Climate Change. Government of the United Kingdom.

Svane, Ö., Weingaertner, C., 2006. MAMMUT—managing the metabolism of urbanization: testing theory through a pilot study of the Stockholm underground. Sustainable Development 14, 312–326.

Tanguay, G.A., Rajaonson, J., Lefebvre, J.-F., Lanoie, P., 2010. Measuring the sustainability of cities: an analysis of the use of local indicators. Ecological Indicators 10, 407–418.

Trigaux, D., Allacker, K., De Troyer, F., 2014. Model for the environmental impact assessment of neighbourhoods. WIT Transactions on Ecology and the Environment 181, 103–114.

United Nations, 1987. World commission of environment and development report. Our Common Future,.

United Nations, 1992. The United Nations Conference on Environment and Development - Earth Summit - Rio de Janeiro.

United Nations, 2001. Indicators of Sustainable Development: Framework and Methodologies. Background Paper No. 3. Commission on Sustainable Development Ninth Session 16-27 April 2001, New York.

United Nations, 2007. Indicators of Sustainable Development: Guidelines and Methodologies, third ed. United Nations Publications, New York.

United Nations, 2018. World Urbanization Prospects - the 2018 Revision.

Verma, P., Raghubanshi, A.S., 2018. Urban sustainability indicators: challenges and opportunities. Ecological Indicators 93, 282–291.

Wang, Y., Zou, Z., 2010. Spatial decision support system for urban planning: case study of harbin city in China. Journal of Urban Planning and Development 136 (2), 147–153.

Sustainable urban design

C H A P T E R

20

Towards sustainable urban redevelopment: urban design informed by morphological patterns and ecologies of informal settlements

Wenjian Pan, Juan Du

Department of Architecture & Urban Ecologies Design Lab, Faculty of Architecture, The University of Hong Kong, Hong Kong Special Administrative Region, China

O U T L I N E

Urban Ecology
https://doi.org/10.1016/B978-0-12-820730-7.00020-3

377

1. Introduction

1.1 Informal settlements in the global

Globally, informal settlements, such as *slum* in Latin America, *gecekondu* in Turkey, *bustee* in India, *desakota* in Southeast Asia, as well as *urban village* in China, were extensively documented urban phenomena during rapid industrialization and urbanization (McGee, 1991; Brillembourg et al., 2005; Saunders, 2010; Sengupta, 2010; Al, 2014). According to UN-Habitat (2013), around 25% of the world's population live in informal settlements, with 213 million informal settlement residents contributed to the increase of global population since 1990. There are three main characteristics/types of informal settlements, including '(1) inhabitants have no security of tenure and land for dwellings, (2) neighbourhood with lacking of basic services and urban infrastructure and (3) the housing may not comply with current planning and building regulations, and is usually situated in geographically and environmentally hazardous areas'. The informal settlements 'can be a form of real estate speculation for all income levels of urban residents, affluent and poor' (Patel, 2013; UN-Habitat et al., 2015). Nevertheless, those informal settlements that are located in city centers, though are identified as 'illegal' status and so-called 'backward' places, usually function as hubs of urban vitality (Hitayezu et al., 2018). In their function as a medium that bridged rural poverty and the urban paradise, these dynamic informal settlements in cities have continuously played a crucial role in supporting the operation and evolution of the formal parts of cities. Saunders (2010) vividly portrayed these informal settlements in *Arrival City*. Even though they are situated in different geographical and sociocultural settings around the world, these arrival cities do share very similar characteristics in terms of high building density and intensity, high population density, as well as socio-institutional spontaneity and autonomy.

Urban informality, typically the informal settlements, is an organic part of the city ecosystems, which maintains buffers, loopholes and niches to digest socio-environmental changes and then adjust the direction of urban growth, so as to create the city equipped with a complete and adaptable operational system. Kovacic et al. (2019) point out that informal settlements represent 'innovative and inventive urban environment, with microscale solutions that can have a positive effect in dealing with interconnected sustainability challenges'. Diversity is a basic nature of cities, and urban density is a very critical factor in sustaining this urban value (Jacobs, 1961). Arrival cities usually possess high-density urban morphological patterns that are associated with high population density. They not only contribute to the construction of the dynamic urban neighbourhood atmosphere but also work as a 'safe harbor' that offers affordable housing, humanized places, employment opportunities and upgraded channel for the low-income groups. They may have a far more positive than negative impact on their respective cities and, therefore, should be maintained in the cities after rehabilitation (Qin, 2012). In reality, most residents have a positive attitude of arrival cities even though they have relatively poor living conditions. This may be because the *arrival city* residents construct their houses, conduct informal small businesses, self-organize daily activities, and spontaneously establish social networks and institutions (Engquist and Lantz, 2009; Saunders, 2010; Li and Wu, 2013). It is the *arrival city* that allows the poorer classes to hope for a better future, which is why 'squatters' or so-called 'illegal constructions' are eventually transformed into 'homes' rather than temporary shelters.

1.2 Environmental effects of informal settlements within cities

As a critical indicator of urban morphology, urban density has been extensively discussed and debated by architectural design and urban planning professionals over the past century. This is related to neighbourhood environmental performance, social equity and well-being, urban accessibility and urban resilience (Boyko and Cooper, 2011; Lehmann, 2016; Wang and Shaw, 2018; Sharifi, 2019; Sim, 2019). While urban practitioners and social reformers have tried to enhance the living conditions of the overcrowded cities by reducing the physical density of the cities (Le Corbusier, 1930; CIAM, 1933), since the 1970s, an urban pattern with high-density and compact characteristics has been advocated as an approach to prevent urban sprawl due to its high connectivity, efficiency and diversity (Locke Science Publishing Company, 1979; Jenks and Burgess, 2000; D'Acci, 2019).

Many urban researchers and social reformers around the world, inspired by the compact and small-scale building layouts with the mix of old and new buildings that are attached with multiple functions, have recognized and emphasized the importance of physical and social diversity since the 1960s (Jacobs, 1961; Lynch, 1984; Owen, 2009; Rowe and Kan, 2014; Shi and Yang, 2015; Powe et al., 2016; Ye, 2017; Sim, 2019). Remarkably, previous studies have demonstrated the dominating effect of urban morphology on the environmental performances of urban neighbourhoods (Steemers and Steane, 2004; Stewart and Oke, 2012; Aflaki et al., 2017). As revealed by many environmental studies around the world, the urban neighbourhoods with compact (high-density) but diverse morphological characteristics can exhibit lower urban heat island intensity and more thermally comfortable outdoor conditions than those planned and regularized urban areas with homogeneous building form (Jr and Rodgers, 2001; Jabareen, 2006; Sharmin et al., 2015; Sobstyl et al., 2018; Wu et al., 2018). The traditional neighbourhoods with small-scale, organic and higher-density urban fabric were also observed to present better environmental performances than the newly built modern urban blocks (Andreou and Axarli, 2012). The informal settlement, a widely observed global urban phenomenon, includes diverse densities and organic urban fabrics that can provide insight into the relationship between urban density and urban resilience. Accordingly, a systematic investigating framework for analyzing the morphologies of informal settlements needs to be explored and established (McCartney and Krishnamurthy, 2018).

Beyond the global south, the city managers of the global north have also advocated for the growth of urban intensity (high-density) by concluding that informal settlements contain a variety of social and environmental resources that have generated many positive effects on cities. These have included urban neighbourhood diversity (social inclusiveness), mixed-use pattern and urban compactness, walkability, accessibility and humanized spatial characteristics (Mosel et al., 2016; Cortinovis et al., 2019). By contrast, the traditional methods of zoning and low-density pattern development have generated many negative effects on residents' quality of life, ranging from high carbon emissions and energy consumption, traffic congestions, air pollution, and a deserted street atmosphere. Although many informal settlements are problematic in terms of both physical environment and social order to some extent (such as lack of basic infrastructure, unstable living status, poor hygiene condition as well as continual events of violence), many scholars do agree that the informal settlements closely match the concept of organic urbanization, new urbanism, smart growth and flexible development (Khalil, 2010; Dovey, 2012; Kamalipour, 2016). This is because of their reasonable high-density morphological patterns and dynamic neighbourhood life (criteria of 'quality and optimal density'; for details, see Lehmann, 2016). What is needed is for the government to provide the necessary support and suitable guidance for self-help upgrading rather than resorting to total demolition to escape the social conflicts.

1.3 Objectives and analytical methods

Taking China's urban villages as informed sites with the potential for dialogue with informal settlements from the global perspective, this chapter analyzes the morphological characteristics of urban informal settlements and examines their role and environmental effects within rapidly transforming urban contexts. It also aims to discuss lessons learnt from the urban villages for sustainable urban planning and design. As China's pioneering special economic zone, Shenzhen possesses a considerable number of urban villages housing over half of the city's total population. It becomes very necessary to better understand the ecological role of such a neighbourhood typology within the city and identify its advantages and deficiencies, fundamentally, the environmental aspects.

Through on-site survey and mapping, the characteristics of building form, spatial geometries and occupants' behaviors were elaborated. The measurement of sky view factor (SVF) and total site factor (TSF), respectively, indicates the openness and solar admittance of each analyzed space. Several intelligent environmental testing instruments were employed for measuring pedestrian urban heat island intensity (PUHII), thermal comfort level (universal thermal climate index [UTCI]) and ventilation capacity (wind speed ratio [VR_w]). Based on these environmental understandings, the adopted strategies and limitations of existing design entries for urban village redevelopment were also analyzed. This chapter reveals the urban villages together with other normal urban building blocks (the scope of urban village neighbourhood) to demonstrate high environmental adaptability, owing to their diverse morphological patterns of the building fabrics, as well as the socio-ecological mechanisms of their self-organization and self-maintenance.

2. Urban villages: China's typical informal settlements

2.1 Overview of urban villages in Chinese cities

The urban village (*chengzhongcun*) is a unique urban phenomenon in China, in which thousands of former rural villages have been rapidly urbanized along with the surrounding construction environment (Wang et al., 2009; Chung, 2010). Since the late 1980s, the urban villages have contributed significantly to urban affordability and social inclusiveness by offering adequate low-rent housing and employment opportunities for numerous migrant workers and the newly graduated youths (Liu et al., 2010; Kochan, 2015; Du, 2020). According to a survey conducted by the Development Research Center of the State Council in China in 2006, there were over 50,000 urban villages throughout the main cities of China, with a residential population of more than 50 million. Even in some megacities of the Pearl River Delta (PRD) region, more than 50% of the population in these cities live in urban villages, accounting for only 20% of the urban land (Sun, 2013). Undoubtedly, urban villages have been one of the most prevailing urban residential patterns in Chinese cities today and should be carefully treated and evaluated during urban design and renewal.

As claimed by Smart and Lin (2007), the pre-existing social connections and townships in the urban villages have contributed to China's economic transformation and takeoff. Even though the urban villages are labeled as 'informal', 'illegal' and even 'backward' places in today's China, they not only highly support and supplement the development of the formal sectors but also absorb the beneficial elements and resources to achieve self-development and self-upgrading in flexible ways (Du, 2020). These urban villages, which are characterized by social inclusiveness, institutional independence, operational flexibility, spatial and environmental diversity and industrial creativity, enable people to gradually adapt to the cities and seek for their position and life value. The ecologies involved in urban villages can inspire the 'formal' decision-makers and urban practitioners to promote a more sustainable urban (re)developmental program.

2.2 Misperception of urban villages' physical environment

During the past two decades, the government, developers and even many urban planners and researchers suggested that urban villages should be demolished in the process of urban renewal, which can be ascribed to their 'high-density' building forms and the so-called very 'poor' daylight, thermal comfort and ventilation conditions. The 'handshake buildings' (*woshou lou*), 'a ray of sky' (*yixiantian*), 'dark and damp' (*yin'an chaoshi*) and 'dirty-messy-bad' (*zang-luan-cha*) have been the most commonly used phrases to describe the physical environment of urban villages in both media and academic publications. These negative discourses have further supported the demolition approach for urban village renewal. However, there are various types of urban villages in the cities of China (Wu, 2016; He and Liu, 2018), each of which possesses specific morphological characteristics and environmental performances (Fig. 20.1). Such variations do require a more systematic assessment that takes into account the local geographical and environmental specifications before the launch of urban renewal.

2.3 Underestimated environmental aspects of urban villages

From the perspective of social diversity, most of the scholars hold a positive attitude and have advocated for the necessity of building the urban villages with productive sociocultural resources and multiple adaptive functions (such as Li, 2002; Bach, 2010). Regarding the economic contribution, there are also many debates on pros and cons between the 'big economic growth' of the formal sector and the 'small business effect' of the informal industries in the urban villages (Sassen, 1997; Huang et al., 2016; Liu and Wong, 2018). However, the environmental aspect of the urban villages has been underestimated or even been neglected due to the negative stereotype of urban villages as well as the lack of a holistic angle. As argued in this chapter, the urban village should be understood as a 'social concept' (place) in relation to multiple dimensions rather than only a 'material space' (area). Whether the urban villages should be treated as a normal urban neighbourhood typology with specific sociocultural heritage or the so-called backward 'Chinese slums' depends on their typological characteristics and the evaluation scope.

Different from the English literature on informal settlements in the Global South (such as Davis, 2006; Kovacic et al., 2019; Saad et al., 2019), the Chinese literature does not mainly focus on the issue of poverty as the residents in urban villages have diverse social backgrounds and can successfully earn their employment opportunities within the settlements or in cities. According to a few anthropogenic surveys (such as Edr-lab, 2016), the real needs and environmental perceptions of migrant residents who live in urban villages are quite different from what government and researchers have observed and realized. In fact, not all urban villages are of the same quality. The migrant residents have high adaptability to the physical environment of urban villages in self-upgrading living conditions, establishing and maintaining diverse social networks and conducting self-employment. Many would say that those low-income migrant workers have no better choice but to tolerate the so-called 'terrible' living conditions. For most low-income renters, affordability is the first factor

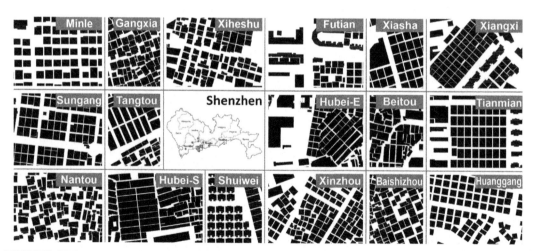

FIGURE 20.1 Urban villages in Chinese cities possess diverse fabrics and morphological patterns (case of Shenzhen City). *Surveyed and drawn by authors.*

to consider, whereas livability is less urgent. However, urban villages where affordability and livability achieve a reasonable balance might have become the optimal choice for them. For the massive migrants, urban villages are not only a place for living and working but a platform that they can own their position in the city.

2.4 Urgency and challenges in the city of Shenzhen

Before the reform and opening up in the late 1970s, there were about 2000 agrarian villages in Shenzhen (Bao'an County in the past) (Du, 2020). A Large number of farmer houses, which were surrounded by the vast areas of farmlands, distributed randomly. Initially, these villages possessed high-density low-rise built form. With the unprecedentedly rapid urbanization in four decades, it had remarkably changed the building form and overall morphologies of these villages (Chung et al., 2001). These rural villages, the later so-called urban villages, have diverse morphological characteristics and different developmental levels, ranging from the burgeoning stage (traditional high-density low-rise pattern), the developing stage (high-density mid-rise pattern), to the mature stage (high-density and mid- to high-rise pattern) and the redeveloping stage (upgrade or reconstruction) (Ma and Wang, 2011).

Significantly, these urban enclaves have also provided their farmlands for urban construction and thereby supported Shenzhen to achieve its miracles of speedy urban development and continuous high economic growth. Nowadays, Shenzhen has 320 urban villages (administrative village; for detailed view of this concept, see Wang, 2006), including 1044 natural village units, which cover an area of 93.49 km^2. Also, there are 352,618 'illegal' residential buildings and 290 million square metre of architectural areas (Hao et al., 2011; Liu et al., 2017a; Wang, 2019, Fig. 20.2). Informal housing in urban villages accounted for 70% of Shenzhen's rental market volume (Zhang, 2011). The urban villages remain a signature of the city in indigenous culture, urban structure and urban landscape (Shane, 2016; O'Donnell, 2019). However, Shenzhen also confronts with a conundrum, namely, whether these urban villages, where 60% of Shenzhen's population (around 12 million) live in the modernizing farmer houses, should be demolished and replaced by typical high-rise and high-end urban blocks. From the past two decades to the present, most of the urban villages located in the urban core of Shenzhen have been involved in the municipal urban renewal plans (Wang, 2019). What is more, many of them have already been demolished and reconstructed, such as Yumin, Tianbei, Caiwuwei, Yunong and Gangxia villages (see the area circled by a dotted box in Fig. 20.2).

3. The case study

3.1 Concept of urban village neighbourhood

It is important that the adoption of strategies for urban development and redevelopment should not only treat each site in isolation, but there should also be an expanded scope of thinking that considers the surrounding area or contexts. Through this mode of thinking, more inspiration can be gotten for efficient strategies and solutions. It should be noted that the so-called 'urban problems' (such as the widely criticized 'environmental problems' of

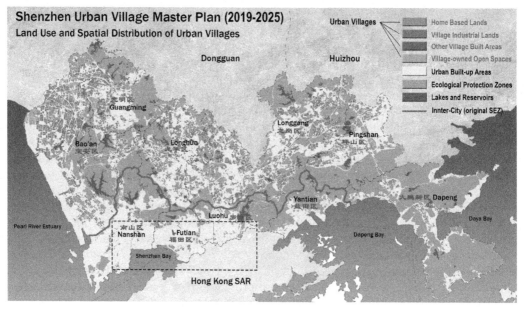

FIGURE 20.2   Land use and spatial distribution of urban villages in Shenzhen (2019). The area within the red line (gray line in printed version) (second line border) is the inner city (original special economic zone) where many urban villages had been demolished during the past decade (Source: SBPNR, 2019). *Translated and reproduced by authors.*

urban villages) should be evaluated and judged within their specific urban contexts since it is possible that the role and effect of the perceived 'problem' under different urban contexts can be altered. Therefore, the observation scope in this chapter contains the site of the urban village as well as its surrounding urban areas, referred to as urban village neighbourhood.

An urban village neighbourhood (UVN) contains the territories of the urban village per se and its surrounding urban built-up areas (within a walkable urban spatial scale), including the modern residential units, urban amenities and infrastructures (Hu et al., 2015). The UVN, which is featured by diverse material and social characteristics, can be recognized as a typology of the urban unit within the city systems (Fig. 20.3). The indigenous villagers, massive migrant renters, nearby citizens, and even transnational small business operators all live and work together in an assembled neighbourhood, forming a mixed cluster of urban communities in the process of urban (re)development (Henri, 1974; Simone, 2014; Streule et al., 2020; Fig. 20.4). Notably, the UVN is a dynamic constructed urban cell, and its physical boundary is blurred, which changes and develops with the sociocultural transformation of the city. The concept of UVN provides an approach to achieve sustainable city governance through reconfiguring the existing urban structures (Simone, 2014; Hu et al., 2015) and re-examing the ecological role of informal settlements in cities, so as to avoid unnecessary demolition and reconstruction. From the perspective of urban planning and sustainable (re) development, the city of Shenzhen is framed and organized in a series of spatial networks comprising many neighbourhoods with a low cost of living (i.e., urban villages), each of which occupies a central location within an urban district (Du, 2010; Duan, 2019; Fig. 20.2).

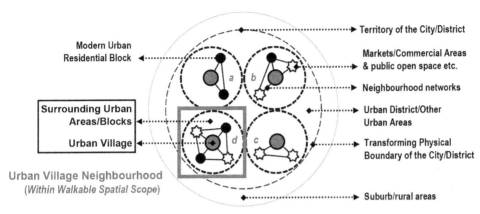

FIGURE 20.3 A conceptual diagram shows the scope and configuration of urban village neighbourhood (UVN) within rapidly transforming urban contexts in China. *a, b* and *c*: developing UVNs, *d*: mature UVN. *Adjusted and drawn by authors,* diagram format took reference and developed from *Hu, J., Li, Z. M., and Shen, Y., 2015. Briefly discuss the design concept under "the equal consideration of both internal and external users" in the urban village community. Architecture and Culture, 139(10), 127-128.*

3.2 Investigation of Hubei village neighbourhood

3.2.1 Continuous history and diverse morphological patterns

Hubei village neighbourhood, which possesses distinct generations of built form at its different stages, is an ideal model of UVN to explore and examine the ecological effect of urban villages within a city. The history of Hubei can be dated back to the age of Emperor Hongwu, Ming Dynasty (AD1368–1398) when the Zhang-Clan built the village in the current location of modern-day Shenzhen 600 years ago (AD 1369–present) (Hubei Village Committee, 2016). Hubei village neighbourhood contains four old sections (i.e., the ancient and old villages, including southern, northern, western and eastern sections), the new village, and a few newer urban blocks (Fig. 20.5). The four old sections cover a land area of 45,000 m^2 with the floor areas of more than 300,000 m^2, which house around 50,000 residents (Edr-lab, 2016). Site survey and mapping identify eight distinct forms of morphological pattern in Hubei for comparative analysis, including the following:

- Pattern A: high-density mid-rise random (old village)
- Pattern B: mid-density mid-rise (work unit)
- Pattern C: low-density mid- to high-rise (modern form I)
- Pattern D: high-density mid-rise uniform (new village)
- Pattern E: open space
- Pattern F: urban park
- Pattern G: low-density high-rise (modern form II)
- Pattern H: high-density low-rise (ancient village)

All of the investigated morphological patterns organically distribute in a mature UVN within an area of 0.48 km^2, yielding a continuous morphological transformation during past centuries, sharing the same background climate and exhibiting a wide range of environmental performances across the tested spaces (Fig. 20.6A,B and C).

FIGURE 20.4 Examples of urban village neighbourhood (UVN) in Shenzhen's urban centres (Luohu and Futian districts) (Based on Baidu Satellite Map 2019) *Drawn by authors.*

3.2.2 *Morphological characteristics of the investigated spaces*

The investigation selects 28 most popular outdoor spaces in Hubei for morphological and environmental analysis (Fig. 20.6A) and mainly investigated a set of widely adopted urban morphological factors that are related to the construction of sustainable cities: building

FIGURE 20.5 Original fabrics of Hubei rural village surrounded by farmlands (Google Satellite Map 1969) versus modern-day urban fabrics (Baidu Satellite Map 2016), and aerial view of Hubei village neighbourhood surrounded by modern high-rise building blocks (photographed by authors in 2016). *Drawn by authors.*

coverage ratio (BCR), mean nearest neighbour distance (MNN), canyon aspect ratio (H/W), mean building height (MBH), SVF, TSF and canyon orientation (Yannas, 2001; Natanian et al., 2019). The factors of SVF and TSF were measured by the software of WinSCANOPY (for details of the measurement method, see Yang, 2009). These selected spaces cover eight morphological patterns in Hubei (patterns A ~ H) (Table 20.1).

Obviously, the areas with high-density characteristics (patterns A, D and H) have a higher BCR (55%—93%) than other areas within Hubei. The ancient village with a high-density low-rise pattern (pattern H) possesses the highest BCR with a value over 90%, whereas it has the lowest mean building height (MBH is around 4.0m). Comparing the situations of high-density mid-rise old and new villages (patterns A and D), they have similar building heights

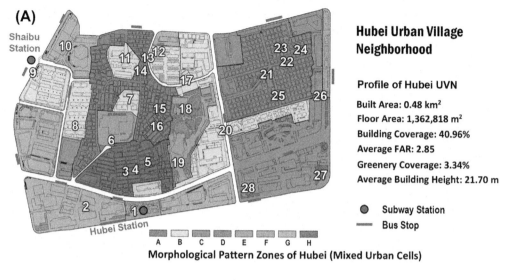

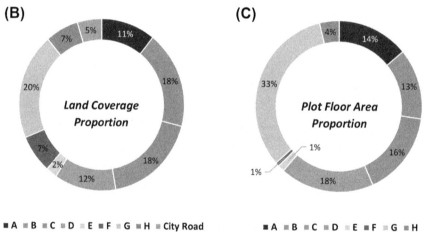

FIGURE 20.6 (A) Spatial distribution of multiple morphological patterns in the Hubei village neighbourhood; (B) land coverage proportions of the eight patterns within Hubei village neighbourhood; (C) plot floor areas proportions of the eight patterns in Hubei village neighbourhood. *Drawn by authors.*

(MBH = 18.0−21.0 m), but the old village (pattern A) has more diverse spatial geometries with multiple building coverage ratios and mean nearest neighbour distances in its different areas (BCR = 35−80%; MNN = 2.0−13.0 m) than the regularized new village (pattern D) and thereby presents a higher degree of spatial porosity (Table 20.1).

This investigation uses SVF to define the degree of openness of a given space, which considers the impact of multiple elements surrounding the space such as vegetation and incidental constructs. Relevantly, TSF indicates solar admittance, but sometimes this factor does not positively relate to SVF, yet, it is significantly affected by canyon orientations

TABLE 20.1 Built information and morphological characteristics of the 28 analyzed outdoor spaces in Hubei village neighbourhood (calculation of each urban morphological and geometric factors for each analyzed space is based on an area of 60 m × 60 m).

Pattern	No.	Year Built	Morphological characteristics/functions	Canyon axis orientation	BCR (%)	MNN (m)	MBH (m)	H/W	SVF (%)	TSF (%)
A	14	1980s	High-density mid-rise (food market street)	NNE–SSW	62.0	2.3	20.3	4.51	8.60	5.87
	15	1980s	High-density mid-rise (narrow canyon)	NW–SE	73.9	1.7	18.5	2.04	7.25	14.65
	16	1980s	High-density mid-rise (narrow canyon)	NE–SW	77.0	1.7	18.5	10.58	5.99	26.18
B	7	1990s	Mid-density mid-rise (public space)	N–S	48.1	13.4	19.5	1.72	21.75	56.93
	8	1990s	Mid-density mid-rise (linear canyon)	W–E	35.8	12.2	21.9	1.31	10.75	33.87
	11	1980s	Public space in old village (food market)	W–E	34.7	12.6	9.3	1.02	25.53	71.91
	12	1990s	Mid-density mid-rise (linear canyon)	W–E	40.7	9.5	20.4	1.44	6.78	44.54
	17	1980s	High density mid-rise (medium-wide)	WSW–ENE	54.9	4.6	17.7	1.84	18.99	42.47
C	20	2000s	Low-density mid- to high-rise (wide canyon)	NNW–SSE	31.2	11.2	14.0	0.90	44.94	85.85
	26	2000s	Low-density mid- to high-rise (wide canyon)	W–E	31.0	5.6	23.8	1.92	31.93	76.13
	28	2000s	Low-density mid- to high-rise (public space)	Open form	26.5	13.1	22.9	1.41	31.02	80.92
D	22	1986	High-density mid-rise (medium-wide canyon)	WSW–ENE	73.7	3.8	21.6	3.23	9.07	43.47
	23	1986	High-density mid-rise (medium-wide canyon)	WSW–ENE	79.6	1.8	21.4	8.81	5.93	29.04
	24	1986	High-density mid-rise (narrow canyon)	NNW–SSE	79.6	1.8	21.4	14.00	2.12	6.73
	25	1986	High-density mid-rise (narrow canyon)	NNW–SSE	72.0	3.8	20.5	2.47	12.69	34.58
E	1	2000s	Urban public square (with vegetation)	Open form	7.60	24.1	3.0	0.14	42.02	64.26
	6	1980s	Seafood market plaza in the central area	Open form	35.2	25.2	17.4	0.28	50.11	95.79
	13	1980s	Public space in old village (food market)	Open form	41.3	6.0	20.1	1.04	27.16	71.04

(Continued)

TABLE 20.1 Built information and morphological characteristics of the 28 analyzed outdoor spaces in Hubei village neighbourhood (calculation of each urban morphological and geometric factors for each analyzed space is based on an area of 60 m × 60 m).—cont'd

Pattern	No.	Year Built	Morphological characteristics/functions	Canyon axis orientation	BCR (%)	MNN (m)	MBH (m)	H/W	SVF (%)	TSF (%)
	19	1986	Public square in urban park	Open form	10.1	26.5	8.3	0.15	56.38	92.49
	21	1986	Public space in new village	Open form	48.9	2.7	21.1	0.66	29.13	67.86
F	9	1990s	Vegetated urban pocket park	Open form	18.0	12.0	11.0	0.82	12.23	7.82
	18	1986	Vegetated urban park (and water bodies)	Open form	16.9	26.5	8.3	0.41	12.48	22.62
	27	2000s	Vegetated urban street park	Open form	14.5	8.4	19.5	0.77	45.41	92.80
G	2	2000s	Low-density high-rise canyon	WNW—ESE	48.6	11.5	41.3	1.61	15.22	21.87
	10	2000s	Low-density high-rise (official towers)	NNE—WSW	48.1	15.2	22.4	2.66	21.62	49.00
H	3	1800s	High-density low-rise (linear narrow canyon)	NNW—SSE	92.8	1.8	4.2	2.08	23.74	40.97
	4	1800s	High-density low-rise (linear narrow canyon)	WSW—ENE	92.8	1.8	4.2	4.09	7.59	57.25
	5	1800s	High-density low-rise (linear narrow canyon)	WSW—ENE	83.4	2.4	4.4	0.99	22.77	76.93

BCR, building coverage ratio; MNN, mean nearest neighbour distance; MBH, mean building height; H/W, canyon aspect ratio; SVF, sky view factor; TSF, total site factor.
Surveyed and analyzed by authors.

such as the situations in Space No. 14 (in pattern A) and Space No. 22 (in pattern D), which are all embedded in the high-density mid-rise areas of Hubei and possess very low degrees of spatial openness (SVF$_{14}$ = 8.60%; SVF$_{22}$ = 9.07%) (Table 20.1). However, these two spaces exhibit distinct performances in terms of solar admittance: the former is an NNE-SSW orientation canyon and the latter is WSW-ENE (nearly W-E), resulting in their dramatic solar admittance differences (TSF$_{14}$ = 5.87%; TSF$_{22}$ = 43.47%) (Fig. 20.7). The space with an orientation of near W-E has a better agreement with the trajectory of direct solar radiation in equatorial regions and low latitudes and, consequently, permitting lesser hours with shadow in the canyon (Allard et al., 2010).

3.2.3 Neighbourhood dynamics

Hubei is located in the east of Dongmen Commercial Zone in Luohu District and also widely known as a hub of the wholesale seafood industry in the PRD region during the past decades. Even today, Hubei is still the major supplier of many seafood restaurants in Shenzhen. Two metro stations (Hubei and Shaibu) and several bus stops are located nearby

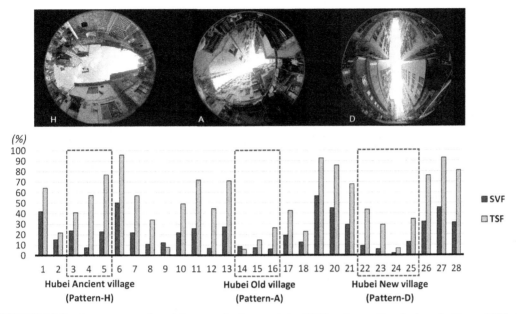

FIGURE 20.7 Comparisons of spatial geometric characteristics (SVF) and capacity of solar admittance (TSF) of the 28 analyzed outdoor spaces in Hubei village neighbourhood . *SVF*, sky view factor; *TSF*, total site factor. *Photographed and drawn by authors.*

Hubei (Fig. 20.6A). These public transport stations not only enable Hubei to act as a destination for urban transit users who work nearby commercial and official areas in Luohu but also distribute Hubei's working population over the city. According to site observation by the authors and a big data report (DT Finance, 2017), Hubei manifested very high accessibility from the public transport system and accounted for much of its patronage in Dongmen Community.

Tracking observation during a typical summer day shows that the eight morphological patterns in Hubei village neighbourhood demonstrated distinct outdoor neighbourhood dynamics (as examined by the number of people staying outside and conducting activities). The residents in Hubei have their preferences to choose specific outdoor spaces for activities, ranging from sitting and chatting, playing cards and chess, playing ball games, singing, dancing as well as just lying leisurely (Fig. 20.8 up). Specifically, much more residents stayed outside during the night-time than the daytime. The number of residents stayed in the old and new villages (patterns A and D) was larger than the other areas in Hubei throughout a day (in particular during the night-time) due to their diverse building forms and outdoor spatial geometries and smaller scale of outdoor spaces. Such spatial diversity can attract more people with different backgrounds as well as activate various small businesses with a more flexible mode of operation and management (Fig. 20.8 down-A and B), and consequently, the phenomenon of 'night city' was formed in these two areas (Song et al., 2016).

Meanwhile, those outdoor spaces, which are embedded in or located near the old village (pattern A), also presented a relatively high degree of vitality. They function as buffer zones that bridge the nearby modern urban areas and the old village. Usually, these areas have diverse spatial characteristics regarding 'size and scale', 'shape and geometry' and 'publicity

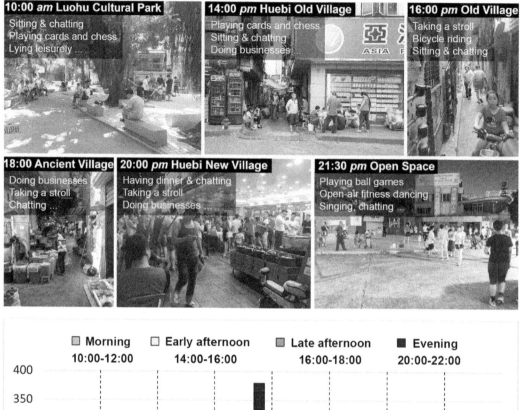

FIGURE 20.8 Up: Usage patterns of outdoor spaces in various sections of Hubei village neighbourhood during different time periods within a typical summer day. Down: The number of people staying outside in the eight patterns in Hubei during four time periods of a typical summer day. *Surveyed and photographed by authors in 2016.*

and privacy' and thus meet occupants' various needs. These spaces also offer low-cost services and amenities for the residents and have more shelters and shadowed areas for preventing the strong direct solar radiation in summer daytime. By contrast, the mid-density mid-rise residential units (pattern B) was observed to have very few people staying there in both the day and the night (Fig. 20.8 down-B). The regularized outdoor spaces and the mode of semi-gated management have strengthened these units' psychological boundaries even though the entrances are actually open for all occupants in Hubei (Grant, 2004).

3.2.4 Environmental diversity

This section examines the performances of PUHII, UTCI and VR_w in Hubei's different types of outdoor spaces (patterns A ~ H) (methods of calculation and evaluation standards of these three environmental indices refer to Liu et al., 2017b; Blazejczyk et al., 2012; Ng et al., 2011, respectively). The mobile traverse monitoring was continuously taken on foot and on a bicycle for three days during the typical summer conditions of Shenzhen (5-7 October 2016, sunny to partly cloudy, weak wind, and clear night) The environmental parameters of pedestrian air temperature (T_a), relative humidity (RH), wind speed (v) and globe temperature (T_g) were collected by several intelligent environmental testing instruments from 10:00 a.m. to 22:00 p.m. in each monitoring day. Data analysis was based on four time-periods, including morning (10:00 a.m.-12:00 p.m.), early afternoon (14:00 p.m.-16:00 p.m.), late afternoon (16:00 p.m.-18:00 p.m.), and evening (20:00 p.m.-22:00 p.m.) (Fig. 20.9).

The results of environmental monitoring demonstrated that the eight morphological patterns in Hubei presented distinct environmental performances throughout a typical summer day in subtropical Shenzhen.

- ### Pedestrian urban heat island effect

The eight morphological patterns in Hubei occurred 'weak' to 'significant' pedestrian urban heat island effect, and many of them exhibited variations during different time periods. The average PUHII of these spaces in the early night-time (20:00 p.m.-22:00 p.m., 1.62°C) was generally stronger than during the daytime (10:00 a.m.-18:00 p.m., 1.37°C). The average PUHII of spaces located in the old and new villages (average $PUHII_A = 1.38$°C; average $PUHII_D = 1.20$°C) presented temperatures lower than 1.50°C (a threshold of significant heat island effect, Guangdong Climate Center, 2014) and was very close to the performance of urban parks (average $PUHII_F = 1.21$°C) (Fig. 20.10). Both patterns A and D presented a significant difference of PUHI effect between the daytime and the early night-time. The strongest average daytime PUHII was recorded in the high-density low-rise ancient village (pattern H) during the early afternoon with the average PUHII range reaching 2.21°C, due to the E—W linear building layout complying with the sun track (see Fig. 20.7). On the other hand, the weakest daytime average PUHII (0.25°C) was observed in the high-density mid-rise new village (pattern D) in the late afternoon; most of the spaces in this pattern are shaped by the uniform building form and have a very low average degree of spatial openness ($SVF_D = 2.12$%--12.69%). Regarding the spaces located in the low-density mid- to high-rise modern blocks (pattern C), they exhibited a higher PUHII than those located in high-density mid-rise old and new villages (patterns A and D) during the daytime by 0.33 and 0.43°C, respectively, but lower PUHII in the evening by 0.58 and 0.21°C, respectively (Fig. 20.11A).

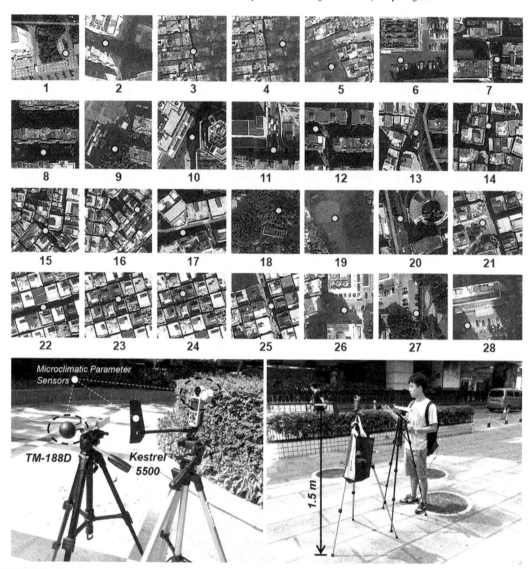

FIGURE 20.9 Urban fabrics and monitoring points of the 28 tested spaces (with a grid area of 60 m × 60 m) in Hubei village neighbourhood, installation and operational test of instruments at the height of 1.5 m on site (down-left image: thermal index meter TM-188D and mobile weather station Kestrel 5500), and data collection process during on-site monitoring in 2016 (down-right image). *Drawn and photographed by authors.*

- *Outdoor thermal comfort (UTCI)*

By examining the UTCI and the hourly variations of UTCI between the eight morphological patterns, it was found that daytime solar radiation played a dominant role in influencing outdoor thermal comfort conditions. The peak of UTCI in the eight patterns occurred in the morning and the early afternoon (10:00 a.m.–16:00 p.m.). Interestingly, both of the high-density midrise areas (patterns A and D) as well as and the vegetated urban park (pattern F) performed

FIGURE 20.10 Thermal images illustrate the daytime (14:00 p.m.–15:00 p.m.) cooling/heating effects of the investigated morphological patterns in Hubei: comparisons of high-density patterns (patterns A, D and H) and open space (pattern E) and urban park (pattern F). *Photographed by authors.*

relatively well in terms of overall outdoor thermal comfort conditions while they possessed vastly different morphological characteristics and land cover types (Fig. 20.7; Table 20.1). Patterns A and D presented a relatively low UTCI during the daytime (day-$UTCI_A = 32.73°C$; day-$UTCI_D = 32.70 < 33.00°C$) and very close to pattern F (day-$UTCI_F = 32.90°C$). By contrast, the open spaces (pattern E) exhibited the highest average daytime UTCI (33.93°C) due to their high solar heat gain from the morning to the late afternoon ($SVF_E = 27.16-56.38\%$; $TSF_E = 64.26-95.79\%$). Entering the night-time, the situation was different. Pattern A presented the highest UTCI in the early night-time (night-$UTCI_A = 31.90°C$); however, it demonstrated the most stable outdoor thermal comfort condition among the eight morphological patterns with a diurnal UTCI difference of only 0.83°C. In comparison, the low-density mid-to high-rise pattern C exhibited the lowest UTCI (31.20°C) among the eight patterns in the evening but presented considerable fluctuations with its diurnal UTCI difference reaching 1.97°C. Although pattern F presented a diurnal UTCI difference of 1.60°C, it still exhibited a relatively low UTCI (31.30 < 32.00°C) during the early night-time and was very close to the performance of open spaces pattern E (31.40°C) and the low-density high-rise pattern G (31.30°C). The high coverage of vegetation and waterscape in the urban parks (pattern F) generated pronounced transpiration heat absorption and evaporative cooling effects to decrease the outdoor air temperature and reduce solar heat gain. This resulted in its relatively low average UTCI throughout a summer day (Fig. 20.11B).

- *Pedestrian ventilation capacity (wind speed ratio)*

In terms of pedestrian ventilation capacity, most of the morphological patterns in Hubei presented very weak wind environments throughout a typical summer day ($VR_w < 0.10$).

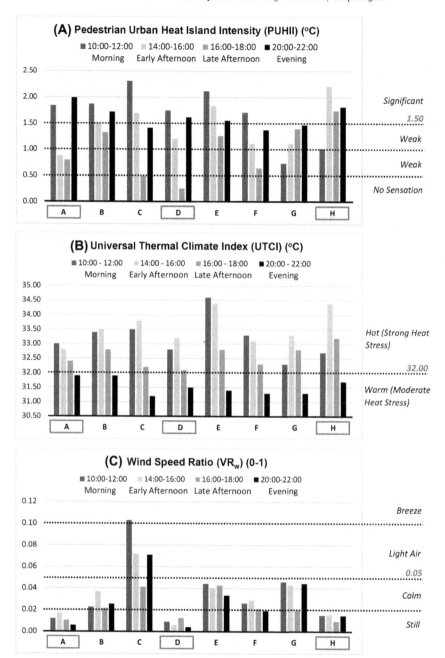

FIGURE 20.11 Average values of the examined environmental indices in the eight morphological patterns throughout a typical summer day: (A) pedestrian urban heat island intensity; (B) universal thermal climate index; (C) wind speed ratio (the red rectangle [gray rectangle in printed version] circles mark three typical patterns of urban villages in Shenzhen). *Drawn by authors.*

Pattern C was observed to have the best pedestrian ventilation capacity, followed by patterns E and G. Among the eight patterns, these three 'open' patterns presented a significantly higher average VR_w (0.07, 0.04 and 0.04 respectively) than the other morphological patterns (average $VR_w < 0.03$). Also, significant fluctuations occurred in pattern C from the morning to the evening, whereas patterns E and H presented very stable wind environment since no dramatic variations were observed between different hours. The diurnal difference range of the tested spaces within patterns E and H was lower than 0.02. Similarly, patterns A and D also exhibited very weak but stable ventilation performance with the average VR_w of 0.00–0.02. This indicates that these two high-density mid-rise patterns have pronounced effects in terms of blocking the urban wind in the city center. Meanwhile, the mid-density mid-rise pattern B was observed to present a better ventilation capacity (avg. $VR_w = 0.03$) than the high-density patterns A, D and H (avg. $VR_w = 0.00–0.02$) as well as the vegetated urban park pattern F (avg. $VR_w = 0.02$) (Fig. 20.11C).

- *Environmental effects of urban village within an UVN*

The above preliminary findings reveal that those urban villages with high-density mid-rise patterns (patterns A and D) generally presented better thermal environmental performances (as examined by PUHII and UTCI) than the newly built large-scale high-rise building blocks under extremely hot weather. It is worthy of note that such morphological patterns can result in very weak pedestrian ventilation capacity (as examined by VR_w). Significantly, none of the eight morphological patterns in Hubei village neighbourhood are always able to exhibit good outdoor environmental performances during all time periods throughout a typical summer day. However, the physical environment of each morphological pattern possesses its merits and drawbacks during the specific time periods, and their timings are asynchronous but coherent (Fig. 20.11). This leads to an understanding that the multiple morphological patterns in an UVN are environmentally complementary and have formed a synergetic network, which enables the residents a continuous outdoor living (Fig. 20.8). This pilot empirical study also inspires further investigation of the relationship between environmental performances and the morphological characteristics of each urban pattern to obtain more in-depth insight into inclusive and sustainable urban design in the future.

4. Reflections of existing urban design practices

The previous section examined the performances of three key environmental indices of urban villages and revealed the positive effects of the high-density mid-rise patterns in these urban informal settlements. Nevertheless, many architects and urban designers who have recognized the sociocultural values of urban villages still hold a negative attitude towards the physical environment in these urban villages, which may have affected their judgement and strategy choice during urban renewal practices. By understanding the environmental diversity of a UVN and the advantages of the high-density mid-rise urban villages within the UVN, this section selects 14 representative urban and architectural design entries (online accessible, Table 20.2) as study materials to discuss what has been emphasized in existing

TABLE 20.2   Information of the 14 analyzed design entries for urban village renewal in China.

No.	Year	Cities	Title of the design entries	Key aspects and main strategies for improvement
01	2003	Shenzhen	Organic urban renewal—An design research of urban village redevelopment	• **Increase ratio of outdoor public space** • 3D functional zoning and 'vertical city' • **Culture conservation** • **Urban acupuncture** • Roof garden/greenery
02	2005	Shenzhen	Overall planning of urban villages (old village) redevelopment in Shenzhen	• **Increase ratio of outdoor public space** • **Culture conservation** • **Urban acupuncture** • Occupant self-help renovation
03	2006	Shenzhen	Design research of urban village—the case of Daxin village, Shenzhen	• Improve daylighting • **Increase ratio of outdoor public space** • **Culture conservation** • **Urban acupuncture** • Mixed-use pattern • Small-scale outdoor spaces
04	2006	Shenzhen	An experimental study of the reconstruction of high-density City village in Shenzhen	• High-density development and diversity • Improve daylighting • **Increase ratio of outdoor public space** • 3D functional zoning and 'vertical city' • **Culture conservation** • **Urban acupuncture** • Roof garden/greenery • Occupant self-help renovation • Mixed-use pattern • Walkability
05	2006	Shenzhen	Village/City—City/Village: Four urban proposals for 'village-amidst-the-city', Shenzhen, China	• **Increase ratio of outdoor public space** • Open the route circulation for fire prevention • 3D functional zoning and 'vertical city' • Add new buildings and new functions • **Culture conservation** • **Urban acupuncture** • Roof garden/greenery • Introduce solar energy • Mixed-use pattern • Flexible residential units
06	2008	Shenzhen	Traditional Chinese medicine therapy applied in urban village reconstruction	• **Increase ratio of outdoor public space** • Open the route circulation for fire prevention • **Culture conservation** • **Urban acupuncture** • Roof garden/greenery • Energy saving materials • Occupant self-help renovation

TABLE 20.2 Information of the 14 analyzed design entries for urban village renewal in China.—cont'd

No.	Year	Cities	Title of the design entries	Key aspects and main strategies for improvement
07	2012	Shenzhen	Baishizhou five urban villages regeneration research	• High-density development and diversity • **Increase ratio of outdoor public space** • 3D functional zoning and 'vertical city' • Add new buildings and new functions • **Culture conservation** • **Urban acupuncture** • Roof garden/greenery • Mixed-use pattern • Flexible residential units
08	2013	Shenzhen	Urban redevelopment of Baishizhou	• 3D functional zoning and 'vertical city' • **Culture conservation** • Roof garden/greenery • Mixed-use pattern • Walkability
09	2013	Shenzhen	The necklace—affordable housing in Shenzhen	• Improve daylighting • Open the wind corridors • **Increase ratio of outdoor public space** • Add new buildings and new functions • **Culture conservation** • **Urban acupuncture** • Roof garden/greenery • Energy saving materials • Mixed-use pattern • Functional change • Flexible residential units • Walkability
10	2013	Shenzhen	Relational urbanism – Baishizhou urban regeneration	• High-density development and diversity • **Increase ratio of outdoor public space** • Add new buildings and new functions • Integrate old and new urban fabrics • **Culture conservation** • Introduce solar energy • Mixed-use pattern
11	2015	Shenzhen	The third way for urban village redevelopment	• **Increase the ratio of outdoor public space** • Open the route circulation for fire prevention • Integrate old and new urban fabrics • **Urban acupuncture** • Walkability
12	2015	Shenzhen	Micro-mega-structure in urban village: designer center design	• 3D functional zoning and 'vertical city' • Add new buildings and new functions • Integrate old and new urban fabrics • **Culture conservation**

(Continued)

VII. Sustainable urban design

TABLE 20.2   Information of the 14 analyzed design entries for urban village renewal in China.—cont'd

No. Year Cities	Title of the design entries	Key aspects and main strategies for improvement
13 2012 Zhengzhou	On design of plan for urban Village renovation—A case of planning design of Chencun	• Introduce solar energy • Energy-saving materials • Improve daylighting • **Increase ratio of outdoor public space** • **Urban acupuncture** • Roof garden/greenery
14 2015 Nanjing, Guangzhou, Shanghai Xi'an	Beyond existence—An exploration to the design of community of urban low-income earners	• Improve daylighting • **Increase ratio of outdoor public space** • 3D functional zoning and 'vertical city' • **Urban acupuncture** • Roof garden/greenery • Mixed-use pattern

Reviewed and analyzed by authors.

urban village renewal design practices in China and what are the limitations of these adopted strategies and approaches in terms of the perception and evaluation of the urban villages' environmental effects.

4.1 Main strategies for urban village redevelopment

Most of these design entries advocated for the necessity of the urban villages as productive social resources and multiple adaptive functions for cities. Many designers called for long-term conservation and improvement of these urban villages in the city, as they were functioning as material carriers of city memory and local culture inheritance (such as Wang and Dai, 2006; Llabres et al., 2013). The proposed or adopted strategies for upgrading include (1) 'reducing building density', (2) 'establishing roof garden', (3) 'adding new functions to activate neighbourhood dynamics' and (4) 'integrating rehabilitation with respect to the original village fabrics' (Table 20.2). In these design entries, the problem of 'lacking public open spaces' along with the 'absence of green cover' in urban villages was highly deemed as the result of extremely 'high building density'. Most of the design entries had a 'common sense' to reduce the urban villages' building density or to group the buildings into a few 'larger' blocks/units while demolishing the buildings with a central location to make an outdoor public courtyard, so as to release the outdoor spaces for public activities (such as URBANUS, 2006).

4.2 Limitations of the strategy of 'reduce density'

However, the strategy of 'reduce building density' mainly relied on the references of mathematical modeling under ideal conditions while lacking the consideration of the complexity of the local situations. On the one hand, gentrifying these urban villages into modern low-density high-rise blocks with large-scale outdoor public spaces might lead to the collapse of long-established sociocultural ecologies and existing economic structures and networks in these urban villages (Li, 2002; Lin et al., 2011), which is essential for the massive migrants' survival in the cities. This can significantly reduce or deprive the 'real income'[1] of those migrant residents (Liu et al., 2018). On the other hand, the lands of urban villages are in collective ownership, and each building in the urban villages possesses its property owner (despite the government has accepted or not) (Wang and Wang, 2014). Thus, the design entries that intended to remove any buildings for 'enhancing' the environmental qualities to meet the 'formal' urban design standards would cause subtle conflicts of interest (due to the ambiguous and complicated property ownership) and, as a result, exacerbate the difficulty of moving a single step.

4.3 Necessity of innovative thinking and a more flexible approach

The urban village is currently a socially and environmentally heterogeneous urban settlement type in China, the form and role of which are changing together with the transformation of its associated, surrounding urban areas. Viewing the urban villages as an organic part of the city ecosystems, perhaps architects and urban designers, should concentrate on how to maintain or sustain 'something' (additive thinking) of these 'informal' urban villages rather than reduce or eliminate 'something' (subtraction thinking). By considering the above obstacles and potential negative impacts, an integrated rehabilitation countermeasure with respect to the original village fabrics (while without significantly reducing the existing building density of these urban villages) should be advocated. A fundamental step is to introduce a precise environmental evaluation mechanism into an urban renewal program to understand multiple aspects of these urban villages' physical environment before putting forward strategies for design interventions. To achieve this goal, decision-makers and urban practitioners should hold a holistic angle to situate the element of 'environment' in the framework of ecologies of urban villages and understand their operational logic, which is elucidated in the next section.

5. Ecologies of urban villages: the 'four shared elements'

Since the ownership of the buildings and the land in the urban villages is complicated, it is difficult and unfair to demolish private-owned buildings to create adequate open spaces within urban villages to meet the normal urban standards. Consequently, an alternative approach is a 'shared' mode of management and utilization of multiple public social and environmental resources in the city to respond and digest the occurring social and environmental changes during the rapid urban transformation (including both 'sharing advantages' and 'sharing problems'). Beyond infrastructural improvement, strategies for rehabilitation and enhancement of environmental quality should be open and can be established not only in the site of urban village per se but also in the surrounding urban elements, typically

---

[1]Titmuss (1962) defined that 'real income' of an individual in the city means his or her command over the society's scarce resources, no matter the resources are paid with monetary incomes or not. Harvey (1973) pointed out that the concept of 'real income' has enriched the connotation of people's income with taking into account not only direct access to resources but also multiple public goods and services that influence urban residents' real income because of their externality effect (such as the informal industry chains, employment opportunities, social networks, shared public spaces and facilities in China's urban villages).

the concept and scope of UVN (for details, see Fig. 20.3). By considering the socio-environmental characteristics of the urban villages, each UVN should contain the 'four shared elements'. These should include 'shared environment' (morphological diversity), 'shared space' (temporal–spatial accessibility), 'shared amenity/welfare' (functional distribution), as well as 'shared social capital' (social inclusiveness). These 'four shared elements' are interrelated and refer to the formation of environmental diversity and energy saving, the inspiration of occupants' outdoor living, the improvement of neighbourhood convenience and quality of life, as well as the interaction and integration of multiple social groups respectively. It is worthy of note that the element of 'shared environment' should function as a base to involve, support and activate the other three elements (Fig. 20.12).

5.1 Shared environment—morphological diversity

Each urban village is a hub of urban vitality, which is closely connected to its surrounding urban areas. The developed UVN should include multiple urban (sub)neighbourhoods with diverse morphological patterns to form a mixed-density pattern. Such a UVN can provide physical foundation for a diverse urban and architectural environments with sharing diverse climate conditions and environmental components. Notably, the zoning of the UVN is not limited to a 'collage' or 'accumulation' of various urban morphological patterns but instead refers to the inheritance and evolution of a pre-existing urban fabrics with their attached sociocultural networks. A UVN is a complete, self-organized and well-operated urban

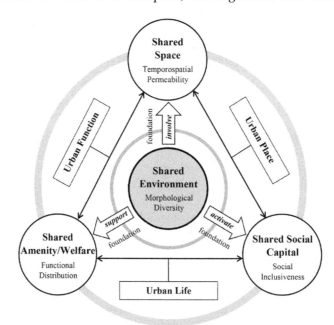

FIGURE 20.12   Interrelationships between the 'four shared elements' in an urban village neighbourhood. *Drawn by authors.*

ecosystem cell. Therefore, all sections in a functional UVN will exhibit a temporally and spatially dynamic and a continuous, coevolutionary process during urban (re)development (e.g., the coexistence and coevolution of the multiple urban morphological patterns as in the case of Hubei village neighbourhood; for details, see Fig. 20.5). Thus, efficient identification of a UVN requires systematic morphological mappings and careful investigation of the targeted urban village and the social—cultural histories of the surrounding areas (Kostof, 1991; Robinson, 2013; McCartney and Krishnamurthy, 2018).

5.2 Shared space—temporospatial permeability

Although gated communities have dominated the urban residential pattern in Chinese cities since the late 1980s (Xu and Yang, 2009), the 'mixed-use open urban neighbourhood' has also been reestablished as a more sustainable and inclusive model for a successful urban living pattern to address the emerging social and environmental issues in contemporary China (Shi and Yang, 2015; Normile, 2016; Li et al., 2018; Wang and Zang, 2018). Each type of urban (sub)neighbourhood possesses its advantages and limitations. The coexistence of multiple (sub)neighbourhood patterns within an urban ecosystem cell (e.g., an urban village neighbourhood) promotes their functional collaboration and shares their spatial resources. The mix of multiple typologies of urban (sub)neighbourhoods within an urban ecosystem cell contributes to making "lived spaces" with diverse characteristics based on size/scale, geometry, surrounding building elements, microclimate characteristics and private—public level, which can inspire residents to participate in various outdoor activities (Henri, 1974). To qualify as an urban ecosystem cell, the outdoor spaces between the sub-neighbourhoods cannot be the boundaries that separate different communities. Instead, they are excellent communication mediums for different social groups. Meanwhile, the spaces within each sub-neighbourhood function as 'internal protection shelters' for each specific social group to allow them to maintain, adjust, adapt and develop their long-established cultural attributes and social networks (Latham and Brown, 2018; Chen, 2019). Examples of this would be the 'Chao-Shan Hailufeng Community' in the southern section of Hubei, the 'Zhejiang Village' in Beijing's Dahongmen and the 'Village of Cabbies' in Shenzhen's Shixia Cun.

Physical accessibility is crucial to make the element of 'Shared Space' since it enables residents to have free mobility through the various sections of a UVN. These open and shared spatial resources allow residents to choose the appropriate place to conduct their social activities and economic businesses. Thus, to be an active UVN, all of these (sub)neighbourhoods along with their associated outdoor spaces should be physically connected and feature an open venue with high accessibility and walkability.

5.3 Shared amenity/welfare—functional distribution

City operation and residents' daily life require both formal and informal city sectors. Urban villages are inclusive residential areas in the cities that contain many small informal businesses and markets. These small businesses are flexible in operation. Some are temporary, and they often change to meet the residents' changing demands. Urban villages provide low-cost amenities (as the urban welfare) for all social groups (for residents within and outside the urban villages) that range from food markets, restaurants, leisure entertainment (e.g., urban park and open space), educational institutions, convenience stores and small workshops. Spatial diversity serves these businesses well, and with that, they prosper. A mature UVN should consider the coexistence of the formal and informal city sectors that are cofunctional and interrelated. For an urban area to be zoned as a UVN, the diversity and distribution of various urban functions and their networks need to be considered in relation to the spatial structures. An investigation of the networks of multiple formal and informal amenities and businesses within and near an urban village is an effective way to understand how the urban village and its surrounding urban built areas benefit each other and cofunction to support a UVN's operation and then understand how can the UVN's evolution and transformation be inspired.

5.4 Shared social capital—social inclusiveness

An urban village neighbourhood has a pattern of 'big mixed residence and small settlement' (*dazaju-xiaojuju*) that provides an excellent platform for residents from all social strata to communicate, educate and influence each other (such as the identified eight distinct patterns in Hubei village neighbourhood; for details, see Fig. 20.5, Fig. 20.6, and a discussion of positive effects of mixed urban communities for social sustainability by Ponce, 2010). This type of physical and social mixing in a UVN can promote the development of a balanced neighbourhood that is based on social structures, physical environment, industries and indigenous culture. On the other hand, clusters of 'small settlements' enable the construction of a community and facilitate a sense of belonging for each social group. Also, the 'small settlements' (subneighbourhoods) within a UVN can be implemented at different stages of development and can feature distinct urban morphological patterns that exhibit diverse sociocultural characteristics (such as the case of Hubei). This creates an adaptive mechanism of 'time' and 'space' that attracts people that have different social identities. Also, integration of a physical form in a UVN is limited to the subneighbourhood level with each subneighbourhood maintaining its sociocultural and morphological distinctions. The benefit of this type of 'limited' combination of morphological patterns and social integration in the new arrivals in a UVN is culturally and psychologically optional, because there are platforms (shared environment, shared space and shared amenity) and time gaps for the interactions of the new arrivals so that they can gradually interact with other classes within the UVN.

A sense of belonging (involvement), freedom (multiple options), interaction (engagement) and collaboration (social trust, mutual benefit and collective efficacy) are the four keys that preconstitute the social capital[2] (Cagney and Wen, 2008) and the urban value that is shared by all the residents in a UVN. The consideration and involvement of these aspects of social capital in a UVN can promote its prosperity and sustainability, which can sustain and enrich the effects of other three 'shared elements'.

---

[2]Social capital refers to those factors that effectively function social groups, including relationships, shared sense of identity, shared understanding, shared norms, shared values, trust, cooperation and reciprocity (definition from Wikipedia). Coleman (1988) argued that 'social capital is defined by its function. It is not a single entity but a variety of different entities, with two elements in common: they all consist of some aspects of social structures, and they facilitate certain actions of actors — whether persons or corporate actors — within the structure'.

6. Conclusion and future outlook

6.1 Summary of environmental analysis

This chapter has investigated the spatially and temporally diverse environmental performances within a model urban village neighbourhood (UVN) in Shenzhen. Preliminary measurement results reveal that the high-density mid-rise urban villages (patterns A and D) presented a weak pedestrian urban heat island effect during the summer daytime (average day-PUHII<1.50°C). Most often, these high-density informal settlements were also found to exhibit a 'hot' outdoor thermal perception during the daytime (average day-UTCI = 32.10−33.20°C > 32.00°C) but were still better than the mid- to low-density modern blocks, open spaces and even urban parks (patterns B, C, E and F, average day-UTCI = 32.20−34.60°C). However, such high-density mid-rise morphological patterns can result in a stagnant pedestrian wind environment with the average value of VR_w being lower than 0.02. Significantly, each morphological pattern in the Hubei village neighbourhood possesses its merits and drawbacks during the specific time periods, and such environmental complementarity mechanism within a UVN properly supports the continuous outdoor living of the residents, which is crucial to saving energy, contacting with nature and forming a healthy lifestyle (Sim, 2019), in particular under the national initiative of 'Healthy Cities' (Summerskill et al., 2018). Along with previously productive sociocultural and economical studies on urban villages, the environmental understandings in this chapter lead to the identification and conceptualization of 'four shared elements' in a UVN.

6.2 Insights for urban design and policy implications

Currently, there is a shift in the burgeoning urban construction in developing countries, which is driving the debate about urban sustainability by criticizing the potential risk of 'western modernism' in contemporary urban development. This criticism is based on the concept of modernization that leads to the desire to clean up urban areas that do not fit the specific concept of use and style (Kwinter, 2010; Young, 2016). However, past land use can influence urban ecological systems and remain cities' characteristics and identities (Mumford, 1961; Kostof, 1991; Ramalho and Hobbs, 2012; Robinson, 2013; Simone, 2014). Cities are evolving, and the connotations and demands of urban (re)development should likewise develop to maintain a balanced urban ecology (Geddes, 1915/1949). To seek more flexible and innovative solution for the emerging urban challenges, some scholars advocate to reconsider the phenomenon of urban informality as a 'site of critical analysis' rather than narrowly as a 'setting, sector or outcome' (Banks et al., 2019). The morphologies, functions, and effects of urban informal settlements can and should inspire urban practitioners to think deeper about the paradigms of 'ecology *of* the city' (system networks and operation of the city) and 'ecology *for* the city' (the way we treat and interact with the city) (Pickett et al., 2016). Roy and AlSayyad (2004) advocated that instead of eliminating or formulizing the informal sectors in the cities, informality should be understood as a 'new way of life' and 'mode of metropolitan urbanization' and involved in the consideration during the formal urban planning. Roy (2005) further argued that 'urban informality is an important epistemology for urban planning'. Even though dealing with the 'unplannable' (exceptions to the order of formal urbanization) was challenging, decision-makers and urban practitioners should learn to work with such 'state or logic of exception' to deal with 'uncertainties' of the cities and achieve a more effective and alternative urban planning and design (Roy, 2005; Schoon and Altrock, 2014; d'Alençon et al., 2018; Lejano and Bianco, 2018; Du, 2020).

In Chinese context, the existence and ecologies of urban villages represent a critical dimension of the 'Chineseness' of urbanization, modernity and urbanity, which contain, mitigate, digest and integrate various confrontations between the rural and urban people, industry and agriculture, the formal and the informal, as well as capital and the state (Zhan and Tong, 2017). In the era of urban renewal, many urban villages in China are struggling with the fate of demolition and gentrification (Chen, 2011; Sun, 2013). This situation has led to a wave of debate on 'sustainable enough' urban design, such as the discussion of the planned, regularized superblocks and the unplanned, spontaneously developing informal neighbourhoods. As also, a recent debate has developed about reevaluating the value of the older and smaller buildings to foster and inherit a city's identity and vitality. There have been many reflections and insights about urban design and city governance from those informally self-developing urban villages. The urban villages function in an open-inclusive manner and feature high accessibility. They offer adequate low-cost, convenient services to both the residents living in these urban enclaves and nearby urban areas.

Indeed, the physical environment of the neighbourhoods and their functional vitality are closely related to and spontaneously formed by the residents' life experience. The heterogeneous spatial patterns of a rational/optimal high-density (mixed-density pattern) contribute to the urban villages' diverse environmental conditions and energetic outdoor atmosphere, which help to form a flexible mode of space utilization. Nonetheless, the diversity of social components, including the differentiation of residents' daily routines, makes each urban village a hub of urban vitality both day and night. The establishment, maintenance and development of the UVN are not simply fully integrating people with multiple social backgrounds into one homogeneous group, but rather about remaining and sustaining their coexistence and codevelopment in the city. Therefore, the 'informal' urban villages should not be arbitrarily deemed as the so-called 'dirty-massive-bad' settlements as the government and most of the media have reported. Instead, they should be given an objective and just evaluation, especially with regard to their role in maintaining or even strengthening urban resilience and adaptability. Alongside the development of the city, the role of urban villages is also changing. With regard to the uniqueness and specification of urban villages, it is essential to clarify an urban direction. Namely, those urban villages are not always the 'problem' of the city, but rather can be the 'opportunity', 'resource' and 'approach' to solve the emerging urban problems in the contemporary Chinese cities, in particular, the intensifying conflicts between increasing urban population and living cost and the deteriorating urban environment and ecosystems. During this course, the making of related urban policies should be based on a holistic understanding of the ecologies and operational logic of the urban villages as well as their short-term and long-term effects on the cities.

Acknowledgements

Portions of the data and environmental analysis of Hubei village neighbourhood in Section 19.3.2 were presented in the sixth ZEMCH International Conference, 29 January–01 February 2018, Melbourne, Australia. The authors thank to Shenzhen Center for Design for their sharing of maps, site information, and related survey reports of Hubei Village to support urban morphological analysis. The authors also thank to those who provided design entries for review and interpretation in this chapter.

References

Aflaki, A., Mirnezhad, M., Ghaffarianhoseini, A., et al., 2017. Urban heat island mitigation strategies: a state-of-the-art review on Kuala Lumpur, Singapore and Hong Kong. Cities 62, 131–145.

Al, S. (Ed.), 2014. Villages in the City: A Guide to South China's Informal Settlements. Hong Kong University Press, Hong Kong.

Allard, F., Ghiaus, C., Szucs, A., 2010. Chapter 11: natural ventilation in high-density cities. In: Ng, E. (Ed.), Designing High-Density Cities for Social and Environmental Sustainability. Earthscan, London; Sterling, VA.

Andreou, E., Axarli, K., 2012. Investigation of urban canyon microclimate in traditional and contemporary environment. Experimental investigation and parametric analysis. Renewable Energy 43, 354–363.

Bach, J., 2010. "They come in peasants and leave citizens": urban villages and the making of Shenzhen, China. Cultural Anthropology 25 (3), 421–458.

Banks, N., Lombard, M., Mitlin, D., 2019. Urban informality as a site of critical analysis. The Journal of Development Studies 1–16.

Blazejczyk, K., Epstein, Y., et al., 2012. Comparison of UTCI to selected thermal indices. International Journal of Biometeorology 56, 515–535.

Boyko, C.T., Cooper, R., 2011. Clarifying and re-conceptualising density. Progress in Planning 76 (1), 1–61.

Brillembourg, A., Feireiss, K., Klumpner, H. (Eds.), 2005. Informal City: Caracas Case. Prestel, Munich.

Cagney, K.A., Wen, M., 2008. Social capital and aging-related outcomes*. In: Kawachi, I., Subramanian, S., Kim, D. (Eds.), Social Capital and Health. Springer, New York, NY.

Chen, W.D., 2011. *Weilai Meiyou Chengzhongcun* (A Future without Urban Villages). China Democracy and Legality Publishing House, Beijing.

Chen, C., 2019. Chengzhongcun: Chengshi Shequ Zhili de Anquanfa (Urban Village: The safety valve of urban community governance). Expanding Horizons (02), 109–115.

Chung, H., 2010. Building an image of villages-in-the-city: a clarification of China's distinct urban spaces. International Journal of Urban and Regional Research 34 (2), 421–437.

Chung, C.H., Inaba, J., Koolhaas, R., Leong, T., 2001. Great Leap Forward: Harvard Design School Project on the City. Harvard Design School, Cambridge.

CIAM (Congrès International d'Architecture Moderne), 1933. The Athens Charter. The Fourth CIAM Conference.

Coleman, J.S., 1988. Social capital in the creation of human capital. American Journal of Sociology 94, S95–S120.

Cortinovis, C., Haase, D., Zanon, B., Geneletti, D., 2019. Is urban spatial development on the right track? Comparing strategies and trends in the European Union. Landscape and Urban Planning 181, 22–37.

D'Acci, L., 2019. A new type of cities for liveable futures. Isobenefit Urbanism morphogenesis. Journal of Environmental Management 246, 128–140.

d'Alençon, P.A., Smith, H., et al., 2018. Interrogating informality: conceptualisations, practices and policies in the light of the new urban agenda. Habitat International 75, 59–66.

Davis, M., 2006. Planet of Slums. Verso Books, London.

Dovey, K., 2012. Informal urbanism and complex adaptive assemblage. International Development Planning Review 34 (3), 371–389.

DT Finance, 2017. Chongxin Renshi Ditie Shang de Shenzhen: 2017 Shenzhen Chengshi Dashuju Huoyue Baogao (Re-understanding Shenzhen through the Subway: A city big data report of Shenzhen's urban vitality). A big data city research report.

Du, J., 2010. Shenzhen: urban myth of a new Chinese city. Journal of Architectural Education 63 (2), 65–66.

Du, J., 2020. The Shenzhen Experiment: The Story of China's Instant City. Harvard University Press, Cambridge, Massachusetts.

Duan, P., 2019. Invisible Shenzhen People (22) – Low-Cost Urbanization (5). Baishizhou Special Interest Group. February 27.

Edr-lab, 2016. A Survey Report Of Hubei Village. 408 Research Group. Tongji University, Shanghai.

Engquist, J., Lantz, M. (Eds.), 2009. Dharavi: Documenting Informalities. Academic Foundation, New Delhi.

Geddes, P., 1949. Cities in evolution, (new and revised edition), 1915. William and Norgate, London.

Grant, J., 2004. Types of gated communities. Environment and Planning B: Planning and Design 31, 913–930.

Guangdong Climate Center, 2014. A Guide for Assessing Urban Heat Island Effect (First Version). Guangdong Meteorological Service.

Hao, P., Sliuzas, R., Geertman, S., 2011. The development and redevelopment of urban villages in Shenzhen. Habitat International 35, 214–224.

Harvey, D., 1973. Social Justice and the City. Edward Arnold, London.

He, L.W., Liu, J., 2018. Why are Chinese "villages in the city" different – distinguishing the differences in the same city and between different cities. Academic Monthly 50 (7), 80–89.

Henri, L., 1974. The Production of Space. Horizon Press, New York.

Hitayezu, P., Rajashekar, A., Stoelinga, D., 2018. The Dynamics of Unplanned Settlements in the City Of Kigali. Final Report, C-38312-RWA-1. International Growth Centre.

Hu, J., Li, Z.M., Shen, Y., 2015. Briefly discuss the design concept under "the equal consideration of both internal and external users" in the urban village community. Architecture & Culture 139 (10), 127–128.

Huang, G.Z., Xue, D.S., Zhang, H.O., 2016. The development of urban informal employment and its effect on urbanization in China. Geographical Research 35 (3), 442–454.

Hubei Village Committee, 2016. Shenzhen (Pengcheng) Hubei Zhangshi Zumai Yuanliu Kaoji (A Historical Record of the Origin and Development of the Zhang Families In Hubei Village, Shenzhen). Unpublished historical record by Hubei Village Committee, Shenzhen.

Jabareen, Y.R., 2006. Sustainable urban forms: their typologies, models, and concepts. Journal of Planning Education and Research 26, 38–52.

Jacobs, J., 1961. The Death and Life of Great American Cities. Random House, New York.

Jenks, M., Burgess, R. (Eds.), 2000. Compact Cities: Sustainable Urban Forms for Developing Countries. Spon Press, London.

Jr, B.S., Rodgers, M.O., 2001. Urban form and thermal efficiency: how the design of cities influences the urban heat island effect. Journal of the Amercian Planning Association 67 (2), 186–198.

Kamalipour, H., 2016. Forms of informality and adaptations in informal settlements. International Journal of Architectural Research 10 (3), 60–75.

Khalil, H.A.E.E., 2010. New Urbanism, Smart Growth and Informal Areas: A Quest for Sustainability. Sustainable Architecture & Urban Development, CSAAR. January.

Kochan, D., 2015. Placing the urban village: A spatial perspective on the development process of urban villages in contemporary China. International Journal of Urban and Regional Research 39 (5), 927–947.

Kovacic, Z., Musango, J.K., et al., 2019. Interrogating differences: a comparative analysis of Africa's informal settlements. World Development 122, 614–627.

Kostof, S., 1991. The City Shaped: Urban Patterns and Meanings Through History. Little, Brown, Boston.

Kwinter, S., 2010. Notes on the third ecology. In: Mostafavi, M., Dothty, G. (Eds.), Ecological Urbanism. Lars Muller Publishers, London, pp. 94–105.

Latham, Z., Brown, R., 2018. Shenzhen's Urban Villages: Dialogic Cultural Landscapes and Resilient Rituals. Socio.Hu, Special Issue 2018.

Le Corbusier, 1930. La Ville Radieuse. [1935]. Editions de l'Architecture d'Aujourd'hui, Boulogne.

Lehmann, S., 2016. Sustainable urbanism: towards a framework for quality and optimal density? Future Cities and Environment 2, 8. https://doi.org/10.1186/s40984-016-0021-3.

Lejano, R.P., Bianco, C.D., 2018. The logic of informality: Pattern and process in a São Paulo favela. Geoforum 91, 195–205.

Li, M.Y., Ma, Y., Sun, X.M., Wang, J.Y., Dang, A.R., 2018. Application of spatial and temporal entropy based on multi-source data for measuring the mix degree of urban functions. City Planning Review 42 (2), 97–103.

Li, P.L., 2002. Tremendous changes: the end of villages – a study of villages in the center of Guangzhou city. Social Sciences in China 01, 168–179.

Li, Z.G., Wu, F.L., 2013. Residential Satisfaction in China's informal settlements: a case study of Beijing, Shanghai, and Guangzhou. Urban Geography 34 (7), 923–949.

Lin, Y.L., de Meulder, B., Wang, S.F., 2011. Understanding the 'village in the city' in Guangzhou: economic integration and development issue and their implications for the urban migrant. Urban Studies 48 (16), 3583–3598.

Liu, L., Lin, Y.Y., Liu, J., et al., 2017b. Analysis of local-scale urban heat island characteristics using an integrated method of mobile measurement and GIS-based spatial interpolation. Building and Environment 117, 191–207.

Liu, R., Wong, T.C., 2018. Urban village redevelopment in Beijing: the state-dominated formalization of informal housing. Cities 72, 160–172.

Liu, S.S., Wang, J., Xu, Y., Zeng, H., 2017a. Study on approaches of buildings carbon mitigation based on spatial-temporal characteristics of civil building energy consumption: a case study on Shenzhen, China. Acta Scientiarum Nauralium Universitaties Pekinensis 54 (1), 125–136.

Liu, Y.T., He, S.J., Wu, F.L., Webster, C., 2010. Urban villages under China's rapid urbanization: unregulated assets and transitional neighbourhoods. Habitat International 34, 135–144.

Liu, Y., Lin, Y.L., Fu, N., Geertman, S., van Oort, 2018. Towards inclusive and sustainable transformation in Shenzhen: urban redevelopment, displacement patterns of migrants and policy implications. Journal of Cleaner Production 173, 24–38.

Llabres, E., et al., 2013. Baishizhou RUM. Rational Urbanism. Available: https://www.relationalurbanism.com/baishizhou-rum. (Accessed 20 June 2017).

Locke Science Publishing Company, 1979. The charter of machu picchu. Journal of Architectural Research 7 (2), 5–9.

Lynch, K., 1984. Good City Form. MIT Press, Cambridge, MA.

Ma, H., Wang, Y.W., 2011. The Space Evolvement and Integration of Villages in Shenzhen. Intellectual Property Publishing House, Beijing.

McCartney, S., Krishnamurthy, S., 2018. Neglected? strengthening the morphological study of informal settlements. SAGE Open 8 (1).

McGee, T.G., 1991. The emergence of desakota regions in Asia: expanding a hypothesis. In: Ginsburg, N., Koppel, B., McGee, T.G. (Eds.), The Extended Metropolils: Settlement Transition in Asia. University of Hawaii Press, Honolulu, pp. 3–26.

Mosel, I., Lucci, P., Doczi, J., Cummings, C., et al., 2016. Urbanisation: Consequences and Opportunities for the Netherlands' Directorate-General for International Cooperation. A Report. Overseas Development Institute.

Mumford, L., 1961. The City in History: Its Origins, Its Transformations, and Its Prospects. Harcourt, Brace and World, New York.

Natanian, J., Aleksandrowicz, O., Auer, T., 2019. A parametric approach to optimizing urban form, energy balance and environmental quality: the case of Mediterranean districts. Applied Energy 254, 113637.

Ng, E., Yuan, C., et al., 2011. Improving the wind environment in high-density cities by understanding urban morphology and surface roughness: a study in Hong Kong. Landscape and Urban Planning 101, 59–74.

Normile, D., 2016. China Rethinks Cities — after decades of reckless growth, the country revises its vision. Science 352 (6288), 916–918.

O'Donnell, M.A., 2019. Heart of Shenzhen: The movement to preserve "Ancient" Hubei Village. In: Banerjee, T., Loukaitou-Sideris, A. (Eds.), The New Companion to Urban Design. Routledge, London, pp. 480–493.

Owen, D., 2009. Green Metropolis: Why Living Smaller, Living Closer, and Driving Less Are the Keys to Sustainability. Penguin Publishing Group, New York.

Patel, K., 2013. Topic Guide: Provision and Improvement of Housing for the Poor. https://doi.org/10.12774/eod_tg11.dec13.patel.

Pickett, S.T.A., Cadenasso, M.L., et al., 2016. Evolution and future of urban ecological science: ecology in, of, and for the city. Ecosystem Health and Sustainability 2 (7) e01229.

Ponce, J., 2010. Affordable housing as urban infrastructure: a comparative study from a European perspective. The Urban Lawyer 42–43 (4–1), 223–245.

Powe, M., Mabry, J., Talen, E., Mahmoudi, D., 2016. Jane Jacobs and the value of older, smaller buildings. Journal of the American Planning Association 82 (2), 167–180.

Qin, H., 2012. Residency of the New Urban Poor: How to Evaluate Shantytowns, Illegal Buildings, Urban Villages, and Low-Rent Houses, vol. 01. Tribune of Social Sciences, pp. 195–219.

Ramalho, C.E., Hobbs, R.J., 2012. Time for a change: dynamic urban ecology. Trends in Ecology and Evolution 27 (3), 179–188.

Robinson, J., 2013. The urban now: Theorising cities beyond the new. European Journal of Cultural Studies 16 (6), 659–677.

Rowe, P.G., Kan, H.Y., 2014. Urban Intensities: Contemporary Housing Types and Territories. Birkhäuser, Basel; Boston.

Roy, A., AlSayyad, N. (Eds.), 2004. Urban Informality. Transnational Perspectives from the Middle East, Latin America, and South Asia. Lexington Books, Lanham.

Roy, A., 2005. Urban informality — toward an epistemology of planning. Journal of the American Planning Association 71 (2), 147—158.

Saad, O.A., Fikry, M.A., Hasan, A.E., 2019. Sustainable upgrading for informal areas. Alexandria Engineering Journal 58, 237—249.

Sassen, S., 1997. Informalization in Advanced Market Economies. Development Policies Department, International Labour Organization, Geneva.

Saunders, D., 2010. Arrival City: How the Largest Migration in History Is Reshaping Our World. William Heinemann, London.

SBPNR (Shenzhen Bureau of Planning and Natural Resources), 2019. Shenzhen Urban Village (Old Village) Master Plan for Integrated Rehabilitation (2019-2025). Shenzhen municipal government.

Schoon, S., Altrock, U., 2014. Conceded informality. Scopes of informal urban restructuring in the Pearl River Delta. Habitat International 43, 214—220.

Sengupta, U., 2010. The hindered self-help: housing policies, politics and poverty in Kolkata, India. Habitat International 34, 323—331.

Shane, D.G., 2016. Notes on villages as a global condition. Architectural Design 86 (4), 48—57.

Sharifi, A., 2019. Resilient urban forms: a macro-scale analysis. Cities 85, 1—14.

Sharmin, T., Steemers, K., Matzarakis, A., 2015. Analysis of microclimatic diversity and outdoor thermal comfort perceptions in the tropical megacity Dhaka, Bangladesh. Building and Environment 94, 734—750.

Shi, B.X., Yang, J.Y., 2015. Scale, distribution, and pattern of mixed land use in central districts: a case study of Nanjing, China. Habitat International 46, 166—177.

Sim, D., 2019. Soft City: Building Density for Everyday Life. Island Press, Washington.

Smart, A., Lin, G.C.S., 2007. Local capitalisms, local citizenship and translocality: rescaling from below in the Pearl River Delta region, China. International Journal of Urban and Regional Research 31 (2), 280—302.

Sobstyl, J.M., Emig, T., Qomi, A., et al., 2018. Role of city texture in urban heat islands at night time. Physical Review Letters 120, 108701.

Song, H.Y., Pan, M.M., Chen, Y.Y., 2016. Nightlife and public spaces in urban villages: a case study of the Pearl River Delta in China. Habitat International 57, 187—204.

Steemers, K., Steane, M.A., 2004. Environmental Diversity in Architecture. Spon Press, New York.

Stewart, I.D., Oke, T.R., 2012. Local climate zones for urban temperature studies. American Meteorological Society (93), 1879—1900.

Summerskill, W., Wang, H., Horton, R., 2018. Healthy cities: key to a healthy future in China. The Lancet 391 (10135), 2086—2087.

Sun, L., 2013. Current Situation and Human Settlements Remediation of Chinese Urban Village. China Building Industries Press, Beijing.

Streule, M., Karaman, O., Sawyer, L., Schmid, C., 2020. Popular urbanization: Conceptualizing urbanization and processes beyond informality. International Journal of Urban and Regional Research. https://doi.org/10.1111/1468-2427.12872.

Titmuss, R.M., 1962. Income Distribution and Social Change: A Study in Criticism. Allen & Unwin, London.

UN-Habitat, 2013. Streets as Public Spaces and Drivers of Urban Prosperity. UN-Habitat, Nairobi.

UN-Habitat, OHCHR, and UNOPS, 2015. Habitat III Issue Papers — 22. Informal Settlements, Nairobi: UN-Habitat.

URBANUS, 2006. Cun Cheng: Cheng Cun (Village/City: City/Village). China Electric Power Press, Beijing.

Wang, P., 2019. Shenzhen chengzhongcun de weilai Jiazhi Yu "weigengxin" moshi tanjiu (the value of urban villages for future Shenzhen and exploration of "micro-upgrade" model). Theoretical Research in Urban Construction 08, 22—25.

Wang, Q., Zang, X.Y., 2018. Origin, features, and adaptive planning strategies of urban block form. City Planning Review 42 (9), 127—134.

Wang, Y., Shaw, D., 2018. The complexity of high-density neighbourhood development in China: intensification, deregulation and social sustainability challenges. Sustainable Cities and Society 43, 578—586.

Wang, Y.P., Wang, Y.L., 2014. China's Urbanisation from below: Village Led Land and Property Development. Working Paper, Lincoln Institute of Land Policy.

Wang, Y.P., Wang, Y.,L., Wu, J.,S., 2009. Urbanization and informal development in China: urban villages in Shenzhen. International Journal of Urban and Regional Research 33 (4), 957—973.

Wang, J.X., 2006. Village Governance in Chinese History. In: Political Economy of Village Governance in Contemporary China, PhD Dissertation. Indiana University, Bloomington, pp. 41–66.

Wang, Y.W., Dai, D.H., 2006. An experimental study of the reconstruction of high-density city village in Shenzhen. Urbanism and Architecture 12, 37–41.

Wu, F.L., 2016. Housing in Chinese urban villages: the dwellers, conditions and tenancy informality. Housing Studies 31 (7), 852–870.

Wu, W., Ren, H.Y., Yu, M., Wang, Z., 2018. Distinct influences of urban villages on urban heat islands: a case study in the Pearl River Delta, China. International Journal of Environmental Research and Public Health 15 (8), 1666.

Simone, A., 2014. Jakarta, Drawing the City Near. University of Minnesota Press, London.

Xu, M., Yang, Z., 2009. Design history of China's gated cities and neighbourhoods: prototype and evolution. Urban Design International 14 (2), 99–117.

Yang, F., 2009. The Effect of Urban Design Factors on the Summertime Heat Islands in High-Rise Residential Quarters in Inner-City Shanghai. PhD Thesis. The University of Hong Kong.

Yannas, S., 2001. Toward more sustainable cities. Solar Energy 70 (3), 281–294.

Ye, J.J., 2017. Managing urban diversity through differential inclusion in Singapore. Environment and Planning D: Society and Space 35 (6), 1033–1052.

Young, R.F., 2016. Modernity, postmodernity, and ecological wisdom: towards a new framework for landscape and urban planning. Landscape and Urban Planning 155, 91–99.

Zhan, Y., Tong, X.X., 2017. "Chengzhongcun" Yu zhongguo chengshihua de Teshu daolu ("Urban village" and the special road of China's urbanization). Journal of Tsinghua University - Philosophy and Social Sciences 6, 33–35.

Zhang, L., 2011. The political economy of informal settlements in post-socialist China: the case of chengzhongcun(s). Geoforum 42 (4), 473–483.

Assessing the role of urban design in a rapidly urbanizing historical city and its contribution in restoring its urban ecology: the case of Varanasi, India

Vidhu Bansal[1], Sunny Bansal[2], Joy Sen[3]

[1]Research Scholar, Department of Architecture and Regional Planning (ARP), Indian Institute of Technology (IIT) Kharagpur, West Bengal, India; [2]Research Scholar, Ranbir and Chitra Gupta School of Infrastructure Design and Management (RCGSIDM), Indian Institute of Technology (IIT) Kharagpur, West Bengal, India; [3]Professor and Head, Department of Architecture and Regional Planning; Joint Faculty, Ranbir and Chitra Gupta School of Infrastructure Design and Management, Indian Institute of Technology (IIT) Kharagpur, West Bengal, India

Cities all around the world are reeling under the waves of rapid urbanization and globalization. This conjoint phenomenon not only puts the city's infrastructure under huge pressure but also induces a significant challenge to its urban landscape. Here, urban design can act as a marker of activities and a catalyst of change, which can alter a city's urban ecology. These changes are evident in both contemporary and historical cities, the latter being more sensitive to these transformations. Hence, this study attempts to assess the urban design of the historical city of Varanasi at multiple spatial and temporal scales and their impact on urban ecology. Principles of 'new urbanism' are applied to the historical interface to identify the elements of design in conjunction with 'space syntax analysis'. The study would conclude in a methodology for generating appropriate urban design strategies that can restore the pristine efficacy of the city's ecosystem

1. Introduction

History has its own way of revealing itself. It does not lie in the fold of a book but maybe at the turn of a street, tucked away in a hidden niche, a story left untold in the nooks and crannies of that old neighbourhood. Cities are among the most important of the history lessons, which are often left unacknowledged in plain sight.

> Dull, inert cities, it is true, do contain the seeds of their own destruction and little else. But lively, diverse, intense cities contain the seeds of their own regeneration, with energy enough to carry over for problems and needs outside themselves. —*Jane Jacobs, The Death and Life of Great American Cities.*

Cities have always been a reflection of the people who inhabited it. The needs dictated the formation of settlements. Piece by piece like a temporal jigsaw puzzle manifested into its tangible form. The values, rituals, culture and traditions were allied with the necessities, directing the generation process. The progression coursing over thousands of years involving multiple trials and errors to find out which form actually suited the dynamic needs of this evolving phenomenon. There have been multiple approaches to understand the developmental changes in the city's expansion and growth. From a very conventional and rigid 'master plan' approach to flexible ideas like 'new paradigm', which talks about the various interactions that are performed among individuals, a central system of planning and governing would have limited influence on cities (Peris et al., 2018). The cumulative effect of these ideas has a diverse impact on all aspects of the city's form, functioning and ecology.

As stated in Sustainable Development Goals (SDGs), more than half of the world's population now live in urban areas. By 2050, that figure will have risen to 6.5 billion people—two-thirds of all humanity. SDG 11 suggests 'Making cities safe and sustainable means ensuring access to safe and affordable housing, and upgrading slum settlements. It also involves investment in public transport, creating green public spaces, and improving urban planning and management in a way that is both participatory and inclusive' (UNDP, 2016). One of the subtasks of this development goal also focuses on strengthening efforts to protect and safeguard the world's cultural and natural heritage. A development strategy with an innovative approach that bridges the gap between traditional urban regeneration process and the dynamism of modern man's urban surroundings might be the need of the hour to achieve the balance of the urban ecosystems. The indicators listed in SDG 11 and the new urban agenda focus on the efficient resource management and maintaining the balance through economic, social and environmental links between urban, periurban and rural areas, which is basically the entire urban ecosystem in spatial sense for a city. Hence policies aligned to this global goal are something required in the local setting too.

In traditional cities, rapid urbanization scars the organic growth with its overwhelming demands of space and infrastructure. A city reeling under its own pressure can succumb to its ill effects and lead to congested cores and sprawling suburbs, which in turn add to the grievances of the citizens. It is a multifaceted phenomenon and can have various perspectives to look, which can broadly be classified into the following categories—spatial (physical), socioeconomic and ecological (environmental) (Newman, 2006). In this scenario, sustainable development cannot be achieved without significantly transforming the way we build and manage our urban spaces (UNDP, 2016). A 2013 International Business Times' article writes about India as one of the major stakeholders of this population growth and is predicted to become the most populous country swapping its position with China by 2050 (Bansal et al., 2017). The predicted population growth will put immense pressure on the urban services provided by the Indian cities. In an absence of adequate infrastructure to service the core and peripheral areas, the cities may succumb to the excruciatingly huge pressure of urbanization. This should not only prompt the 'new and contemporary' Indian cities to get in tune with latest tools and techniques so as to empower their arsenal with competent urban services, but it should also be a reminder for 'traditional cities' to come at par with their fresher counterparts in enhancing their urban services.

In the following section, the major problem and the research question is addressed, which would give an overview of the issue that is being dealt with in this study. This is followed by in-depth literature reviews dealing with 'Living Historical Cities', 'Concept of Planning and Evolutionary Changes', 'Urban Design and Urban Ecology', and 'Space Syntax Analysis' among other allied concepts. This leads to an introduction to the case study that is Varanasi and a detailed description of the workings of the city and the problem, which is currently being faced by it. The later part of this chapter deals with the methodology and analysis portion. It is then followed by a discussion on the results obtained and a final portion of conclusion.

2. Problem and research question

Indian traditional cities are complex entities. They have strong cultural relations embedded into its urban fabric, which makes them multifarious urban opportunities and problems for the planners to deal with. On one hand, they have the rich legacy of architectural history that can take back anyone into ancient times by the virtue of their indigenous built styles, whereas on the other hand, they struggle to deal with the needs of modern man and the fast-paced modern life. The rapid inflow of people into the urban zones with not so rapid improvement in infrastructure render these cities inefficient, and they become a ghost of their own glorious past.

The effect of excessive pressure on the infrastructure is much more evident in public areas than the residential zones. This may sometimes lead to a complete lack of urban places in these cities, or even if they are present, then also they have a negligible place in the daily life of people. Now, the presence of public spaces and their essentiality have been asserted as the universal urban trait throughout history. From the ancient Greek polis to the 14th century Renaissance city and to the present 21st-century postindustrial city, the transformation of public life in these spaces has reflected the continuous evolution of urban environments (Adhya, 2007). The public realm is an important part of our town and cities. This is the physical setup, an urban theatre where the interrelationships occur among people and between people and various components of urban life (Short, 1998) as cited in (Adhya, 2007). It can be seen that there is a renewed interest in public space with a growing belief that while modern societies no longer depend on the town square or the piazza for basic needs, good public space is required for the social and psychological health of modern communities (Mehta, 2014).

Indian cities had a rich legacy of the vibrant social culture ingrained in their public spaces in ancient times, but with the wave of urbanization, these spaces are gradually succumbing to a state of decline and nonexistence. These spaces are not only essential for social interaction and as breathing spaces but also a necessity to maintain a healthy mental balance of the citizens. It is extremely important at this point to seek a balance between nature and built environment to maintain the overall ecological balance of the city. This could only be possible if these urban places are restored and reingrained in the daily milieu of a citizen's life.

> There is a common tendency to focus on natural features (e.g. rivers and trees) rather than the processes that shape and structure them (e.g. flow of air, water and materials; plant reproduction and growth). Ignoring natural processes leads to harmful consequences, including the failure of planners to accommodate dynamic change, their failure to make connections among seemingly unrelated issues and phenomena and to realize opportunities. *Spirn, (2012)*

The above-written quote precisely specifies the issue here. As an urban designer and planner, the vision of the city becomes confined to a formal ordered approach. It is essentially required to not think of a solution that has been procreated in solitude on the blank pages by an isolated individual. Rather, it should be a solution based on the indigenous idea, stemming from the inherent needs and age-old wisdom. Often, when focusing on the issue of maintaining the natural balance of built and unbuilt, the significance of the process that shapes the urban ecology of the city is lost. Moreover, the tools and techniques that would facilitate this transformation should act as a bridge that preserves the heterogeneity and essence of Indian cities and not a tunnel that beckons an identity crisis. Here, urban design might act as a corrective procedure to restore this lost significance. This leads to the research question for this study: 'Can ancient urban patterns become markers of balanced urban ecology?' If so, then how to assess the current conditions?

2.1 Living historical cities: brief about historical cities

The traditional (historical) cities have deep-seated cultural values associated with them, which render a unique aura to these regional composites. Here, the built environment is the most salient manifestation of heritage (Chhabra et al., 2003). History serves as the base upon which the city thrives, and the historical built environment or HBE is the principal spatial element on which the interplay between time and space is played out. These historical elements are both iterative and partial in their existence. Although historic preservation has played an increasingly important role in shaping urban form yet demolition and removal of historic structures and the haphazard mushrooming of new built forms, are all too common in the cities worldwide, and Indian cities are no different. The juxtaposition of new development with once historic environments can sometimes empty the latter of their deep and cultural significance (Newman, 2016). It is evident that urbanization spree is the causative agent of urban sprawls on the periphery and urban congestion in the core for many Indian cities. This has resulted in haphazard growth and is a reason for creating a sense of chaos and urban planning anomalies throughout. Earlier, the town belonged to everyone, and a sense of ownership was opaque, but gradually a shift can be seen in the scenario (Kush, 2015; Curry, 2005). Slowly and steadily parks are giving ways to parkways, and green lungs are getting smaller day by day. Populations' pressure on space has never been so inevitable (Kush, 2015). Here, it is required that those in power of decision-making and designing, such as urban planners and designers, must do a better job preserving and modifying existing built forms for new development schemes, rather than the typical *Tabula Rasa* approach.

Indian traditional cities can be considered as a vibrant microcosm that has evolved gradually over time, amalgamating culture and modernity. These cities are evolved as traditional settlement types, developing their form owing to the needs of the inhabitants governed by climate and culture as the driving forces. The form that we see today is a composite result of trials and errors in the built and unbuilt context trailing through hundreds and thousands of years. Varanasi, Ujjain, Kancheepuram, Puri, Ayodhya, Udaipur and so on can be considered as examples of Indian traditional cities. They have an inherent organization, which is engulfed in an outer layer of chaos, invisible to the observer who is outside the Indian system (Singh, 2017). However, when we put them in the course of rapid urbanization, then the situation gets drastically modified. These cities are comparatively more sensitive to the issues of urbanization from their modern counterparts because they are simultaneously trying to balance an age-old traditional heritage with the dynamic needs of modern civilization.

Earlier, one can distinguish in each historic city on the basis of specific urban patterns or features such as the nature and density of land uses, the height of buildings, width and pattern of circulation routes (roads, alleys and footpaths), building typologies, as well as specific infrastructure components. These form the components of the 'urban tissue' (Steinberg, 1996). Within the tangible manifestation, that is the 'urban tissue' of the city, lies the intangible expressions. Be it special occasions such as festivals, pilgrimage, ritualistic activities, or carrying out day-to day urban activities, these intangible values are embedded in the urban landscape. They shape the culture, value and the spirit of the place (M, 2008). Any

VII. Sustainable urban design

distortion in the tangible form thus has a direct implication on the intangible values. When rapid urbanization comes into the picture, it is observed that it takes a huge toll on the existing urban conditions and starts affecting the physical spaces, tending to haphazard urban development. One of the major shortcomings of the phenomenon is the gradual degradation of urban form. The form existed earlier by the virtue of urban patterns, which helped to distinguish one city from the other. For example, on comparing old historical neighbourhoods of Varanasi (Northern India) with Ujjain (Central India) and Kancheepuram (Southern India), it can be observed that the urban grain differed from each other to a large extent. This quality is getting diminished as the cities are expanding into newer realms, with apartments of various typologies dotting their skyline.

Generally, mitigation of this urban issue is done through charting urban policies, which mostly overlook the indigenous needs and aspirations of the stakeholders on one hand and cause identity crisis by lacking the natural order of spaces on the other hand. The study attempts to look into the case area through the lens of spaces as a concept and a physical entity since they have gone through a multitude of changes in the past few decades, and it is important to manage them in the magnitude they are present. This study draws its inspiration from the works of Christopher Alexander (CA), viz., *A Pattern Language* technique that would be forming the basis of new urbanism design movement and is applied along with allied techniques for observing the various urban spaces against a larger continuum of the sustainable traditional urban environment.

2.2 Concept of planning

The basic idea behind urban planning is to produce some kind of explicit rationality driven by collective purposes that are capable of regulating the multitude of individual decisions that build the city over time. How this rationality is constructed and what are the underlying criteria and analytical interpretations remain the most critical questions in urban planning. A strict diagnosis of the current scenario and an efficient prognosis of the outcomes are the basis of competent planning. However, it can be seen that physical and morphological dimensions of urban planning and this kind of approach in methods and models remain largely dubious, if not simply absent from the regular practice (Serra et al., 2011). Thus, the need of the hour is to invoke the prerequisites for a process that can take a holistic view of current urban scenarios and might lead to some plausible conclusion on the steps that could be taken for battling this issue at a grass-root level.

Taking the case of planning in Indian traditional (historical) cities, then it can be said that they are complex entities. They have strong cultural relations embedded into its urban fabric, which makes them simultaneously a case of multifarious urban opportunities and problems for the planners to deal with. On one hand, it has the rich legacy of architectural history, which can take back anyone into ancient times by the virtue of their indigenous built styles, whereas on the other hand, it struggles to deal with the needs of modern man and the fast-paced modern life. The rapid inflow of people into the urban zones with not so rapid improvement in infrastructure render these cities inefficient, and they become a ghost of their own glorious past.

Mental maps of Indian cities can categorize a city into many powerful typologies. Going by the colloquial genre, we can identify three major typologies (north Indian style) as *ghars* (residences), *nukkads* (small urban gathering areas formed on the junction/crossing of streets or lanes) and *galis* (alleys/lanes). The inhabitants of these cities find comfort and a sense of belongingness with the organic layouts of *mohallas (neighborhoods in the Indian context), galis* and *nukkads* with which they have started to associate their own identity, and any idea to disturb this symbiosis is generally reckoned as an intrusion to the culture of indigenous inhabitants.

Moreover, the general notion propagates 'planned environments' as a thing of elites and a domain that does not include the everyday man into the loop of the processes. This, in turn, can be attributed to the fact that the said schemes are never presented in a style or format, which is easily accepted or comprehended by the citizens. The living structure that is needed to sustain a city and nurture it and that did exist to some degree in the traditional societies, rural communities, and early urban settlements has disappeared. The knowledge to create it or generate it is lost. This was not the case long ago, earlier with the help of the shared pattern languages, which existed in traditional society, when citizens were able to generate a complete living structure (Saligaros, 2005).

2.2.1 Evolutionary changes in planning

The conceptual level understanding of city planning has been an ever-evolving journey. The myriad concepts from Ebenezer Howard's 'Garden City' to Le Corbusier's 'Radiant City' and 'Superblocks'. On one hand, one can see concepts like 'megaregion' that incorporate multiple facets of transportation, economy and environment to the rural—urban gradation of 'Transects' used by new urbanists (Badger, 2012).

When talking about traditional (historical) cities, 'urban regeneration' as a concept comes at the forefront. Regeneration in biological terms is the process of renewal, restoration and growth that makes genomes, cells, organisms and ecosystems resilient to natural fluctuations or events that cause disturbance or damage. The analogy can as well be true for a city that is considered as a living organism exhibiting each and every nuance of its life cycle. Oxford dictionary defines urban regeneration as the process of improving derelict or dilapidated districts of a city, typically through redevelopment. It is called urban renewal (in the United Kingdom) and urban revitalization (in the United States) and is a program of land redevelopment in areas of moderate- to high-density urban land use. Historically the concept of urban renewal as a method for social reform emerged in England as a reaction to the increasingly cramped and unsanitary conditions of the urban poor in the rapidly industrializing cities of the 19th century (Worldbank, 2007) (Fig. 21.1).

Of recent, with visible effects of environmental degradation becoming more obvious in their daily lives, citizens are becoming more aware to be ecologically sensitive with their

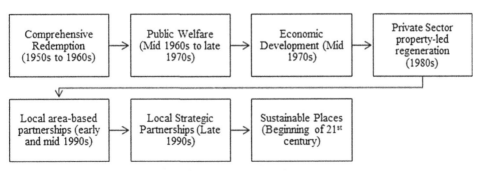

FIGURE 21.1   Evolution in urban regeneration strategies.

surroundings. Many innovative planning and design strategies are being conceptualized to make the cities more sensitized towards the looming environmental crisis.

2.2.2 Urban design, urban ecology and its relationship

Urban design can be defined as the process of creating functional and aesthetically pleasant spaces. Urban design as a tool for creating functional spaces has been used for a long time. Yet most of the time, aesthetics overshadow the functionality. There are multiple perspectives for looking at urban design and its complexities. 'Some see it as a "glass box" a transparent, self-explicable process, whereas others view it as a "black box", a phenomenon obscured by the fathomless complexities and depths of the design imagination' (Carmona, 2014). It is thus required to understand the background, historically and politically motivated actions, and consequential decisions in context with urban design. It is only then that we can arrive at an integrated thought process to understand the spectrum between the 'glass box' and the 'black box' of the urban design process. Another view on the subject by renowned urban designer Kevin Lynch is that the city is first a human habitat and a 'good city form' can only exist if it sustains human life. He explored the role that natural features play in enhancing the identity, legibility, coherence and immediacy of urban form from the scale of the street to that of the region (Lynch, 1981). The cities are nothing but an ecosystem in itself acting on the interface of the larger ecosystem. Here, urban design is a bridge between the natural and built; hence, ecology as a parameter of functionality can be of great significance during generating or evaluating any urban design scenario. In this context, the three main qualities of urban design can be called utility, durability and the ability to bring to the user a sense of well-being and emotional satisfaction (Moughtin et al., 1999).

2.2.3 Role of an urban ecosystem in the process of transformation

Ecological urbanism considers the city as a part of the natural environment. It considers the city as an ecosystem in itself. With each city evolving from its own deep and long-term context, the urban design could be positioned as a manifestation of all the human actions, adaptations and interaction with its surroundings. In cities, these human activities result in a typical kind of climatic conditions, soil and hydrology and urban flora and fauna in conjunction with the material and energy flow cycles (Spirn, 2012). These local conditions majorly governed the form of urban grain in traditional societies, consequently leading to its final form that is organically generated. In this scenario, architecture and urban design were not defined as a designed environment but were something generated through the collective actions of the inhabitants. They build according to their needs and means, which was inherently sustainable. This gave birth to concepts of 'vernacular architecture' and 'traditional urban design' of any place. This is well evident in the physical format of traditional cities (older settlement patterns) all around the world, as they differ in their urban grain and form, which evolved according to their local context.

Nowadays, the sustainable urban form that is the product of urban design strives to have following components in the design, namely, decreased energy use, reduced waste and pollution, reduced automobile use, preservation of open space and sensitive ecosystems and livable and community-oriented human environments (Jabareen, 2006). Be it Copenhagen in Denmark that aims to become carbon neutral by 2025 to Stockholm in Sweden, which prides itself in being climate-smart. The urban form is governed by the services and a larger vision imagined by the city authorities. This type of system may be suitable for the case of contemporary cities but for more sensitive cases like those of traditional cities or historical interfaces, which need a rethinking.

2.2.4 Space syntax analysis as an analyzing tool in the process of transformation

Spaces as an entity in themselves are unique mediums to understand this essence. Spaces are also the canvass for urban design to manifest itself and augment the life of the people. But can spaces be the prime drivers of ecological design? The spaces in the form of land are a vital resource in the urban ecological perspective. In this particular study, the intention is to take the case of ancient water bodies and adjacent land–space systems and to understand their present condition with respect to their locational virtues. The study explores the case area of Varanasi at particular zones to understand this space–land system. Spaces are the main protagonist in the human's life. The attachment, the possession, the nostalgia and the belongingness can all be linked to spatial–temporal components of life. When we talk about urban ecology, humans become the major driver of the change. Each action from the side of mankind is reciprocated by a consequential answer from nature. The trick is to find a balance in this action–reaction phenomenon that we often lose.

The space syntax modeling approach is used to understand the physical relationships like visual connectivity of the patterns found on-site with their surrounding patterns. The characteristics of this analysis are granular, with highly disaggregate data at the level of street segment, useful for detail urban design; predictive, with the results integrated in a multivariate statistical model; relational across scales—the same model works for local and global analysis; relational across themes—the same model describes urban morphology, centrality, hierarchy or structure (Serra et al., 2011). Space syntax is a set of analytical computer-based techniques to analyze any spatial configuration such as built spaces and urban environments. The technique has been developed at the University College London through research on quantitative analysis of natural movements in environments spearheaded by Bill Hillier and Julienne Hanson and published in their work 'Social Logic of Space' which was completed in 1984 (Adhya, 2007). It is also a set of theories linking space and society. Space syntax addresses where people are, how they move, how they adapt, how they develop and how they talk about it (Space Syntax Limited, 2019).

3. Case study—the Indian context

3.1 Introduction to Varanasi, details of the urban fabric (physical built environment at the neighbourhood and city level)

3.1.1 A living history

Varanasi is the story of 3000 years of continuity, the incessant cycle of death and rebirth, a city of myths and concrete facts. Every nook and cranny has a narrative hidden underneath be it the buildings, the open spaces, the leftover ponds or the *ghats* (stepped banks of a river). Though many explorations have happened and keep happening to unravel its mystery, there are a plethora of stories that need to be brought to the light of the day to understand the full essence of this saga called Varanasi.

Varanasi or Kashi started as a microversion of itself confined to the sacred boundaries, which were the sacred *'kshetra'* (territory). The city grew around its center, the luminous 'linga' (*symbol of Hindu deity Lord Shiva/Shaivism*). The periphery was enacted by pilgrim's feet, and the topography determined the placement of temples. There were actually multiple centers each attracting a constellation of neighbourhoods. This was the old city to which were added cantonment and civil lines in the colonial period in a very different pattern (Sinha, 2019). Slowly and gradually, multiple layers of urban patterns were added on this interface to give its present form. Mark Twain puts it himself:

> Benares is older than history, older than tradition, older even than legend and looks twice as old as all of them put together.

Varanasi is one of the oldest continuously living cities in the world whose fabric is woven across the web of interconnecting lanes, serving myriad services. It is in its *ghats*, along which thrives, trade, commerce and life. Varanasi has been a hub of pilgrims, scholars, educationists, artists and craftsmen for long. The image of Varanasi as a diversified community structure while still maintaining its regional characteristics can be attributed to different immigrants who came to this city for solace, peace and sacred merit, (Sanskrit) education and also as a consequence of various invasions. In this manner, Varanasi has evolved as a mosaic of social−cultural spaces, representing the whole of India (Singh, 2009). It is already a creative city that inspires the sharing of culture and knowledge. The spiritual scene of the city renders it with a unique faith-based economy, which in turn has led to the evolution of various allied activities and services sprouting alongside 'ghats'.

Here, the niches and the alleys hold the precise life of this vibrant city. Multiple ecosystems exist in the microareas that need to be explored in greater detail as a prerequisite before framing any reform mechanism for this city. However, rapid and haphazard urbanization has disturbed the delicate balance between the city's various facets. It is an ancient city trying to bear the burden of the needs of a modern man. It serves as a perfect example of an Indian city with the charm of a tourism center (with its heritage factor) and the problems of any other city, thereby making it a living laboratory for observing the urban ecological process and assessing the role of urban design in this context. This would be an endeavor, which might help in making a more robust and sustainable system that could be resilient towards the vulnerabilities posed by urbanization of the 21st century. Some basic information about the city is as follows:

Location and boundaries: Varanasi is located in the eastern part of the north Indian state of Uttar Pradesh. The city is on the western bank of the river Ganga (82 degrees56′E−83 degrees03′E and 25 degrees14′N−25 degrees23.5′N). It is 80.71 m above the mean sea level. It is bound by river Varuna on north and Asi Nala on the south.

Population: Varanasi is one of the largest urban centers and fast-growing cities in the state. According to Census (2011), the city had a population of 11.98 lakhs. The population of Varanasi city grew from 10.9 lakhs in 2001 to 11.9 lakhs in 2011 at a growth rate of 10%.

Climate: Varanasi experiences a humid subtropical climate. Summer: 22−46°C winter: up to 5°C and rainfall is 1110 mm annually.

Connectivity: Varanasi is well connected by road, rail and airways with other parts of the country. The road distance from major cities is 820 km from Delhi, 286 km from Lucknow and 122 km from Allahabad. There are three national highways, i.e., NH-2, NH-56 and NH-29 and four state highways, i.e., SH-87, SH-73, SH-74 and SH-98 passing through the heart of the city. Varanasi junction is located in the cantonment area. Three other smaller stations also exist there (CRISIL, 2015).

3.1.2 Architectural expression

Amid all the problems and expectations, the city still tries to charm the citizens and tourists alike with its manifold expressions. As one moves in the alleys and lanes of the vibrant microcosm, it is amazing to see the type of people one meets and the kind of spaces they inhabit. These spaces are diverse in their nature and have a different interpretation in everybody's life and get transformed by the virtue of their usage. The traditional Indian way of habitation makes mixed usage a very inherent part of urban living. Its prevalence in the urban structure makes it difficult to segregate it into an absolute residential or a commercial zone. There is always an interplay of both, and only a higher dominant character would signify the major characteristic of that area. Some of the spaces that form the urban fabric of Varanasi and that come under both of these characters are discussed subsequently. The typology of these spaces is deliberated as follows.

Gallies and the hidden spaces: Varanasi has a peculiar urban profile that varies from core areas to peripheral areas. In core areas, there is a mix of commercial character along the main arterial roads, which adds to the vibrancy of the city and extends up to the *ghats*. These arterial roads branch off to more narrow *gallies* (very narrow alleys), which form the web of connector streets in the residential zone. The urban character here becomes more introvert as well as densely packed. These urban pockets are devoid of dedicated open spaces that might have existed in earlier times. Around the turn of the 19th century, most of the parks were paved, and *talaabs* and *kunds* (ponds) were drained to make way for the houses and apartments. Yet on further probing, we can find hidden expanses of relatively open spaces embedded in this dense urban fabric. While wandering in the lanes, we can counter many such spaces that could be accessed from the narrow alleys. One such lane lies in one of the densest zones of the city starting from *Thatheri Bazaar* and leads up to the *Panchganga Ghat*. *Sher-wali Kothi* is one of these spaces that can be found in this alley. The humble entrance paves the way to a palatial house with a private shrine. *Kath ki Haveli* is another hidden space in this dense neighbourhood, which is in a state of dilapidation. Many shrines such as *Gopal Mandir* and *Tailang Swami temple* are situated on this route, which can be considered of significance in terms of religious congregation yet are known only to a few locals.

Ghats: Another signature characteristic of Varanasi are the *ghats* that form one of the longest continuous urban spaces for the city. These are 84 in number, and there is a continuous connection from *Assi Ghat* till up to *Prahalad Ghat*. These *ghats* form the pious platforms for performing religious rituals and also as grounds for funeral pyres. They add up as a backdrop for musical concerts and also as a stage for classical dance. They form cricket grounds for budding sportsmen and also offer quietness for an urban ascetic. They have diverse usage. The character of these *ghats* differs from the northernmost *Adi Keshav*, which is set in the low-density urban area and then moves towards more southern *ghats*, which are set in a dense urban environment. It is also observed that the northern *ghats* are steeper in their profile, and as you move south, the slope becomes gentler. Each *ghat* has its own historical or social significance, which is visible in its usage whether it is in every day or for a particular event (pulsar).

Havelis: As one tries to look up images of Varanasi, the signature *Havelis* (palatial mansions) prop up in the search. These mansions are a testimony to the efforts of different kingdoms that came from all over India to claim a part of Varanasi and to establish themselves in this exuberant city. Most of them are now being maintained by trusts of erstwhile princely states, whereas the rest have been converted to heritage hotels offering a slice of royal life in the hustle and bustle of the 21st century.

Sacred spaces: Varanasi's cultural landscape is dotted with many sacred spaces forming a very dominant religious overlay. Temples form the major portion of these spaces, whereas mosques, churches and Buddhist stupas form the other components. For Hindus all over India, it is the city of Shiva. It is said that each pebble of this city is a manifestation of Shiva, and it gets affirmed with millions of *shivling*, which dot its sacred space. The temples here can be found in a variety of shapes and sizes and belong to popular deities as well as the *gram devtas* (village deities). There are shrines in the corner of the street, under the steps of the *ghats* and also well-established temple complexes such as the *Vishwanath Temple, Tridev Mandir, Tulsi Manas* temple and others. The other significant sacred spaces in Varanasi are the mosques, and the most popular among them are the *Gyan Vapi* mosque and the *Alamgir mosque*. *Gyan Vapi* Mosque that sits adjacent to *Vishwanath Temple* takes its name from the *Gyan Vapi* well or the well of light. It is believed that the original *shivling* of the *Kashi Vishwanath* temple was hidden in this well when the temple was demolished for building the mosque. Original Kashi Vishwanath mandir has been demolished and built numerous times. Another major mosque that is of significance in the Varanasi's urbanscape is Alamgiri Mosque. The history of mosque is controversial, as the residents claim it to be on the same site as the ancient Bindu Madhav ka Dharara (Bindu Madhav's Shrine). Presently, the temple of Bindu Madhav is located in the vicinity of the mosque as a small shrine, and the location is often thronged by cows and buffaloes.

4. Methodology

4.1 Extracting and analysing the spaces on the basis of new urbanism approach

The hypothesis for this study lies in the fact that Indian traditional cities have been inhabited for long, and the very essence of these cities lies in their unique urban fabric. However, this factor is slowly diminishing, as urbanization is taking a toll on these cities. The study takes upon the quest to reveal the urban patterns, which are most suitable in retaining and creating the unique form of the city. The patterns evolved by CA place their origin to the design virtues, which are the core of any sensible human pattern language (Alexander et al., 1977), similar to those found in organically evolved cities. A reverse approach will entirely be true, that is, the cities that have evolved organically might have some of these patterns, which have been used in the pattern language. The study takes upon the city and tries to unearth such evidence, which lies scattered in the city just hidden under the layers of rapid urbanization and its consequence, the haphazard development. The study would take up the patterns developed by CA as the starting point for the surface level search. The patterns searched are similar in the description but may have a contextually different vocabulary.

CA completed his work *A Pattern Language* in 1977. After finishing this work, he produced *The Timeless Way of Building* in 1979, which came as an information manual for designers and architects on how to use the urban patterns. Its central argument is that the patterns that are formed by the daily usage of the people should be discerned and improved upon for further usage. The pattern language gives us a collection of urban design elements that when used in relation to each other than tend to make up places that are healthy and sustainable and will lead to satisfied citizens. Its central argument is that patterns developed simply by living life allow unconscious cognitive relationships with space to be discerned, consciously recognized and further improved upon (Park and Newman, 2017).

The process for pattern mining starts by doing a reconnaissance survey of the spots in the city, which might have patterns (tangible and intangible) embedded into the urban fabric. Three broad survey zones have been identified in the inner belt of the city. These zones are part of the sacred cluster, which is identified as 'Panch teerths' or the five sacred spots in the historical texts. These include Ghats of Assi, Dashashwamedh and Panchganga. They have been selected on the basis of their historicity and hence a high probability of finding the urban patterns that are the true native of this city. The neighbourhoods that are in direct contact with these ghats are selected.

A detailed photographic and video study of the spots was conducted. CA's *A Pattern Language* is taken as a base reference for this study. The patterns similar to CA patterns are added into the new list for that area and anything that is not present in the base reference but recurs in multiple places is also recorded. The connectivity rule of CA's pattern language is also considered while finding patterns (Alexander et al., 1977; Park and Newman, 2017). Some of the patterns of significance in the context of Varanasi are promenades, individually owned shops, shopping streets, pedestrian streets, activity nodes and building complex. These would be specific ghats, stand-alone specialized Dukaan (shops); bazaar streets (specialized markets); gullies (the narrow pedestrian alleyways); nukkads (points of activity on the cross-roads of small streets); kunds (man-made waterbodies with religious significance); paanshops (a mouth freshener and a local delicacy made from betel leaf); Pakka Mahal and Havelis (houses in old neighbourhoods); respectively. These are some of the signature elements of Varanasi that are instrumental in making the cognitive image of Varanasi as seen today.

5. Analysis

5.1 Analysis based on space syntax technique

The city of Varanasi has a spirit embodied in its urban fabric, which is ubiquitous in its essence. The transcendental urban mood vibrates in every sphere of this vibrant city. During the analysis, it is important to consider the sensitivity of the place so that the holistic essence of the city is reflected. The space syntax theories and techniques perform analysis, which is nonnormative in nature, i.e., why a particular structure is in that form. The idea behind it is based on mathematical formalism and proposes several methods for representing space and spatial systems that can be represented into graphs, making possible the analysis of even very large spatial networks, like those of cities or even city regions (Serra et al., 2011). This makes it the perfect technique to analyze complex spatial interface like that of Varanasi.

Space Syntax is founded on two fundamental propositions (Hilier and Stonor, 1990):

- Space is not a background to human activity but is intrinsic to it.
- Space is first and foremost configurational. In other words, what happens in any individual space—a room, corridor, street or public space—is fundamentally influenced by the relationships between that space and the network of spaces to which it is connected.

For the specific case of this study that is a part of a larger study, 'visibility graph analysis (VGA)' is used to analyze the three zones to understand physical relationships. Since visibility and visual perception significantly influence how people behave, appreciate and experience the environment, it could serve as determining factors while assessing the urban design and landscape layout of the spaces (Mahmouda and OmarOmar, 2015). VGA technique was developed to perform integrative analysis of several positions inside an environment by calculating the intervisibility of positions regularly distributed over the entire environment. It was also strongly correlated that visual perception affects the movement of people and becomes an integral part of the social fabric (Turner, 2001).

Results: Ancient urban patterns, urban ecology and hidden ecosystems?

5.1.1 Assi Ghat and Pushkar Kund

Assi Ghat is the southernmost ghat of the city and has relatively new platforms due to various developmental activities happening in recent years. Dedicated parking lots, wider streets and accessibility, in general, have made it one of the favourite hotspots among the residents of younger age bracket. The ghat is among the most thronged after a place in terms of public usage. Among the less used public places in this particular zone are the kunds. One such kund is the Pushkar Kund that lies very close to the ghats but in a relatively introvert zone. Here, the placement of kund and ghat is analyzed using VGA to understand the basic spatial situation, which makes all the difference.

The VGAs convert the zone map into a heat map visualization. The red-orange region shows higher visual integration (high visually connected places), and the blue-green region shows low visual integration (low visually connected places).

In Hinduism, nature is sacred, be it trees, water bodies, animals or birds. Sacredness may have ensured that they might be taken care of and would be sustained for future generations. For example, kunds or ponds were the focal points during worship, as the main water source for daily household activities and as a water source for domestic animals. They were an ecosystem in itself. It has been documented in ancient texts that kunds in Varanasi were interconnected at subterranean level with each other and with River Ganga. During the times of flood, these kunds acted as sinks for the overflowing water of Ganga, hence maintaining the balance (Cunha, 2018). Since the water of kunds was the same as that of Ganga, they had almost equal importance as the sacred river. These were revered and were the main part of any worship held in the community. But a gradual disconnect from ancient rituals and changing religious dynamics, draining of kunds during the time of British colonization and rapid urbanization have led to a change in the status quo of these entities. Taking the case of Assi Ghat and Pushkar Kund (Fig. 21.2), then it can be well observed that Assi still remains a hot spot of activity, whereas Pushkar has been gradually pushed towards the periphery in the citizen's minds. In the visibility graph of Assi Ghat zone, the road leading to ghat and the ghat itself lies in orange-red zone showing high visual connectivity, whereas the access road towards Pushkar Kund and the adjoining area lies in the blue region showing very low visual connectivity. Here the global visual integration score of the zone is 4.61, whereas local scores are 4.73 for the ghat area and 2.88 for the kund area (Fig. 21.3).

5.1.2 Dhashashwamedh Ghat, main road and market area

Dhashashwamedh Ghat (Fig. 21.4) is famous for its Aarti (eitualistic prayers offered to River Ganga choreographed into synchronized incantations). Here the main street connecting

FIGURE 21.2   Assi Ghat zone—Assi Ghat and Pushkar Kund.

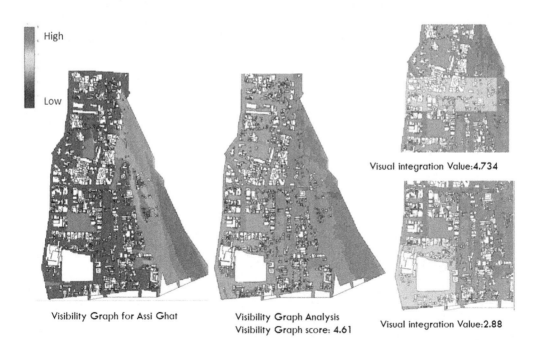

FIGURE 21.3   Assi Ghat—visibility graph analysis.

Dashashwamedh ghat zone

FIGURE 21.4 Dashashwamedh Ghat zone—Ghat and main market area.

from the nearby crossroads (Godowliya chauraha) towards the main worship ghat is lined with many varieties of shops, major being that of fabric and dress materials, food outlets, worship items, trinkets and other knick-knacks. This forms as nuclei of attraction and hence a very high range of movement. The global score for the area is 1.42 (Fig. 21.5), whereas for ghat and the main street, the score is 1.72 (Fig. 21.6).

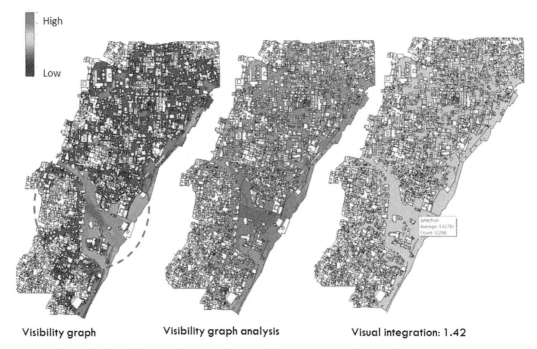

Visibility graph   Visibility graph analysis   Visual integration: 1.42

FIGURE 21.5 Dhashwamedha Ghat—visibility graph analysis.

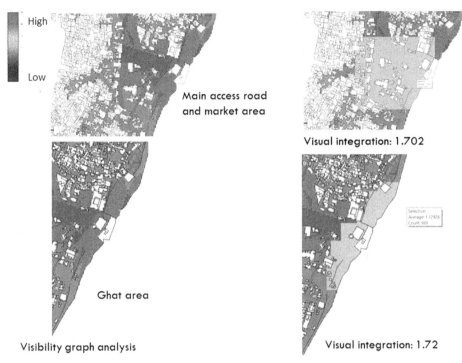

High

Low

Main access road
and market area

Visual integration: 1.702

Ghat area

Visibility graph analysis

Visual integration: 1.72

FIGURE 21.6   Visibility graph analysis—Ghat area and main market road.

5.1.3 Panchganga Ghat, Thatheri Bazaar and Chaukhamba

The Panchganga Ghat zone is among the northernmost ghats of the city (Fig. 21.7). The city has a very steep profile in this zone. The ghat has very long staircases, and the gradual rise of the slope can be observed in the neighbourhoods. The urban fabric of this particular zone is very dense as compared with the previous two zones. This is just one axis of the market that can be called as the main street. It is called as *Thatheri Bazaar*. This road merges into another market, which is in the interior of this zone and is known as the *Chaukhamba* area. The area is devoid of any green space and is mostly residential neighbourhoods. The global score for the area is 1.30 (Fig. 21.8), whereas local scores are 1.09 for *Thatheri Bazaar* street and 1.54 for *Chaukhamba* area. A few important sites that lie in this area are the Alamgir mosques with score 1.48 and Shri Gopal Mandir with a score of 1.54 (Fig. 21.9).

6. Discussion

The analysis conducted here has given the visual integration scores that has helped to understand the spaces that are highly integrated or highly visually connected. VGA investigates the properties of a visibility graph derived from a spatial environment. The VGA can be

FIGURE 21.7   Panchganga Ghat zone.

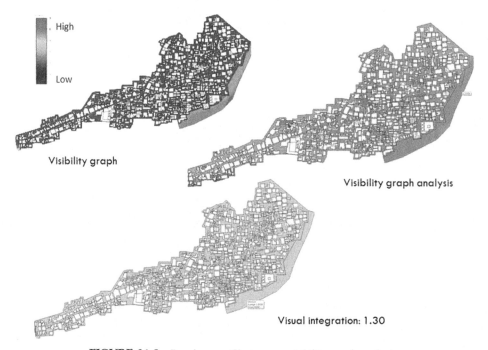

FIGURE 21.8   Panghganga Ghat zone—visibility graph analysis.

Visual integration score: 1.48
Alamgiri mosque and Ghat's Access

Visual integration score: 1.54
End of Chaukhamba area

Visual integration score: 1.09
Thatheri Baazaar

Visual integration score: 1.50
Shri Gopal Mandir

FIGURE 21.9 Visibility graph analysis: Thatheri Baazar, Chaukhamba and Shri Gopal Mandir.

applied to two levels, eye level for what people can see and knee level for how people can move, which is critical to understand spatial layouts. Through the above examples, it can be observed that places, which are more significant, generally have high visual integration.

The zones taken under this study are from southern, central and northern parts of the ghats' armature so as to understand how the physical setting is changing at an intracity level. In the southern part, the global visual integration scores are higher showing more visual connectedness among the zone's components, and hence the people movement is also higher. This translates to a high level of social interrelationship happening in this zone as compared with its northern counterpart. It is also seen that more green spaces and open spaces are observed in the southern zone, whereas the northern zone is devoid of any open space. These zones with the virtue of their layout are giving certain urban patterns, which are long-term identifiers for this city. They are functional but may be lost under the layers of urbanization. Through techniques such as VGA, one can understand the inherent logic of space objectively and sieve out such urban design patterns.

7. Conclusion

The study started as an investigation for assessing the role of ancient urban design as markers of urban ecology. This is a case study—based research; hence the testbed for this study was Varanasi. Varanasi as a city has complex layers of religious, social, cultural and economic intertwined with its urban fabric, which gives a very unique character to this city. The study takes the case of three zones and tries to understand its spatial structure through visual graph analysis, which is a technique under space syntax analysis (umbrella term). At the southernmost zone that is Assi Ghat, the study is conducted to understand the lost relevance of kunds with respect to gaining popularity of ghats. It is established that urbanization and its consequential effect, that is, haphazard development, could be termed as the leading causes that slowly took a toll on the spatial significance of these places and pushed them to the background of the city's interface. Similarly, the other two points also look into urban places at the respective locations and try to understand the visual scores and their significance. For example in the middle zone, that is, Dashashwamedha Ghat, a comparative understanding is developed among central market spine, chief ghat area. Both of these zones are located in a highly visually connected zone that gives them high visual integration scores. Contrasting this case will be that of the northern zone that is Panchganga Ghat that has an extremely dense urban fabric of intertwined lanes. The public places are hidden deep inside the zone and are only possible to explore them through local knowledge, word of mouth information. The urban places in all of these zones govern certain kind of urban ecosystems that need to be sensitively handled. Be it kund and the role it plays as a cultural catalyst and as a water resource for the urban community, or a market lane with its different scale of economies interacting with resource management cycles, each and every space is a marker of the process of urban ecology. Hence, by extracting the urban patterns from the larger area and then reiterating them for modern usage, the author intends to reclaim the significance of urban design as an efficient marker for urban ecology. These can not only be used as components of urban design that can create a culturally and functionally sustainable city that is in coherence with its historical identity but will also help in restoring the urban ecological balance.

In traditional societies, the man was more in sync with the natural environment. The space he used was appropriate to his need, and hence a balance of nature and built was preferably tilted towards nature. This tilt has slowly shifted to the built side, and that has generated a scenario of concern. Thus, it is required to focus on studies that take up the challenges of managing cities and its urban ecologies from various perspectives to evolve unique solutions.

The typology of spaces discussed is as follows:

Gallies: These are the narrow lanes in the city of Varanasi that act as the connector among dense residential neighbourhoods. They are the lifelines of the citizens by acting as facilitators of various day-to-day activities.

Hidden spaces: These spaces are mostly open spaces with an introvert character. These spaces could be accessed through *gallies* and generally have small entry points. This normally takes them off from the conventional tourist map yet they can be of major significance when talking about the holistic urban landscape of the city.

Ghats: These are the stepped embankments. They can be found in many north Indian cities but have a special place in Varanasi due to its prominence in the religious activities. They also act as the biggest urban spaces in Varanasi.

Havelis: These are the palatial mansions that were built by kings and rich nobles in the earlier era. Varanasi had a very strategic position in terms of trade in the bygone era, and all the kingdoms wanted a slice of it. So they came and acquired area for themselves and their people. The havelis acted as markers of various kingdoms.

Sacred spaces: In this article, sacred spaces of two major religions are discussed that are temples and mosques. Temples are discussed because the whole city is dotted extensively by temples and shrines of various sizes. Mosques are discussed because of their locational significance.

References

Adhya, A., 2007. Report on University Research, vol. 3. AIA.

Alexander, C., Silverstein, M., Ishikawa, S., 1977. A Pattern Language.

Badger, E., November 9, 2012. The Evolution of Urban Planning in 10 Diagrams. Retrieved from Citylab: https://www.citylab.com/design/2012/11/evolution-urban-planning-10-diagrams/3851/.

Bansal, S., Pandey, V., Sen, J., 2017. Redefining and exploring the smart city concept in Indian perspective: Case study of Varanasi. In: Seta, F., Sen, J., Biswas, A., Khare, A., Seta, F., Sen, J., Biswas, A., Khare, A. (Eds.), From Poverty, Inequality to Smart City: Proceedings of the National Conference on Sustainable Built Environment 2015. Springer, Singapore, pp. 93–107. https://doi.org/10.1007/978-981-10-2141-1_7.

Carmona, M., 2014. The place-shaping continuum: a theory of urban design process. Journal of Urban Design 19 (1), 2–36. https://doi.org/10.1080/13574809.2013.854695.

Chhabra, D., Healy, R., Sills, E., 2003. Staged authenticity and heritage tourism. Annals of Tourism Research 30 (3).

CRISIL, 2015. City Development Plan for Varanasi-2041.

Cunha, D.D., January 2018. Water Urbanism Studio-Varanasi (Joint Studio by IIT Kharagur and Columbia University). I. K. University, Interviewer).

Curry, N., 2005. Countryside Recreation, Access and Land Use Planning.

Hilier, B., Stonor, T., 1990. UCL Space Space Online Training Platform. Retrieved from otp.spacesyntax.net: http://otp.spacesyntax.net/.

Jabareen, Y.R., 2006. Sustainable urban forms- their typologies, models, and concepts. Journal of Planning Education and Research 38–52.

Kush, D., 2015. People and Their Spaces. IIA NATCON.

Lynch, K., 1981. A Theory of Good City Form. MIT Press.

M, O.P., 2008. Preservation of the Spirit of Place. ICOMOS Quebec 2008, Quebec.

Mahmouda, A.H., OmarOmar, R.H., 2015. Planting design for urban parks: space syntax as a landscape design assessment tool. Frontiers of Architectural Research 4 (1), 35–45. https://doi.org/10.1016/j.foar.2014.09.001.

Mehta, V., 2014. Evaluating public space. Journal of Urban Design 19 (No. 1), 53–88. Routeledge, Taylor and Francis.

Moughtin, C., Cuesta, R., Sarris, C., Signoretta, P., 1999. Urban Design: Methods and Techniques. Routledge.

Newman, G.D., 2016. The eidos of urban form: a framework for heritage-based place making. Journal of Urbanism: International Research on Placemaking and Urban Sustainability 9 (4), 388–407.

Newman, P., 2006. The Environmental Impact of Cities. Environment and Urbanization.

Park, Y., Newman, G.D., 2017. A framework for place-making using Alexander's patterns. Urban Design International 22 (4), 349–362. https://doi.org/10.1057/s41289-017-0040.

Peris, A., Meijers, E., Ham, M.v., 2018. The Evolution of the Systems of Cities Literature since 1995: Schools of Thought and Their Interaction. Networks and Spatial Economics. Springer.

Saligaros, N.A., 2005. The structure of pattern languages. In: Principles of Urban Structure. Techne Press, Amsterdam.

Serra, M., Gil, J., Pinho, P., 2011. Challenges of Portuguese urban planning. In: Proceedings of 7VCT, pp. 505–509 (Lisbon, Portugal).

Short, J., 1998. The Urban Order: An Introduction to Cities,Culture, and Power. Blackwell Publishers, MA.

Singh, R.P., 2009. Banaras: Making of India's Heritage City. Cambridge Scholars Publishing.

Singh, R.P., 2017. The Liveability of India's Cultural Capital, Banaras (Varanasi/Kashi). (V. Bharne, Interviewer). April-May.

Sinha, A., August 2019. Primary Interview Done by Author in Relation with Varanasi. (B. Vidhu, Interviewer).

Space Syntax Limited, August 2019. UCL Space Syntax. Retrieved from otp.spacesyntax.ne. http://otp.spacesyntax.net/overview-2/.

Spirn, A.W., 2012. Ecological Urbanism: a Framework for the Design of Resilient Cities.

Steinberg, F., 1996. Conservation and Rehabilitation of Urban Heritage in Developing Countries.

Turner, A., 2001. Depthmap: A Program to Perform Visibility Graph Analysis. 3rd International Symposium on Space Syntax. Georgia Institute of Technology, Atlanta.

UNDP, 2016. Sustainable Development Goals. United Nations.

Worldbank, 2007. URBACT II. Worldbank.

'Green building' movement in India: study on institutional support and regulatory support

Ravindra Pratap Singh[1], A.S. Raghubanshi[2]

[1]Research Scholar, Integrative Ecology Laboratory, Institute of Environment and Sustainable Development, Banarasi Hindu University, Varanasi, Uttar Pradesh, India; [2]Integrative Ecology Laboratory (IEL), Institute of Environment & Sustainable Development (IESD), Banaras Hindu University (BHU), Varanasi, Uttar Pradesh, India

OUTLINE

1. Introduction

In order to draw the attention of the world and address the global environmental problems, the United Nation appointed a commission known as 'Brundtland Commission' to study and report the various issues related to environment and sustainable development in 1980s. It was a leap forward to address the issue of sustainability. The report was submitted in 1987 and was given the name 'Our Common Future'. This commission set the common goals for world community to save the earth, and to minimize the pace of damage to the environment, the commission gave the definition of sustainable development which had been accepted by UN and world community (WCED, 1987).

The commission report led to the United Nation to hold the Earth Summit at Rio de Janeiro, Brazil, in 1992, where the world community set the plan of actions called 'Agenda 21' to contain the global environmental problems due to rapid development. Under the 'Agenda 21', the plans were set to control the CO_2 emission, switching over to renewable source of energy, control the emission of other gases creating a greenhouse effect, global warming, ozone layer depletion, etc.

Urban growth is a global phenomenon that has resulted from intense human activities and development. Urbanization in the form of cities has resulted in accelerated economic activities and social changes, while also influencing the ecological balance at multiple scales (Grimm et al., 2008). This rapid urban growth has also led to an exponential increase of land use & land cover changes especially in the form of an increase in built spaces (Verma and Raghubanshi, 2019). The built spaces are the places for living and work, commuting and storage of goods and where various other services required by society are rendered. Both public and private buildings and constructions have seen an increase, which raises certain questions about the resource consumption and waste generation from these constructed structures.

The world is facing acute challenge of climate change on a global scale. The construction industry has been identified as a major source of environmental pollution in the form its resource and energy-intensive nature as well as waste generation (Thanu and Rajasekaran, 2018). Many countries, including India, depend on nonrenewable source of electricity generation and fossil fuels to meet their energy demand. Such energy consumption by buildings is a major cause for concern from the perspective of mitigating climate change.

According to the International Energy Agency, the residential sector consumes about 10% of the total world energy delivered, which is increasing at the rate of 1.5% per year. Non-OECD countries account for 28% electricity consumption out of their total energy consumption, while in India, residential electricity consumption is 22% of total energy, second only to the industrial sector (Chandel et al., 2016). Economic development, climate change due to greenhouse gas emissions, energy security and accessibility have been identified as major concerns for urbanization in India (Thanu and Rajasekaran, 2018). Economic development and demographic changes demand a high level of efficient and sustainable built space to accommodate the rising demand. Thus, the building construction has received some attention in the recent years especially in developing countries like India, in order to create sustainable urban areas.

The building construction is regulated by a number of building laws that have evolved over the years. However, from the perspective of sustainability, they lack any definite focus. These regulatory codes and standards are made according to the local conditions, for example, demographics and topography, and vary for different countries. In the era of global climate change, a need for incorporating energy efficiency and sustainable resource utilization has risen, which has resulted in evolution of certain green building norms. These regulatory norms aim to standardize the building construction so as to create green infrastructure that would contribute to reducing the effect of urbanization of the ecological balance and mitigate the climate change phenomena. They essentially focus on the environmental, economic and social aspects of building construction (Chandel et al., 2016).

The objective of this chapter is to critically evaluate the two of these green building norms, LEED (Leadership in Environment and Energy Design) India and GRIHA, in the backdrop of the existing bylaws and propose changes in the existing framework. These green building norms would result in strong policy support to help in creating ecologically sensitive urbanization.

These facts and factors compelled the thinker, policymakers, scientists, environmentalists, architects and engineers to think about designing and constructing buildings that minimize the adverse impact of the construction industry on the environment. This gave birth to the concept of green building.

2. Green building concept — meaning and initiatives

The sustainable construction, 'Green Building' or high-performance building are synonymous. Broadly green building can be defined as a 'Building design and construction using methods and materials that are resource-efficient and that will not compromise the health of the environment or the associated health of occupants, construction workers and the general public' (Landmam, 1999).

The international movement for sustainable construction/green building started in the 1990s after the Earth Summit. The United States and the United Kingdom took the lead in this movement, and the US Green Building Council was established in 1993 to promote sustainable construction. In order to evaluate the green building effort, a rating system was established, by USGBC, called LEED. In the last 30 years, many countries developed their own green rating systems for certification of buildings. Some important green rating systems of the world are shown in Table 22.1.

TABLE 22.1   Green rating systems.

Sl. No.	Green rating system	Country of origin
1	BREEAM	United Kingdom
2	LEED	United States
3	GB Tool	United States
4	Green Globe	United States
5	Green Star	Australia
6	CASBEE	Japan
7	GBES 3 Star	China
8	GRIHA	India

Green building initiative in India was started in 2001 with the introduction of LEED India as a green rating certification system, and later on in 2005, GRIHA rating system was developed and introduced in India.

3. Methodology

In order to identify the literature related to LEED India and GRIHA certification, peer-reviewed literature and Government of India reports were assessed by the authors. The Scopus database was accessed (11 September 2019) to identify relevant peer-reviewed literature from the last 10 years (2010–20). This literature was identified was applying the method followed by Verma and Raghubanshi (2018). 'LEED', 'LEED India' and 'GRIHA' were the keywords used, after which all the documents were shortlisted based on their abstract. It was found that 2786 documents were found globally, which were based on LEED certification. This literature saw a consistent decrease in during the 2010–20 decade (Fig. 22.2). Most of the studies based on LEED certification were done in the United States (Fig. 22.1).

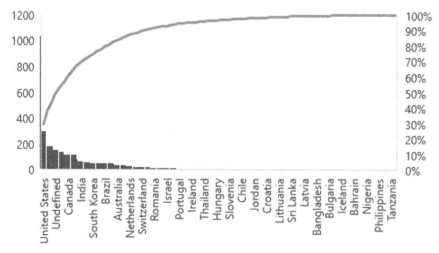

FIGURE 22.1   Studies on LEED certification in different countries.

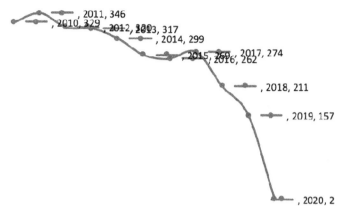

FIGURE 22.2   Number of studies on LEED certification.

LEED India and GRIHA search resulted in only 31 articles. 23 studies were from India, 4 from United States, while 1 each from Malaysia, Thailand, Turkey and United Kingdom.

4. LEED India

This rating system was introduced in India in the year 2001. It is associated with US Green Building Council's LEED certification system and managed by Indian Green Building Council (IGBC), a voluntary and private organization (IGBC 2013a,b,c,d). The rating system is based on the following broad principles:

- Sustainable site development
- Water saving
- Energy efficiency
- Material selection
- Indoor environment quality
- Innovation

Under this rating system, the building is certified under four categories, i.e., platinum, gold and silver and certified, based on credit achieved by the project.

5. GRIHA

The GRIHA, another rating system, is India's national rating system developed in the country and introduced in 2005. It is backed by Ministry of New and Renewable Energy, Government of India (ADaRSH, 2013a,c). It claims to be based on nationally accepted energy and environmental principles. It is based on five 'R' philosophy of sustainable development:

- **Refuse** to broadly adopt the international trend, material, technologies, products, etc., especially in an area where local substitutes/equivalents are available.

- **Reduce** the dependency on high energy products, systems, processes, etc.
- **Reuse** materials, products and traditional technologies, so as to reduce the cost incurred in designing dealings as well as in operating them.
- **Recycle** all possible waste generated from the building site, during construction, operation and demolition.
- **Reinvent** the engineering systems, designs and practices such that India creates a global example that the world can follow.

It has tried to quantify the important elements of sustainable construction:

- Energy/power consumption
- Water consumption
- Waste generation
- Renewal energy integration

GRIHA rates a building based on 34 criteria and awards the points on a scale of 100. The building is certified in five categories, i.e., Star 1, Star 2, Star 3, Star 4 and Star 5, based on points gained (ADaRSH, 2013d).

6. Impact of built environment on ecology/environment

When a building is constructed in the ecosystem, we create a mini ecosystem also called built environment. The main purpose of the built environment is to separate human from the natural system, providing space for human functions protected from elements like the sun, wind and rain on one hand and from physical danger on the other. This build environment affects the microclimate (heat island), hydrology (runoff), soil and plants. It creates false natural habitats in the natural ecosystem. But, the question arises whether human habitat, its planning, content and functioning are analogous to natural system. And if it is not, then how much is it different from natural ecosystem, its impact on environment and what are the measures to minimize it. The green building philosophy originates from this point. The built environment should be planned, designed and constructed in such a way that its adverse effluence on natural ecosystem is minimized. The planners, architects and engineers have to learn the behaviour and functioning of natural system. The built environment in which we live and the infrastructure that supports our lives demand huge share of the natural resources we consume.

The natural ecosystem exhibits some remarkable features:

- The sole source of power is solar energy, which is renewable.
- Concentrated toxic materials are generated and ultimately consumed in the system locally.
- Efficiency and productivity are in dynamic balance with resiliency.
- Ecosystem remains resilient in the face of change through high biodiversity of species.
- In the ecosystem, cooperation and competition are interlinked and held in balance.

The natural processes are predominantly cyclic rather linear (Kibert et al., 2002). The storages are organic, promote resilience within each range of scales by diversifying the execution of function into arrays of narrow niches, promote efficient use of materials by developing cooperative waves of interactions among members of complex communities and sustain sufficient diversity of information and functions to adapt and evolve in response to changes in their external environment (Kibert et al., 2002).

Build environment, contrary to the natural ecosystem, is an artificially created false ecosystem of human habitats. The processes are linear, create a lot of waste, consume natural resources, like fossil fuels (mostly nonrenewable source of energy), and cause damage to the environment disturbing the balance of natural ecosystem and reducing the life-supporting capability of the ecosystem for all species.

With the global population of mankind going up every year substantially, the scale of the built habitat occupies a significant fraction of earth surface. The long timescale of the built habitat about 75–100 years is another disturbing factor (Kibert et al., 2002). The changes in living habits and living styles have pushed the energy demand globally very high. It requires huge quantities of raw materials such as earth, stone, steel and other materials. For processing and converting these materials into a suitable and useable form, huge energy is required daily and yearly. The source of energy for these activities is on predominantly nonrenewable fossil fuels.

The concentration of buildings affects microclimate, hydrology (runoff), soils (top fertile layer) and plants and disturbs the natural ecosystem of the locality. The built environment interacts with the natural environment at different levels. The individual structure may affect only its local environment, but cities can have an impact on the regional environment by affecting the weather through changes in the earth's albedo (Wermick and Ausubel, 1995) and other surfaces characteristics, altering natural hydrologic cycles and degrading air, water and land via the emissions of their energy systems as well as through the behaviour of their inhabitants.

The construction industry is the major user of the conventional energy on one hand and biggest polluter on the other hand. The built environment impacts enormously to global environmental change, effects on biodiversity, influences water and air quality, is biggest culprit of CO_2 emission and is major source of solid waste.

7. Demographic growth and urbanization

The global urban population is more than 4 billionm and more than 50% of people are living in urban area (UN Report Revised, 2018). Though the rate of growth of urban population is declining globally, the same is on the rise in developing and populous countries like India (Table 22.2).

The urban population growth is stabilizing globally (UN Report 2018/1, 2018), whereas in India, it is still growing consistently (Figs. 22.3–22.5).

TABLE 22.2   Global urban population and growth rate.

Year	Urban population (in billion)	% of urban population w.r.t. total population	Urban population growth (percentage)
1961	1.048	33.609	2.902
1971	1.382	36.748	2.626
1981	1.794	39.772	2.857
1991	2.329	43.398	2.531
2001	2.919	47.137	2.270
2011	3.957	52.103	2.068
2018	4.196	55.271	1.936

7.1 Urban population of India

The urban population in India is 31.15% of (Table 22.3) the total population, but the actual number of people is 377.10 million, which is more than the population of many countries. The country is facing many challenges like maintenance of civic amenities, drinking water, traffic regulations, high energy demand, disposal of waste, high volume of rainwater during rainy season and floods like situation due to blockage of channels, hygiene, health and shortage of urban houses, etc. (MOUD compendium, 2013, MOHUA Annual Report, 2017–18).

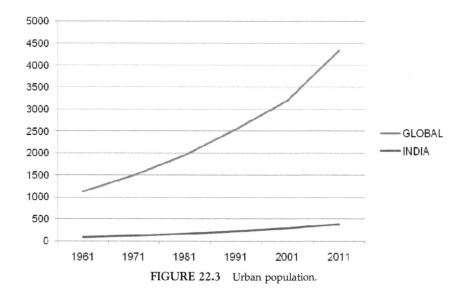

FIGURE 22.3   Urban population.

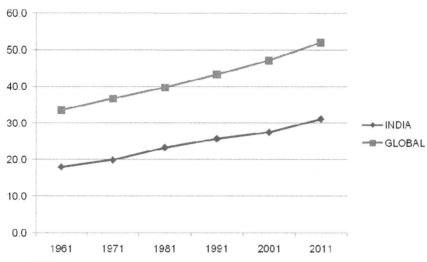

FIGURE 22.4 Percentage of urban population with respect to total population.

As per estimates, according to the current growth rate, the urban population on India will be 575 million in year 2030 and 875 million in year 2050. The urban population percentage will be more than 50% by the year 2050.

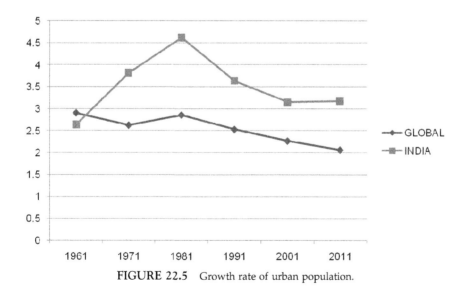

FIGURE 22.5 Growth rate of urban population.

TABLE 22.3 Urban Population of India and growth rate.

Year	Urban population (in billion)	% of urban population w.r.t. total population	Urban population growth (percentage)
1961	0.0789	18.0	2.64
1971	0.1091	19.9	3.83
1981	0.1595	23.3	4.62
1991	0.2176	25.7	3.64
2001	0.2861	27.8	3.15
2011	0.3771	31.15	3.18

8. Green Building Certification in India- Under LEED India

The LEED India registered 4980 projects for green building certification since 2001 to 2018. The actual registration for green building certification started in 2004. The data are presented in Table 22.4.

TABLE 22.4 Registered projects under LEED India.

Sl. No.	Year	Number of projects each year	Cumulative
1	2004	5	5
2	2005	2	7
3	2006	9	16
4	2007	15	31
5	2008	132	163
6	2009	101	264
7	2010	386	650
8	2011	435	1085
9	2012	535	1620
10	2013	390	2010
11	2014	661	2771
12	2015	628	3399
13	2016	559	3958
14	2017	492	4450
15	2018	530	4980

9. Green Building certification - Under GRIHA

The GRIHA registered 1200+ projects for green building certification from 2004 to 2018. Though the GRIHA was founded in 2004, actual registration for green building certification started in 2009. The data are presented in Table 22.5.

The combined data of projects registered under LEED India and GRIHA are graphically presented showing the growth of green building certification scenario in the country in the last 17 years (Fig. 22.6).

TABLE 22.5 Registered projects under GRIHA.

Sl. No	Year	Number of projects each year	Cumulative
1	2009	9	9
2	2010	63	70
3	2011	31	101
4	2012	113	214
5	2013	133	347
6	2014	132	479
7	2015	131	610
8	2016	135	745
9	2017	163	908
10	2018	298	1205

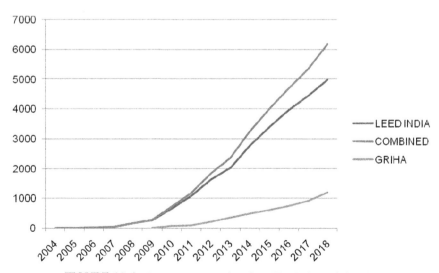

FIGURE 22.6 Projects registered under LEED India and GRIHA.

10. Green building certification

10.1 Regulatory Framework of the Country & Regulatory framework of the country

The Green Building certification started in the country in a structured way with the opening of the LEED India chapter in the year 2001. Thereafter, GRIHA certification system was introduced in 2005 backed by the Central Government. Though the GRIHA has been backed by the Central Government, the certification of any building is voluntary. The IGBC and GRIHA council made significant progress by expanding the outreach to different states and districts of the country through awareness programme, workshops, seminars at regional and national levels, newsletter, webinars and other programmes.

The council developed a pool of experts and evaluators as design and implementation of green building concepts requires a large number of qualified professionals in all parts of the country. In the last 17 years, the GRIHA and LEED India have registered 6181 projects for green building certification when there are 110.14 million houses in urban areas of the country. The share of certified green building in the pie of total urban building stock is insignificant.

In developing countries, the role of government is predominant and key for any change in society. The Central Government took lead by way of preparing and circulating Model Building Byelaws (MBBL) in 2004 and incorporating some elements of sustainable construction in MBBL 2004.

The MBBL 2004 was circulated and all the states were advised to amend their building bylaws with the liberty to incorporate modifications as per local requirement. The States and UTs have been slow in amending their bylaws. The Central Government has further amended the MBBL 2016, and this time green building got a substantial focus. A new chapter has been added in MBBL. The States have been advised to amend their building bylaws. Some of the States and UTs who amended their building bylaws after the release of MBBL 2016, incorporating the important and key elements in their building bylaws.

In the country, there is a well-structured framework (Fig. 22.7) to implement and enforce the building bylaws and policies of Central Government, State Government and urban local bodies (ULBs).

The public building is being constructed by government departments/government construction agencies. These government construction agencies of Central Government, as well as State Government, are setting the standard and specification for the construction industry in the country. They have a good pool of engineers, designers, architects and researchers. Society perceives them as pioneers. They are well equipped with economic, infrastructure and human resources. The public building should be made mandatorily to follow green building concepts. The Central Government and States have taken many initiatives to adopt green building concept and further can play a very crucial role by way of making all the public buildings to be constructed with certified green building. The policies and legislatures need to be amended. It will have a far-reaching impact on rapid propagation of green building concepts. It will have great impact on awareness in the society, availability of green building technology, identification of green building products, setting standards and specifications, green label certification, etc.

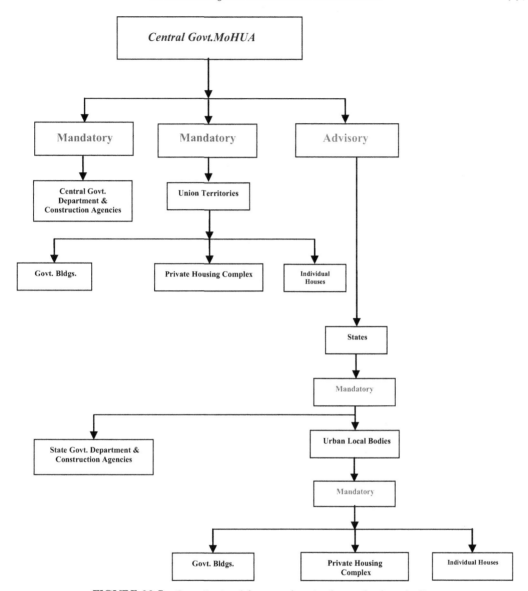

FIGURE 22.7 Organizational framework to implement byelaws/policy.

11. Green building effort and urbanization: critical assessment

As per 2011 census (DG census of India, 2011), there are 13 states in India having an urban population more than 10 million (Table 22.6). The growth of certification of Green Building Projects is not uniform throughout the country. The more developed states of West and South India have a better environmental status than less developed states of North and East India (Table 22.7).

TABLE 22.6   States with urban population more than 10 million.

State/UTs	Urban population in 2001	Urban population in 2011
1. Maharashtra	41.10	50.83
2. Uttar Pradesh	34.54	44.47
3. Tamil Nadu	27.48	34.95
4. West Bengal	22.43	29.13
5. Andhra Pradesh (+Telangana)	20.81	28.35
6. Gujarat	18.93	25.71
7. Karnataka	17.96	23.58
8. Madhya Pradesh	15.97	20.06
9. Rajasthan	13.21	17.08
10. Delhi	12.91	16.33
11. Kerala	8.27	15.93
12. Bihar	8.68	11.73
13. Punjab	8.26	10.39
India	286.12	377.11

TABLE 22.7   Green building projects registered in states with an urban population more than 10 million.

State/UTs	Number of projects registered under LEED India
1. Maharashtra	1333
2. Uttar Pradesh	441
3. Tamil Nadu	402
4. West Bengal	305
5. Andhra Pradesh (+Telangana)	381
6. Gujarat	280
7. Karnataka	407
8. Madhya Pradesh	36
9. Rajasthan	92
10. Delhi	223
11. Kerala	117
12. Bihar	8
13. Punjab	60

12. Elements of green building concept in building byelaws

The MoHUA (Ministry of Housing and Urban Affairs), Central Government, is the apex organization of the country to prepare the policies and legislatures, issue advice and direction to State Government and most importantly orient and shape the urban affairs of the country in the desired direction. Its role is very important as far as propagation of green building concepts in the country is concerned. It has taken a lead role by preparing the MBBL, 2016 as well as incorporating the elements of green building in the bylaws.

The implementation and management of urban policies and matters fall under the purview of State Government. Every state has its own urban affair ministry and respective departments. The urban affairs ministry of State Government is responsible for preparation and approval of legislatures, building bylaws and other guidelines applicable for the entire State. In the State, the entire territory has been divided into districts. In each district, there are ULBs responsible for implementation, execution, supervision and management of building bylaws and policies adopted by the State Government. The ULBs have some autonomy and powers to modify the provisions of building bylaws with the concurrence of the State Government as per local conditions. The ULBs supervise and enforce the implementation of building bylaws and regulations in respective urban areas.

12.1 Model building Byelaws 2004 and 2016

Some elements of green building concept have been incorporated in MBBL 2004 and the same is shown in Table 22.8.

After 14 years since first MBBL in 2004, the Central Government felt need to amend the MBBL and it was amended in 2016. The green building concept got greater and ample space in MBBL 2016. In the preface, it is mentioned that growing environmental concerns is one of the main vision for revising the bylaws. It is also mentioned that the MoHUA in 2015 felt that the MBBL 2004 needs to be reviewed and updated keeping in view the emerging issues:

1. Norms for rooftop solar PV installations.
2. Sustainability and green building.

TABLE 22.8   Elements of Green building in MBBL 2004.

Sl. No.	Paragraph	Elements
1.	2.10.6(ii)a	Separate conveying system to facilitate reuse of sullage water for gardening and washing purposes are to be provided within the plot.
2.	2.10.6(ii)b	For recharging groundwater, rainwater harvesting provisions are to be provided within the plot.
3.	5.4.3(4)	Low capacity cistern to be provided in place of 12.5 L capacity.
4.	5.4.3(5)	Water harvesting through storing of water runoff including rainwater in all new buildings on plot of 100 sqm. and above will be mandatory.
5.	5.4.3(6)	All buildings having a minimum discharge of 10,000 L and above per day shall incorporate a wastewater recycling system. The recycled water should be used for horticulture purpose.
6.	5.4.3(7)	Installation of solar-assisted water heating system in buildings.

Separate chapters have been added for green buildings and sustainability, provision and rainwater harvesting.The elements of the green building concept incorporated in MBBL 2016 have been identified as given in Table 22.9.

The provision has been made to give incentive to builders/individual for the adoption of green building concept in the fom of FAR and fee waiver. Some States/UTs have followed it and amended their bylaws in line of model building bylaws. The incentives provisions were also made in the building bylaws.

12.2 Opportunities for green building elements: ecological balance

12.2.1 Energy conservation and use of nonconvention energy

The new building byelaws have given focus on energy conservation. The building sector consumes around 40% of the total energy produced globally. The green building concept mandates to reduce the energy demand from the building sector from construction to life cycle. Presently, the energy is predominantly produced from the burning of fossil fuels, which results in emission of CO_2 and other greenhouse gases into the environment. The use of low energy consumption light fixtures and efficient electric appliances has been made mandatory. The use of solar photovoltaic cells and solar heater has been provisioned in the byelaws, thereby reducing the energy from the conventional sources resulting in saving of fossil fuels.

12.2.2 Water conservation and recycling of wastewater

The new building byelaws have made provisions to use water and sanitary fixtures that are efficient and use less water without compromizing the functional requirement. The low capacity flushing cistern coupled with efficient water closet results in considerable saving of freshwater. The treatment of wastewater and reuse of the treated water for nonpotable uses have been provisioned. The use of treated wastewater reduces the demand of freshwater as well as reduces the pollution of the river water and water bodies. It will have far-reaching impact on improving the ecology of the water bodies and surroundings. It will also reduce the energy consumption as required in drawing water from the groundwater/other water source.

12.2.3 Rainwater Harvesting

The present pattern of urbanization is leading to deficient recharging of the natural aquifer due to hard surface and buildings in major area of the city as well as destruction of water bodies in the urban area. The groundwater is withdrawn in most of the cities without taking care of recharge of aquifer. The over exploitation of ground water and deficient recharge of aquifer is producing distress in the aquifer, and it has been found that many cities of the worlds are sinking. The new building byelaws have made mandatory provisions for installation of rainwater harvesting.

13. Status of amendment byelaws by the states/UTs after the MBBL 2004

The building byelaws have been enacted by the State Government of various States from time to time. In 2004, the Central Government prepared MBBL 2004 and advised all States and UTs to formulate and amend building bylaws in line of MBBL with the liberty to modify

TABLE 22.9 Elements of green building in MBBL 2016.

Sl. No.	Paragraph	Elements
1.	9	Rainwater harvesting (RWH) has been made mandatory for plot size 100 sqm. or more
		• RWH techniques have been dealt within detail.
		• RWH made mandatory for group housing, commercial building, institutional building, etc.
		• RWH provisions for open space in cities have been included.
2.	9.6	Enforcement and monitoring − A provision has been added for proper enforcement and monitoring of RWH provisions. A rain harvesting cell in ULB which shall be responsible for enforcement and monitoring of RWH provisions.
3.	10	Green building and sustainability provision − A separate chapter has been added.
	10.2(1)	Water conservation and management:
		• Rainwater harvesting
		• Wastewater recycle and reuse
		• Reduction of hardscape
	10.2(2)	Solar energy utilization:
		(a) Installation of solar photovoltaic panel
		(b) Installation of solar-assisted water heating system.
	10.2(3)	Energy efficiency (the concept of passive solar design of a building):
		(a) Low-energy consumption of lighting fixtures (electrical appliances with BEE star, etc.)
		(b) Energy-efficient HVAC system
		(c) The lighting of common area by solar energy/LED devices.
	10.2(4)	Waste management:
		(a) Segregation of waste
		(b) Organic waste management
4.	10.2.1	**(i)** Provision of minimum one tree for every 80 sqm. of the plot area.
		(ii) Compensatory plantation at ratio 1:3.
		(iii) The unpaved area shall be more than or equal to 20% of recreational open space.
5.	10.2.2	Mandatory provision of water reuse and recycling for all buildings having a minimum discharge of 10,000 L and above per day.
6.	10.2.3	Mandatory provision of rooftop solar energy installation for certain category of building.
7.	10.2.4	Mandatory provision of installation of assisted water heating system in a certain category of buildings.
8.	10.2.5	Mandatory provision of sustainable waste management for certain category of buildings.
9.	10.3	Various guidelines for certification of building as per green rating system.
10.		Provision of incentive for certified GRIHA rating buildings.

VII. Sustainable urban design

the provision according to local requirements. Some States amended its building bylaws promptly and some States took considerable time. Some of the States have still not amended their bylaws. The data regarding amendments of their building bylaws have been studied and tabulated in Table 22.10.

14. Challenges to the application of green building norms

In developing countries like India, there are many challenges in the application of such new concept in society. The building technique has been evolved in different societies from the ancient period, and accordingly, traditional techniques have been in use and the people have developed expertise in those traditional techniques. The rapid urbanization has given new challenges to the society to change and adopt building construction techniques as per new requirements of the time. Some of the important challenges to the applicability and the rapid growth of green building techniques have been identified as under. However, green building certification continues to face challenges due to the following factors:

     (i) Lack of awareness.
    (ii) High perceived first cost.
   (iii) Perceived lack of demand from the end users.
   (iv) The perception that green building design and construction is difficult.
    (v) Lack of green building experts in the vicinity.
   (vi) The inertia of the society to leave the traditional method of construction techniques and material use.
  (vii) Lack of awareness of environmental factors.
 (viii) Reluctance on the part of users/occupants to change the living and behavioural pattern.
   (ix) Green building certification projects envisaged many restrictions and measure to design and execute projects in a streamlined manner thereby protecting the environment.
    (x) Reluctance on the part of construction agency, government agencies and institutions with the perception that it is cumbersome, costly and time-consuming.

15. Conclusion

Creating 'green' built environments are the need of the hour for sustainable urbanization. The standards and building codes will determine how the built infrastructures interact with the surrounding environment and the associated trade-offs for a healthy living and balanced ecological functioning. This chapter explores this aspect for the urban ecological studies and brings out relevant information in respect of the building constructions and the need for environmentally sound regulatory framework. The authors have critically assessed two such certification programmes, LEED India and GRIHA, in respect of a rapidly urbanizing country.

Apart from creating publicity about the ease of adopting such standards and creating environmental awareness, private builders also need to be paid attention. The private builders

TABLE 22.10 States and Year of amendment of building bylaws.

Sl. No.	State/UTs	Whether amended	Year of amendment
1	Andhra Pradesh	Yes	2017
2	Andaman and Nicobar (UT)	No	Last in 1999
3	Arunachal Pradesh	Yes	2009
4	Assam	Yes	2014
5	Bihar	Yes	2014
6	Chandigarh (UT)	Yes	2018
7	Chhattisgarh	Yes	2012*
8	Dadra and Nagar Haveli (UT)	Yes	2014
9	Daman and Diu	Yes	2018
10	Delhi	Yes	2004
11	Goa	Yes	2010
12	Gujarat	Yes	2017
13	Haryana	Yes	2017
14	Himachal Pradesh	Yes	2017
15	Jammu and Kashmir	Yes	2011
16	Jharkhand	Yes	2016
17	Karnataka	Yes	2017
18	Kerala	Yes	2011
19	Lakshadweep (UT)	Yes	2016
20	Madhya Pradesh	Yes	2012
21	Maharashtra	Yes	2018
22	Manipur	Yes	2013
23	Meghalaya	Yes	2011
24	Mizoram	Yes	2012
25	Nagaland	Yes	2012
26	Odisha	Yes	2008
27	Pondicherry (UT)	Yes	2012
28	Punjab	No	Draft in 2017
29	Rajasthan	Yes	2017
30	Sikkim	No	Last 1991

(Continued)

TABLE 22.10 States and Year of amendment of building bylaws.—cont'd

Sl. No.	State/UTs	Whether amended	Year of amendment
31	Tamil Nadu	No	Draft 2018
32	Telangana	Yes	2017
33	Tripura	—	—
34	Uttar Pradesh	Yes	2008, 2011 and 2016
35	Uttarakhand	Yes	2015
36	West Bengal	Yes	2009 and 2015

with multistoried colonies are growing rapidly in all parts of the country in the last 20 years. Rising urban population and high land cost are the factors pushing the demand for multistoried buildings. This group can be attracted and moulded with the help of some incentives and mandatory provisions in the building bylaws to adopt green building concepts in their projects.

The individual private houses are most difficult target groups since the majority of the people do not engage architects or designers/engineers to plan and supervise their buildings. They can be targeted in the last with some good incentives. However, before targeting this group, studies have to be conducted for their behaviour changes for region. India is a vast country with different climates, a lot of variation in culture, behaviour, social and environment. The economy also plays a very important role in construction of individual houses. Other factors like size of the plot, location, behaviour, etc., are the challenges for adoption of green building initiatives by the society and individuals. There may need to formulate green building rating system according to regions and climate zones.

References

ADaRSH (Association for Development and Research of Sustainable Habitats), 2013a. About GRIHA. Website. http://www.grihaindia.org.
ADaRSH (Association for Development and Research of Sustainable Habitats), 2013b. GRIHA Rating. Website. http://www.grihaindia.org.
ADaRSH (Association for Development and Research of Sustainable Habitats), 2013c. SVA GRIHA. Website. http://www.grihaindia.org.
Chandel, S.S., Sharma, A., Marwaha, B.M., 2016. Review of energy efficiency initiatives and regulations for residential buildings in India. Renewable and Sustainable Energy Reviews 54, 1443–1458.
DG Census of India, 2011. Census Report (New Delhi, India).
Grimm, N.B., Faeth, S.H., Golubiewski, N.E., Redman, C.L., Wu, J., Bai, X., Briggs, J.M., 2008. Global change and the ecology of cities. Science 319 (5864), 756–760.
IGBC (Indian Green Building Council), 2013a. Annual Review: 2012-2013. Website. https://igbc.in.
IGBC (Indian Green Building Council), 2013b. Certification. Website. https://igbc.in.
IGBC (Indian Green Building Council), 2013c. LEED India. Website. https://igbc.in.
IGBC (Indian Green Building Council), 2013d. Vision. Website. https://igbc.in.
Kibert, C.J., Sendzimir, J., Bradley Guy, G., 2002. Construction Ecology. New Fetter Lane London.
Landman, M., 1999. Breaking through the Barriers to Sustainable Buildings. Tufts University, Boston USA.

Ministry of Housing and Urban Affair, Annual Report 2017-18. (New Delhi, India).

Ministry of Housing and Urban Development, State of Housing in India, A Statistical Compendium 2013.New Delhi, India.

Thanu, H.P., Rajasekaran, C., October 2018. Critical study on performance of building assessment tools with respect to Indian context. IOP Conference Series: Materials Science and Engineering 431 (8), 082011. IOP Publishing.

The UN Population Report, 2018/1, 2018. Website. https://data.worldbank.org.

The UN Report [2018], Revised, World Urbanization Prospect [Website https://data.worldbank.org].

Verma, P., Raghubanshi, A.S., 2018. Urban sustainability indicators: challenges and opportunities. Ecological Indicators 93, 282–291.

Verma, P., Raghubanshi, A.S., 2019. Rural development and land use land cover change in a rapidly developing agrarian South Asian landscape. Remote Sensing Applications: Society and Environment 14, 138–147.

WCED, 1987. Brundtland, G.H. Report of the World Commission on Environment and Development: "Our Common Future". Oxford University Press, Oxford, UK.

Wermick, Ausuben, 1995. National Materials Flows and the Environment (New York, USA).

Challenges and innovations of transportation and collection of waste

Piotr Nowakowski[1], Mariusz Wala[2]

[1]Silesian University of Technology, Katowice, Poland; [2]PST Transgór S.A., Rybnik, Poland

1. Municipal waste generation—global indicators and local approach

In 2018, approximately 55% of the world's population lived in urban areas. It is estimated that this number can grow by up to 68% by 2050. From the beginning of this century, we can observe a significant increase in the residence of the human population to urban areas (UN, 2018). Today, urbanization rate has reached 82% in Northern America, 81% in Latin America,

74% in Europe and 68% in Oceania. Another highly populated continent, Asia has about 50% and Africa almost 43% of the population living in urban areas with some variations between individual countries (CIA, 2019). Fig. 23.1 indicates the percentage of the population living in urban areas for the countries with the highest world economies and the most populated countries. Only India's urbanization rate is below 50% with some countries reaching more than 80% (CIA, 2019; Statista, 2019). The growing number of urban areas and populations has a direct impact on various economic, social and environmental issues.

Many countries around the world have increased economic growth. As a result, their gross domestic product and purchase power have risen. The economic growth is significant in highly populated developing countries including China where the growth for the 2015—2017 year was estimated at over 6.5% annually, in India about 7% and in Indonesia 5% annually (CIA, 2019). As we observe growth in the global population, the growing number of waste is a natural consequence of that. The global waste stream is estimated at approximately 1.3 billion tons per year and is expected to increase to approximately 2.2 billion tons per year by 2025 (Hoornweg and Bhada-Tata, 2012). The main contributors in a waste generation are the population growth and purchasing power of individuals. On average, each resident in African region generates 0.65 kg/cap/day, in Asia regions 0.95—1.1 kg/cap/day, and OECD 2.2 kg/cap/day (Hoornweg and Perinaz, 2012; OECD, 2018). Gross domestic product purchasing power parity (GDP-PPP) has increased in developed and developing countries, and it contributed to change of living style. The purchasing power parity in the leading world economies was, in 2017, 21.7 trillion $ in China, 18 trillion $ in the United States, 9.6 trillion $ in India, 4.7 trillion $ in Japan and 3.7 trillion $ in Germany (WorldBank, 2018; CIA, 2019). On one hand, we have a growing volume of waste, but on the other hand, there is also growing pressure from international organizations, nongovernment institutions and governments to put effort into the protection of the natural environment and to minimize negative impacts on the environment by human activity and economic growth.

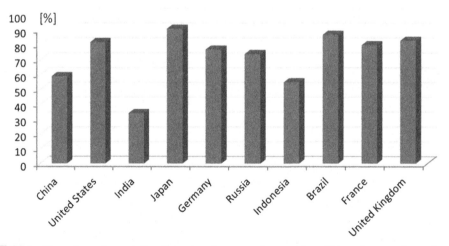

FIGURE 23.1 Percentage of population living in urban areas in countries of biggest economies and number of population (CIA, 2019; Statista, 2019).

Growing costs of extraction of raw materials and the negative influence on environmental had an impact on changing a model of the economy. Additional problems with increasing mass of waste required a shift from a linear economy, where a majority of waste and secondary raw materials ended up dumped or landfilled, to the circular approach where raw materials are intended to circulate in a closed loop (Andersen, 2007). Fig. 23.2 presents a circular economy approach indicating the main sequence after the extraction of raw materials, production, distribution and use, with highlighted waste collection and transportation element as the main subject in this chapter.

After the use of durable products, several options emerge including reusing products, recycling or recovering. The products intended for disposal must be properly collected to facilitate recycling and recovery. Other wastes including consumables and packaging of products should be collected separately to facilitate the recycling of secondary raw materials or energy recovery (Ning, 2001). In the areas with a larger number of the population, a new concept of a process for reclaiming of secondary raw materials was introduced—urban mining. The main task in urban mining is to extract secondary raw materials not only deposited in landfills but also kept in the household or disposed of. It includes wastes, products and buildings. It has the potential to achieve numerous elements and materials after recovery (Brunner, 2011). In any urban area in the world, the waste management system must include legislation, waste collection and transportation rules, waste treatment and processing and social involvement to operate efficiently.

This chapter focuses on key elements of wastes collection and transportation. It is one of the most important factors contributing to the effectiveness of a closed-loop economy. Properly operating waste management system including waste separation contributes to the improvement of social involvement in using secondary raw materials and creating environmental responsibility by individuals. It is an important challenge for developing countries where a significant portion of waste stream ends up in dumps or landfills. Environmental policy in each country must comply with agreements signed up and imposed globally by international organizations such as the United Nations or in Europe—the European Union. Such regulations include a framework for local legislation in minimizing environmental impacts from industry and citizens (Conca, 1995; Jordan, 2012).

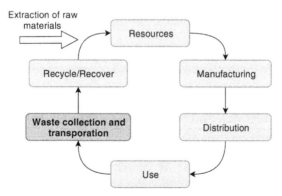

FIGURE 23.2   A circular economy concept indicating waste collection and transportation process.

FIGURE 23.3   Key players and responsibility levels in the waste management system.

The legislation defines the rules on how the waste management system operates and indicates key indicators to be achieved (e.g., recycling or recovery ratio). Therefore, key players of the waste management system have assigned responsibilities (Fig. 23.3). The local or regional authority is responsible for introducing, verifying and supervising the system including controlling the operation imposing fees and fines (Mazzanti and Montini, 2009). Also, it can be defined details of the system operation such as a number of separated categories of waste, recycling ratio for a type of raw materials or more detailed including standards of waste collection bins or requirements of waste collection vehicles. Other players in the system are waste collection companies, which are obliged to prepare operational part of the physical collection of wastes, including a sufficient number and type of collection bins, employees, vehicles, collection schedules and cooperating with waste treatment and processing plants (Janz et al., 2011). Finally, the residents must comply with regulations and participate in waste separation. A contribution of each resident that has properly disposed of waste is crucial to achieve a higher quality of each waste category and facilitate further treatment (Pichtel, 2014). Local authorities and educational institutions must participate in instructing and educating individuals about the waste collection system. Different methods can be applied including school education and direct information for residents using various communication channels (leaflet, posters, Internet) about separated waste collection in the local area.

2. Waste categories and fate of waste streams from households

Collecting different categories of waste requires effective source separation of wastes depending on the secondary raw materials contents and further processing of wastes. The main types of waste coming from households can be divided into several categories (Edjabou et al., 2015). They depend on raw materials contents and the number of waste streams necessary to be managed from households. The main categories of separated waste include plastic, glass, metals, paper and biodegradable waste. The hazardous waste category includes mainly waste electrical and electronic equipment (WEEE or e-waste), batteries and other minor groups such as painting cans, plants and garden care substances, dissolvent substances and others. Additional two categories including a larger amount of waste are bulky waste and construction and demolition waste (European Commission, 2008; European Union, 2011; European Commission, 2012). Biodegradable waste—in the composting process, organic waste is converted into a fertilizer-like product and could be considered for use in residents' gardens, parks and landscaping. Composting facilities do not require sophisticated technology and equipment, and therefore, the investment can be done at relatively low cost. Another method of processing biodegradables is anaerobic digestion and requires a good-quality organic waste stream with sufficient volume to operate. In the process, biogas is produced, and also liquid fertilizer can be dried. The biogas usually is used for generating heat and electricity. The anaerobic digestion plants require advanced technology of processing, collecting and generating energy. Each process needs to be controlled depending on changeable parameters of the waste stream, temperature, installed power, etc. (Hoornweg and Bhada-Tata, 2012).

Another option for the treatment of municipal solid waste is energy recovery. The wastes can be burned at special waste-to-energy plants to produce heat or/and generate electricity. In such a case, a profit is from generated power and reduction of the material intended for landfills. In the case of incineration, the waste of the volume can be reduced by over 80% (Stehlík, 2009).

There are various methods of municipal solid waste management and the possible destination of the waste stream. Although some of them are intended for maximal utilization of secondary raw materials in a recycling process or energy recovery in many countries, the most common is landfilling. Landfills are necessary for the disposal of wastes that cannot be recovered. They require a location properly constructed, maintained and monitored. If any of these activities fails, it can become simply dumpsite with a negative impact on the environment and society (Kaza and Bhada-Tata, 2018).

In numerous countries, separated waste collection is promoted. The separate collection system requires preparation of waste bins, containers or bags for storage, different types of vehicles capable to collect each category of waste and design of schedules of the collection including routing frequency. Depending on environmental policy and waste treatment facilities, availability of waste processing technologies various systems can be compared, e.g., in Italy, Brazil and China (Vaccari et al., 2013; Ibáñez-Forés et al., 2018; Xiao et al., 2018).

Waste composition is influenced by factors such as culture, economic development, climate and energy sources; composition impacts how often waste is collected and how it is disposed of. Low-income countries have the highest proportion of organic waste (Aleluia and Ferrão, 2016; Ikhlayel, 2018). Another possibility in the separation of waste at source could be into 'wet' (food waste, organic components) and 'dry' (all recyclables) and all other waste as a mixed stream or residue. Waste that is collected unsegregated could be separated into organic and recycling streams at a waste sorting facility. Mixed or general waste is a category without separation and as an additional category to separated collection scheme where all materials classified as other or difficult to identify can be disposed of. This option is the least preferred way of wastes disposal from households in countries where the separated system had been introduced, and therefore, the municipalities try to discourage residents by imposing higher collection costs for a household or an enterprise. Although the mixed waste stream is intended for a landfill in practice in developed countries, it undergoes a special treatment to select some fractions that can be used in incineration plants to produce heat or energy (Malinauskaite and Jouhara, 2019).

The annual waste stream varies depending on the country and many factors. In the European Union, the lowest value was in Romania 260 kg/cap/year—and the highest in Denmark 780 kg/cap/year (Eurostat, 2019a). In the United States, the average waste stream is 744 kg/cap/year (OECD, 2018). India generates 226 kg/cap/year in urban areas whereas much lower waste stream in rural settlements (Kumar et al., 2017), and the amount in China is similar 237 kg/cap/year (NBC-China, 2015; Mian et al., 2017). Fig. 23.4 shows the difference in the distribution of five categories of waste between countries of high income and low income. Significantly, the recyclables (paper, plastic, glass and metal) are dominant in high-income countries, and an organic fraction is the main contributor in the waste stream in low-income countries (Kaza and Bhada-Tata, 2018).

In the European Union, there are variations considering the treatment of waste stream from households. Some of the countries including Finland, Sweden and Denmark recycle approximately half of the waste stream and incinerate the rest. Landfill dominates in Bulgaria, Romania, Croatia, Slovakia and Greece (Fig. 23.5). The highest rate of recycling and composting is observed in Germany, Austria and Belgium (Eurostat, 2019b).

Separation of waste by households, firms and institutions is a preliminary stage for using the waste for recycling purposes. It means the individuals or enterprises are obliged to use different containers and to dispose of various categories of waste. In general, several categories of recyclable waste were selected: paper and cardboard, plastics, glass and metal. The separation for categories of waste had been introduced in highly developed countries, and sometimes, there are also some subcategories such as different colors of glass and types of plastics (bottles, foil, etc.) (Tanskanen and Kaila, 2001). The separated categories of waste

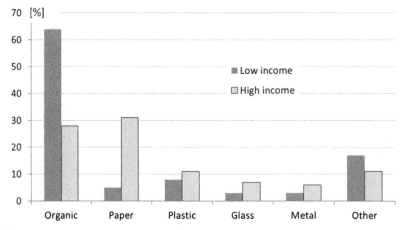

FIGURE 23.4 Distribution of separated waste stream categories depending on income (Kaza and Bhada-Tata, 2018).

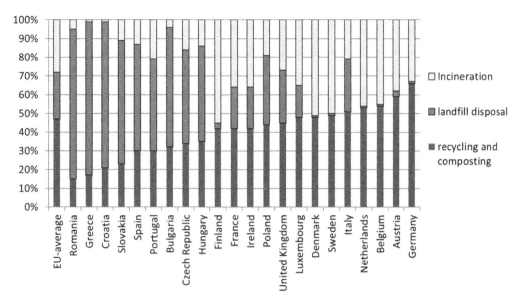

FIGURE 23.5 Waste treatment in selected countries in the European Union (Eurostat, 2019b).

are mainly coming from everyday consumption, packaging of goods and some simple equipment or items from households or commercial activity. There are also other categories of waste intended for separate collection but collected less frequently or on-demand of a resident or an enterprise representative. In these categories is included bulky waste—all kinds of large-dimension objects such as furniture, carpets, big toys and so on. These waste items require sometimes support with handling. Another category of waste is construction and demolition waste. As it can be also a category from industrial or commercial sources of large volumes, for households, it is limited for setting up one or more containers for pickup by the waste collection company.

One of the most valuable concerning the contents of raw materials but also containing hazardous substances is waste electrical and electronic equipment. This category of waste will be discussed in a separate section of this chapter.

3. Storage and transportation of separated waste

The storage, collection and transportation of waste are the main cost contributors in the waste management system. It is estimated that the collection and transportation of waste reach approximately 70% of costs in the waste management system (Boskovic et al., 2016). For the source separation in households, it is necessary to provide waste bags, bins and containers for each category of waste to be collected separately from households. The collection companies have to accordingly prepare collection schedules for the vehicles and staff to collect each type of waste. Also, additional attributes are necessary to take into consideration when designing location and placement of refuse containers such as distance from

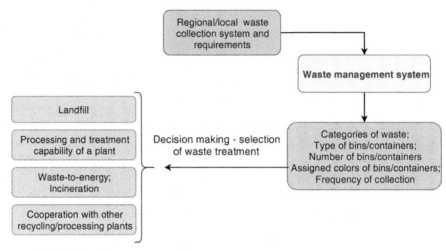

FIGURE 23.6   Local waste management policy and waste collection framework.

households, convenience in access a bin or container and identification of containers such as color, labels and so on (Leeabai et al., 2019). Fig. 23.6 shows the local requirements of waste collection systems depending on the capability of recycling or recovery facilities.

Usually, bags and containers have unified dimensions and variants, and also it is necessary to differentiate them by various colors and printed labels to allow an individual to dispose of exactly each category of waste and to prevent against mixing one type of waste with other categories.

The color system of waste bins and containers has been widely applied in different countries. The color meaning has not been unified globally, but the main pallet in use is green, blue, red, yellow, black and brown. It helps the residents where to dispose of each type of separated waste such as plastics, paper/cardboard, glass, metal, biodegradable waste and a mixed fraction (Lane and Wagner, 2013).

The location of the containers is necessary to be close to residential areas for the convenience of residents but also to assure easy access for waste-collecting companies. Fig. 23.7 presents several solutions to waste separate collections—including plastic bags (in Italy and Poland) and buckets. Fig. 23.7C and D shows stationary containers and bins—and the additional waste bin is intended for beverages cartons (as they include metal, plastic and paper).

4. Waste collection vehicles with reduced emissions and electric vehicles

A big variety of vehicles are necessary to collect the waste. Light commercial vehicles, trucks and special vehicles must be used for the collection. It includes vehicles equipped with a hydraulic compactor on a body and vehicles with lifts to empty or load a container. Waste collection companies are interested in minimal transportation costs. For this purpose, it is necessary to design routes of the collection vehicles to minimize the cost of collection and exhausts emissions (Di Maria and Micale, 2013; Jatinkumar Shah et al., 2018). The vehicles

(A)   **(B)**

(C)   **(D)**

FIGURE 23.7   Examples of bags, bins and containers for separate collection in (A) Italy, (B) Poland, (C) Czech Republic, (D) Spain.

should be suitable to run along various streets including some areas hardly accessible. Fig. 23.8 presents various arrangements and city streets design including modern city, historical city center suburban blocks or residential houses.

   The requirements of employing sufficient bins and containers for a collection area and the frequency of the collection need to be correlated to waste stream and other sanitary and social issues. It depends on the category of waste, climate, temperature and location. For the transportation of the waste, payload capacity and maximal volume are important parameters for assigning each vehicle for a collection area. Depending on the summarized capacity of waste bins and containers, the designed schedules must include these factors.

   Fig. 23.11 presents two categories of collection vehicles. Fig. 23.11A and B shows examples of vehicles for waste collection with gross vehicle mass up to 3500 kg. The first presented vehicle (in service in Italy) is suitable for the collection of waste in narrow streets in old cities and towns for collection of separated waste, and the second one is equipped with hydraulic press and also can be applied in suburbs characterized by difficult maneuvering or in residential areas with difficult access to houses (Poland). Other vehicles with a payload capacity 17 and 23 m^3 (Fig. 23.9C and D) are suitable to collect a larger volume of waste and are equipped with a hydraulic waste compactor.

   Additional factor having an impact on human health and environmental pollution during waste collection is exhaust emissions. The emissions contribute to urban air quality; therefore, local administration requires the application of vehicles with limited emissions, in compliance

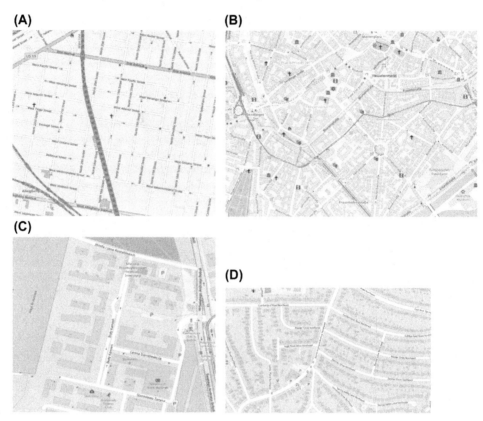

FIGURE 23.8 Types of urban city streets design (A) modern city, (B) historical city center, (C) suburban blocks, (D) residential houses. *maps—© OpenStreetMap contributors.*

with standards, e.g., Euro 5 or Euro 6 in the European Union, or other depending on regulations outside the European Union. In the European Union, a family of standards imposed permissible levels of exhaust emissions by various types of vehicles and other transportation machinery. The exhaust emissions include nitrogen oxides (NOx), carbon monoxide (CO), particulate matter, total hydrocarbon and nonmethane hydrocarbons. Table 23.1 presents limit values of emissions for light commercial vehicles for both diesel and petrol engine, and Table 23.2 includes limit values for heavy vehicles in compliance with standard Euro 5 and Euro 6 (European Parliament, 2007; European Commission, 2016).

Table 23.2 indicates emission limits for heavy-duty engines applied in heavy vehicles including trucks and other waste collection vehicles. The values shown in the table are defined by engine energy output in g/kW.

The main parameters required for the selection of the refuse collection vehicle are payload volume, gross vehicle mass, power, average fuel consumption, number of employees and the possibility of working on one tank for one shift. Table 23.3 shows the technical data of selected vehicle types commonly used in Poland. The vehicles must comply with minimum Euro 5 standard or can be powered by natural gas, and it is required by local legislation.

FIGURE 23.9 Waste collection vehicles from households: (A) separate collection (bags) 3.5t vehicle, (B) mixed waste with hydraulic press 3.5t vehicle, (C) two axes refuse truck, (D) three axes refuse truck.

In many cities, local governments restrict access to vehicles that do not meet emission standards. Also, any vehicle in public service including waste collections must meet these emission standards. It is a policy for supporting clean air zones in cities. As an example, clean air zones have been proposed in the United Kingdom. Local authorities have brought measures into place to improve the air quality. The clean air zones would apply only to buses, taxis and

TABLE 23.1 European emission standards (Euro 5 and Euro 6) for light commercial vehicles.

Vehicle type	Tier (first registration)	CO (g/km)	NOx (g/km)	HC + NO$_x$ (g/km)	Particulate matter (g/km)
Light commercial vehicle (diesel)	Euro 5a (2012)	0.74	0.28	0.35	0.005
	Euro 5b (2013)	0.74	0.28	0.35	0.0045
	Euro 6d (2022)	0.74	0.125	0.215	0.0045
Light commercial vehicle (petrol)	Euro 5a (2012)	2.27	0.082	—	0.005
	Euro 5b (2013)	2.27	0.082	—	0.0045
	Euro 6d (2021)	2.27	0.082	—	0.0045

TABLE 23.2   European emission standards for heavy-duty diesel engines.

Tier	Test cycle	CO	HC	NOx	NH$_3$ (ppm)	Particulate matter
Euro V	Steady-State cycle	1.5	0.46	2	—	0.02
Euro VI	World Harmonized Stationary Cycle	1.5	0.13	0.4	10	0.01
	World Harmonized Transient Test Cycle	4	0.16	0.46	10	0.01

TABLE 23.3   Technical data of refuse collection vehicles commonly used in collection companies in Poland.

Type of vehicle	Gross vehicle mass (GVM) (kg)	Payload capacity (kg)	Payload volume (m^3)	Engine power (kW)	Average fuel consumption (l/ 100 km)	Working day on one tank/one charge
Diesel powered vehicles						
Three axes vehicle	26,000	10,200	23	235	60	Yes
Two axes vehicle	19,000	8140	17	175	45	Yes
Two axes vehicle up to GVM 3.5 t	3500	360	5.5	107	17	Yes
Natural gas—powered vehicles						
Three axes vehicle	26,000	9500	19.5	250	112	No
Three axes vehicle	26,000	8500	18.6	243	84	Yes

heavy goods vehicle, and they will be introduced by 2020. Later the number of vehicle groups would be extended (RAC, 2018). There are numerous initiatives for monitoring air quality levels (Aqicn, 2019; EEA-AQI, 2019; EPA-AirNow, 2019). The emissions monitoring sensors are mainly placed in urban areas, and it is possible to check air quality in most of the cities around the world.

Replacing older diesel vehicles with all-electric versions would be very helpful in reducing unhealthy air pollution in cities. For such purposes, electric vehicles are becoming a new category offered by manufacturers of heavy vehicles. It includes light commercial vehicles for the collection of categories of waste in minor quantities and volume. The electric VW e-Crafter is the 3.5-tonne cargo van with has 170 km of range and uses a 35-kW lithium ion battery. Charging time to 80% of battery capacity is approximately 45 min. The top speed is limited to 90 km/h. In urban settings, e-Crafter may be driven 65—95 km a day including many stops (Volkswagen, 2019). Another example of light commercial vehicle is the Mercedes-Benz eSprinter all-electric van with a range of up to 150 km and a maximum payload of 1000 kg. The van is equipped with two battery pack options: 55 kW for the maximum range of 150 km and 41 kW for a range of 115 km. The rating power of an electric motor is 84 kW, the maximal range is 120 km, maximal speed is 120 km/h and load capacity is 10.5 m^3 (Daimler, 2019).

Recently, automotive manufacturers have presented novel electric waste collection trucks. The first example is the Zoeller E-PTO—an electric drive system that replaces the conventional auxiliary drive of a truck. The vehicle itself is driven by a diesel engine but can eliminate emissions from the chassis motor after reaching a waste collection point. There are two versions of the E-PTO with energy storage 43 and 24 kW. The electric auxiliary drive is powered by an electric motor supplied by a battery pack of charging time 4–6 h and is sufficient to operate a full working day (Zoeller, 2019). New, completely electric refuse vehicles including the drive of the truck have been announced by Volvo with a model Mack LR BEV. The vehicle is powered by two 130-kW motors producing a combined 496 peak horse power. Power is sent through a two-speed transmission to rear axles. All of the equipment including the hydraulic systems and accessories are electrically driven through 12, 24 and 600 V circuits. The vehicle includes four lithium-ion batteries, and full real-world testing will begin in 2020 (Volvo, 2019). Refuse collection vehicles' manufacturer—Dennis Eagle from Terberg Group proposes a novel construction of refuse collection vehicle includes low-entry chassis, refuse compaction body, bin lift and telematics system. The power module is equipped with 300-kW battery packs and control systems and a 200-kW electric motor driving a conventional axle. The capacity of the compartment for waste will be 19 m^3 with an automatic split bin lift. The electric refuse collection vehicle is going to be produced by the end of 2019 (TerbergGroup, 2019).

Real-world operation of waste collection electric vehicles will make it possible to evaluate usability in various climatic conditions including temperature and humidity. An additional factor would depend on traffic conditions, a number of stops and so on.

5. Novel solutions in the automation of waste management and application of information technologies in waste collection

Selection of a separated waste collection model, designing collection schedules and providing various sizes of waste bins and containers is a challenge for many developing countries where large-scale initiatives have not been taken yet. But even in developed countries, many innovations potentially can improve the waste management system operation and increase the economic efficiency of the collection.

A study of Hannan et al. (2015) presents an overview of ICT (information and communication technology) technologies and their use in solid waste monitoring and management systems. The rapid development of information and communication technologies allowed improving the planning and design of modern solid waste management systems. Most recent applications help in planning, monitoring and supervising the collection of solid waste. ICT in waste management has been divided into four categories, such as spatial technologies, identification technologies, data collection technologies and data transfer technologies. Spatial technologies include geographic information system (GIS), global positioning system and remote sensing (RS). Identification technologies include using barcodes and radio frequency identification. Data acquisition technologies include detection and imaging technologies, and data transmission technologies include both short-range and long-range communication technologies (Hannan et al., 2011). New solutions in the improvement of waste collection and transportation have been focused on the application of IT in different activities. Several tools have been used in supporting fleet management systems and also waste bins and containers. One of the applications was the use of GIS in the location of the vehicles with additional features typical for logistics management such as details of vehicles, drivers and routes (Rada et al., 2013). Fig. 23.10 shows the registered routing of waste collection vehicles in the South of Poland. It can be used not only for evaluation of the efficiency of the collection but also to estimate costs including fuel and working hours. After the wide application of fleet management systems in the real world, the managers of the waste collection companies have the availability of monitoring routes of a vehicle to apply route optimization systems to lower transportation costs.

Applications of RS-based systems include the choice of storage location, environmental features and monitoring of the impact on the solid waste landfill and environmental impact assessment. Artificial intelligence algorithms have been widely applied in solving vehicle routing problems and focusing on shortening the routes of collection vehicles and reducing collection cost (Das and Bhattacharyya, 2015; Ehmke et al., 2018; Jatinkumar Shah et al., 2018). These algorithms are essential in software used in the design of a sequence of driving the streets in case of curbside collection (Zbib and Wøhlk, 2019). The optimal sequence of routing is important in street cleaning and snow removal. Other tasks of waste collection include collecting from a set of companies in different locations in a city or waste collection on-demand, when the waste collection is ordered by a company or individual requires

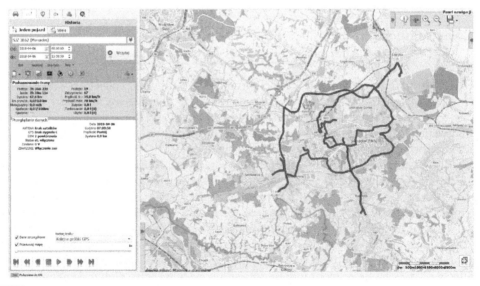

FIGURE 23.10 Example of routing of a waste collection vehicle in urban municipality in Poland. *map—© OpenStreetMap contributors.*

solving vehicle routing problem to find optimized routes (Nowakowski et al., 2017, 2018). Waste collection vehicles can be equipped in additional sensors for online transmission of the data.

Other innovations have been introduced in waste collection containers, creating an intelligent system equipped with sensors indicating a level of filling waste bins or containers. For data collection technologies, various sensors are used in the measurement of the level of waste in a bin or container, the waste mass inside the bin, monitoring of environmental conditions, waste sorting and monitoring of combustion. Real-world solutions in supporting waste collection across the United States and the European Union have been applied by Enevo (2019). In this system, containers equipped with sensors and communication modules are controlled online, and routing plan for the collection vehicles is assigned dynamically.

Research proposals (Aziz et al., 2015, 2018) consider a waste collection schedule based on image processing for waste collection containers. Containers can be empty, partly full or full of waste objects of various shapes and sizes. The proposed methods can identify and determine the level of waste in a container. Hannan presents a content-based image retrieval system to explore the possibility of detecting the filling level of a solid waste container from images with an extracted texture. For feature extraction, the texture is used as an image function for the basket level detection system (Hannan et al., 2016). The proposals, researches and real-world applications indicate that the broad implementation of automated, sensor-equipped components of waste collection system supported by information and communication technologies have great potential for development.

6. Case study of separate collection—waste electrical and electronic equipment

For several years in numerous countries in the world, a serious problem emerged with a growing number of waste electrical and electronic equipment (WEEE) also described as e-waste (Ongondo et al., 2010; Shah Khan et al., 2014). Consequently, several countries began to work out legislation to create a waste management system including all key players in electrical and electronic equipment (EEE) supply chain, including institutions and companies involved in the collection, handling and treatment of WEEE. Initially, e-waste management systems were put in practice in Switzerland, South Korea, Taiwan, Japan and then European Union (Chung and Murakami-Suzuki, 2008; Wäger et al., 2011; European Commission, 2012). Other countries followed these concepts or proposed similar solutions, e.g., the United States, China, India and other developing countries (Shinkuma and Managi, 2010; Wath et al., 2011).

The main purpose of the WEEE management system is to assign responsibilities for each stakeholder in the supply chain including producer of EEE, municipalities, collection companies and disassembling and recycling plants and individuals. The residents need to be aware of how to dispose of waste equipment and where the nearest collection point is and do not store the waste items in households (Jafari et al., 2017; Nowakowski, 2019).

The producers of EEE are responsible to collect and recycle the equipment they produced. As it is a difficult task to create and keep the collection and recycling network in each country by a producer, there are established WEEE organizations that are capable of helping to collect the appliances put on the market when they reach end-of-life. After an agreement is signed, the WEEE organizations take the responsibility to collect WEEE and employ different collection companies to provide the collection and transportation from collection sites to disassembly and processing plants. The key element to collect efficiently WEEE is the creation of a network including the appropriate number and methods of collection and social involvement in environmentally responsible disposal of WEEE (Goodship and Stevels, 2012).

In developed countries, the WEEE organizations and collection companies created an efficient system of collecting and processing e-waste. In developing countries especially in India, China and South America also, informal sector of collectors plays an important role in a pickup of waste from residents and then delivering to treatment and disassembling plants. To avoid any disposal of hazardous substances into the environment by extracting only precious components, including metals, the informal sector should be supervised to prevent potential harm to the environment by the disposal of hazardous substances (Gerxhani, 2004; Streicher-Porte et al., 2005). Therefore, there are initiatives to include also informal collectors in a formalized chain or organization leaving some independence but providing education and information campaigns and a possible collection of all waste appliances they collect (Gu et al., 2016; Li et al., 2017).

In urban municipalities, an individual has many choices to dispose of WEEE. The methods of disposal can be stationary or mobile including variations and modifications. The stationary collection points can be located in the municipal collection centers, shops and supermarkets with EEE and local storage places in the vicinity of urban and residential areas. Small equipment can be disposed of in minor bins or boxes located in frequently visited places. Fig. 23.11 shows two choices of e-waste disposal including mobile and stationary collections. The challenges and novel applications to improve the collection rate are shown in the curly bracket in Fig. 23.11. Waste collection bins or containers can be equipped with sensors to facilitate information exchange and communication with waste collection companies. Mobile collection includes important optimization issues with vehicle routing. One of the ways to dispose of e-waste is the replacement of old equipment by a new one purchased. In this case, a delivery cost includes also pickup of a waste item, before the final shipment of all collected appliances to e-waste disassembly plant and treatment facility. Collection of waste appliances—mostly large home appliances during delivery of new ones, is the most efficient taking into consideration transportation costs. In such a case, a delivery vehicle can be also used as a waste collection vehicle. Additional perspectives on the application of Internet-based collection services are growing interest of residents in new methods of requesting a waste pickup. In such a case, a resident who has any waste can call or use a mobile app to request waste collection (Nowakowski et al., 2018).

Numerous types of waste electrical and electronic appliances are small in size. For the convenience of residents and customers of supermarkets, many small e-waste collection bins have been set up. Fig. 23.12 shows examples of designs such as bins in Poland (Fig. 23.12A) and in Spain (Fig. 23.12B).

Example of intelligent small WEEE container capable of checking the level of waste inside and sending information to a waste collection company has been developed in the Czech Republic. Fig. 23.12C shows the container placed in a residential area with blocks in Poland. The location of all collection points can be supported by online maps (e.g., Asekol, 2019; LIFE WEEE − RAEE, 2019). An individual can find the nearest location of a stationary collection

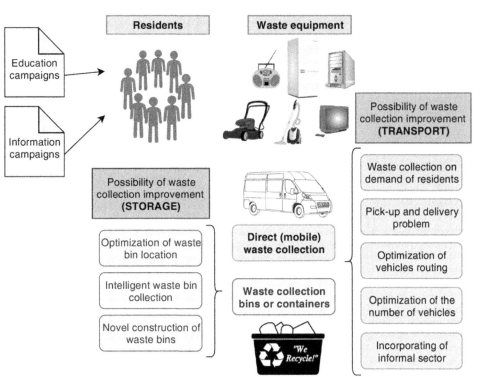

FIGURE 23.11 Relations between individuals and ways of disposal e-waste and potential improvements.

point or a schedule for mobile collection. Alternatively, waste equipment for the collection can be reported online (Nowakowski et al., 2018). Broad application of artificial intelligence supporting systems with the novel approach with bin and container design including communication modules has brought a gradual increase in the e-waste collection rate.

7. Conclusions

Waste management is a key element in urban ecology. It has a significant influence on the environment and society. One of the most important tasks is the collection and transportation of waste. This chapter indicated that the growing number of wastes in the cities would be a challenge for local governments, waste collection companies and residents. Gradual implementation of separate waste collection is a necessary step in many developing countries. Some initiatives have been successfully taken in different cities by introducing the separate collection. In separate collection, waste streams from households can be divided into several categories. Larger number of categories requires not only separate waste bins and containers but also social responsibility. It requires adaptation of curricula on different levels of education to provide sufficient knowledge about proper ways of waste disposal and also negative environmental impacts especially when the waste stream is mixed or worse—dumped in

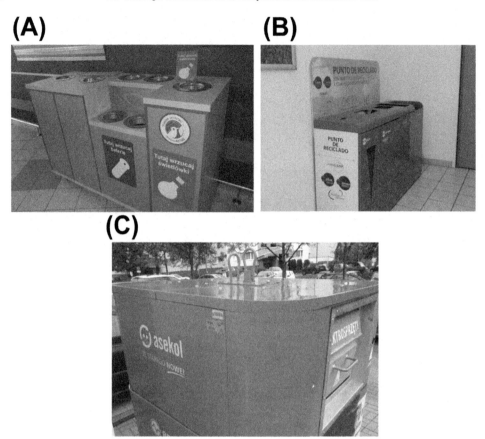

FIGURE 23.12 Small e-waste collection bins located in supermarkets in Poland (A) and Spain (B), and a novel container with a communication module (C).

green areas. Effective education and information campaigns need to be provided for different age groups.

Many novel methods and systems have been implemented in the infrastructure—including collection vehicles, intelligent waste bins, containers equipped with communication modules and sensors and advanced treatment and technologies in waste processing plants. The collection companies use artificial intelligence systems to cut the collection costs by minimizing route length, number of vehicles and employees. Environmental issues having an impact on poor air quality in cities also have been addressed by the necessity of using vehicles meeting standards of Euro 5 and 6 or powered by natural gas. For reducing exhaust emissions to zero in waste collection, numerous electric vehicles including trucks have been recently manufactured. By the end of 2019 and the beginning of 2020, fully electric waste collection trucks are going to put into service.

We can conclude and summarize that the main challenges in waste collection in urban areas require concurrent development of the selected activities concerning:

- Society—better education and understanding of waste separation and proper waste disposal including information about environmental and health hazards
- Technology—by a new design of waste collection bins and containers with the possibility of measuring the amount of waste and communicating with the collection company, implementation of information technologies in collection vehicles for monitoring routing and process of emptying bins and containers
- Environment and health—providing zero-emission electric waste collection vehicles with reduced noise level during collection
- Management—setting up waste sufficient number of bins and containers depending on categories and cost reduction in the optimal number of employees and vehicles

References

Aleluia, J., Ferrão, P., 2016. Characterization of urban waste management practices in developing Asian countries: a new analytical framework based on waste characteristics and urban dimension. Waste Management 58, 415–429. https://doi.org/10.1016/j.wasman.2016.05.008.

Andersen, M.S., 2007. An introductory note on the environmental economics of the circular economy. Sustainability Science 2, 133–140.

Aqicn, 2019. Chennai Air Pollution: Real-Time Air Quality Index. http://aqicn.org/city/chennai/alandur/. (Accessed 8 October 2019).

Asekol, 2019. Collection Points < ASEKOL | Asekol. https://www.asekol.cz/en/asekol/collection-points/. (Accessed 5 August 2019).

Aziz, F., Arof, H., Mokhtar, N., et al., 2015. Rotation invariant bin detection and solid waste level classification. Measurement 65, 19–28. https://doi.org/10.1016/j.measurement.2014.12.027.

Aziz, F., Arof, H., Mokhtar, N., et al., 2018. Waste level detection and HMM based collection scheduling of multiple bins. PLoS One 13, e0202092. https://doi.org/10.1371/journal.pone.0202092.

Boskovic, G., Jovicic, N., Jovanovic, S., Simovic, V., 2016. Calculating the costs of waste collection: a methodological proposal. Waste Management and Research 34, 775–783. https://doi.org/10.1177/0734242X16654980.

Brunner, P.H., 2011. Urban mining A contribution to reindustrializing the city. Journal of Industrial Ecology 15, 339–341. https://doi.org/10.1111/j.1530-9290.2011.00345.x.

Chung, S.-W., Murakami-Suzuki, R., 2008. A Comparative Study of e-Waste Recycling Systems in Japan, South Korea and Taiwan from the EPR Perspective: Implications for Developing Countries. http://www.ide.go.jp/English/Publish/Download/Spot/pdf/30/007.pdf. (Accessed 19 October 2011).

CIA, 2019. The World Factbook - Central Intelligence Agency. https://www.cia.gov/library/publications/resources/the-world-factbook/. (Accessed 5 October 2019).

Conca, K., 1995. Greening the United Nations: environmental organisations and the UN system. Third World Quarterly 16, 441–458.

Daimler, 2019. eVito and eSprinter: local zero-emissions mobility. Daimler. https://www.daimler.com/innovation/case/electric/driving-event-evito-und-esprinter.html. (Accessed 8 September 2019).

Das, S., Bhattacharyya, B.K., 2015. Optimization of municipal solid waste collection and transportation routes. Waste Management 43, 9–18. https://doi.org/10.1016/j.wasman.2015.06.033.

Di Maria, F., Micale, C., 2013. Impact of source segregation intensity of solid waste on fuel consumption and collection costs. Waste Management 33, 2170–2176. https://doi.org/10.1016/j.wasman.2013.06.023.

Edjabou, M.E., Jensen, M.B., Götze, R., et al., 2015. Municipal solid waste composition: sampling methodology, statistical analyses, and case study evaluation. Waste Management 36, 12–23. https://doi.org/10.1016/j.wasman.2014.11.009.

EEA-AQI, 2019. European Environment Agency - European Air Quality Index. https://airindex.eea.europa.eu/. (Accessed 8 October 2019).

Ehmke, J.F., Campbell, A.M., Thomas, B.W., 2018. Optimizing for total costs in vehicle routing in urban areas. Transportation Research Part E: Logistics and Transportation Review 116, 242–265. https://doi.org/10.1016/j.tre.2018.06.008.

Enevo, 2019. Enevo-Better Waste Services and Advanced Tech, All at a Lower Cost. https://www.enevo.com/waste-solutions-services. (Accessed 8 September 2018).

EPA-AirNow, 2019. Environmental Protection Agency (USA) - AirNow. https://airnow.gov/. (Accessed 1 October 2019).

European Commission, 2008. Directive 2008/1/EC of the European Parliament and of the Council of 15 January 2008 Concerning Integrated Pollution Prevention and Control (Codified version)(Text with EEA Relevance).

European Commission, 2012. Directive 2012/19/EU of the European Parliament and of the Council of 4 July 2012 on Waste Electrical and Electronic Equipment (WEEE)Text with EEA Relevance - LexUriServ.do.

European Commission, 2016. Commission Regulation (EU) 2016/646 of 20 April 2016 Amending Regulation (EC) No 692/2008 as Regards Emissions from Light Passenger and Commercial Vehicles (Euro 6) (Text with EEA Relevance).

European Parliament, 2007. Regulation (EC) No 715/2007 of the European Parliament and of the Council of 20 June 2007 on Type Approval of Motor Vehicles with Respect to Emissions from Light Passenger and Commercial Vehicles (Euro 5 and Euro 6) and on Access to Vehicle Repair and Maintenance Information.

European Union, 2011. Directive 2011/65/EU of the European Parliament and of the Council of 8 June 2011 on the Restriction of the Use of Certain Hazardous Substances in Electrical and Electronic Equipment. http://eur-lex.europa.eu/LexUriServ/LexUriServ.do?uri=OJ:L:2011:174:0088:0110:EN:PDF. (Accessed 28 October 2011).

Eurostat, 2019a. Waste generated by households by year and waste category. In: Waste Generated by Households by Year and Waste Category. https://ec.europa.eu/eurostat/web/environment/waste/main-tables. (Accessed 2 October 2019).

Eurostat, 2019b. Waste Management Indicators - Statistics Explained. https://ec.europa.eu/eurostat/statistics-explained/index.php/Waste_management_indicators#Recycling. (Accessed 7 September 2019).

Gerxhani, K., 2004. The informal sector in developed and less developed countries: a literature survey. Public Choice 120, 267–300.

Goodship, V., Stevels, A., 2012. Waste Electrical and Electronic Equipment (WEEE) Handbook. Elsevier Science.

Gu, Y., Wu, Y., Xu, M., et al., 2016. The stability and profitability of the informal WEEE collector in developing countries: a case study of China. Resources, Conservation and Recycling 107, 18–26. https://doi.org/10.1016/j.resconrec.2015.12.004.

Hannan, M.A., Abdulla Al Mamun, Md, Hussain, A., et al., 2015. A review on technologies and their usage in solid waste monitoring and management systems: issues and challenges. Waste Management 43, 509–523. https://doi.org/10.1016/j.wasman.2015.05.033.

Hannan, M.A., Arebey, M., Begum, R.A., et al., 2016. Content-based image retrieval system for solid waste bin level detection and performance evaluation. Waste Management 50, 10–19. https://doi.org/10.1016/j.wasman.2016.01.046.

Hannan, M.A., Arebey, M., Begum, R.A., Basri, H., 2011. Radio Frequency Identification (RFID) and communication technologies for solid waste bin and truck monitoring system. Waste Management 31, 2406–2413. https://doi.org/10.1016/j.wasman.2011.07.022.

Hoornweg, D., Bhada-Tata, P., 2012. What a Waste: a Global Review of Solid Waste Management.

Hoornweg, D., Perinaz, B.T., 2012. Urban Dev Ser Knowl Pap. What a Waste: a Global Review of Solid Waste Management, vol. 15, pp. 87–88.

Ibáñez-Forés, V., Coutinho-Nóbrega, C., Bovea, M.D., et al., 2018. Influence of implementing selective collection on municipal waste management systems in developing countries: a Brazilian case study. Resources, Conservation and Recycling 134, 100–111. https://doi.org/10.1016/j.resconrec.2017.12.027.

Ikhlayel, M., 2018. An integrated approach to establish e-waste management systems for developing countries. Journal of Cleaner Production 170, 119–130. https://doi.org/10.1016/j.jclepro.2017.09.137.

Jafari, A., Heydari, J., Keramati, A., 2017. Factors affecting incentive dependency of residents to participate in e-waste recycling: a case study on adoption of e-waste reverse supply chain in Iran. Environment, Development and Sustainability 19, 325–338. https://doi.org/10.1007/s10668-015-9737-8.

Janz, A., Günther, M., Bilitewski, B., 2011. Reaching cost-saving effects by a mixed collection of light packagings together with residual household waste? Waste Management and Research 29, 982–990. https://doi.org/10.1177/0734242X11416156.

Jatinkumar Shah, P., Anagnostopoulos, T., Zaslavsky, A., Behdad, S., 2018. A stochastic optimization framework for planning of waste collection and value recovery operations in smart and sustainable cities. Waste Management 78, 104–114. https://doi.org/10.1016/j.wasman.2018.05.019.

Jordan, A., 2012. Environmental policy in the European Union: actors, institutions, and processes. Earthscan.

Kaza, S., Bhada-Tata, P., 2018. Decision Maker's Guides for Solid Waste Management Technologies. World Bank.

Kumar, S., Smith, S.R., Fowler, G., et al., 2017. Challenges and Opportunities Associated with Waste Management in India. Royal Society Open Science.

Lane, G.W.S., Wagner, T.P., 2013. Examining recycling container attributes and household recycling practices. Resources, Conservation and Recycling 75, 32–40. https://doi.org/10.1016/j.resconrec.2013.03.005.

Leeabai, N., Suzuki, S., Jiang, Q., et al., 2019. The effects of setting conditions of trash bins on waste collection performance and waste separation behaviors; distance from walking path, separated setting, and arrangements. Waste Management 94, 58–67. https://doi.org/10.1016/j.wasman.2019.05.039.

Li, Y., Xu, F., Zhao, X., 2017. Governance mechanisms of dual-channel reverse supply chains with informal collection channel. Journal of Cleaner Production 155, 125–140. p.2. https://doi.org/10.1016/j.jclepro.2016.09.084.

LIFE WEEE − RAEE, 2019. LIFE WEEE − RAEE: Tesori da recuperare! - Apps on Google Play. https://play.google.com/store/apps/details?id=org.disit.lifeweee&hl=en. (Accessed 5 August 2019).

Malinauskaite, J., Jouhara, H., 2019. The trilemma of waste-to-energy: a multi-purpose solution. Energy Policy 129, 636–645. https://doi.org/10.1016/j.enpol.2019.02.029.

Mazzanti, M., Montini, A., 2009. Waste and Environmental Policy. Taylor & Francis.

Mian, M.M., Zeng, X., Nasry, A. al NB., Al-Hamadani, S.M.Z.F., 2017. Municipal solid waste management in China: a comparative analysis. Journal of Material Cycles and Waste Management 19, 1127–1135. https://doi.org/10.1007/s10163-016-0509-9.

NBC-China, 2015. National Bureau of Statistics of China (2015) Annual data.

Ning, D., 2001. Cleaner production, eco-industry and circular economy. Research of Environmental Sciences 6.

Nowakowski, P., 2019. Investigating the reasons for storage of WEEE by residents − a potential for removal from households. Waste Management 87, 192–203. https://doi.org/10.1016/j.wasman.2019.02.008.

Nowakowski, P., Król, A., Mrówczyńska, B., 2017. Supporting mobile WEEE collection on demand: a method for multi-criteria vehicle routing, loading and cost optimisation. Waste Management 69, 377–392. https://doi.org/10.1016/j.wasman.2017.07.045.

Nowakowski, P., Szwarc, K., Boryczka, U., 2018. Vehicle route planning in e-waste mobile collection on demand supported by artificial intelligence algorithms. Transportation Research Part D: Transport and Environment 63, 1–22. https://doi.org/10.1016/j.trd.2018.04.007.

OECD, 2018. Waste - municipal waste - OECD data. theOECD. http://data.oecd.org/waste/municipal-waste.htm (Accessed 6 September 2019).

Ongondo, F., Williams, I., Cherrett, T., 2010. How are WEEE doing? A global review of the management of electrical and electronic wastes. Waste Management 31, 714–730. https://doi.org/10.1016/j.wasman.2010.10.023.

Pichtel, J., 2014. Waste Management Practices: Municipal, Hazardous, and Industrial, second ed. Taylor & Francis.

RAC, 2018. Clean Air Zones − what Are They and where Are They? | RAC Drive. https://www.rac.co.uk/drive/advice/emissions/clean-air-zones/. (Accessed 8 October 2019).

Rada, E.C., Ragazzi, M., Fedrizzi, P., 2013. Web-GIS oriented systems viability for municipal solid waste selective collection optimization in developed and transient economies. Waste Management 33, 785–792. https://doi.org/10.1016/j.wasman.2013.01.002.

Shah Khan, S., Aziz Lodhi, S., Akhtar, F., Khokar, I., 2014. Challenges of waste of electric and electronic equipment (WEEE): toward a better management in a global scenario. Management of Environmental Quality: An International Journal 25, 166–185. https://doi.org/10.1108/MEQ-12-2012-0077.

Shinkuma, T., Managi, S., 2010. On the effectiveness of a license scheme for E-waste recycling: the challenge of China and India. Environmental Impact Assessment Review 30, 262–267. https://doi.org/10.1016/j.eiar.2009.09.002.

Statista, 2019. Indonesia - urbanization 2007-2017. Statista. https://www.statista.com/statistics/455835/urbanization-in-indonesia/. (Accessed 6 October 2019).

Stehlík, P., 2009. Contribution to advances in waste-to-energy technologies. Journal of Cleaner Production 17, 919—931. https://doi.org/10.1016/j.jclepro.2009.02.011.

Streicher-Porte, M., Widmer, R., Jain, A., et al., 2005. Key drivers of the e-waste recycling system: Assessing and modelling e-waste processing in the informal sector in Delhi. Environmental Impact Assessment Review 25, 472—491.

Tanskanen, J.-H., Kaila, J., 2001. Comparison of methods used in the collection of source-separated household waste. Waste Management and Research 19, 486—497. https://doi.org/10.1177/0734242X0101900604.

TerbergGroup, 2019. 0% Emissions. 100% Electric. The All-New eCollect - Terberggroup-Dennis Eagle. https://newsmedia.terberggroup.com/en/trrg/overview/press-releases/dennis-eagle-uk/ecollect—fully-electric-rcv/. (Accessed 8 October 2019).

UN, 2018. 68% of the world population projected to live in urban areas by 2050, says UN. In: UN DESA | United Nations Department of Economic and Social Affairs. https://www.un.org/development/desa/en/news/population/2018-revision-of-world-urbanization-prospects.html. (Accessed 25 September 2019).

Vaccari, M., Bella, V.D., Vitali, F., Collivignarelli, C., 2013. From mixed to separate collection of solid waste: benefits for the town of Zavidovići (Bosnia and Herzegovina). Waste Management 33, 277—286. https://doi.org/10.1016/j.wasman.2012.09.012.

Volkswagen, 2019. Modelle | Volkswagen Nutzfahrzeuge - E-Crafter. https://www.volkswagen-nutzfahrzeuge.de/de/innovationen/elektromobilitaet/modelle.html. (Accessed 2 October 2019).

Volvo, 2019. Mack Trucks Unveils Fully Electric Refuse Demonstration Model. https://www.volvogroup.com/en-en/news/2019/may/news-3292007.html. (Accessed 8 October 2019).

Wäger, P., Hischier, R., Eugster, M., 2011. Environmental impacts of the Swiss collection and recovery systems for waste electrical and electronic equipment (WEEE): a follow-up. The Science of the Total Environment.

Wath, S.B., Dutt, P.S., Chakrabarti, T., 2011. E-waste scenario in India, its management and implications. Environmental Monitoring and Assessment 172, 249—262. https://doi.org/10.1007/s10661-010-1331-9.

WorldBank, 2018. World Development Indicators | DataBank. https://databank.worldbank.org/indicator/NY.GDP.MKTP.KD.ZG/1ff4a498/Popular-Indicators. (Accessed 6 August 2019).

Xiao, S., Dong, H., Geng, Y., Brander, M., 2018. An overview of China's recyclable waste recycling and recommendations for integrated solutions. Resources, Conservation and Recycling 134, 112—120. https://doi.org/10.1016/j.resconrec.2018.02.032.

Zbib, H., Wøhlk, S., 2019. A comparison of the transport requirements of different curbside waste collection systems in Denmark. Waste Management 87, 21—32. https://doi.org/10.1016/j.wasman.2019.01.037.

Zoeller, 2019. MAGNUM_XXL_E-PTO - Zoeller. https://www.zoeller-kipper.de/wp-content/uploads/MAGNUM_XXL_E-PTO_Final_01-GB.pdf. (Accessed 2 October 2019).

Critical assessment and future dimensions for the urban ecological systems

Pramit Verma[1], Rishikesh Singh[1], Pardeep Singh[2], A.S. Raghubanshi[1]

[1]Integrative Ecology Laboratory (IEL), Institute of Environment & Sustainable Development (IESD), Banaras Hindu University (BHU), Varanasi, Uttar Pradesh, India; [2]Department of Environmental Studies, PGDAV College, University of Delhi, New Delhi, India

1. Introduction

This chapter indicates that there is a dire need for a comprehensive book on urban ecological studies, which holistically addresses emerging concepts of urbanization and compiles an all-inclusive account of development in its various themes. Several previous works have dealt with different aspects of urban ecological research focusing on climate change, flood risk, green and open space, and environmental pollution and global environmental change, urbanization process and environmental justice (Klauer et al., 2016), energy (Bridge et al., 2018), certain hotspots of urbanization, like China (Loo, 2018). Others have highlighted the process of urbanization, analysing the challenges associated with it (Kleer and Nawrot, 2018). Urban boundary and basic concepts of urban ecology, phenomena of ex situ resource mobilization for urban growth due to trade and globalization, anthropogenic subsidization of urban material balance, spatial modelling using crowd-sourced and freely available data, rural ecology and urbanization of periurban areas, social—ecological systems and good governance, anthropogenic and natural nutrient fluxes from urban landscape, urban energy and socioeconomic driving factors, human health, climate change and urbanization, sustainable cities and sustainability monitoring need a transdisciplinary approach, combining all these aspects.

2. Transdisciplinary nature

The transdisciplinary nature of urban ecology research can be seen from the diverse publications and relationships in the next section. The need for integrative frameworks that collate the biophysical and social aspects has been highlighted by many authors (Norris et al., 2016; Simon et al., 2018; Grove and Pickett, 2019; Rademacher et al., 2019). The social—ecological relationships that can be used to create such a framework are poorly understood between the urban and ecological systems (Rademacher et al., 2019). The same author group has suggested that a coproduction approach can be a better answer for integrating these aspects into a transdisciplinary nature. Coproduction simply means advancing research in all aspects simultaneously that would help in understanding the relationships between the biophysical and social systems and thereby help in creating an integrative framework. However, the transdisciplinary approach has been criticized as inclusive of certain 'wicked problems' related to the complexity between social and ecological systems (Carroll et al., 2007; Norris et al., 2016). A detailed description of the problems of a transdisciplinary approach in social—ecological systems has been described by Norris et al. (2016). Due to limited space, we would like to highlight certain aspects of such a transdisciplinary approach. Firstly, since these systems and their subsidiaries differ greatly, 'systematic thematic comparison requires great care and methodological rigour' (Simon et al., 2018). Secondly, the complexity inherent in transdisciplinary research makes it time-consuming and sometimes unpredictable (Ibid). Simon et al. (2018) also identify coproduction as the way forward; however, it should be understood that validation at each step, understanding about the complexity and unpredictability and sound methodological basis need to be constructed at the outset of such research. This chapter discusses the major research fields in urban ecology domain, integration with other themes, integration with emerging fields and major challenges and opportunities. We have taken help of the bibliometric analysis to identify the emerging fields and their relationships. In the results of the bibliometric analysis, we will see that the most productive journal in this field in terms of the number of articles is the one with a transdisciplinary approach.

3. Methodology—bibliometric analysis

A bibliometric analysis was performed using the bibliometrix package in R (ver. 3.5.1) (Aria and Cuccurullo, 2017). The keyword cooccurrence plot and the conceptual structure and dendrogram were utilized to describe the bibliometric results. The first evaluates the role of scientific actors and their quantitative impact, whereas the other explains the structural and dynamic features, respectively (Ibid). The conceptual structure plot uses correspondence analysis (CA), multiple correspondence analysis (MCA) or metric multidimensional scaling (MDS) and clustering into two nodes with only two connections allowed for each node (Aria and Cuccurullo, 2017). Word cooccurrence plot represents the major topics and the integration between them. Conceptual structure thematic dimensions of the different fields comprising urban ecology extracted from keywords, title and abstract (Aria and Cuccurullo, 2017). The thematic strategic plot represents the function of the major themes in the field of urban ecology, their emerging or declining nature, etc. (Cobo et al., 2011). The Web of Science database was searched on 21 December 2019 with the keyword term 'urban ecology'. The results were not refined further to understand the full extent of urban ecology research. The search years included 2009−2019. We followed Cobo et al. (2011) in conducting the bibliometric analysis. The analysis has considered the Keywords-Plus field of the Web of Science results. These are used to tag the articles with predefined research areas according to the references cited in the article. These are different from the author keywords and provide a comparable platform for analysis.

4. Urban ecology: state of research and associations

4.1 Major research fields in urban ecology domain

The most frequently used author keyword was 'urbanization', after 'urban ecology', followed by 'ecosystem services' and 'biodiversity'. However, the keyword-plus category from the Web of Science database can be considered more reflective of the major research fields in urban ecology. Author keyword-plus is determined based on an algorithm, which classifies articles according to the references cited in the article. Hence, the keyword-plus provides a comprehensive and comparative universal index. Biodiversity was the most researched aspect, followed by urbanization, conservation diversity and ecology (Table 24.1). Ecosystem services were the least cited aspect out of the top 10 author keywords-plus.

On the basis of the number of articles published during the study period, the top 10 most relevant journals in the field of urban ecology are given in Table 24.2. Urban Ecosystems followed by Landscape and Urban Planning are the top two most relevant journals accounting for the bulk of articles published during 2009−19, 77 and 72, respectively. This was followed by Urban Forestry and Urban Greening with 29 articles. PeerJ, with 14 articles in 2009−20, came at the 10th place.

TABLE 24.1 Author keywords and keywords-plus frequency in the bibliometric results in 2009–19.

Author keywords (DE)	Articles	Keywords-plus (ID)	Articles
URBAN ECOLOGY	798	BIODIVERSITY	203
URBANIZATION	80	URBANIZATION	192
ECOSYSTEM SERVICES	54	CONSERVATION	142
BIODIVERSITY	51	DIVERSITY	135
URBAN PLANNING	28	ECOLOGY	119
SUSTAINABILITY	21	PATTERNS	83
URBAN BIODIVERSITY	21	LANDSCAPE	79
GREEN INFRASTRUCTURE	19	MANAGEMENT	75
INVASIVE SPECIES	19	URBAN	75
SPECIES RICHNESS	19	ECOSYSTEM SERVICES	74

TABLE 24.2 Top 10 journals in the field of urban ecology and the number of articles published during the search period.

Journals	Number of articles	Impact factor (2018)
URBAN ECOSYSTEMS	77	2.49
LANDSCAPE AND URBAN PLANNING	72	5.14
URBAN FORESTRY and URBAN GREENING	29	3.04
LANDSCAPE ECOLOGY	18	4.35
SUSTAINABILITY	17	2.59
BIOLOGICAL CONSERVATION	16	4.45
ECOLOGICAL INDICATORS	16	4.49
ECOSPHERE	15	2.75
JOURNAL OF APPLIED ECOLOGY	15	5.78
PEERJ	14	2.35

5. Integration with other themes and emerging fields

The keyword cooccurrence figure describes the integration between different fields of urban ecology (Fig. 24.1). The circles represent the concepts, and the distance and lines connecting the concepts show the linkages between them. Size of the concept is indicative of its frequency, whereas closer concepts are found with more interconnecting lines and less distance between them. Some concepts found overlapping with each other indicate their string

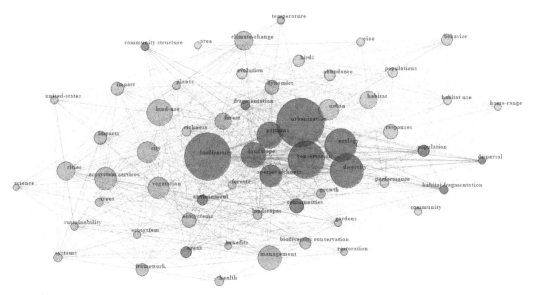

FIGURE 24.1 Keyword cooccurrence plot showing the keywords-plus as nodes and the relationship between them for urban ecology articles.

relationship. Urbanization and biodiversity occupied the most prominent position with the most frequent occurrence. The three different colours represented three clusters. We followed a scheme of naming the three clusters according to the most frequent keyword-plus. The red cluster was dominated by urbanization, the green cluster by urban, and the blue cluster had landscape and management as the major concepts. The urbanization cluster had biodiversity, conservation, diversity and ecology as some of the dominant themes. Urbanization, patterns, conservation and landscape were present quite closeer to each other, and sometimes overlapping reflecting their strong association and cooccurrence pattern. Ecology and diversity were also found overlapping with each other. The lesser researched concepts included dispersal, habitat fragmentation and community structure in this cluster. The urban cluster had most of the keywords-plus distributed almost evenly. It had habitat and responses as some of the major themes. The lesser linked concepts included community, area, home range, behaviour and size. The landscape and management cluster was dominated by ecosystem services, vegetation, climate change, city and cities, whereas health, restoration, framework, united states, systems, science, temperature and so on were found less frequently. The most dominant concepts in each cluster may serve as a keyword for research related to any concept in a cluster, thus serving as a helpful search word. However, the lesser researched concepts from the perspective of urban ecology also indicate the need to include them in future research related particularly with the higher-frequency concepts. Depending on the number of concepts, it can be said that there is a large number of related concepts on the fringes of urban ecology, which need to be focussed, and the higher-frequency larger concepts can help as a window in bringing them out. For example, research related to urban and habitat concepts can include home range, habitat use, richness, forests or evolution, to address these lesser-explored areas from the perspective of urban ecology. Similarly, research related with urbanization and biodiversity can also focus on habitat fragmentation, community structure, dispersal or population, and that related to management and landscape can expand to include health, systems, science, restoration, gardens, trees, plants or temperature

6. Evolution of major research fields and challenges

Conceptually the urban ecology research was divided into four clusters, with two major clusters and two minor clusters (Fig. 24.2). The first cluster, represented by the red colour in the figure, was predominantly about themes related to communities, landscapes, biodiversity conservation, habitat fragmentation, urbanization, ecosystem services, forests, land use, management, trees, landscapes, fragmentation and so on. We found that most of the concepts were related to ecology of cities broader theme. The *ecology of cities* is described as the study of urban areas, which incorporates the interaction between biological, built and social components in an urban ecosystem (Pickett et al., 2016), whereas the ecology in cities is concerned more with the study of ecosystem forms in an urban ecosystem treating them as 'analogues of their nonurban counterparts' (Pickett et al., 2016). The ecology in cities was found to be the broader theme in the second cluster shown in blue colour in the figure. This was dominated by birds, behaviour, abundance, community, performance, growth, gradient, evolution, climate change, population, size, community, evolution, etc.

The two minor clusters were characteristic of the management and sustainability of the urban ecosystems. One of these was represented by 'benefits' alone, highlighting its niche in this research field. The other minor cluster was represented by sustainability, science, framework,

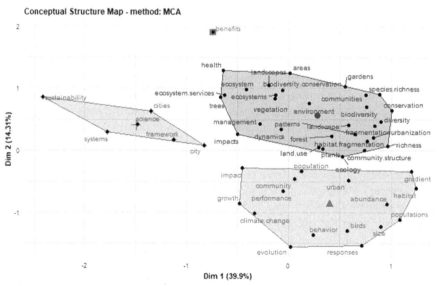

FIGURE 24.2 Conceptual structure plot showing the conceptual arrangement of urban ecology fields based on their relationships with each other.

cities and city, supporting our conclusion that most of the concepts in the cluster were related to sustainability. Based on this conceptual cluster map of the bibliometric results, we could discern three major thematic areas in the urban ecological research at present: (1) ecology of cities; (2) ecology in cities and (3) sustainability and systems science. These are broad and interrelated categories with subject fields merging and collating with other depending on the research question. To find out the thematic structure of the identified keyword-plus, a thematic structure plot was constructed. The results are explained in the following paragraphs.

The thematic strategic plot gives information about the state of various themes in terms of their density and centrality (Fig. 24.3). Density and centrality can be understood as two metrics, which determine the overall development of a focused theme. High density represents a high frequency of occurrence of a keyword, and a high centrality indicates its strong relationship with other themes. This gives more information than a keyword-cooccurrence plot as the keywords are arranged based on their frequency as well as the relationship to each other in four quadrants of the plot. The upper right quadrant represents themes that are well developed as well as 'important for the structuring of a research field' (Cobo et al., 2011). These are called motor themes. The upper left quadrant represents themes that have well-developed internal linkages but are isolated and not so important for the research field. The lower left quadrant represented themes that are emerging or disappearing in nature. The lower right quadrant represents themes that are transverse, having high density but low centrality. These are not so well-developed themes but found frequently in the research field. The presence of motor themes indicates their well-developed conceptual basis as well as the structural integrity of the research field.

Climate change and impact were found in the upper right quadrant. These motor themes were composed of climate change, evolution, temperature and impact. These were related with the influence of urban growth on other systems, such as the urban heat island effect

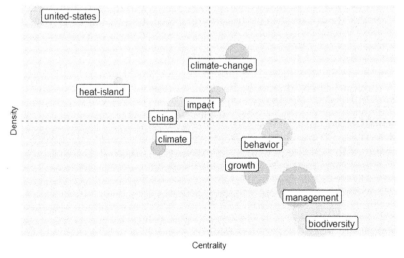

FIGURE 24.3 Thematic strategic plot showing the arrangement of major themes of urban ecology according to density and centrality of the concept.

(Wang et al., 2019); natural phenomenon, such as changes in bird songs (Marín-Gómez et al., 2020); and evolutionary occurrences, such as loss of Darwin's finches (De Leon et al., 2019), or trait variation in lizards (Baxter-Gilbert et al., 2019), etc. Climate change and the impact of urbanization are two subject areas with high centrality and density. The themes in the upper left quadrant were China, heat island and the United States. These represented well-researched but isolated themes. The lower left quadrant had the only climate as its constituent that represented the emerging or disappearing themes. The lower right quadrant had biodiversity, management, behaviour and growth; the most frequent keywords found in the keyword cooccurrence plot. However, their location in this quadrant signified that their conceptual basis needed stronger integration and research to include them in the motor themes. Since the most frequent keywords were not in the motor theme quadrant, we could hypothesize that more well-developed conceptual basis is yet to be created for the urban ecology as a distinct discipline. However, the integration of a large number of themes in the cooccurrence plot showed that urban ecology has achieved a certain level of a structured approach.

Based on the broad thematic areas identified for urban ecology, we utilized the bibliometric analysis and the other works presented in this book, to describe some of the major challenges and opportunities encountered in urban ecological studies. A transdisciplinary approach is needed to assess the role of humans as components of urban ecosystems, the relationship between the biotic and abiotic functions and the urban areas in 'altered environmental conditions' (Groffman et al., 2017). However, there are some major challenges faced in carrying out such a transdisciplinary study, which have been described in the next section.

7. Challenges and opportunities

In the previous sections, a background about the state of research in urban ecology was established. Certain broad categories of such research were also recognised. In this section, we have described the major challenges faced with respect to the research on different aspects of the urban ecology. As mentioned earlier, these broad themes were (1) ecology of cities; (2) ecology in cities; and (3) sustainability and management (Table 24.3). The list in Table 24.3 though not exhaustive could provide a way forward.

An urban area has as many aspects as there are perspectives. From the perspective of human health, even smart cities appear to be unsustainable (Colding and Barthel, 2017); from the view of ecosystem services, the urban green cover provides a much-needed filter from air pollution, and even the natural water bodies act as networks to ward off excess water in times of floods (Eck, 1999). These perspectives have been discussed in detail in this section, highlighting some of the challenges and opportunities of their respective fields.

8. Ecology of cities

Major challenges include energy balance and urban design, ornamental horticulture, urban greening, social diversity, equity and education, air and water pollution, waste management, cost—benefit analysis, transport and mobility. These include the human—ecosystem interaction in urban areas.

TABLE 24.3 Important challenges faced in major themes of urban ecology, their opportunities and broad objectives.

	Objectives	Challenges	Opportunities	References
1. Ecology of cities				
Energy and material balance	Energy performance	# Urban design of the built envelope # Location and geography of the buildings	# Green buildings and green building material # Zero-emission buildings # Efficient lighting # Sufficient roof area # Solar photovoltaic	Natanian et al. (2019), AL-Dabbagh (2020)
Invasive alien species (IAPs)	Urban biodiversity	# Ornamental horticulture # Disturbed habitats	# Conserve natural and seminatural habitat remnants # Remove IAP # Preborder import restrictions, postborder bans, industry codes of conduct, consumer education	Hulme et al. (2018)
Social diversity	Cultural ecological services Healthy lifestyle	# Loss of green spaces # Spatial, social and economic inequality # Human appreciation—ecosystem relationship	# Ecosystem appreciation # Social networks # Virtual mobility # Shared economies # Innovation	Ngiam et al. (2017), Alberti (2017)
Urban agriculture	Food security	# Energy flow # Life cycle assessment and urban metabolism # Biogeochemical cycles # Unexplored urban peripheries	# Rooftop gardens # Ecological network analysis # Renewable energy # Increasing recycling rates through the supply chain # Local socioeconomic and ecological issues	Piezer et al. (2019), Geneletti et al. (2017)
Air pollution	Human and ecosystem health	# Urban heat island effect # Environmental risk # Wind speed # Landscape factors	# Green spaces # Ecosystem services # Air quality standards # Landscape indicators	Łowicki (2019)
Water pollution	Human and ecosystem health	# Excessive water consumption # High pollution load # Rainstorm waterlogging # Governance	# Telecoupling #Water footprint exploration # Simulation and modelling # Land use land cover change # Decentralized sewage systems # Sponge city # Natural water bodies # Ecological restoration	Ren et al. (2017), Friis et al. (2016)

(Continued)

TABLE 24.3 Important challenges faced in major themes of urban ecology, their opportunities and broad objectives.—cont'd

	Objectives	Challenges	Opportunities	References
Waste management	Human health and ecosystem protection	# Unregulated waste disposal # Attenuation # Infrastructure	# Recovery, reuse of organic fraction of MSW # Urban infrastructure innovation and symbiosis # Emerging ICT technologies	Dong et al. (2018);
Decoupling economic growth and carbon emissions	Low carbon cities	# Coal consumption # Coal purification # Outsourcing emissions # Ecological modernization # Investment	# Change in assessment methodologies # Identifying cobenefits of the low carbon strategies # Institutions as knowledge and case examples # Innovation in legal, institutional, financial and governance framework	Dhakal and Ruth (2017), Li et al. (2019)
2. Ecology in cities				
Monitoring and assessment	Data collection and temporal variations	# Continuous monitoring # Spatial and spectral resolution of remotely sensed data # Sampling and scale	# Technological advancements	Verma and Raghubanshi (2019)
Biogeochemical models	Material and energy flow	# Restricted to the C, N, P or energy flow through food systems # Natural ecosystem information alone is not sufficient # Effects and changes are spread ecosystems	# Set points of control, like government at a city level # Cross-city comparisons # Interdisciplinary and systems approach	Bai (2016)
Invasive plant species (IAPs)	Biodiversity	# Proximity to large propagule sources # Ecosystem alteration # Environmental disturbance	# Role of IAPs in phytorem ediation # Ecosystem restoration # Integrated approach	Rai and Kim (2019), Bai (2016), Potgieter et al. (2019)
Hydrology	Water bodies and the hydrological cycle	# Flooding # Mapping and monitoring water flows # Urban water management	# Green spaces # Studies at watershed scales	Nguyen et al. (2019)
Biodiversity	Ecosystem health	# Biodiversity at the city and regional scale # Spatial metrics # Temporal dynamics	# Ecosystem services # Hierarchical multi-scale landscape ecology models	Norton et al. (2016)

TABLE 24.3   Important challenges faced in major themes of urban ecology, their opportunities and broad objectives.—cont'd

	Objectives	Challenges	Opportunities	References
Ecosystem services	Ecosystem health and sustainability	# Incorporation of ecosystem services into decision-making # Valuation and measurement # Demand and supply # Indicators # Databases # Comparable metrics and scales	# Place-based empirical knowledge # Appropriate measures and policies # Integrated approach # Social scientists, ecologists and environmental experts # Integrate formal and informal institutions	Larondelle and Lauf (2016), Kaczorowska et al. (2016)
3. Management and sustainability				
Displacement and migration	Disaster-induced displacement and ecological burden	# Lack of any data # Future management of shelters # Basic survival and accessibility	# Comprehensive shelter surge # System dynamic model for temporary operation and management in urban areas	Uddin et al. (2018)
Governance	Local governance	# Effective communication # Management tools # Technological tools # Green infrastructure # Local and regional frameworks	# Strategic management tools # Efficient devices # Recovery and recycling of resources # Rooftop gardens, urban agriculture	Wu (2017)
Economic disparity	Economic sustainability	# Lack of resources # Resource distribution # Employment # Slums # Inefficient resource use # Lack of access to public services # Biodiversity in social and economic disparity	# Income # Education # Traditional knowledge base # Financial institutions # Services penetration # 'New urban sociology'	MacGregor-Fors and Escobar-Ibáñez (2017), Gottdiener (2019)
Indicators and targets	Sustainability framework	# Data availability # Differences in the scale of operations # Lack of targets and thresholds # Differences in the definition of sustainability	# Integrated and flexible sustainability framework # Data availability at the scale of implementation # Identification of local and global indicators and thresholds	Verma and Raghubanshi (2018)

(Continued)

TABLE 24.3 Important challenges faced in major themes of urban ecology, their opportunities and broad objectives.—cont'd

	Objectives	Challenges	Opportunities	References
Urban design	Energy and cost efficiency	# Uniform building codes and rating systems # Social cohesion	# Green infrastructure # Building materials # Planning and implementation # Initiative by public and institutional building # Transit-oriented development # Waste management	AL-Dabbagh (2020)
Urban renewal	Sustainable development	# Neighbourhood scale # Social, economic and environmental conflicts # High density and historical cities	# Sustainability assessment # Building condition # Decision-making framework # Public awareness and communication # Economics	Zheng et al. (2017)
Sustainable transport	Energy efficiency and human health	# High-density cities and demographic load # Unplanned urbanization # Social factors # Economic aspects	# Policy instruments # Social acceptance and encouragement for cleaner modes of transport # Harmony of infrastructure and social practices # Sustainable transitions	
Smart cities	Human health and sustainable built environments	# Cost # Planning and implementation # Local and regional requirements # Disconnected from the real urban fabric of people # 'Frankenstein urbanism' # Human health # People's connection with nature	# Innovation # Governance # Efficiency # Transit-oriented # Ecological urbanization	Cugurullo (2018), Colding and Barthel (2017)

The energy performance of urban ecosystems is an essential area of research. It is determined to a large extent by the type of building design and geoclimatic region (Natanian et al., 2019). Green buildings and green building material, zero-emission buildings, utilization of roof area for agriculture or gardening and the use of solar photovoltaic panels in energy generation are some of the major opportunities, which can help in increasing the energy performance of cities. Rooftop gardening or urban greenery is important for urban areas, especially where the urbanization is unfolding rapidly and in an unplanned manner. They offer opportunities to resuscitate the ecological fabric of cities. Urban greenery, however, also poses certain challenges in the form of introduction of nonnative plant species, which prove invasive for the local varieties. Ornamental horticulture and disturbed habitats can be congenial for the spread of invasive plant species (IAPs). In urban areas, planning, with a focus on conserving the natural and native vegetation component, should be undertaken along with some policy support such as restriction on the export of IAPs and consumer education (Hulme et al., 2018). Thus, informed governance and education play an important role in regulating the biodiversity, its protection and betterment. Diversity in the form of social and cultural ecosystem services is determined by informed governance. It is found that social disparity is responsible for unhealthy living conditions for the lower socioeconomic urban population (Ngiam et al., 2017). Social networks, traditional knowledge systems and shared economies are essential to protect the cultural ecosystem services (Alberti, 2017). Local socioeconomic and ecological conditions can be determining factors for the local and regional food security too. Since urban areas depend on different resource bases, such as food imports, which are sometimes found away from the source of their consumption, urban agriculture can serve to strengthen the local food security and sustainability, as the cost of bringing in food from far away locations is reduced. However, research into the actual cost and benefit of urban agriculture has been able to show that, in case of rooftop gardens, the energy flow through the system is not sustainable, as the system has to depend on external sources of energy, like fossil fuels (Piezer et al., 2019). In the same research, it was recommended that to understand the energy flow, comparison between different components of the system needs to be made. Utilising a higher share of renewable energy would be able to tip the balance towards sustainability in energy flow. Similar studies need to be conducted utilising 'Ecological Network Analysis', with the aim of increasing the recycling rates through the supply chain (Piezer et al., 2019). Urban design can play a vital role in greening and utilising urban peripheries for agriculture (Geneletti et al., 2017). Green space also plays an essential part in mitigating the effects of air pollution and urban heat island effect. Certain geographical factors are also important in determining the effect of air pollution such as wind speed and direction, landscape, etc. (Łowicki, 2019). Air pollution has been identified as the single largest factor for many human health issues (Singh et al., 2019). Air, along with water, is essential for life to thrive, but humans have negatively affected both. Overexploitation of water resources, high pollution load, lack of urban planning leading to urban floods and so on are some of the challenges faced by cities. They are a reflection in part of the high consumption of water resource and its wastage, but the majority of them arise due to lack of governance and policies based on scientific and local perspective. Groundwater has receded around the globe in urban areas, and the rate of consumption exceeds the rate of recharge (de Graaf et al., 2019). This has led to telecoupling, the interaction of socioeconomic and environmental factors over a long distance (Deines et al., 2016). There is a need to adopt a watershed approach that includes the remote sensing analysis and simulation modelling into actual decision-making, and create policies that can address the renewal of natural and existing water bodies (Ren et al., 2017). The concept of Sponge city might act as a guiding concept in this regard (Friis et al., 2016).

Urban ecosystem runs on subsidized material and energy balance, which means that by using energy, from fossil fuels and other sources, more materials are consumed within the urban areas than are exported back to the ecosystem. However, a large quantity of waste is exported, which has become important because of unregulated practices at the local scale and its potential economic value. The basic needs of waste management lie in regulating the waste flow, infrastructure to manage the waste generation and education of the common people who are at the receiving end of bad waste management policies, especially in case of hazardous waste (Dong et al., 2018; Adeola, 2000). The emerging technologies, however, also can be used in turning waste-to-economy (Dong et al., 2018). Since economic growth is inextricably linked to the urban resource utilization and waste generation, there is a need to decouple economic growth from carbon emissions. Carbon emissions are caused at every step of energy utilization, in the transport of people and materials, electricity generation, water treatment plants, etc. The resource utilization by areas is heavily dependant on energy, and energy consumption leads to carbon emissions. Coal consumption and purification, outsourcing of emission, investment to adopt cleaner and zero-emission technologies are some of the major challenges faced to decouple carbon emissions from economic growth (Dhakal and Ruth, 2017). However, there is a need to identify the co-benefits of low carbon strategies and change the assessment methodologies to include those cobenefits (Li et al., 2019). Innovation in legal, institutional, financial and governance would be essential to implementing these in practice.

9. Ecology in cities

Ecology in cities has been traditionally regarded as research into urban areas regarding them as analogues of natural ecosystems (Grimm et al., 2008). The human has been neglected; however, it is very difficult to continue with this approach due to the scale and unsustainability associated with urbanization. In this section, we have considered those themes that focus on the ecological aspect of urban areas alone.

Monitoring the expansion of cities helps in planning for its future growth, and in this regard, the role of remote sensing comes into play. Among all the monitoring strategies, remote sensing allows for the monitoring of large expanses of land and temporal advantages. Remote sensing is also increasingly used in monitoring urban vegetation, water exploration, land cover changes, etc. Thus, it provides a useful tool to monitor natural as well as human-dominated landscapes. High-resolution data availability and revisit time of satellites is one of the major issues with remote sensing. The vegetation mapping is also used to understand the natural terrestrial carbon flux and storage. The carbon cycle is one of the most important of the biogeochemical cycles. Biogeochemical cycles are restricted to the carbon, phosphorus, nitrogen or energy flows within ecosystems. The natural ecosystem information alone is not sufficient and more cross-city comparisons with interdisciplinary and systems approach (Bai, 2016). Biodiversity monitoring is also related to IAPs. IAPs causing ecosystem alteration and environmental disturbances are a major concern, especially when a large propagule source is present in the vicinity (Rai and Kim, 2019). However, the role of IAPs in phytoremediation and ecological restoration warrants an integrated approach towards IAPs management and local biodiversity preservation (Bai, 2016). Apart from the biogeochemical cycles, the hydrological cycle determines urban habitats and vegetation to a large extent. Climatic and geographic factors come into play determining the type of precipitation, water bodies and the water run-off. Urban flooding, mapping and monitoring water flows are an integral part of urban water management. Green spaces and studies at the watershed scale help in the restoration of natural water bodies and preserving the hydrological cycles (Nguyen et al., 2019). The type of climate, geography and hydrology determine the type of biodiversity at different scales, which in turn influences the ecosystem health. Spatial metrics are frequently used in determining landscape indicators to quantify the effect of the urban landscape on biodiversity (Goddard et al., 2010); however, their use in identifying biodiversity niches is also prevalent (Walz, 2011). A major challenge is to determine a set of universally agreed-upon metrics out of a very large number of available metrics (Skidmore et al., 2015). Hierarchical multiscale landscape models are also helpful in determining biodiversity and ecosystem health (Norton et al., 2016). Ecosystem health is essentially the state of an ecosystem where it can function sustainably. It has been defined by terms such as 'system resilience, organization and vigour' by Costanza (1992). It also includes ecosystem services, which play an essential role not only in human survival but also for the functioning of all components in an ecosystem. Incorporation of ecosystem services into the decision- and policy-making level is found scarce, especially in developing countries (Kenter et al., 2011); there is a lack of universal indicators, databases and comparable metrics and scales. Place-based empirical knowledge, a collaboration between social scientists, ecologists and environmental experts, institutional leadership and education are important aspects that can help to strengthen the ecosystem health (Larondelle and Lauf, 2016; Kaczorowska et al., 2016).

10. Management and sustainability

Among the three themes of urban ecology, management and sustainability are the overarching objectives for most of the studies concerned with urban ecology. Since it is connected to all aspects of ecology *in* and *of* cities, a clear understanding and measure of sustainability is the first important step.

Indicators and targets help in creating the backbone of sustainability framework (Verma and Raghubanshi, 2018). Conceptual knowledge about sustainability is necessary to undertake this. Data availability, targets and thresholds of how much is sustainable, applicability at different geographical and administrative scales are the major challenges met in this approach. However, an integrated and flexible mechanism, like the United Nations Sustainable Development Goals 2015, is a set of such goals. There are 230 indicators in the SDGs that can be adopted by nations according to their region specificity. Data availability at multiple scales is required to implement the sustainability frameworks, along with certain thresholds for determining system sustainability. The urban ecological sustainability is subject to many challenges, some are sudden and less perceptible. One of these is the disaster-induced displacement (DID). Climate and geography have been determining factors for population distribution and urban habitats (Berlemann and Steinhardt, 2017), and DID is an effect of climate change—induced migration. Climate change may lead to an increase in the frequency, magnitude or both of natural disasters making the phenomenon of DID more important. The migrated population puts large pressure on the existing resources, leading to basic survival and accessibility problems (Uddin et al., 2018) A system dynamics model for temporary operation and management in urban areas is recommended for managing DID (Uddin et al., 2018). The role of local governance becomes important in managing DID and other sustainability issues. Effective communication, infrastructure, preparedness and framework for sustainability and future directions can lead to efficient local governance.

Governance leads to policy-making; however, it is vital to consider the social and economic fabric of the urban areas. Distribution of resources, employment, access to public services, even biodiversity in socially and economically poor areas need to be considered in policymaking at different scales. Here, income, education, traditional knowledge base, financial institutions and a 'new urban ecology' approach are the need of the hour (MacGregor-Fors and Escobar-Ibanez, 2017; Gottdiener, 2019). Governance and local resource planning can also lead to efficient energy and material utilization. This could be achieved by urban planning considering the social, economic and environmental need. For example, transit-oriented development would be beneficial for urban areas housing the central business district; building material and green infrastructure can provide health, environmental and aesthetic benefits. However, historical cities with a lack of accessibility and rapidly urbanizing urban areas, especially in the fringes, can offer considerable difficulty in sustainability and management, for example, Varanasi city in India. Urban renewal, aimed at neighbourhood development, in high density and cities, requires public awareness and communication, finance and sustainable urban design (Zheng et al., 2017). Sustainable transport is one such aspect of urban planning that can reduce air pollution as well as lead to greater accessibility in these high-density areas. Cycling has been identified as one such aspect of sustainable transport that can be encouraged as the choice of transport for masses. A harmony in infrastructure (design) and social practices would need to be a part of this urban renewal. Urban renewal, grounded in ecological sensitivity, would also have a positive impact on human health and built environments. Smart cities have been projected as the culmination of such urban renewable. A smart city basically functions on the basis of efficiency and application of information and technology in delivering services (Zhuhadar et al., 2017). 'Planning, designing, buildings and operating a city's infrastructure' are based on using technology and innovations (Zhuhadar et al., 2017). However, the smart city paradigm has been criticised on account of being disconnected from natural fabric, heavy dependence on a digital framework that depends on high-end infrastructure and lack of focus on human health (Cugurullo, 2018; Colding and Barthel, 2017). Many smart cities have been criticised as being a somewhat heterogeneous mixture of isolated urban fragments put together to form a 'Frankenstein urbanism' (Cugurullo, 2018). The lack of people's connection with nature, especially children, is an interesting topic of exploration (Colding and Barthel, 2017). Smart cities may serve as incubation centres of innovation and science and efficiency and need to consider ecological urbanization as an integral part.

11. Conclusions

Considering all these different perspectives in a single narrative is quite difficult; however, this multitude of perspectives is what defines urban ecology and the social—ecological systems. In the development of urban ecology as a distinct discipline, certain major challenges and problems have been recognized. Urbanization can be considered the culmination of all processes human. It has led to the creation of many problems but is also responsible to offer solutions to the problems. However, given the imbalance between the two, a comprehensive sustainability approach is yet to be delivered. This would only be possible when the diverse subject areas develop with common objectives while following different pathways, i.e., co-production. Urban ecology has developed as a distinct discipline, and in the future, we can expect to see more transdisciplinary developments in this field.

References

Adeola, F.O., 2000. Cross-national environmental injustice and human rights issues: a review of evidence in the developing world. American Behavioral Scientist 43 (4), 686–706.

AL-Dabbagh, R.H., 2020. Toward green building and eco-cities in the UAE. In: Renewable Energy and Sustainable Buildings. Springer, Cham, pp. 221–233.

Alberti, M., 2017. Grand challenges in urban science. Frontiers in Built Environment 3, 6.

Aria, M., Cuccurullo, C., 2017. Bibliometrix: an R-tool for comprehensive science mapping analysis. Journal of Informetrics 11 (4), 959–975.

Bai, X., 2016. Eight energy and material flow characteristics of urban ecosystems. Ambio 45 (7), 819–830.

Baxter-Gilbert, J., Riley, J.L., Whiting, M.J., 2019. Bold New World: urbanization promotes an innate behavioral trait in a lizard. Behavioral Ecology and Sociobiology 73 (8), 105.

Berlemann, M., Steinhardt, M.F., 2017. Climate change, natural disasters, and migration—a survey of the empirical evidence. CESifo Economic Studies 63 (4), 353–385.

Bridge, G., Barr, S., Bouzarovski, S., Bradshaw, M., Brown, E., Bulkeley, H., Walker, G., 2018. Energy and Society: A Critical Perspective. Routledge.

Carroll, M.S., Blatner, K.A., Cohn, P.J., Morgan, T., 2007. Managing fire danger in the forests of the US Inland Northwest: a classic "wicked problem "in public land policy. Journal of Forestry 105 (5), 239–244.

Cobo, M.J., López-Herrera, A.G., Herrera-Viedma, E., Herrera, F., 2011. An approach for detecting, quantifying, and visualizing the evolution of a research field: A practical application to the fuzzy sets theory field. Journal of informetrics 5 (1), 146–166.

Colding, J., Barthel, S., 2017. An urban ecology critique on the "Smart City" model. Journal of Cleaner Production 164, 95–101.

Costanza, R., 1992. Toward an Operational Definition of Ecosystem Healtch. Ecosystem Health: New goals for environmental management, pp. 239–256.

Cugurullo, F., 2018. Exposing smart cities and eco-cities: frankenstein urbanism and the sustainability challenges of the experimental city. Environment and Planning: Economy and Space 50 (1), 73–92.

de Graaf, I.E., Gleeson, T., van Beek, L.R., Sutanudjaja, E.H., Bierkens, M.F., 2019. Environmental flow limits to global groundwater pumping. Nature 574 (7776), 90–94.

De León, L.F., Sharpe, D.M., Gotanda, K.M., Raeymaekers, J.A., Chaves, J.A., Hendry, A.P., Podos, J., 2019. Urbanization erodes niche segregation in Darwin's finches. Evolutionary Applications 12 (7), 1329–1343.

Deines, J.M., Liu, X., Liu, J., 2016. Telecoupling in urban water systems: an examination of Beijing's imported water supply. Water International 41 (2), 251–270.

Dhakal, S., Ruth, M., 2017. Challenges and opportunities for transition to low carbon cities. In: Creating Low Carbon Cities. Springer, Cham, pp. 1–4.

Dong, L., Wang, Y., Scipioni, A., Park, H.S., Ren, J., 2018. Recent progress on innovative urban infrastructures system towards sustainable resource management. Resources, Conservation and Recycling 128, 355–359.

Eck, D.L., 1999. Banaras, City of Light. Columbia University Press.

Friis, C., Nielsen, J.Ø., Otero, I., Haberl, H., Niewöhner, J., Hostert, P., 2016. From teleconnection to telecoupling: taking stock of an emerging framework in land system science. Journal of Land Use Science 11 (2), 131–153.

Geneletti, D., La Rosa, D., Spyra, M., Cortinovis, C., 2017. A review of approaches and challenges for sustainable planning in urban peripheries. Landscape and Urban Planning 165, 231–243.

Goddard, M.A., Dougill, A.J., Benton, T.G., 2010. Scaling up from gardens: biodiversity conservation in urban environments. Trends in Ecology and Evolution 25 (2), 90–98.

Gottdiener, M., 2019. New Urban Sociology. The Wiley Blackwell Encyclopedia of Urban and Regional Studies, pp. 1–5.

Grimm, N.B., Faeth, S.H., Golubiewski, N.E., Redman, C.L., Wu, J., Bai, X., Briggs, J.M., 2008. Global change and the ecology of cities. Science 319 (5864), 756–760.

Groffman, P.M., Cadenasso, M.L., Cavender-Bares, J., Childers, D.L., Grimm, N.B., Grove, J.M., et al., 2017. Moving towards a new urban systems science. Ecosystems 20 (1), 38–43.

Grove, J.M., Pickett, S.T., 2019. From transdisciplinary projects to platforms: expanding capacity and impact of land systems knowledge and decision making. Current Opinion in Environmental Sustainability 38, 7–13.

Hulme, P.E., Brundu, G., Carboni, M., Dehnen-Schmutz, K., Dullinger, S., Early, R., et al., 2018. Integrating invasive species policies across ornamental horticulture supply chains to prevent plant invasions. Journal of Applied Ecology 55 (1), 92–98.

Kaczorowska, A., Kain, J.H., Kronenberg, J., Haase, D., 2016. Ecosystem services in urban land use planning: integration challenges in complex urban settings—case of Stockholm. Ecosystem Services 22, 204–212.

Kenter, J.O., Hyde, T., Christie, M., Fazey, I., 2011. The importance of deliberation in valuing ecosystem services in developing countries—evidence from the Solomon Islands. Global Environmental Change 21 (2), 505–521.

Klauer, B., Manstetten, R., Petersen, T., Schiller, J., 2016. Sustainability and the Art of Long-Term Thinking. Routledge.

Kleer, J., Nawrot, K.A., 2018. The Rise of Megacities: Challenges, Opportunities and Unique Characteristics. World Scientific Publishing Co. Pte. Ltd.

Larondelle, N., Lauf, S., 2016. Balancing demand and supply of multiple urban ecosystem services on different spatial scales. Ecosystem Services 22, 18–31.

Li, L., Shan, Y., Lei, Y., Wu, S., Yu, X., Lin, X., Chen, Y., 2019. Decoupling of economic growth and emissions in China's cities: a case study of the Central Plains urban agglomeration. Applied Energy 244, 36–45.

Loo, B.P., 2018. Unsustainable Transport and Transition in China. Routledge.

Łowicki, D., 2019. Landscape pattern as an indicator of urban air pollution of particulate matter in Poland. Ecological Indicators 97, 17–24.

MacGregor-Fors, I., Escobar-Ibáñez, J.F., 2017. Birds from urban Latin America, where economic inequality and urbanization meet biodiversity. In: Avian Ecology in Latin American Cityscapes. Springer, Cham, pp. 1–10.

Marín-Gómez, O.H., Dáttilo, W., Sosa-López, J.R., Santiago-Alarcon, D., MacGregor-Fors, I., 2020. Where has the city choir gone? Loss of the temporal structure of bird dawn choruses in urban areas. Landscape and Urban Planning 194, 103665.

Natanian, J., Aleksandrowicz, O., Auer, T., 2019. A parametric approach to optimizing urban form, energy balance and environmental quality: the case of Mediterranean districts. Applied Energy 254, 113637.

Ngiam, R.W.J., Lim, W.L., Collins, C.M., 2017. A balancing act in urban social-ecology: human appreciation, ponds and dragonflies. Urban Ecosystems 20 (4), 743–758.

Nguyen, T.T., Ngo, H.H., Guo, W., Wang, X.C., Ren, N., Li, G., et al., 2019. Implementation of a specific urban water management-Sponge City. The Science of the Total Environment 652, 147–162.

Norris, P.E., O'Rourke, M., Mayer, A.S., Halvorsen, K.E., 2016. Managing the wicked problem of transdisciplinary team formation in socio-ecological systems. Landscape and Urban Planning 154, 115–122.

Norton, B.A., Evans, K.L., Warren, P.H., 2016. Urban biodiversity and landscape ecology: patterns, processes and planning. Current Landscape Ecology Reports 1 (4), 178–192.

Pickett, S.T., Cadenasso, M.L., Childers, D.L., McDonnell, M.J., Zhou, W., 2016. Evolution and future of urban ecological science: ecology in, of, and for the city. Ecosystem Health and Sustainability 2 (7), e01229..

Piezer, K., Petit-Boix, A., Sanjuan-Delmás, D., Briese, E., Celik, I., Rieradevall, J., et al., 2019. Ecological network analysis of growing tomatoes in an urban rooftop greenhouse. The Science of the Total Environment 651, 1495–1504.

Potgieter, L.J., Gaertner, M., O'Farrell, P.J., Richardson, D.M., 2019. Perceptions of impact: invasive alien plants in the urban environment. Journal of Environmental Management 229, 76–87.

Rademacher, A., Cadenasso, M.L., Pickett, S.T., 2019. From feedbacks to coproduction: toward an integrated conceptual framework for urban ecosystems. Urban Ecosystems 22 (1), 65–76.

Rai, P.K., Kim, K.H., 2019. Invasive alien plants and environmental remediation: a new paradigm for sustainable restoration ecology. Restoration Ecology 28 (1), 3–7.

Ren, N., Wang, Q., Wang, Q., Huang, H., Wang, X., 2017. Upgrading to urban water system 3.0 through sponge city construction. Frontiers of Environmental Science and Engineering 11 (4), 9.

Simon, D., Palmer, H., Riise, J., Smit, W., Valencia, S., 2018. The challenges of transdisciplinary knowledge production: from unilocal to comparative research. Environment and Urbanization 30 (2), 481–500.

Singh, N., Mhawish, A., Ghosh, S., Banerjee, T., Mall, R.K., 2019. Attributing mortality from temperature extremes: A time series analysis in Varanasi, India. Science of The Total Environment 665, 453–464.

Skidmore, A.K., Pettorelli, N., Coops, N.C., Geller, G.N., Hansen, M., Lucas, R., Mücher, C.A., et al., 2015. Environmental science: agree on biodiversity metrics to track from space. Nature 523 (7561), 403–405.

Uddin, M.S., Ahmad, M.M., Warnitchai, P., 2018. Surge dynamics of disaster displaced populations in temporary urban shelters: future challenges and management issues. Natural Hazards 94 (1), 201–225.

Verma, P., Raghubanshi, A.S., 2018. Urban sustainability indicators: Challenges and opportunities. Ecological indicators 93, 282–291.

Verma, P., Raghubanshi, A.S., 2019. Rural development and land use land cover change in a rapidly developing agrarian South Asian landscape. Remote Sensing Applications: Society and Environment 14, 138–147.

Walz, U., 2011. Landscape structure, landscape metrics and biodiversity. Living Reviews in Landscape Research 5 (3), 1–35.

Wang, J., Zhou, W., Wang, J., 2019. Time-series analysis reveals intensified urban heat island effects but without significant urban warming. Remote Sensing 11 (19), 2229.

Wu, F., 2017. Emerging Chinese cities: Implications for global urban studies 1. In: The Post-Urban World. Routledge, pp. 157–175.

Zheng, H.W., Shen, G.Q., Song, Y., Sun, B., Hong, J., 2017. Neighborhood sustainability in urban renewal: an assessment framework. Environment and Planning B: Urban Analytics and City Science 44 (5), 903–924.

Zhuhadar, L., Thrasher, E., Marklin, S., de Pablos, P.O., 2017. The next wave of innovation—review of smart cities intelligent operation systems. Computers in Human Behavior 66, 273–281.

Index

'*Note*: Page numbers followed by "f" indicate figures and "t" indicate tables.'

Printed in the United States
By Bookmasters